Lecture Notes in Computer Science 4374

Commenced Publication in 1973
Founding and Former Series Editors:
Gerhard Goos, Juris Hartmanis, and Jan van Leeuwen

Editorial Board

David Hutchison
Lancaster University, UK

Takeo Kanade
Carnegie Mellon University, Pittsburgh, PA, USA

Josef Kittler
University of Surrey, Guildford, UK

Jon M. Kleinberg
Cornell University, Ithaca, NY, USA

Friedemann Mattern
ETH Zurich, Switzerland

John C. Mitchell
Stanford University, CA, USA

Moni Naor
Weizmann Institute of Science, Rehovot, Israel

Oscar Nierstrasz
University of Bern, Switzerland

C. Pandu Rangan
Indian Institute of Technology, Madras, India

Bernhard Steffen
University of Dortmund, Germany

Madhu Sudan
Massachusetts Institute of Technology, MA, USA

Demetri Terzopoulos
University of California, Los Angeles, CA, USA

Doug Tygar
University of California, Berkeley, CA, USA

Moshe Y. Vardi
Rice University, Houston, TX, USA

Gerhard Weikum
Max-Planck Institute of Computer Science, Saarbruecken, Germany

James F. Peters   Andrzej Skowron
Ivo Düntsch   Jerzy Grzymała-Busse
Ewa Orłowska   Lech Polkowski (Eds.)

Transactions on Rough Sets VI

Commemorating the Life and Work
of Zdzisław Pawlak, Part I

 Springer

Editors-in-Chief

James F. Peters
University of Manitoba, Winnipeg, Manitoba R3T 5V6, Canada
E-mail: jfpeters@ee.umanitoba.ca

Andrzej Skowron
Warsaw University, Banacha 2, 02-097 Warsaw, Poland
E-mail: skowron@mimuw.edu.pl

Volume Editors

Ivo Düntsch
Brock University, St. Catharines, Ontario L2S 3A1, Canada
E-mail: duentsch@brocku.ca

Jerzy Grzymała-Busse
University of Kansas, Lawrence, KS 66045, USA
E-mail: jerzy@ku.edu

Ewa Orłowska
National Institute of Telecommunications, ul. Szachowa 1, 04-894 Warsaw, Poland
E-mail: e.orlowska@itl.waw.pl

Lech Polkowski
University of Warmia and Mazury and
Polish-Japanese Institute of Information Technology Warsaw
10560 Olsztyn, Poland
E-mail: polkow@pjwstk.edu.pl

Library of Congress Control Number: 2007922187

CR Subject Classification (1998): F.4.1, F.1, I.2, H.2.8, I.5.1, I.4

LNCS Sublibrary: SL 1 – Theoretical Computer Science and General Issues
ISSN   0302-9743 (Lecture Notes in Computer Science)
ISSN   1861-2059 (Transactions on Rough Sets)
ISBN-10   3-540-71198-8 Springer Berlin Heidelberg New York
ISBN-13   978-3-540-71198-8 Springer Berlin Heidelberg New York

This work is subject to copyright. All rights are reserved, whether the whole or part of the material is concerned, specifically the rights of translation, reprinting, re-use of illustrations, recitation, broadcasting, reproduction on microfilms or in any other way, and storage in data banks. Duplication of this publication or parts thereof is permitted only under the provisions of the German Copyright Law of September 9, 1965, in its current version, and permission for use must always be obtained from Springer. Violations are liable to prosecution under the German Copyright Law.

Springer is a part of Springer Science+Business Media

springer.com

© Springer-Verlag Berlin Heidelberg 2007

Typesetting: Camera-ready by author, data conversion by Scientific Publishing Services, Chennai, India
Printed on acid-free paper   SPIN: 12028375   06/3142   5 4 3 2 1 0

Preface

Volume VI of the *Transactions on Rough Sets* (TRS) commemorates the life and work of Zdzisław Pawlak (1926-2006)[1]. His legacy is rich and varied. Professor Pawlak's research contributions have had far-reaching implications inasmuch as his works are fundamental in establishing new perspectives for scientific research in a wide spectrum of fields.

From a very early age, Zdzisław Pawlak devoted his life to scientific research. The pioneering work by Prof. Pawlak included research on the design of computers, information retrieval, modeling conflict analysis and negotiation, genetic grammars, and molecular computing. His research led to the introduction of knowledge representation systems during the early 1970s and the discovery of rough sets during the early 1980s. Added to that was Prof. Pawlak's lifelong interest in painting, photography, and poetry. During his lifetime, he nurtured worldwide interest in approximation, approximate reasoning, and rough set theory and its applications[2]. Evidence of the influence of Prof. Pawlak's work can be seen in the growth in the rough-set literature that now includes over 4000 publications by more than 1600 authors in the rough set database[3] as well as the growth and maturity of the International Rough Set Society[4]. Numerous biographies of Zdzisław Pawlak have been published[5].

This volume of the TRS presents papers that reflect the profound influence of a number of research initiatives by Zdzisław Pawlak. In particular, this volume introduces a number of new advances in the foundations and applications of artificial intelligence, engineering, logic, mathematics, and science. These advances have significant implications in a number of research areas such as the foundations of rough sets, approximate reasoning, bioinformatics, computational intelligence, cognitive science, data mining, information systems, intelligent systems, machine intelligence, and security. In addition, it is evident from the papers included in this volume that rough set theory and its application form a very active research area worldwide. A total of 41 researchers from 8 countries are represented in this volume, namely, Canada, India, France, Norway, Poland, P.R.

---

[1] Prof. Pawlak passed away on 7 April 2006.

[2] See, *e.g.*, Pawlak, Z., Skowron, A.: Rudiments of rough sets, *Information Sciences* 177 (2007) 3-27; Pawlak, Z., Skowron, A.: Rough sets: Some extensions, *Information Sciences* 177 (2007) 28-40; Pawlak, Z., Skowron, A.: Rough sets and Boolean reasoning, *Information Sciences* 177 (2007) 41-73.

[3] http://rsds.wsiz.rzeszow.pl/rsds.php

[4] http://roughsets.home.pl/www/

[5] See, *e.g.*, Peters, J.F. and Skowron, A., Zdzisław Pawlak: Life and Work. *Transactions on Rough Sets* V, LNCS 4100 (2006) 1-24. See, also, R. Słowiński, Obituary, Prof. Zdzisław Pawlak (1926-2006), *Fuzzy Sets and Systems* 157 (2006) 2419-2422.

China, Sweden, Russia, Thailand, and the USA. Evidence of the vigor, breadth and depth of research in the theory and applications of rough sets can be found in the articles in this volume.

Most of the contributions of this commemorative volume of the TRS are on an invitational basis and every paper has been refereed in the usual way. This special issue of the TRS contains 23 papers and extended abstracts that explore a number of research streams that are either directly or indirectly related to research initiatives by Zdzisław Pawlak. These research streams are represented by papers on propositional logics (Mohua Banerjee and Md. Aquil Khan), intuitionistic rough sets for database applications (Theresa Beaubouef and Fred Petry), missing attribute value problem (Jerzy W. Grzymała-Busse and Witold J. Grzymała-Busse), Zdzisław Pawlak's contributions to the study of vagueness (Mihir Chakraborty), data mining (Alicja Wakulicz-Deja and Grzegorz Ilczuk), approximation of concepts (Anna Gomolińska), intelligent systems (Andrzej Jankowski and Andrzej Skowron), acoustics (Bozena Kostek), rule evaluation (Jiye Li, Puntip Pattaraintakorn, and Nick Cercone), rough sets in China (Qing Liu and Hui Sun), four-valued logic (Jan Małuszyński, Andrzej Szałas and Aida Vitória), crisp and fuzzy information systems (Alicja Mieszkowicz-Rolka and Leszek Rolka), artificial intelligence and rough sets (Tosiharu Munakata), topology and information systems (Piero Pagliani and Mihir K. Chakraborty), conjugate information systems (Maria Semeniuk-Polkowska), incomplete transactional databases (Grzegorz Protaziuk and Henryk Rybinski), classifiers, rule induction and rough sets (Jerzy Stefanowski), approximation spaces (Jarosław Stepaniuk), relevant attributes in high-dimensional data (Julio J. Valdés and Alan J. Barton), knowledge discovery in databases (Anita Wasilewska, Ernestina Menasalvas, Christelle Scharff), information quanta and approximation operators (Marcin Wolski), lattice theory for rough sets (Jouni Järvinen).

The editors of this volume extend their hearty thanks to reviewers of papers that have been submitted to the TRS during the past 12 months: Manuel Ojeda-Aciego, Mohua Banerjee, Jan Bazan, Mihir Chakraborty, Anna Gomolińska, Etienne Kerre, Pawan Lingras, Victor Marek, Piero Pagliani, Sheela Ramanna, Dominik Ślęzak, Jerzy Stefanowski, Jarosław Stepaniuk, Piotr Synak, Piotr Wasilewski and Yiyu Yao.

This issue of the TRS has been made possible thanks to the laudable efforts of a great many generous persons and organizations. The editors and authors of this volume also extend an expression of gratitude to Alfred Hofmann, Ursula Barth, Christine Günther and the LNCS staff at Springer for their support in making this volume of the TRS possible. In addition, the editors extend their thanks to Marcin Szczuka for his consummate skill and care in the compilation of this volume. The editors have been supported by the State Committee for Scientific Research of the Republic of Poland (KBN), research grant No. 3T11C00226, and the Natural Sciences and Engineering Research Council of Canada (NSERC) research grant 185986.

December 2006

Ivo Düntsch
Jerzy W. Grzymała-Busse
Ewa Orłowska
James F. Peters
Lech Polkowski
Andrzej Skowron

LNCS Transactions on Rough Sets

This journal subline has as its principal aim the fostering of professional exchanges between scientists and practitioners who are interested in the foundations and applications of rough sets. Topics include foundations and applications of rough sets as well as foundations and applications of hybrid methods combining rough sets with other approaches important for the development of intelligent systems.

The journal includes high-quality research articles accepted for publication on the basis of thorough peer reviews. Dissertations and monographs up to 250 pages that include new research results can also be considered as regular papers. Extended and revised versions of selected papers from conferences can also be included in regular or special issues of the journal.

Honorary Editor: Zdzisław Pawlak – deceased
Editors-in-Chief: James F. Peters, Andrzej Skowron

Editorial Board

M. Beynon
G. Cattaneo
M.K. Chakraborty
A. Czyżewski
J.S. Deogun
D. Dubois
I. Düntsch
S. Greco
J.W. Grzymała-Busse
M. Inuiguchi
J. Jrvinen
D. Kim
J. Komorowski
C.J. Liau
T.Y. Lin
E. Menasalvas
M. Moshkov
T. Murai

M. do C. Nicoletti
H.S. Nguyen
S.K. Pal
L. Polkowski
H. Prade
S. Ramanna
R. Słowiński
J. Stefanowski
J. Stepaniuk
Z. Suraj
R. Świniarski
M. Szczuka
S. Tsumoto
G. Wang
Y. Yao
N. Zhong
W. Ziarko

Table of Contents

Contributed Papers

Monographs

Propositional Logics from Rough Set Theory

Mohua Banerjee* and Md. Aquil Khan

Department of Mathematics and Statistics,
Indian Institute of Technology,
Kanpur 208 016, India
{mohua,mdaquil}@iitk.ac.in

Abstract. The article focusses on propositional logics with semantics based on rough sets. Many approaches to rough sets (including generalizations) have come to the fore since the inception of the theory, and resulted in different "rough logics" as well. The essential idea behind these logics is, quite naturally, to interpret well-formed formulae as rough sets in (generalized) approximation spaces. The syntax, in most cases, consists of modal operators along with the standard Boolean connectives, in order to reflect the concepts of lower and upper approximations. Non-Boolean operators make appearances in some cases too.

Information systems ("complete" and "incomplete") have always been the "practical" source for approximation spaces. Characterization theorems have established that a rough set semantics based on these "induced" spaces, is no different from the one mentioned above. We also outline some other logics related to rough sets, e.g. logics of information systems – which, in particular, feature expressions corresponding to attributes in their language. These systems address various issues, such as the temporal aspect of information, multiagent systems, rough relations.

An attempt is made here to place this gamut of work, spread over the last 20 years, in one platform. We present the various relationships that emerge and indicate questions that surface.

1 Introduction

A "logic of rough sets" would, in the natural sense, represent a formal system, statements in the language of which would be interpreted as rough sets in some approximation space. Thus "models" in the semantics of such a system would be approximation spaces, equipped with a meaning function that assigns rough sets to well-formed formulae (wffs) of the language.

Rough sets have been defined in more than one way for a Pawlak approximation space (X, R) – [1] lists five definitions, all of which are equivalent to each other. One of these is most commonly used:

(*) a rough set in (X, R), is the pair $(\underline{A}, \overline{A})$, for each $A \subseteq X$,

where $\underline{A}, \overline{A}$ denote the lower and upper approximations of A respectively. Another is a definition given by Pawlak in [2], and of interest to us in this paper:

---

* The author acknowledges the support of the Department of Computer Science, University of Regina, Canada, during a visit to which the paper was finalized.

J.F. Peters et al. (Eds.): Transactions on Rough Sets VI, LNCS 4374, pp. 1–25, 2007.
© Springer-Verlag Berlin Heidelberg 2007

(**) $A \subseteq X$ is a rough set in (X, R), provided the boundary of A, $BnA \neq \emptyset$. For generality's sake, we could remove the restriction in (**) and consider definable sets (i.e. subsets with empty boundary) as special cases of rough sets.

Thus, in the semantics based on approximation spaces, the meaning function defining models, assigns to wffs either subsets of the domain, or pairs of subsets in accordance with (*) [3,4,5,6,7,8,9,10]. This is true even for semantics based on generalized approximation spaces, where different relations (may be more than one in number, with operations on them) are considered [6,11]. The logics invariably involve modalities to express the concepts of lower and upper approximations – some are simply known normal modal logics, or have non-Boolean connectives (and no modalities) in the language, but there are translations into modal logics. We make a study of this group of systems in Section 2. It may be remarked that the "rough logic" proposed by Pawlak [3] (the first system to be called so) makes an appearance here (cf. Section 2.6).

The "practical" source of Pawlak approximation spaces are *complete / deterministic* information systems. These have the form $\mathcal{S} \equiv (U, A, Val, f)$, where U is a set of objects, A a set of *attributes*, Val a set of *values* for the attributes, and f a function from $U \times A$ to Val. An equivalence relation $R_{\mathcal{S}}$ is induced on U (thus giving the approximation space $(U, R_{\mathcal{S}})$), as

$$x \ R_{\mathcal{S}} \ y \text{ in } U, \text{ if and only if } f(x, a) = f(y, a), \text{ for all } a \in A.$$

The converse also holds: given any approximation space (U, R), one can define an information system $\mathcal{S} \equiv (U, A, Val, f)$ such that the induced equivalence $R_{\mathcal{S}}$ is just the relation R. So, in effect, a semantics based on approximation spaces induced by complete information systems, is identical to the one discussed above.

Generalized information systems, termed *incomplete/nondeterministic*, are those where f is a function from $U \times A$ to $\mathcal{P}(Val)$, and yields different kinds of binary relations (e.g. similarity, inclusion – cf. Section 3.1) apart from equivalences, on U. Thus any information system (complete or incomplete) on a domain U, induces a relational system or a (generalized) approximation space on U, i.e. the (non-empty) set U together with a set of binary relations. This is called a *standard structure* on U [12,13,14]. For example, for the complete information system (U, A, Val, f) above, $(U, R_{\mathcal{S}})$ is a standard structure on U. In Section 3.1, $(U, sim_{\mathcal{S}}, in_{\mathcal{S}})$ is a standard structure for the incomplete information system $\mathcal{S} \equiv (U, A, Val, f)$, with similarity and inclusion relations $sim_{\mathcal{S}}, in_{\mathcal{S}}$. (Different sets of relations can give different standard structures on the same set U.)

The induced relations in the standard structure may be characterized by a set of properties. As we know, equivalences are characterized by the properties of reflexivity, symmetry and transitivity. The similarity and inclusion relations considered in Section 3.1 are characterized by the properties $(S1), (S2), (S4) -$ $(S6)$ given there. By a *general structure* on U [12,13,14], one means *any* relational system comprising a non-empty set, along with binary relations that satisfy the set of properties characterizing the induced relations in the standard structure. Again, for the complete information system (U, A, Val, f) above, any Pawlak approximation space (U, R) is a general structure. A general structure for $\mathcal{S}$ of

Section 3.1, would be of the form (U, sim, in), where sim, in are binary relations on U satisfying $(S1), (S2), (S4) - (S6)$.

One finds logics with semantics defined on incomplete information systems, for instance, in [15], or with semantics defined on general structures [16]. However, Vakarelov [12,13,14,17] has established a series of characterization results, enabling an identification of semantics based on general and standard structures (as in case of the Pawlak approximation space and complete information system above). In case of [15] too, we demonstrate here that the logic in question is equivalent to a normal modal logic with certain generalized approximation spaces defining models. These systems are discussed in Section 3.

In another line, there are "logics of information systems", which accommodate in their language, expressions corresponding to objects and attributes [18,19,4,20]. Amongst these is a system that addresses the temporal aspect of information (cf. [4]), while [20] presents a logic for multiagent systems. There are also treatises on "rough relations" – a logic has been proposed [21] on the one hand, and on the other, we have the proposal of a logic programming language in "rough datalog" [22]. In Section 4, we briefly sketch these and other approaches, such as rough mereology [23]. It will be seen that, some of the logics [4,16,20] have atomic propositions as (or built from) *descriptors*, the key feature of *decision logic* [2]. Decision logic is well-known, and not presented in this article.

One should mention that a few of the logics described here, have also been used as a base to express various concepts involving rough sets. For instance, Yao and Lin [6] have defined graded and probabilistic rough sets, using graded and probabilistic modal operators in the language of normal modal systems. Common and distributed knowledge operators have been interpreted in generalized approximation spaces by Wong [24]. In [25], another modal system (inspired by [3]) has been used to propose postulates for rough belief change.

A comparative study of the presented logics is made in Section 5. The paper concludes by indicating possible future directions of investigation in Section 6.

2   Logics with Semantics Based on Approximation Spaces

In this section, we take a look at logics with approximation spaces defining models. We find six kinds of systems.

For a logic $\mathcal{L}$, "α is a theorem of $\mathcal{L}$" shall be indicated by the notation $\vdash_\mathcal{L} \alpha$.

2.1   Normal Modal Systems

The modal nature of the lower and upper approximations of rough sets was evident from the start. Hence, it is no surprise that normal modal systems were focussed upon, during investigations on logics for rough sets. In particular, in case of Pawlak rough sets, the two approximations considered as operators clearly obey all the $S5$ laws. The formal connection between the syntax of $S5$ and its semantics in terms of rough sets is given as follows [26].

According to the Kripke semantics for $S5$, a wff α is interpreted by a function π as a subset in a non-empty domain U, the subset representing the extension of the formula – i.e. the collection of situations/objects/worlds where the wff holds. Moreover, in an $S5$-model $\mathcal{M} \equiv (U, R, v)$ (say), the accessibility relation R is an equivalence on U. Further, if $\Box$, $\Diamond$ denote the necessity and possibility operators respectively then for any wff α, $v(\Box\alpha) = \underline{v(\alpha)}$ and $v(\Diamond\alpha) = \overline{v(\alpha)}$.

A wff α is *true* in $\mathcal{M}$, if $v(\alpha) = U$. Now it can easily be seen that all the $S5$ theorems involving $\Box$ and $\Diamond$ translate into valid properties of lower and upper approximations.

Taking a cue from this connection, similar links have been pointed out (e.g. in [6,27]) between "rough sets" on generalized approximation spaces, and different normal modal systems. The basic idea is to define generalized approximation operators corresponding to any binary relation R on the domain U – this has been done by many (e.g. for tolerance relations in [28] and others – cf. [29]). More explicitly, a map $r : U \rightarrow \mathcal{P}(U)$ is defined as $r(x) \equiv \{y \in U : xRy\}$. Then the operators $\underline{apr}, \overline{apr} : \mathcal{P}(U) \rightarrow \mathcal{P}(U)$ are given by

$$\underline{apr}(A) \equiv \{x : r(x) \subseteq A\}, \text{ and } \overline{apr}(A) \equiv \{x : r(x) \cap A \neq \emptyset\}.$$

The rough set operators then satisfy various properties, depending upon the nature of R. Now let $\mathcal{L}$ denote a normal modal language, and $\mathcal{M} \equiv (U, R, v)$ be a model for $\mathcal{L}$. v, as before, interprets a wff as a subset in U. Then it is straightforward to observe that for any wff α of $\mathcal{L}$,

$$v(\Box\alpha) = \underline{apr}(v(\alpha)), \text{ and dually, } v(\Diamond\alpha) = \overline{apr}(v(\alpha)).$$

By the above interpretation, the modal logics like $KB, KT, K4, S5$ etc. could be said to capture the properties of rough sets in generalized approximation spaces based on different R (symmetric, reflexive, transitive, equivalence etc.).

As remarked in the Introduction, this link has been made use of further. Considering graded and probabilistic modal operators on the above systems, *graded* and *probabilistic* rough sets have been defined in [6]. Wong [24] has interpreted common and distributed knowledge operators (as defined in logic of knowledge) in generalized approximation spaces with an indexed set of indiscernibility relations (corresponding to the knowledge operator of each agent).

2.2   DAL

[11] considers generalized approximation spaces containing a family of equivalence relations instead of just one. The logic DAL that is defined in [11], has models based on these spaces. Further, the set of equivalence relations is assumed to be closed with respect to the operations of intersection and transitive closure of union of relations.

The language of DAL, expectedly, includes a family of modal operators intended to correspond to the indiscernibility relations on the domains of the models. Formally, this is done by having a set $\mathcal{R}$ (say) of *relational variables* apart from the set $\mathcal{P}$ of propositional ones. There are binary operations $\cap$, $\uplus$, and a collection REL of *relational expressions* is built inductively out of the members of $\mathcal{R}$ with these operations. Apart from the classical Boolean connectives, a modal connective $[R]$ is then introduced in the language for each $R \in REL$.

A *DAL*-model is a structure $\mathcal{U} \equiv (U, \{\rho_R\}_{R \in REL}, m)$, where, (i) for any $R \in REL, \rho_R$ is an equivalence relation in the set U; (ii) $\rho_{R \sqcap S}$ is the greatest equivalence relation in U included in both ρ_R and ρ_S; (iii) $\rho_{R \sqcup S}$ is the least equivalence relation including both ρ_R and ρ_S; and (iv) m is the meaning function from $\mathcal{P} \cup \mathcal{R}$ to $\mathcal{P}(U) \cup \{\rho_R\}_{R \in REL}$ such that $m(p) \subseteq U$, for $p \in \mathcal{P}$, and $m(R) \equiv \rho_R$, for $R \in REL$.

For evaluating truth of wffs in *DAL*-models, one defines a function v that is determined by the meaning function m:

$v(p) \equiv m(p)$, for $p \in \mathcal{P}$,

$v([R]\alpha) \equiv \{x \in U : y \in v(\alpha),$ for all y such that $x\ m(R)\ y\}$,

the Boolean cases being defined in the standard way.

Definitions of truth and validity then are as usual: α is true in $\mathcal{U}$, provided $v(\alpha) = U$, and valid if it is true in all *DAL*-models.

DAL has been axiomatized as follows. The connective $\langle\rangle$ is the dual of $[]$.

> A1. All classical tautologies,
> A2. $[R](\alpha \rightarrow \beta) \rightarrow ([R]\alpha \rightarrow [R]\beta)$,
> A3. $[R]\alpha \rightarrow \alpha$,
> A5. $\langle R \rangle \alpha \rightarrow [R]\langle R \rangle \alpha$,
> A5. $[R \uplus S]\alpha \rightarrow [R]\alpha \wedge [S]\alpha$,
> A6. $(([P]\alpha \rightarrow [R]\alpha) \wedge ([P]\alpha \rightarrow [S]\alpha)) \rightarrow ([P]\alpha \rightarrow [R \uplus S]\alpha)$,
> A7. $[R]\alpha \vee [S]\alpha \rightarrow [R \wedge S]\alpha$,
> A8. $(([R]\alpha \rightarrow [P]\alpha) \wedge ([S]\alpha \rightarrow [P]\alpha)) \rightarrow ([R \wedge S]\alpha \rightarrow [P]\alpha)$.

The only rules of inference are Modus Ponens and Necessitation (corresponding to the connective $[R]$ for each $R \in REL$).

The axiomatization yields a completeness result with respect to the aforementioned semantics.

Theorem 1. *For any DAL-wff α, $\vdash_{DAL} \alpha$, if and only if α is valid.*

2.3   Pre-rough Logic

Following in the footsteps of Rasiowa, the algebra of rough sets was investigated in [7] in order to arrive at a logic for the theory. An algebraic structure called *pre-rough algebra* was proposed – this is a *quasi Boolean algebra* [30] along with a topological operator satisfying all the properties of an *interior*, and more. A corresponding logic *PRL* was framed, and observed to be sound and complete with respect to a semantics based on rough sets.

The language of *PRL* has the primitive logical symbols $\neg, \sqcap, \square$. $\sqcup, \Diamond$ are duals of $\sqcap, \square$, while $\Rightarrow$ is defined as:

$$\alpha \Rightarrow \beta \equiv (\neg \square \alpha \sqcup \square \beta) \sqcap (\neg \Diamond \alpha \sqcup \Diamond \beta),$$

for any wffs α, β of *PRL*.

As in the case of $S5$, a model for PRL is of the form $\mathcal{M} \equiv (U, R, v)$, where the departure from the $S5$-semantics lies in the definition of the meaning function v with respect to the connectives of conjunction $\sqcap$ and implication $\Rightarrow$. For any α, β in PRL, $S, T \subseteq U$,

$$v(\alpha \sqcap \beta) \equiv v(\alpha) \sqcap v(\beta), \text{ and}$$
$$v(\alpha \Rightarrow \beta) \equiv v((\neg\Box\alpha \sqcup \Box\beta) \sqcap (\neg\Diamond\alpha \sqcup \Diamond\beta)), \text{ where}$$
$$S \sqcap T \equiv (S \cap T) \cup (S \cap \overline{T} \cap (\overline{S \cap T})^c) \text{ (^{c} denoting complementation).}$$

Definition of truth of a wff α in $\mathcal{M}$ remains the same: this is if and only if $v(\alpha) = U$. It may then be noticed that $\Rightarrow$ reflects *rough inclusion*: a wff $\alpha \Rightarrow \beta$ is true in (U, R, v) provided $v(\alpha)$ is roughly included in $v(\beta)$. Further, $\sqcap / \sqcup$ are operations that reduce to ordinary set intersection / union only when working on definable sets.

α is *valid* (written $\models_{RS} \alpha$), if and only if α is true in every PRL-model.

Following are the axiom schemes for PRL:

1. $\alpha \Rightarrow \alpha$
2a. $\neg\neg\alpha \Rightarrow \alpha$ 2b. $\alpha \Rightarrow \neg\neg\alpha$
3. $\alpha \sqcap \beta \Rightarrow \alpha$ 4. $\alpha \sqcap \beta \Rightarrow \beta \sqcap \alpha$
5a. $\alpha \sqcap (\beta \sqcup \gamma) \Rightarrow (\alpha \sqcap \beta) \sqcup (\alpha \sqcap \gamma)$ 5b. $(\alpha \sqcap \beta) \sqcup (\alpha \sqcap \gamma) \Rightarrow \alpha \sqcap (\beta \sqcup \gamma)$
6. $\Box\alpha \Rightarrow \alpha$
7a. $\Box(\alpha \sqcap \beta) \Rightarrow \Box(\alpha) \sqcap \Box(\beta)$ 7b. $\Box(\alpha) \sqcap \Box(\beta) \Rightarrow \Box(\alpha \sqcap \beta)$
8. $\Box\alpha \Rightarrow \Box\Box\alpha$ 9. $\Diamond\Box\alpha \Rightarrow \Box\alpha$
10a. $\Box(\alpha \sqcup \beta) \Rightarrow \Box\alpha \sqcup \Box\beta$ 10b. $\Box\alpha \sqcup \Box\beta \Rightarrow \Box(\alpha \sqcup \beta)$

Rules of inference :

1. $\dfrac{\alpha \quad \alpha \Rightarrow \beta}{\beta}$ 2. $\dfrac{\alpha \Rightarrow \beta \quad \beta \Rightarrow \gamma}{\alpha \Rightarrow \gamma}$

    modus ponens   *hypothetical syllogism*

3. $\dfrac{\alpha}{\beta \Rightarrow \alpha}$ 4. $\dfrac{\alpha \Rightarrow \beta}{\neg\beta \Rightarrow \neg\alpha}$

5. $\dfrac{\alpha \Rightarrow \beta \quad \alpha \Rightarrow \gamma}{\alpha \Rightarrow \beta \sqcap \gamma}$ 6. $\dfrac{\alpha \Rightarrow \beta, \beta \Rightarrow \alpha \quad \gamma \Rightarrow \delta, \delta \Rightarrow \gamma}{(\alpha \Rightarrow \gamma) \Rightarrow (\beta \Rightarrow \delta)}$

7. $\dfrac{\alpha \Rightarrow \beta}{\Box\alpha \Rightarrow \Box\beta}$ 8. $\dfrac{\alpha}{\Box\alpha}$

9. $\dfrac{\Box\alpha \Rightarrow \Box\beta \quad \Diamond\alpha \Rightarrow \Diamond\beta}{\alpha \Rightarrow \beta}$

One can then prove, for any PRL-wff α,

Theorem 2. $\vdash_{PRL} \alpha$, *if and only if* $\models_{RS} \alpha$.

We shall meet this logic and its semantics again in the coming sections.

2.4 3-Valued Łukasiewicz Logic $\mathcal{L}3$

The connection of rough sets with 3-valuedness, also came up in the context of algebraic investigations. For example, in [31,32,33], an equivalence of 3-valued Łukasiewicz (Moisil) algebras with rough set structures was observed. In terms of logic, the way we can set up a formal link between the intensely studied $\mathcal{L}3$ and a rough set semantics – in fact, the semantics just outlined in Section 2.3, is as follows.

Let us recall Wajsberg's axiomatization of $\mathcal{L}3$ (cf. [34]). The logical symbols $\neg, \rightarrow$ are taken to be primitive.

Axiom schemes:

> 1. $\alpha \rightarrow (\beta \rightarrow \alpha)$.
> 2. $(\alpha \rightarrow \beta) \rightarrow ((\beta \rightarrow \gamma) \rightarrow (\alpha \rightarrow \gamma))$.
> 3. $((\alpha \rightarrow \neg\alpha) \rightarrow \alpha) \rightarrow \alpha$.
> 4. $(\neg\alpha \rightarrow \neg\beta) \rightarrow (\beta \rightarrow \alpha)$.

The only rule of inference is Modus Ponens.

$\mathcal{L}3$ is known to be sound and complete with respect to the class of 3-valued Łukasiewicz (Moisil) algebras, as well as with respect to the semantics on $\mathbf{3} \equiv \{0, 1/2, 1\}$, with Łukasiewicz negation and implication [34].

Now a logic L_1 is said to be *embeddable* into a logic L_2, provided there is a translation * of wffs of L_1 into L_2, such that $\vdash_{L_1} \alpha$ if and only if $\vdash_{L_2} \alpha^*$ for any wff α of L_1. We use the denotation $L_1 \rightharpoonup L_2$. $L_1 \rightleftharpoons L_2$ denotes existence of embeddings both ways.

[31] establishes the following. There are translations $^\circ$ from $\mathcal{L}3$ into PRL and * from PRL into $\mathcal{L}3$ given by

$$(\neg\alpha)^\circ \equiv \neg\alpha^\circ,$$
$$(\alpha \rightarrow \beta)^\circ \equiv (\Diamond\neg\alpha^\circ \sqcup \beta^\circ) \sqcap (\Diamond\beta^\circ \sqcup \neg\alpha^\circ);$$
$$(\neg\alpha)^* \equiv \neg\alpha^*,$$
$$(\alpha \sqcup \beta)^* \equiv (\alpha^* \rightarrow \beta^*) \rightarrow \beta^*,$$
$$(\alpha \sqcap \beta)^* \equiv \neg(\neg\alpha^* \sqcup \neg\beta^*),$$
$$(\Diamond\alpha)^* \equiv \neg\alpha^* \rightarrow \alpha^*.$$

(One may notice that for any α, $(\alpha^\circ)^*$ and $(\alpha^*)^\circ$ are logically equivalent to α in the respective systems.)

It is then shown that $\mathcal{L}3 \rightleftharpoons PRL$. Thus

Theorem 3
> (a) $\vdash_{\mathcal{L}3} \alpha$, *if and only if* $\models_{RS} \alpha^\circ$, *for an $\mathcal{L}3$-wff α and*
> (b) $\vdash_{\mathcal{L}3} \alpha^*$, *if and only if* $\models_{RS} \alpha$, *for a PRL-wff α.*

2.5   Logic for Regular Double Stone Algebras

Another line of algebraic investigation has resulted in linking rough set structures with the class of regular double Stone algebras [35]. A *double Stone algebra* (DSA) is a structure $(L, \sqcup, \sqcap, ^*, ^+, 0, 1)$ such that

$(L, \sqcup, \sqcap, 0, 1)$ is a bounded distributive lattice,
$y \leq x^*$ if and only if $y \sqcap x = 0$,
$y \geq x^+$ if and only if $y \sqcup x = 1$ and
$x^* \sqcup x^{**} = 1$, $x^+ \sqcap x^{++} = 0$.

The operations $^*, ^+$, as evident, are two kinds of complementation on the domain.
    The DSA is *regular* if, in addition to the above, for all $x \in L$,

$$x \sqcap x^+ \leq x \sqcup x^*$$

holds. This is equivalent to requiring that $x^* = y^*$, $x^+ = y^+$ imply $x = y$, for all $x, y \in L$.
    Considering the definition (*) of rough sets (cf. Introduction), one finds that the collection $\mathcal{RS}$ of rough sets $(\underline{X}, \overline{X})$ over an approximation space (U, R) can be made into a regular DSA. The zero of the structure is the element $(\emptyset, \emptyset)$, while the unit is (U, U). The operations $\sqcup, \sqcap, ^*, ^+$ are defined as

$$(\underline{X}, \overline{X}) \sqcup (\underline{Y}, \overline{Y}) \equiv (\underline{X} \cup \underline{Y}, \overline{X} \cup \overline{Y}),$$
$$(\underline{X}, \overline{X}) \sqcap (\underline{Y}, \overline{Y}) \equiv (\underline{X} \cap \underline{Y}, \overline{X} \cap \overline{Y}),$$
$$(\underline{X}, \overline{X})^* \equiv (\overline{X}^c, \overline{X}^c),$$
$$(\underline{X}, \overline{X})^+ \equiv (\underline{X}^c, \underline{X}^c).$$

For the converse, Comer shows that any regular DSA is isomorphic to a subalgebra of $\mathcal{RS}$ for some approximation space (U, R).
    Using these facts, a logic $\mathcal{L_D}$ for rough sets is defined by Düntsch [8] as follows.
    The language of $\mathcal{L_D}$ has two unary connectives $^*, ^+$ (for two kinds of negation), apart from the binary connectives $\vee, \wedge$ and constant symbol $\top$. We write α^*, α^+ instead of $^*\alpha, ^+\alpha$, just to keep parity with the algebraic notation used above.
    A model of $\mathcal{L_D}$ is a pair (W, v), where W is a (non-empty) set and v is the meaning function assigning to propositional variables, pairs in $\mathcal{P}(W) \times \mathcal{P}(W)$ such that if $v(p) = (A, B)$ then $A \subseteq B$. $v(p) = (A, B)$ is to express that "p holds at all states of A and does not hold at any state outside B". For $\top$, we have $v(\top) \equiv (W, W)$.

v is extended to the set of all wffs recursively:

if $v(\alpha) = (A, B)$ and $v(\beta) = (C, D)$ then
$v(\alpha \vee \beta) \equiv (A \cup C, B \cup D)$,
$v(\alpha \wedge \beta) \equiv (A \cap C, B \cap D)$,
$v(\alpha^*) \equiv (B^c, B^c)$,
$v(\alpha^+) \equiv (A^c, A^c)$.

A wff α is true in a model (W, v), provided $v(\alpha) = (W, W)$.

We would now like to make explicit, how v interprets the wffs of $\mathcal{L_D}$ as rough sets over some approximation space. One refers to [8], and [35].

Consider the range $ran(v)$ of the map v in $\mathcal{P}(W) \times \mathcal{P}(W)$. It can be shown that it forms a regular DSA through the operations $\sqcup, \sqcap, ^*, ^+$:

$$v(\alpha) \sqcup v(\beta) \equiv v(\alpha \vee \beta),$$
$$v(\alpha) \sqcap v(\beta) \equiv v(\alpha \wedge \beta),$$
$$v(\alpha)^* \equiv v(\alpha^*),$$
$$v(\alpha)^+ \equiv v(\alpha^+).$$

$v(\top^*)$ (or $v(\top^+)$) is the zero $((\emptyset, \emptyset))$ of the algebra, while $v(\top) = (W, W)$ is the unit.

In fact, the variety of regular DSA's is just the one generated by regular DSA's of the kind $ran(v)$, where v ranges over all meaning functions for all models.

Using the correspondence between classes of algebras and logic [36], [8] concludes, amongst other properties of $\mathcal{L}_\mathcal{D}$, that

Theorem 4. $\mathcal{L}_\mathcal{D}$ *has a finitely complete and strongly sound Hilbert style axiom system.*

Through Comer's representation result, $ran(v)$ corresponding to any model (W, v) of $\mathcal{L}_\mathcal{D}$, is isomorphic to a subcollection of $\mathcal{RS}$ for some approximation space (U, R). We can now say that $v(\alpha)$ for a wff α, can be identified with a rough set over some (U, R) in precisely the following manner.

Let U consist of all the *join irreducible* elements of $ran(v)$, i.e. $v(\alpha) \in U$, if and only if $v(\alpha) \neq (\emptyset, \emptyset)$, and for all wffs β, γ, if $v(\alpha) = v(\beta) \sqcup v(\gamma)$ then either $v(\alpha) = v(\beta)$ or $v(\alpha) = v(\gamma)$. An equivalence relation R on U can then be obtained, where R is given by:

$$v(\alpha) \; R \; v(\beta) \text{ if and only if } v(\alpha^{**}) = v(\beta^{**}),$$

i.e. if and only if $B = D$, where $v(\alpha) = (A, B)$ and $v(\beta) = (C, D)$.
Now define $f : ran(v) \rightarrow \mathcal{P}(U)$ such that for $v(\alpha) = (A, B)$,

$$f(A, B) \equiv \{v(\beta) = (C, D) \in U : C \subseteq A, D \subseteq B\}.$$

Finally, define the map $g : ran(v) \rightarrow \mathcal{P}(U) \times \mathcal{P}(U)$ as:

$$g(A, B) \equiv (f(A, A), f(B, B)), \text{ where } v(\alpha) = (A, B).$$

(Note that $(A, A), (B, B) \in U$, as $v(\alpha^{++}) = (A, A)$, and $v(\alpha^{**}) = (B, B)$.)
It can then be shown that (a) g is injective, and (b) g preserves $\sqcup, \sqcap, ^*, ^+$.
Moreover, if $v(\alpha) = (A, B)$,

$$g(v(\alpha)) = (\; \underline{f(A, B)}, \overline{f(A, B)} \;),$$

a rough set in the approximation space (U, R).

[8] does not present an explicit proof method for the logic $\mathcal{L}_\mathcal{D}$ – the only comment on the matter is vide Theorem 4. Recently, Dai [9] has presented a sequent calculus for a logic (denoted $RDSL$) with a semantics based on the regular DSAs formed by collections of rough sets of the kind $\mathcal{RS}$ over some approximation space (U, R) (defined earlier in the section). The language of $RDSL$ is the same as that of $\mathcal{L}_\mathcal{D}$, except that the constant symbol $\perp$ (dual for $\top$) is included amongst the

primitive symbols. Models are of the form $(\mathcal{RS}, v)$, where v, the meaning function, is a map from the set of propositional variables to $\mathcal{RS}$. Thus $v(p)$, for a propositional variable p, is a pair $(\underline{X}, \overline{X})$ in the approximation space (U, R). v is extended to the set of all wffs in the same way as for models of $\mathcal{L_D}$.

We note that an $RDSL$-model $(\mathcal{RS}, v)$ may be identified with the $\mathcal{L_D}$-model (U, v). On the other hand, due to Comer's representation result, given any $\mathcal{L_D}$-model (W, v), there is an isomorphism f from $ran(v)$ to a subalgebra $(\mathcal{S}$, say$)$ of $\mathcal{RS}$ on some approximation space. One can thus find an $RDSL$-model $(\mathcal{RS}, v')$ such that $ran(v')$ is $\mathcal{S}$, i.e. $v'(p) \equiv f(v(p))$, for every propositional variable p. So, in this sense, the classes of models of the two logics are identifiable.

As in classical sequent calculus, for finite sequences of wffs $\Gamma \equiv (p_1, p_2, \ldots p_m)$ and $\Delta \equiv (q_1, q_2, \ldots q_n)$ in $RDSL$, the sequent $\Gamma \Rightarrow \Delta$ is said to be valid in a model $(\mathcal{RS}, v)$ if and only if

$$v(p_1) \sqcap \ldots \sqcap v(p_m) \leq v(q_1) \sqcup \ldots \sqcup v(q_n).$$

$\sqcup, \sqcap$ are the operations in the regular DSA $(\mathcal{RS}, \sqcup, \sqcap, ^*, ^+, < \emptyset, \emptyset >, < U, U >)$.
$\Gamma \Rightarrow \Delta$ is said to be valid (in notation, $\models_{RDSA} \Gamma \Rightarrow \Delta$) if and only if $\Gamma \Rightarrow \Delta$ is valid in every $RDSL$-model.

The standard classical axiom $p \Rightarrow p$ and rules for the connectives $\wedge, \vee$ and constant symbols $\top, \bot$ are considered to define derivability $(\vdash_{RDSL})$. In addition, the axioms and rules for the two negations $^*, ^+$ are as follows.

$$1. \ p \Rightarrow p^{**}.$$
$$2. \ p^* \Rightarrow p^{***}.$$
$$3. \ p \Rightarrow p^{++}.$$
$$4. \ p^+ \Rightarrow p^{+++}.$$

$$(R^*) \ \frac{\Gamma \Rightarrow \Delta}{\Delta^* \Rightarrow \Gamma^*} \qquad (R^+) \ \frac{\Gamma \Rightarrow \Delta}{\Delta^+ \Rightarrow \Gamma^+}$$

Soundness and completeness are then proved, with respect to the semantics sketched.

Theorem 5. $\vdash_{RDSL} \Gamma \Rightarrow \Delta$, if and only if $\models_{RDSA} \Gamma \Rightarrow \Delta$.

2.6 Logic for Rough Truth or of Rough Consequence

In [3], a logic R_l (the first in literature to be called "rough logic") was proposed, along with a very appealing notion of *rough truth*. The language of R_l consists of the standard Boolean connectives, and models $\mathcal{M} \equiv (U, R, v)$ are based on approximation spaces. v assigns subsets of the domain U to wffs in the usual manner. Five logical values of "truth", "falsity", "rough truth", "rough falsity" and "rough inconsistency" are considered in this work, with truth and falsity representing the limit of our partial knowledge.

As we know, a wff α is *true* in $\mathcal{M}$, if $v(\alpha) = U$. α is said to be *surely/possibly* true on $x \in U$, if $x \in \underline{v(\alpha)}$ $(\overline{v(\alpha)})$ respectively. α is *roughly* true in $\mathcal{M}$, if it is possibly true on every x in U, i.e. $\overline{v(\alpha)} = U$, or in other words, $v(\alpha)$ is *externally indiscernible* [37] in (U, R). On the other hand, α is *roughly false*, when $v(\alpha) = \emptyset$ ($v(\alpha)$ is *internally* indiscernible), and α is *roughly inconsistent*, if it is $\overline{\text{both}}$ roughly true and false ($v(\alpha)$ is *totally* indiscernible).

Let us consider the modal system $S5$. Note that models of $S5$ and R_l are identical. We can then effect a translation of the above concepts into $S5$. In (U, R, v), a wff α can be termed roughly true if $\overline{v(\alpha)} = v(\Diamond\alpha) = U$, roughly false if $\underline{v(\alpha)} = v(\Box\alpha) = \emptyset$, and roughly inconsistent if both hold.

In [10], a logic L_r having the same models as above was proposed, with the speciality that the syntax-semantics relationships are explored with *rough truth* replacing truth and *rough validity* replacing validity. The notion of consistency is replaced by one of *rough consistency* too. The consequence relation defining the logic is also non-standard. These ideas were first mooted in [5,26], and L_r is a modified version of the formal system discussed there.

L_r has a normal modal language. A model $\mathcal{M} \equiv (U, R, v)$ is a *rough model* of Γ, if and only if for every $\gamma \in \Gamma$, $v(\Diamond\gamma) = U$, i.e. γ is roughly true in $\mathcal{M}$. α is a *rough semantic consequence* of Γ (denoted $\Gamma \approx\!\!\!\mid \alpha$) if and only if every rough model of Γ is a rough model of α. If Γ is empty, α is said to be *roughly valid*, written $\approx\!\!\!\mid \alpha$.

There are two rules of inference:

$$R_1. \qquad \frac{\alpha}{\beta} \qquad\qquad R_2. \qquad \frac{\Diamond\alpha}{\Diamond\beta}$$
$$\text{if } \vdash_{S5} \Diamond\alpha \to \Diamond\beta \qquad \frac{\Diamond\alpha \quad \Diamond\beta}{\Diamond\alpha \wedge \Diamond\beta}$$

The consequence relation is defined as follows. Let Γ be any set of wffs and α any wff in L_r.

α is a *rough consequence* of Γ (denoted $\Gamma \mid\!\sim \alpha$) if and only if there is a sequence $\alpha_1, ..., \alpha_n$ ($\equiv \alpha$) such that each α_i ($i = 1, ..., n$) is either (i) a theorem of $S5$, or (ii) a member of Γ, or (iii) derived from some of $\alpha_1, ..., \alpha_{i-1}$ by R_1 or R_2. If Γ is empty, α is said to be a *rough theorem*, written $\mid\!\sim \alpha$.

A kind of "rough Modus Ponens" is then derivable, in the form: if $\Gamma \mid\!\sim \alpha$, $\vdash_{S5} \alpha' \to \beta$ with $\vdash_{S5} \alpha \approx \alpha'$ then β. Here $\approx$ reflects the notion of "rough equality", $\alpha \approx \beta \equiv (\Box\alpha \leftrightarrow \Box\beta) \wedge (\Diamond\alpha \leftrightarrow \Diamond\beta)$. One also obtains soundness of L_r with respect to the above semantics: if $\Gamma \mid\!\sim \alpha$ then $\Gamma \approx\!\!\!\mid \alpha$.

It is clear that in the face of an incomplete description of a concept p, p and "not" p (in the classical sense) may not always represent conflicting situations. To accommodate this possibility, a set Γ of wffs is termed *roughly consistent* if and only if the set $\Diamond\Gamma \equiv \{\Diamond\gamma : \gamma \in \Gamma\}$ is $S5$-consistent.

With the help of this notion, one obtains

Theorem 6. *(Completeness)*
    (a) Γ *is roughly consistent if and only if it has a rough model.*
    (b) *For any L_r-wff α, if $\Gamma \approx\!\!\!\mid \alpha$ then $\Gamma \mid\!\sim \alpha$.*

Thus, L_r appears as another system that is able to address rough sets and related notions. We shall remark on its relationship with other well-known systems in Section 5. It may be mentioned that L_r has been used as the base logic for a proposal of *rough belief change* in [25].

3   Logics with Semantics Based on Information Systems

We now present logics, the models of which are defined on approximation spaces induced by information systems. We find one pioneering system NIL that has inspired the proposal of many others in the same line. The section also includes a logic by Nakamura, the models of which are directly defined on information systems.

3.1 NIL

Recall that an incomplete information system is of the form $\mathcal{S} \equiv (U, A, Val, f)$, where U is a set of objects, A a set of attributes, Val a set of values for the attributes, and f a function from $U \times A$ to $\mathcal{P}(Val)$.

The logic NIL proposed by Orłowska and Pawlak [16] works on incomplete information systems, in which the function f satisfies an additional condition:

$\quad (\diamond) \quad f(x, a) \neq \emptyset$, for all $x \in U$, $a \in A$.

One observes that, given $\mathcal{S} \equiv (U, A, Val, f)$, two particular kinds of binary relations on the domain U are induced – these dictate the formulation of NIL. Let $x, y \in U$.

Similarity (sim_S): $x \ sim_S \ y$ if and only if $f(x, a) \cap f(y, a) \neq \emptyset$, for all $a \in A$.
Inclusion (in_S): $x \ in_S \ y$ if and only if $f(x, a) \subseteq f(y, a)$, for all $a \in A$.

It can be shown that for every incomplete information system $\mathcal{S} \equiv (U, A, Val, f)$ and $x, y, z \in U$, the following hold.

$\quad (S1)$ $x \ in_S \ x$.
$\quad (S2)$ if $x \ in_S \ y$ and $y \ in_S \ z$ then $x \ in_S \ z$.
$\quad (S3)$ if $x \ sim_S \ y$ for some y, then $x \ sim_S \ x$.
$\quad (S4)$ if $x \ sim_S \ y$ then $y \ sim_S \ x$.
$\quad (S5)$ if $x \ sim_S \ y$, $x \ in_S \ u$, $y \ in_S \ v$ then $u \ sim_S \ v$.

Further, if the condition $(\diamond)$ is satisfied by f then sim satisfies

$\quad (S6)$ $x \ sim_S \ x$.

Thus a *standard structure* (cf. Introduction) corresponding to an incomplete information system $\mathcal{S} \equiv (U, A, Val, f)$ with condition $(\diamond)$, would be (U, sim_S, in_S). On the other hand, a *general structure* for $\mathcal{S}$ would be of the form (U, sim, in), where sim, in are binary relations on U satisfying $(S1), (S2), (S4) - (S6)$. For brevity, we refer to these as standard and general NIL-structures respectively.

NIL could be termed as a modal version of decision logic introduced by Pawlak [2], an association similar to that of rough logic [3] and $S5$ (cf. Section 2.6). The atomic propositions of NIL are the descriptors of decision logic – of the form (a, v), where a is an "attribute constant", and v a constant representing "value of attribute".

Apart from the standard Boolean connectives $\neg, \vee$, the language contains modal connectives $\square, \square_1, \square_2$ corresponding to sim, in and the *inverse* in^{-1} of in respectively. Wffs are built, as usual, out of the atomic propositions (descriptors) and the connectives. Note that there are no operations on the attribute or value constants.

A NIL-model $\mathcal{M} \equiv (U, sim, in, m)$ consists of a general structure (U, sim, in) as above, along with a meaning function m from the set of all descriptors to the set $\mathcal{P}(U)$.

m is extended recursively to the set of all NIL-wffs in the usual manner. In particular,
$$m(\Box \alpha) \equiv \{x \in U : y \in m(\alpha) \text{ for all } y \text{ such that } x \ sim \ y\}.$$
Similarly one defines $m(\Box_1 \alpha)$, and $m(\Box_2 \alpha)$.

α is true in the model $\mathcal{M}$, if $m(\alpha) = U$.

The following deductive system for NIL was proposed in [16].

Axiom schemes:

$A1.$ All classical tautologies,

$A2.$ $\Box_2(\alpha \rightarrow \beta) \rightarrow (\Box_2 \alpha \rightarrow \Box_2 \beta)$,

$A3.$ $\Box_1(\alpha \rightarrow \beta) \rightarrow (\Box_1 \alpha \rightarrow \Box_1 \beta)$,

$A4.$ $\Box(\alpha \rightarrow \beta) \rightarrow (\Box \alpha \rightarrow \Box \beta)$,

$A5.$ $\alpha \rightarrow \Box_1 \neg \Box_2 \neg \alpha$,

$A6.$ $\alpha \rightarrow \Box_2 \neg \Box_1 \neg \alpha$,

$A7.$ $\Box_2 \alpha \rightarrow \alpha$,

$A8.$ $\Box_1 \alpha \rightarrow \alpha$,

$A9.$ $\Box \alpha \rightarrow \alpha$,

$A10.$ $\Box_2 \alpha \rightarrow \Box_2 \Box_2 \alpha$,

$A11.$ $\Box_1 \alpha \rightarrow \Box_1 \Box_1 \alpha$,

$A12.$ $\alpha \rightarrow \Box \neg \Box \neg \alpha$,

$A13.$ $\Box \alpha \rightarrow \Box_2 \Box \Box_1 \alpha$.

Rules of inference:

$$(R1) \ \frac{\alpha, \ \alpha \rightarrow \beta}{\beta} \quad (R2) \ \frac{\alpha}{\Box_2 \alpha}$$

$$(R3) \quad \frac{\alpha}{\Box_1 \alpha} \quad (R4) \ \frac{\alpha}{\Box \alpha}$$

It has been proved that

Theorem 7. *For any NIL-wff α, $\vdash_{NIL} \alpha$ if and only if α is true in all NIL-models.*

3.2   Logics by Vakarelov

Vakarelov addresses the issue of completeness of various logics, the models of which are based on standard structures corresponding to some information system. For instance, in the case of NIL, the question would be about a completeness theorem with respect to the class of NIL-models defined on standard NIL-structures (cf. Section 3.1). In [12], such a theorem is proved, via a key characterization result. In fact, this result set the ground for a series of similar observations when the binary relations involved are changed.

Proposition 1. *(Characterization) Let (U, sim, in) be a general NIL-structure. Then there exists an information system $S \equiv (U, A, Val, f)$ with f satisfying $(\diamond)$, such that $sim_S = sim$ and $in_S = in$.*

In other words, the classes of NIL-models based on standard and general NIL-structures are identical. Hence one obtains the required completeness theorem.

The condition $(\diamond)$, viz. $f(x, a) \neq \emptyset$ for all $x \in U$, $a \in A$, is a restrictive one. However, it is observed by Vakarelov that even if this condition is dropped, a characterization result similar to Proposition 1 can be obtained. Instead of reflexivity of sim (cf. property $(S6)$, Section 3.1), we now have just the condition of *quasireflexivity* – cf. property $(S3)$: if $x \, sim \, y$ for some y, then $x \, sim \, x$. The corresponding logic can be obtained from NIL by replacing the axiom $A9$ by

$$\neg\Box(p \wedge \neg p) \rightarrow (\Box\alpha \rightarrow \alpha).$$

Following this approach, one handles the cases of incomplete information systems inducing different binary relations. For example, [14,13,17] consider these relations amongst others, for $S \equiv (U, A, Val, f)$:

Indiscernibility (ind_S): $x \, ind_S \, y$ if and only if $f(x, a) = f(y, a)$, for all $a \in A$,
Weak indiscernibility (ind_S^w): $x \, ind_S^w \, y$ if and only if $f(x, a) = f(y, a)$, for some $a \in A$,
Weak similarity (sim_S^w): $x \, sim_S^w \, y$ if and only if $f(x, a) \cap f(y, a) \neq \emptyset$, for some $a \in A$.
Complementarity (com): $x \, com \, y$ if and only if $f(x, a) = (VAL_a \setminus f(y, a))$, for all $a \in A$, where Val_a is the value set for the particular attribute a, and $Val \equiv \cup\{Val_a : a \in A\}$.

The characterization result for each has been obtained, the corresponding logical system is defined and the completeness theorem with respect to models on the intended standard structures is proved.

3.3   Logic by Nakamura

[15] discusses a logic with models on incomplete information systems. We recall (cf. Introduction) that given a complete information system $S \equiv (U, A, Val, f)$, one can define the equivalence relation R_S. The lower approximation of $X(\subseteq U)$ under this relation is denoted as $\underline{X}_S$, and its upper approximation as $\overline{X}_S$.

Nakamura defines a *completion* S_0 of an incomplete information system S as a complete information system that can be constructed from S by selecting any one value from $f(x, a)(\subseteq Val)$, for each $x \in U, a \in A$. If $f(x, a) = \emptyset$, one selects a special symbol ϵ. The relationship of S_0 and S is expressed as $S_0 \geq S$.

Now the "lower" and "upper approximations" $\underline{X}, \overline{X}$ of $X \subseteq U$ in an incomplete information system $S \equiv (U, A, Val, f)$ are defined as follows:

$$(*) \quad \underline{X} \equiv \cap_{S_0 \geq S}\underline{X}_{S_0}, \quad \overline{X} \equiv \cup_{S_0 \geq S}\overline{X}_{S_0}.$$

With this background, a logic $INCRL$ is proposed, having the standard Boolean connectives, and two modal operators $[], \langle\rangle$ (corresponding to "surely" and "possibly" respectively).

An $INCRL$-model is an incomplete information system $\mathcal{S} \equiv (U, A, Val, f)$ along with a meaning function $v_{\mathcal{S}}$ from the set of propositional variables of the language to $\mathcal{P}(U)$. $v_{\mathcal{S}}$ is extended as usual for the wffs involving Boolean connectives. For wffs with modal operators, one makes use of completations $\mathcal{S}_0$ of $\mathcal{S}$ and the preceding definitions of lower and upper approximations given in ($*$).

$$v_{\mathcal{S}}([\,]\alpha) \equiv \cap_{\mathcal{S}_0 \geq \mathcal{S}} \underline{v_{\mathcal{S}}(\alpha)}_{\mathcal{S}_0} = \underline{v_{\mathcal{S}}(\alpha)},$$
$$v_{\mathcal{S}}(\langle\rangle\alpha) \equiv \cup_{\mathcal{S}_0 \geq \mathcal{S}} \overline{v_{\mathcal{S}}(\alpha)}_{\mathcal{S}_0} = \overline{v_{\mathcal{S}}(\alpha)}.$$

Truth and validity of wffs are defined again as for most of the previous systems. Nakamura points out relationships of $INCRL$ with the modal system KTB, in particular that all theorems of KTB are valid wffs of $INCRL$. We shall take a further look at the two logics in Section 5.

4   Other Approaches

This section outlines a few proposals of logics related to rough sets, the models of which are based on structures that are even more generalized than the ones already presented. As we shall see, these logics have dimensions not accounted for in the systems presented so far.

4.1   Temporal Approach

Orłowska (cf. [4]), defines a logic $\mathcal{L}_T$ with models on *dynamic information systems*, in order to deal with the temporal aspect of information. A set T of moments of time, and a suitable relation R on the set T are considered along with the set U of objects and A of attributes. Formally, a dynamic information system is a tuple $\mathcal{S} \equiv (U, A, Val, T, R, f)$, where $Val \equiv \cup\{Val_a : a \in A\}$, ($Val_a$, as in Section 3.2, being the value set for the particular attribute a) and the information function $f : U \times T \times A \rightarrow Val$ satisfies the condition that $f(x, t, a) \in VAL_a$, for any $x \in U$, $t \in T$, $a \in A$.

In the language of $\mathcal{L}_T$, atomic statements are descriptors of decision logic, together with an object constant x – so these are triples (x, a, v), and are intended to express: "object x assumes value v for attribute a". There are modal operators to reflect the relations R and R^{-1}. The truth of all statements of the language is evaluated in a model based on a dynamic information system, with respect to moments of time, i.e. members of the set T.

An $\mathcal{L}_T$-model is a tuple $\mathcal{M} \equiv (\mathcal{S}, m)$ where $\mathcal{S}$ is a dynamic information system, and m a meaning function which assigns objects, attributes and values from U, A, Val to the respective constants.

The satisfiability of a formula α in a model $\mathcal{M}$ at a moment $t(\in T)$ of time is defined inductively as follows:

$$\mathcal{M}, t \models (x, a, v) \text{ if and only if } f(m(x), t, m(a)) = m(v).$$

For the Boolean cases, we have the usual definitions. For the modal case,

$$\mathcal{M}, t \models [R]\alpha \text{ if and only if for all } t' \in T, \text{ if } (t,t') \in R \text{ then } \mathcal{M}, t \models \alpha.$$

A wff is true in $\mathcal{M}$, provided it is satisfied in $\mathcal{M}$ at every $t \in T$. $\mathcal{L}_T$ is complete with respect to this class of models, for the axioms of linear time temporal logic, and an axiom which says that the values of attributes are uniquely assigned to objects.

4.2 Multiagent Systems

[20] describes a logic, that takes into account a (finite) collection of *agents* and their *knowledge bases*. We denote the logic as $\mathcal{L}_{MA}$. The language of $\mathcal{L}_{MA}$ has "agent constants" along with two special constants 0,1. Binary operations $+, \cdot$ are provided to build the set $\mathcal{T}$ of *terms* from these constants. Wffs of one kind are obtained from terms, and are of the form $s \Rightarrow t$, $s, t \in \mathcal{T}$, where $\Rightarrow$ is a binary relational symbol. $s \Rightarrow t$ is to reflect that "the classification ability of agent t is at least as good as that of agent s".

Furthermore, there are attribute as well as attribute-value constants. Descriptors formed by these constants constitute atomic propositions, and using connectives $\wedge, \neg$ and modal operators I_t, $t \in \mathcal{T}$ (representing "partial knowledge" of each agent), give wffs of another kind.

$\mathcal{L}_{MA}$-models are not approximation spaces, but what could be called "partition spaces" on information systems. Informally put, a model consists of an information system $\mathcal{S} \equiv (U, A, Val, f)$, and a family of *partitions* $\{E_t\}_{t \in \mathcal{T}}$ on the domain U – each corresponding to the knowledge base of an agent. The family is shown to have a lattice structure, and the ordering involved gives the interpretation of the relational symbol $\Rightarrow$. Wffs built out of descriptors are interpreted in the standard way, in the information system $\mathcal{S}$. The partial knowledge operator I_t for a term t reflects the lower approximation operator with respect to the partition E_t on U. An axiomatization of $\mathcal{L}_{MA}$ is presented, to give soundness and completeness results.

In the context of multiagent systems, it is worth mentioning the approach followed in [38], even though a formal logic based on it has not been defined yet. *Property systems* (*P*-systems) are defined as triples of the form $(U, A, \models)$, where U is a set of objects, A a set of *properties*, and $\models$ a "fulfilment" relation between U and A. For each P-system $\mathcal{P}$, a collection $\mathcal{P}^{op}$ of *interior* and *closure* operators satisfying specific properties are considered. These operators could be regarded as generalizations of lower and upper approximations. Now given a family $\{\mathcal{P}_k\}_{k \in K}$ of P-systems (each for an agent, say) over some index set K and over the same set U of objects, one obtains a *multiagent pre-topological approximation space* as a structure $(U, \{\mathcal{P}_k^{op}\}_{k \in K})$. It is to be seen if such a generalized structure could form the basis of a semantics of some formal logical framework.

4.3 Rough Relations

Discussion about relations on approximation spaces, started from [39]. We find two directions of work on this topic.

Logic of Rough Relations: [40] considers another generalization of the notion of an approximation space – taking systems of the form $AS \equiv (U, I, v)$, where U is a non-empty set of objects, $I : U \to \mathcal{P}(U)$ an *uncertainty function*, and $v : \mathcal{P}(U) \times \mathcal{P}(U) \to [0, 1]$ is a *rough inclusion* function satisfying the following conditions:

$v(X, X) = 1$ for any $X \subseteq U$,
$v(X, Y) = 1$ implies $v(Z, Y) \geq v(Z, X)$ for any $X, Y, Z \subseteq U$,
$v(\emptyset, X) = 1$ for any $X \subseteq U$.

For any subset X of U, we then have the *lower* and *upper approximations*:

$L(AS, X) \equiv \{x \in U : v(I(x), X) = 1\}$,
$U(AS, X) \equiv \{x \in U : v(I(x), X) > 0\}$.

A 'rough set' in AS is the pair $(L(AS, X), U(AS, X))$.

The above is motivated from the fact that any Pawlak approximation space (U, R) is an instance of a generalized space as just defined. Indeed, we consider the function I that assigns to every object its equivalence class under R, and the inclusion function v as:

$$v(S, R) \equiv \begin{cases} \frac{card(S \cap R)}{card(S)} & \text{if } S \neq \emptyset \\ 1 & \text{if } S = \emptyset \end{cases}$$

For an approximation space $AS \equiv (U, I, v)$ with $U = U_1 \times U_2$ and v as in the special case above, [21] discusses relations $R \subseteq U_1 \times U_2$. The lower and upper approximation of R in AS are taken, and a *rough relation* is just a rough set in AS.

A decidable multimodal logic is proposed – for reasoning about properties of rough relations. The modal operators correspond to a set of relations on the domain of the above generalized approximation spaces, as well as the lower and upper approximations of these relations. An axiomatization for the logic is given, and completeness is proved with respect to a Kripke-style semantics.

Rough Datalog: Just as *decision tables* [2] are (complete) information systems with special attributes, viz. the *decision attributes*, [22] considers a *decision system* $(U, A \cup \{d\})$ – but with a difference. Each attribute a in A is a *partial map* from U to a value set V_a, and d, the decision attribute, is a partial map from U to $\{0, 1\}$. It is possible that for some $x \in U$, all attribute values (including the value of d) are undefined. A 'rough set' X is taken to be a pair (X^+, X^-), where X^+ is the set of elements of U that may belong to X, while X^- contains those elements of U that may not belong to X. d indicates the information about membership of an object of U in X.

Formally, let $A \equiv \{a_1, ..., a_n\}$, $A(x) \equiv (a_1(x), ..., a_n(x))$ for each $x \in U$, and $A^{-1}(t) \equiv \{x \in U : A(x) = t\}$, for $t \in V_{a_1} \times ... \times V_{a_n}$. (Note that for some $x \in U$, $A(x)$ could be undefined). Then

$X^+ \equiv \{x \in U : A \text{ is defined for } x, \text{ and } d(x') = 1, \text{ for some } x' \in A^{-1}(A(x))\}$,
and
$X^- \equiv \{x \in U : A \text{ is defined for } x, \text{ and } d(x') = 0, \text{ for some } x' \in A^{-1}(A(x))\}$.

This definition implies that X^+ and X^- may not be disjoint, allowing for the presence of conflicting (contradictory) decisions in the decision table. On the other hand, X^+ and X^- may not cover U either, allowing for the possibility that there is no available information about membership in X.

With these definitions, 'rough relations' are considered in [22]. Standard relational data base techniques, such as relational algebraic operations (e.g. union, complement, Cartesian product, projection) on crisp relations, are extended to the case of rough relations. A declarative language for defining and querying these relations is introduced - pointing to a link of rough sets (as just defined) with logic programming.

4.4   Logics with Attribute Expressions

As we have seen, $\mathcal{L}_T$ and $\mathcal{L}_{\mathcal{MA}}$ (cf. Sections 4.1 and 4.2 respectively) have attribute expressions in the language that are interpreted in information systems. NIL (cf. Section 3.1), also has attribute constants in the language. But unlike the models of $\mathcal{L}_T$ and $\mathcal{L}_{\mathcal{MA}}$, the standard or general NIL-structures defining NIL-models do not accommodate attributes, and the wffs (which are built using the attribute constants) point to collections of objects of the domain.

A class of logics with attribute expressions are also defined in [18,19]. Models are based on structures of the form $(U, A, \{ind(P)\}_{P \subseteq A})$, where the "indiscernibility" relation $ind(P)$ for each subset P of the attribute set A, has to satisfy certain conditions. For the models of one of the logics, for example, the following conditions are stipulated for $ind(P)$:

$(U1)$ $ind(P)$ is an equivalence relation on U,
$(U2)$ $ind(P \cup Q) = ind(P) \cap ind(Q)$,
$(U3)$ if $P \subseteq Q$ then $ind(Q) \subseteq ind(P)$, and
$(U4)$ $ind(\emptyset) = U \times U$.

Other logics may be obtained by changing some of $(U1) - (U4)$. The language of the logics has a set of variables each representing a set of attributes, as well as constants to represent all one element sets of attributes. Further, the language can express the result of (set-theoretic) operations on sets of attributes. The logics are multimodal – there is a modal operator to reflect the indiscernibility relation for each set of attributes as above. A usual Kripke-style semantics is given, and a number of valid wffs presented. However, as remarked in [19], we do not know of a complete axiomatization for such logics.

4.5   Rough Mereology

This is an approach inspired by the theory of *mereology* due to Leśniewski (1916). Leśniewski propounds a theory of sets that has *containment* as the primitive relation, rather than membership. Drawing from this classical theory, *rough mereology* has been proposed [23], providing a useful notion of *rough containment*, of "being a part, *in a degree*".

Formally, this can be defined as a real binary function μ on the domain with values in [0,1], satisfying certain conditions (abstracted from the properties of classical containment). A given information system (U, A, Val, f), a partition of A into, say $A_1, ..., A_n$ and a set of weights $\{w_1, ..., w_n\}$, can generate $\mu(x, y)$, $x, y \in U$. It is assumed that $w_i \in [0, 1]$, $i = 1, ..., n$, and $\sum_{i=1}^{n} w_i = 1$.

A *pre-rough* inclusion μ_o is first defined:

$$\mu_o(x, y) \equiv \sum_{i=1}^{n} w_i.(|ind_i(x, y)|/|A_i|),$$

where $ind_i(x, y) \equiv \{a \in A_i : f(x, a) = f(y, a)\}$. μ_o can then be extended to *rough inclusion* μ over $\mathcal{P}(U)$ by using t-norms and t-conorms. Rough inclusion can be used, for instance, in specifying approximate decision rules.

It may be remarked that predicate logics corresponding to rough inclusions have been proposed recently in [41].

5   Comparative Study

We now discuss some relationships between the logics presented in Sections 2 and 3.

5.1   Embeddings

Let us recall the notion of an embedding of logics – cf. Section 2.4. We consider the logics PRL, $\mathcal{L}3$, $\mathcal{L}_\mathcal{D}$, $RDSL$ presented in Sections 2.3, 2.4 and 2.5 respectively, and point out interrelationships, as well as relations with other known logics.

(1) $\mathcal{L}3 \rightleftharpoons PRL$: This has already been seen in Section 2.4.

(2) $\mathcal{L}3 \rightleftharpoons \mathcal{L}_\mathcal{D}$: As summarized in [1] and observed by Düntsch and Pagliani, regular double Stone algebras and 3-valued Łukasiewicz algebras are equivalent to each other via suitable transformations. Passing on to the respective logics, we would thus find embeddings both ways, between $\mathcal{L}_\mathcal{D}$ and $\mathcal{L}3$.

(3) $\mathcal{L}_\mathcal{D} \rightleftharpoons RDSL$: We can define, in $RDSL$, that *a wff α is a theorem (valid)*, if and only if the sequent $\top \Rightarrow \alpha$ is derivable (valid). Using the formal argument made in Section 2.5 to show that the classes of models of the logics $\mathcal{L}_\mathcal{D}$ and $RDSL$ are identifiable and Theorems 4, 5, one gets the result with the identity embedding.

(4) $\mathcal{L}3 \rightleftharpoons \mathcal{L}_{SN}$: $\mathcal{L}_{SN}$ denotes constructive logic with strong negation [30]. We note that semi-simple Nelson algebras are the algebraic counterparts for $\mathcal{L}_{SN}$. The equivalence of semi-simple Nelson algebras and 3-valued Łukasiewicz algebras through suitable translations has also been observed e.g. by Pagliani. Hence the stated embedding.

(5) $PRL \rightarrowtail S5$: One observes [31] a translation * of wffs of PRL into $S5$ that assigns the operations of negation $\neg$ and necessity $\square$ in PRL those same operations of $S5$. Further, $\sqcap$ is translated in terms of the conjunction $\wedge$ and disjunction $\vee$ of $S5$ as:

$$(\alpha \sqcap \beta)^{\star} \equiv (\alpha^{\star} \wedge \beta^{\star}) \vee (\alpha^{\star} \wedge M\beta^{\star} \wedge \neg M(\alpha^{\star} \wedge \beta^{\star})).$$

Then it can be shown that $\vdash_{PRL} \alpha$ if and only if $\vdash \alpha^{\star}$, for any wff α of PRL.

(6) $S5 \rightleftharpoons L_r$: The logic L_r for rough truth is able to capture, as the class of its theorems, exactly the "$\Diamond$-image" of the class of $S5$-theorems, i.e. $\vdash_{S5} \Diamond\alpha$ if and only if $\vdash\!\!\sim\!\alpha$ [5,10]. Note that the languages of L_r and $S5$ are the same. We translate α in $S5$ to $\alpha^* \equiv L\alpha$. Then $\vdash \alpha$ if and only if $\vdash\!\!\sim\!\alpha^*$. For the other direction, we consider the translation $\alpha^{\circ} \equiv M\alpha$.

(7) $J \rightleftharpoons L_r$: In 1948, Jaśkowski proposed a "discussive" logic – he wanted a formalism to represent reasoning during a discourse. Each thesis, a *discussive assertion* of the system, is supposed either to reflect the opinion of a participant in the discourse, or to hold for a certain "admissible" meaning of the terms used in it. Formally, any thesis α is actually interpreted as "it is possible that α", and the modal operator $\Diamond$ is used for the expression. The logic J (cf. [42]) is such a system. The J-consequence, defined over $S5$, is such that:

$$\vdash_J \alpha \text{ if and only if } \vdash_{S5} \Diamond\alpha.$$

Because of the relationship between L_r and $S5$ noted in (6) above, we have $J \rightleftharpoons L_r$ with the identity embedding. In the whole process, one has obtained an alternative formulation of the paraconsistent logic J (proposed in a different context altogether), and established a link between Pawlak's and Jaśkowski's ideas.

5.2 KTB and Nakamura's Logic $INCRL$

We refer to Section 3.3, and present a connection between $INCRL$, and the normal modal system KTB. KTB, as we know, is sound and complete with respect to the class of reflexive and symmetric Kripke frames.

Let $\mathcal{S} \equiv (U, A, Val, f)$ be an incomplete information system, and let us consider the relation $\Re$ on U defined as follows:

$x \Re y$ if and only if there exists a completion $\mathcal{S}_0$ of $\mathcal{S}$ such that $x R_{\mathcal{S}_0} y$.

Clearly $\Re$ is reflexive and symmetric, but not transitive. From the definitions of $v_{\mathcal{S}}([\ ]\alpha)$ and $v_{\mathcal{S}}(\langle\rangle\alpha)$, we see that

$x \in v_{\mathcal{S}}([\ ]\alpha)$ if and only if, for all $y \in U$ such that $x \Re y, y \in v_{\mathcal{S}}(\alpha)$, and

$x \in v_{\mathcal{S}}(\langle\rangle\alpha)$ if and only if, there exists $y \in U$ such that $x \Re y$ and $y \in v_{\mathcal{S}}(\alpha)$.

So all provable wffs of the modal logic KTB are valid in $INCRL$. What about the converse – are all valid wffs of $INCRL$ provable in KTB? [15] makes a cryptic comment about this, we establish the converse here.

KTB provides an axiomatization for $INCRL$: We show that if α is not provable in KTB then it is not valid in $INCRL$. It suffices then, to construct an incomplete information system $\mathcal{S} \equiv (U, A, \{Val_a\}_{a \in A}, f)$ for any given KTB-frame (W, R), such that $\Re$ is identical with R.

Let g be a function from R ($\subseteq W \times W$) to some set C of constants, satisfying the following conditions:
(i) $g(x, y) = g(y, x)$, (ii) $g(x, y) = g(t, z)$ implies that either $x = t$ and $y = z$, or $x = z$ and $y = t$.

(g essentially assigns, upto symmetry, a unique constant from C to every pair in R.)

Now consider $U \equiv W$, $A \equiv \{a\}$, where a is a *new* symbol. Further, define $f(x, a) \equiv \{g(x, y) : y \in U \text{ and } (x, y) \in R\}$, so that $Val_a \subseteq C$.

We claim that xRy if and only if $x \Re y$. Suppose xRy. Then $g(x, y) \in f(x, a) \cap f(y, a)$ and hence $x \Re y$. Conversely, if $x \Re y$, there exists $d \in f(x, a) \cap f(y, a)$. Now $d \in f(x, a)$ implies that $d = g(x, z)$, for some $z \in U$ such that $(x, z) \in R$, and $d \in f(y, a)$ implies that $d = g(y, t)$, for some $t \in U$ such that $(y, t) \in R$. From the property of g, it follows that either $x = y$ or $x = t$, whence by reflexivity and symmetry of R, we get xRy.

The proof above, in fact, yields a characterization theorem, viz. given any reflexive, symmetric frame (W, R), there exists an incomplete information system $\mathcal{S} \equiv (U, A, \{Val_a\}_{a \in A}, f)$ satisfying the condition $(\diamond)$ (cf. Section 3.1) such that $R = \Re = sim_{\mathcal{S}}$.

5.3 Normal Modal Systems and Vakarelov's Logics

Vakarelov has proved the characterization theorem for incomplete information systems with respect to different sets of relations [12,14,13,17]. As we have re-marked in the Introduction, a special case would be obtained with respect to the indiscernibility relation on the Pawlak approximation space. One finds that if we restrict the logics presented in [14,13,17] to take a modal operator corresponding only to the indiscernibility relation, the resulting system would be just the modal logic $S5$.

As noted at the end of Section 5.2, if an incomplete information system satisfies the condition $(\diamond)$, then the similarity relation $sim_{\mathcal{S}}$ is the same as the relation $\Re$. So it follows that if we restrict the logic NIL to take only the modality $\square$ in the language then the corresponding logic will be just $INCRL$, or, in other words, KTB.

5.4 *DAL* Again

Observing Vakarelov's strain of work, it may be tempting to look for a kind of characterization result in the case of DAL (cf. Section 2.2) as well. Consider a general DAL-structure $\mathcal{U} \equiv (U, \{R_i\}_{i \in I})$, where the family $\{R_i\}_{i \in I}$ of equivalence relations is closed under intersection and transitive closure of union. Can one find an incomplete information system $\mathcal{S} \equiv (U, A, Val, f)$ such that the standard structure for $\mathcal{S}$ is just $\mathcal{U}$? Let us assume that the standard structure is obtained "naturally" from $\mathcal{S}$, viz. that the equivalence relations in it are the ones induced by the *subsets* of A. As it turns out, this is a hard question.

However, we can find an information system, such that the standard structure obtained from it in the above manner cannot be a general DAL-structure.

Suppose for some incomplete information system $\mathcal{S} \equiv (U, A, Val, f)$, R and P are the equivalence relations induced by subsets R', P' of A respectively – we denote this as $ind(R') = R$ and $ind(P') = P$. For the equivalence relation

$R \cap P$, $R' \cup P' \subseteq A$ is such that $ind(R' \cup P') = R \cap P$. But in the case of $R \uplus P$, there may not be any $Q \subseteq A$ such that $ind(Q) = R \uplus P$. Consider the following example [11].

Example 1. $U \equiv \{o1, o2, o3, o4, o5, o6, o7\}$, where each o_i consists of circles and squares. Let $A \equiv \{\text{number of circles } (\bigcirc), \text{ number of squares } (\square)\}$. The information function is given by the following table:

	$\bigcirc$	$\square$
$o1$	1	1
$o2$	1	2
$o3$	2	1
$o4$	2	2
$o5$	3	3
$o6$	3	4
$o7$	3	4

Equivalence classes of indiscernibility relations $ind(\bigcirc)$ and $ind(\square)$ are:
$$ind(\bigcirc): \{o1, o2\}, \{o3, o4\}, \{o5, o6, o7\},$$
$$ind(\square): \{o1, o3\}, \{o2, o4\}, \{o5\}, \{o6, o7\}.$$
The transitive closure of these relations gives the following equivalence classes:
$$ind(\bigcirc) \uplus ind(\square): \{o1, o2, o3, o4\}, \{o5, o6, o7\}.$$
Clearly there is no $Q \subseteq A$ such that $ind(Q) = ind(\bigcirc) \uplus ind(\square)$.

6   Summary and Questions

We have tried to present the various proposals of logics with semantics based on rough sets, including some generalizations. Two main approaches emerge, discussed in Sections 2 and 3. One of these considers logics, the models of which are approximation spaces, while the other considers approximation spaces, but those induced by information systems. However, it is found through characterization results, that both lines of study converge, in that the two semantics for a particular system are identical. This actually reflects on the apt description of the properties of the relations defining the approximation spaces.

The only exception is the logic DAL of the first category. As remarked in Section 5.4, given a general DAL-structure $\mathcal{U} \equiv (U, \{R_i\}_{i \in I})$, it does not seem easy to construct an information system "naturally" to obtain $\mathcal{U}$ back as its standard structure. In case of the logics with attributes as expressions (cf. Section 4.4), one encounters a problem even earlier. The models here are based on structures of the form $(U, A, \{ind(P)\}_{P \subseteq A})$, and there does not appear easily a corresponding "general" structure of the kind $\mathcal{U} \equiv (U, \{R_i\}_{i \in I})$, with appropriate closure conditions on $\{R_i\}_{i \in I}$. These logics have not been axiomatized, though the language can express a lot about attributes – that few of the other systems are able to do.

An interesting picture is obtained from the logics of Section 2, leaving out DAL and other systems with models based on generalized spaces. Most of the logics are embeddable into each other (cf. Section 5). We have

$$\mathcal{L}_\mathcal{D} \;\rightleftharpoons\; \mathcal{L}3 \;\rightleftharpoons\; PRL \;\rightharpoonup\; S5 \;\rightleftharpoons\; L_r \;\rightleftharpoons\; J. \tag{1}$$

In one sense then, the embeddings in (1) establish that no 'new' logic surfaces with the kind of rough set semantics defined. But in another sense, well-known systems have been imparted a rough set interpretation. It should be noted that though the embeddings are defined with respect to *theoremhood*, the relationships would hold in some cases (e.g. $\mathcal{L}3 - PRL$ and $L_r - J$) if derivability of wffs from non-empty premise sets is considered [31,10]. One could attempt to settle the question for the rest. (1) indicates another interesting future line of work, viz. an investigation for logics and interrelations, that may result on replacing $S5$ by other non-modal systems (as in [6]).

All the systems presented other than $\mathcal{L}_\mathcal{T}$ (cf. Section 4.1), deal with static information. The semantics of $\mathcal{L}_\mathcal{T}$ essentially gives rise to a family of approximation spaces on the same domain, the indiscernibility relations changing with moments of time. One could further enquire about the behaviour of rough sets in such a dynamic information system.

As remarked in Section 4.2, another open direction relates to a study of logics that may be obtained from the generalized approach in [38].

Overall, one may say that it has been a remarkable journey in the exploration of logics, beginning with a deceptively simple proposal of "rough sets". We have seen the introduction of novel concepts – e.g. of "rough truth", "rough modus ponens", "rough consistency", "rough mereology". The journey has, by no means, ended. Pawlak's theory has just opened up the horizon before us, to reveal a number of yet unexplored directions in the study of "rough logics".

References

1. Banerjee, M., Chakraborty, M.K.: Algebras from rough sets. In Pal, S.K., Polkowski, L., Skowron, A., eds.: Rough-neuro Computing: Techniques for Computing with Words. Springer Verlag, Berlin (2004) 157–184
2. Pawlak, Z.: Rough Sets. Theoretical Aspects of Reasoning about Data. Kluwer Academic Publishers, Dordrecht (1991)
3. Pawlak, Z.: Rough logic. Bull. Polish Acad. Sc. (Tech. Sc.) **35** (1987) 253–258
4. Orłowska, E.: Kripke semantics for knowledge representation logics. Studia Logica **XLIX** (1990) 255–272
5. Chakraborty, M.K., Banerjee, M.: Rough consequence. Bull. Polish Acad. Sc.(Math.) **41(4)** (1993) 299–304
6. Yao, Y., Lin, T.Y.: Generalization of rough sets using modal logics. Intelligent Automation and Soft Computing **2** (1996) 103–120
7. Banerjee, M., Chakraborty, M.K.: Rough sets through algebraic logic. Fundamenta Informaticae **28(3,4)** (1996) 211–221
8. Düntsch, I.: A logic for rough sets. Theoretical Computer Science **179** (1997) 427–436

9. Dai, J.H.: Logic for rough sets with rough double Stone algebraic semantics. In Slezak, D., Wang, G., Szczuka, M.S., Düntsch, I., Yao, Y., eds.: Proc. RSFSDM-GrC(1), Canada, LNCS 3641, Springer Verlag (2005, 141-148)

10. Banerjee, M.: Logic for rough truth. Fundamenta Informaticae **71(2-3)** (2006) 139–151

11. Farinas Del Cerro, L., Orłowska, E.: *DAL* – a logic for data analysis. Theoretical Computer Science **36** (1997) 251–264

12. Vakarelov, D.: Abstract characterization of some knowledge representation systems and the logic *NIL* of nondeterministic information. In Jorrand, P., Sgurev, V., eds.: Artificial Intelligence II. North–Holland (1987) 255–260

13. Vakarelov, D.: A modal logic for similarity relations in Pawlak knowledge representation systems. Fundamenta Informaticae **15** (1991) 61–79

14. Vakarelov, D.: Modal logics for knowledge representation systems. Theoretical Computer Science **90** (1991) 433–456

15. Nakamura, A.: A rough logic based on incomplete information and its application. Int. J. Approximate Reasoning **15** (1996) 367–378

16. Orłowska, E., Pawlak, Z.: Representation of nondeterministic information. Theoretical Computer Science **29** (1984) 27–39

17. Vakarelov, D., Balbiani, P.: A modal logic for indiscernibilty and complementarity in information systems. Fundamenta Informaticae **50** (2002) 243–263

18. Orłowska, E.: Logic of nondeterministic information. Studia Logica **1** (1985) 91–100

19. Orłowska, E.: Logic of indiscernibility relations. In Goos, G., Hartmanis, J., eds.: Proc. Symposium on Computation Theory, Zabrów, 1984, LNCS 208, Springer Verlag (1985, 177–186)

20. Rauszer, C.M.: Rough logic for multiagent systems. In Masuch, M., Polos, L., eds.: Knowledge Representation and Reasoning under Uncertainty, LNAI 808. Springer-Verlag (1994) 161–181

21. Stepaniuk, J.: Rough relations and logics. In Polkowski, L., Skowron, A., eds.: Rough Sets in Knowledge Discovery 1: Methodology and Applications. Physica-Verlag (1998) 248–260

22. Małuszyński, Vitória, A.: Toward rough datalog: embedding rough sets in prolog. In Pal, S.K., Polkowski, L., Skowron, A., eds.: Rough-neuro Computing: Techniques for Computing with Words. Springer Verlag, Berlin (2004) 297–332

23. Polkowski, L., Skowron, A.: Rough mereology: a new paradigm for approximate reasoning. Int. J. Approximate Reasoning **15(4)** (1997) 333–365

24. Wong, S.K.M.: A rough set model for reasoning about knowledge. In Orłowska, E., ed.: Incomplete Information: Rough Set Analysis. Studies in Fuzziness and Soft Computing, vol. 13. Physica-Verlag (1998) 276–285

25. Banerjee, M.: Rough belief change. Transactions of Rough Sets V **LNCS 4100** (2006) 25–38

26. Banerjee, M., Chakraborty, M.K.: Rough consequence and rough algebra. In Ziarko, W.P., ed.: Rough Sets, Fuzzy Sets and Knowledge Discovery, Proc. Int. Workshop on Rough Sets and Knowledge Discovery (RSKD'93). Workshops in Computing, London, Springer Verlag (1994, 196–207)

27. Yao, Y.: Constructive and algebraic methods of the theory of rough sets. Information Sciences **109** (1998) 21–47

28. Pomykała, J.: Approximation, similarity and rough construction. preprint CT–93–07, ILLC Prepublication Series, University of Amsterdam (1993)

29. Komorowski, J., Pawlak, Z., Polkowski, L., Skowron, A.: Rough sets: a tutorial. In Pal, S.K., Skowron, A., eds.: Rough Fuzzy Hybridization: A New Trend in Decision-Making. Springer Verlag, Singapore (1999) 3–98
30. Rasiowa, H.: An Algebraic Approach to Non-classical Logics. North Holland, Amsterdam (1974)
31. Banerjee, M.: Rough sets and 3-valued Łukasiewicz logic. Fundamenta Informaticae **32** (1997) 213–220
32. Pagliani, P.: Rough set theory and logic-algebraic structures. In Orłowska, E., ed.: Incomplete Information: Rough Set Analysis. Studies in Fuzziness and Soft Computing, vol. 13. Physica-Verlag (1998) 109–190
33. Iturrioz, L.: Rough sets and three-valued structures. In Orłowska, E., ed.: Logic at Work: Essays Dedicated to the Memory of Helena Rasiowa. Studies in Fuzziness and Soft Computing, vol. 24. Physica-Verlag (1999) 596–603
34. Boicescu, V., Filipoiu, A., Georgescu, G., Rudeano, S.: Łukasiewicz-Moisil Algebras. North Holland, Amsterdam (1991)
35. Comer, S.: Perfect extensions of regular double Stone algebras. Algebra Universalis **34** (1995) 96–109
36. Andréka, H., Németi, I., Sain, I.: Abstract model theoretic approach to algebraic logic. CCSOM working paper, Department of Statistics and Methodology, University of Amsterdam (1992)
37. Pawlak, Z.: Rough sets. Int. J. Comp. Inf. Sci. **11(5)** (1982) 341–356
38. Pagliani, P., Chakraborty, M.K.: Information quanta and approximation spaces I: non-classical approximation operators. In: Proc. 2005 IEEE Conf. on Granular Computing, IEEE Press (2005, 605–610)
39. Pawlak, Z.: Rough relations. ICS PAS Reports 435 (1981)
40. Skowron, A., Stepaniuk, J.: Tolerance approximation spaces. Fundamenta Informaticae **27** (1996) 245–253
41. Polkowski, L.: Rough mereological reasoning in rough set theory: recent results and problems. In Wang, G., Peters, J.F., Skowron, A., Yao, Y., eds.: Proc. Rough Sets and Knowledge Technology (RSKT 2006), China, 2006, LNAI 4062, Springer Verlag (2006, 79–92)
42. da Costa, N.C.A., Doria, F.A.: On Jaśkowski's discussive logics. Studia Logica **54** (1995) 33–60

Intuitionistic Rough Sets for Database Applications

Theresa Beaubouef[1] and Frederick E. Petry[2]

[1] Southeastern Louisiana University
Dept. of Computer Science & Ind. Technology
Hammond, LA 70402, USA
tbeaubouef@selu.edu
[2] Center for Intelligent and Knowledge-Based Systems
Tulane University
New Orleans, LA 70118, USA
fep@eecs.tulane.edu

1 Introduction

We introduce the intuitionistic rough set and intuitionistic rough relational and object oriented database models. The intuitionistic rough set database models draw benefits from both the rough set and intuitionistic techniques, providing greater management of uncertainty for databases applications in a less than certain world. We provide the foundation for the integration of intuitionistic rough sets into modeling of uncertainty in databases. This builds upon some of our previous research [2,3] with integrating fuzzy and rough set techniques for uncertainty management in databases.

2 Intuitionistic Rough Sets

An intuitionistic set [1] (intuitionistic fuzzy set) is a generalization of the traditional fuzzy set. Let set X be fixed. An intuitionistic set A is defined by the following:

$$A = \{\langle x, \mu_A(x), \nu_A(x) \rangle : x \in X\}$$

where $\mu_A(x) \mapsto [0,1]$, and $\nu_A(x) \mapsto [0,1]$. The degree of membership of element $x \in X$ to the set A is denoted by $\mu_A(x)$, and the degree of nonmembership of element $x \in X$ to the set A is denoted by $\nu_A(x)$. A is a subset of X. For all $x \in X, 0 \le \mu_A(x) + \nu_A(x) \le 1$. A hesitation margin, $\pi_A(x) = 1 - (\mu_A(x) + \nu_A(x))$, expresses a degree of uncertainty about whether x belongs to X or not, or uncertainty about the membership degree. This hesitancy may cater toward membership or nonmembership.

We next define the intuitionistic rough set, which incorporates the beneficial properties of both rough set [5] and intuitionistic set techniques. Intuitionistic rough sets are generalizations of fuzzy rough sets that give more information about the uncertain, or boundary region. They follow the definitions for partitioning of the universe into equivalence classes as in rough sets, but instead

J.F. Peters et al. (Eds.): Transactions on Rough Sets VI, LNCS 4374, pp. 26–30, 2007.
© Springer-Verlag Berlin Heidelberg 2007

of having a simple boundary region, there are basically two boundaries formed from the membership and nonmembership functions. Let U be a universe, Y a rough set in U, defined on a partitioning of U into equivalence classes.

Definition 1. *An intuitionistic rough set Y in U is $\langle Y, \mu_Y(x), \nu_Y(x) \rangle$, where $\mu_Y(x)$ is a membership function which associates a grade of membership from the interval [0,1] with every element (equivalence class) of U, and $\nu_Y(x)$ associates a degree of non membership from the interval [0,1] with every element (equivalence class) of U, where $0 \leq \mu_Y(x) + \nu_Y(x) \leq 1$, where x denotes the equivalence class containing x. A hesitation margin is $\pi_Y(x) = 1 - (\mu_Y(x) + \nu_Y(x))$.*

Consider the following special cases $\langle \mu, \nu \rangle$ for some element of Y:

$\langle 1, 0 \rangle$ denotes total membership. This correspond to elements found in $\underline{R}Y$.

$\langle 0, 1 \rangle$ denotes elements that do not belong to Y. Same as $U - \overline{R}Y$.

$\langle 0.5, 0.5 \rangle$ corresponds to traditional rough set boundary region.

$\langle p, 1 - p \rangle$ corresponds to fuzzy rough set in that there is a single boundary. In this case we assume that any degree of membership has a corresponding complementary degree of non membership.

$\langle p, 0 \rangle$ corresponds to fuzzy rough set.

$\langle 0, q \rangle$ This case can not be modeled by fuzzy rough sets. It denotes things that are not a member of $\underline{R}Y$ or $\overline{R}Y$. It falls somewhere in the region $U - \overline{R}Y$.

$\langle p, q \rangle$ Intuitionistic set general case , has membership and nonmembership.

Let Y' denote the complement of Y. Then the intuitionistic set having $\langle \mu_Y(x), \mu_{Y'}(x) \rangle$ is the same as fuzzy rough set. The last two cases above, $\langle 0, q \rangle$ and $\langle p, q \rangle$, cannot be represented by fuzzy sets, rough sets, or fuzzy rough sets. These are the situations which show that intuitionistic rough sets provide greater uncertainty management than the others alone. Note, however, that with the intuitionistic set we do not lose the information about uncertainty provided by other set theories, since from the first few cases we see that they are special cases of the intuitionistic rough set. Although there are several various way of combining rough and fuzzy sets, we focus on those fuzzy rough sets as defined in [2,3] and used for fuzzy rough databases, since our intuitionistic rough relational database model follows from this. The intuitionistic rough relational database model will have an advantage over the rough and fuzzy rough database models in that the non membership uncertainty of intuitionistic set theory will also play a role, providing even greater uncertainty management than the original models.

3   Intuitionistic Rough Relational Database Model

The intuitionistic rough relational database, as in the ordinary relational database, represents data as a collection of *relations* containing *tuples*. Because a relation is considered a set having the tuples as its members, the tuples are unordered. In addition, there can be no duplicate tuples in a relation. A tuple $\mathbf{t}_i$ takes the form $(d_{i1}, d_{i2}, \ldots, d_{im}, d_{i\mu}, d_{i\nu})$, where d_{ij} is a *domain value* of a particular *domain set* D_j and $d_{i\mu} \in D_\mu$, where D_μ is the interval $[0,1]$, the domain for intuitionistic membership values, and Dv is the interval $[0,1]$, the domain for intuitionistic nonmembership values. In the ordinary relational database, $d_{ij} \in D_j$. In the intuitionistic rough relational database, except for the intuitionistic membership and nonmembership values, however, $d_{ij} \in D_j$, and although d_{ij} is not restricted to be a singleton, $d_{ij} \neq \emptyset$. Let $P(D_i)$ denote any non-null member of the powerset of D_i.

Definition 2. *A intuitionistic rough relation R is a subset of the set cross product $P(D_1) \times P(D_2) \times \ldots \times P(D_m) \times D_\mu \times D_{nu}$. An intuitionistic rough tuple $\mathbf{t}$ is any member of R. If $\mathbf{t}_i$ is some arbitrary tuple, then $\mathbf{t}_i = (d_{i1}, d_{i2}, \ldots, d_{im}, d_{i\mu}, d_{i\nu})$ where $d_{ij} \in D_j$ and $d_{i\mu} \in D_\mu, d_{i\nu} \in D_\nu$.*

Let $[d_{xy}]$ denote the equivalence class to which d_{xy} belongs. When d_{xy} is a set of values, the equivalence class is formed by taking the union of equivalence classes of members of the set; if $d_{xy} = \{c_1, c_2, ..., c_n\}$, then $[d_{xy}] = [c1] \times [c2] \times \ldots \times [c_n]$.

Definition 3. *Tuples $\mathbf{t}_i = (d_{i1}, d_{i2}, \ldots, d_{in}, d_{i\mu}, d_{i\nu})$ and $\mathbf{t}_k = (d_{k1}, d_{k2}, \ldots, d_{kn}, d_{k\mu}, d_{k\nu})$ are redundant if $[d_{ij}] = [d_{kj}]$ for all $j = 1, \ldots, n$.*

In [3], we defined several operators for the rough relational algebra, and in [2] demonstrated the expressive power of the fuzzy rough versions of these operators in the fuzzy rough relational database model. In an extension of this work we do the same for the rough intuitionistic database.

4   Intuitionistic Rough Object-Oriented Database (IROODB) Model

We next develop the intuitionistic rough object-oriented database model. We follow the formal framework and type definitions for generalized object-oriented databases proposed by [4] and extended for rough sets in [3]. We extend this framework, however, to allow for intuitionistic rough set indiscernibility and approximation regions for the representation of uncertainty as we have previously done for relational databases [2,3]. The intuitionistic rough object database scheme is formally defined by the following type system and constraints.

The type system, $TS = [T, P, f_{impl}^{type}]$, where T can be a literal type $T_{literal}$, which can be a base type, a collection literal type, or a structured literal type. It also contains T_{object}, which specifies object types, $T_{reference}$, the set of specifications for reference types, and a *void* type. In the type system, each domain $dom_{ts} \in D_{ts}$, the set of domains. This domain set, along with a set of operators O_{ts} and a set of axioms A_{ts}, capture the semantics of the type specification. The type system is then defined based on these type specifications, the set of all programs P, and the implementation function mapping each type specification for a domain onto a subset of $\rho(P)$ – the powerset of P that contains all the implementations for the type system:

$$f_{impl}^{type} : T \mapsto \rho(P) \text{ giving } ts \mapsto \{p_1, p_2, \ldots p_n\}.$$

We are particularly interested in object types, and specify a class t of object types as Class $id(id_1 : s_1; \ldots; id_n : s_n)$ or Class $id : \overline{\overline{id}}_1, \ldots, \overline{\overline{id}}_n (id_1 : s_1; \ldots; id_n : s_n)$ where id, an identifier, names an object type, $\{\overline{\overline{id}}_i : 1 \leq i \leq m\}$ is a finite set of identifiers denoting parent types of t, and $\{id_i : s_i : 1 \leq i \leq n\}$ is the finite set of characteristics specified for object type t within its syntax. This set includes all the attributes, relationships and method signatures for the object type. The identifier for a characteristic is id_i and the specification is s_i for each of the $id_i : s_i$. See [4] for details of how rough set concepts are integrated in this OO model, and how changing the granularity of the partitioning affects query results. In that paper the OO model is extended for fuzzy and rough set uncertainty.

If we extend the rough OODB further to allow for intuitionistic types, the type specifications T can be generalized to a set $\check{T}$ as in [4], so that the definitions of the domains are generalized to intuitionistic sets. For every $ts \in T$, having domain ts being dom_{ts}, the type system $ts \in T$ is generalized to $\overline{ts} \in \check{T}$, where domain of $\overline{ts}$ is denoted by $dom_{\overline{ts}}$ and is defined as the set $\overline{p}(dom_{ts})$ of intuitionistic sets on dom_{ts}, and O_{ts} is generalized to $O_{\overline{ts}}$, which contains the generalized version of the operators.

The generalized type system then is a triple $GTS = [\check{T}, P, \overline{f}_{impl}^{type}]$, where $\check{T}$ is the generalized type system, P is the set of all programs, and $\overline{f}_{impl}^{type}$ maps each $\overline{ts} \in \check{T}$ onto that subset of P that contains the implementation for $\overline{ts}$. An instance of this GTS is a generalized type $\overline{t} = [\overline{ts}, \overline{f}_{impl}^{type}(\overline{ts})], \overline{ts} \in \check{T}$.

A generalized object belonging to this class is defined by $\overline{o} = [oid, N, \overline{t}, \overline{f}_{impl}^{type}(\overline{ts}), \nu]$, where ν draws values from the generalized domain that allows an object to contain intuitionistic membership and nonmembership values as part of the state of the object. Both intuitionistic and rough set uncertainty management can be used in this generalized OODB model.

We extended a formal framework of object-oriented databases to allow for modeling of various types of imprecision, vagueness, and uncertainty that typically occur in spatial data. The model is based on a formal type system and specified constraints, thus preserving integrity of the database, while at the same time allowing an OODB to be generalized in such a way as to include both intuitionistic and rough set uncertainty, both well-developed methods of uncertainty management. Incorporation of intuitionistic and rough set uncertainty into the OODB model is essential for representing imprecision and uncertainty in spatial data entities and in their interrelationships.

5   Conclusion

We introduced the intuitionistic rough set, then discussed how the intuitionistic rough set generalizes each of traditional rough, fuzzy, fuzzy-rough, and intuitionistic sets. The intuitionistic rough relational database model was introduced. This model allows for both rough and intuitionistic modeling of uncertainty. Because real world applications involve uncertainty, this model can more accurately represent data and relationships than traditional relational databases. We have also introduced our model for intuitionistic object-oriented databases and shown the significance of both rough sets and intuitionistic sets for uncertainty management.

References

1. Atanassov, K.: Intuitionistic Fuzzy Sets. Fuzzy Sets and Systems 20 (1986) 87–96
2. Beaubouef T., Petry F.: Fuzzy Rough Set Techniques for Uncertainty Processing in a Relational Database. International Journal of Intelligent Systems 15(5) (2000) 389–424
3. Beaubouef, T., Petry, F., Buckles, B.: Extension of the Relational Database and its Algebra with Rough Set Techniques. Computational Intelligence 11(2) (1995) 233–245
4. De Tré, G., De Caluwe, R. A Generalized Object-Oriented Database Model with Generalized Constraints. In Proc. of NAFIPS'99, New York (1999) 381–386.
5. Pawlak, Z.:Rough Sets. International Journal of Man-Machine Studies 21 (1984) 127–134

An Experimental Comparison of Three Rough Set Approaches to Missing Attribute Values

Jerzy W. Grzymala-Busse[1,2] and Witold J. Grzymala-Busse[3]

[1] Department of Electrical Engineering and Computer Science University of Kansas,
Lawrence, KS 66045, USA
[2] Institute of Computer Science, Polish Academy of Sciences,
01–237 Warsaw, Poland
[3] Touchnet Information Systems, Inc.,
Lenexa, KS 66219, USA

Abstract. In this paper we present results of experiments conducted to compare three types of missing attribute values: lost values, "do not care" conditions and attribute-concept values. For our experiments we selected six well known data sets. For every data set we created 30 new data sets replacing specified values by three different types of missing attribute values, starting from 10%, ending with 100%, with increment of 10%. For all concepts of every data set concept lower and upper approximations were computed. Error rates were evaluated using ten-fold cross validation. Overall, interpreting missing attribute values as lost provides the best result for most incomplete data sets.

Keywords: missing attribute values, incomplete data sets, concept approximations, LERS data mining system, MLEM2 algorithm.

1 Introduction

Real-life data are frequently incomplete, i.e., values for some attributes are missing. Appropriate handling of missing attribute values is one of the most important tasks of data mining.

In this paper we assume that missing attribute values have three different interpretations. The first possibility is that missing attribute values are *lost*. Such values are interpreted as originally specified, but currently unavailable since these values were incidentally erased, forgotten to be recorded, etc. A rough set approach to incomplete data sets in which all attribute values were lost was presented for the first time in [12], where two algorithms for rule induction, modified to handle lost attribute values, were introduced.

The next possibility are "do not care" conditions. Such missing attribute values were irrelevant during collection of data. Simply, an expert decided that the attribute value was irrelevant for a classification or diagnosis of the case. For example, a data set describing flu patients may contain, among other attributes, an attribute *Color of hair*. Though some scrupulous patients may fill in this value, other patients may assume that this attribute is irrelevant for the flu diagnosis

J.F. Peters et al. (Eds.): Transactions on Rough Sets VI, LNCS 4374, pp. 31–50, 2007.
© Springer-Verlag Berlin Heidelberg 2007

and leave it unspecified. If we suspect that this attribute does matter, the best interpretation for missing attribute values is replacing them by all possible existing attribute values. A rough set approach to incomplete data sets in which all attribute values were "do not care" conditions was presented for the first time in [4], where a method for rule induction was introduced in which each missing attribute value was replaced by all values from the domain of the attribute.

The third possibility is a missing attribute value interpreted as an *attribute-concept* value. It is a similar case to a "do not care" condition, however, it is restricted to a specific concept. A concept (class) is a set of all cases classified (or diagnosed) the same way. Using this interpretation, we will replace a missing attribute value by all values of the same attribute typical for the concept to which the case belongs. Let us consider a patient, sick with flu, from the flu data set, with a missing attribute value for *Color of hair*. Other patients, sick with flu, filled in values *brown* and *grey* for this attribute. On the other hand, healthy patients characterized the color of their hair as *blond* and *brown*. Using attribute-concept value interpretation, this missing attribute value is replaced by *brown* and *grey*. If we would use "do not care" condition interpretation, the same missing attribute value should be replaced by *blond*, *brown*, and *grey*. This approach was introduced in [10].

In general, incomplete decision tables are described by characteristic relations, in a similar way as complete decision tables are described by indiscernibility relations [7,8,9].

In rough set theory, one of the basic notions is the idea of lower and upper approximations. For complete decision tables, once the indiscernibility relation is fixed and the concept (a set of cases) is given, the lower and upper approximations are unique.

For incomplete decision tables, for a given characteristic relation and concept, there are three important and different possibilities to define lower and upper approximations, called singleton, subset, and concept approximations [7]. Singleton lower and upper approximations were studied in [14,15,19,21,22]. Note that similar definitions of lower and upper approximations, though not for incomplete decision tables, were studied in [16,24,25].

Note that some other rough-set approaches to missing attribute values were presented in [4,11,13,23] as well.

2   Blocks of Attribute-Value Pairs—Complete Data

We assume that the input data sets are presented in the form of a *decision table*. An example of a decision table is shown in Table 1. Rows of the decision table represent *cases*, while columns are labeled by *variables*. The set of all cases will be denoted by U. In Table 1, $U = \{1, 2, ..., 6\}$. Independent variables are called *attributes* and a dependent variable is called a *decision* and is denoted by d. The set of all attributes will be denoted by A. In Table 1, $A = \{Temperature, Headache, Cough\}$. Any decision table defines a function ρ that maps the direct product of U and A into the set of all values. For example, in

Table 1. A complete decision table

	Attributes			Decision
Case	Temperature	Headache	Cough	Flu
1	high	yes	yes	yes
2	very_high	yes	no	yes
3	high	no	no	no
4	high	yes	yes	yes
5	normal	yes	no	no
6	normal	no	yes	no

Table 1, $\rho(1, Temperature) = high$. A decision table with completely specified function ρ will be called *completely specified*, or, for the sake of simplicity, *complete*. In practice, input data for data mining are frequently affected by missing attribute values. In other words, the corresponding function ρ is incompletely specified (partial). A decision table with an incompletely specified function ρ will be called *incomplete*. Function ρ describing Table 1 is completely specified.

An important tool to analyze complete decision tables is a block of the attribute-value pair. Let a be an attribute, i.e., $a \in A$ and let v be a value of a for some case. For complete decision tables if $t = (a, v)$ is an attribute-value pair then a *block* of t, denoted $[t]$, is a set of all cases from U that for attribute a have value v.

Rough set theory [17], [18] is based on the idea of an indiscernibility relation, defined for complete decision tables. Let B be a nonempty subset of the set A of all attributes. The indiscernibility relation $IND(B)$ is a relation on U defined for $x, y \in U$ as follows

$$(x, y) \in IND(B) \quad if \ and \ only \ if \ \rho(x, a) = \rho(y, a) \ for \ all \ a \in B.$$

The indiscernibility relation $IND(B)$ is an equivalence relation. Equivalence classes of $IND(B)$ are called elementary sets of B and are denoted by $[x]_B$. For example, for Table 1, elementary sets of $IND(A)$ are $\{1, 4\}$, $\{2\}$, $\{3\}$, $\{5\}$, $\{6\}$. Additionally, IND(B) = $\{(1, 1), (1, 4), (2, 2), (3, 3), (4, 1), (4, 4), (5, 5), (6, 6)\}$. The indiscernibility relation $IND(B)$ may be computed using the idea of blocks of attribute-value pairs. Let a be an attribute, i.e., $a \in A$ and let v be a value of a for some case. For complete decision tables if $t = (a, v)$ is an attribute-value pair then a block of t, denoted $[t]$, is a set of all cases from U that for attribute a have value v. For Table 1,

[(Temperature, high)] = $\{1, 3, 4\}$,
[(Temperature, very_high)] = $\{2\}$,
[(Temperature, normal)] = $\{5, 6\}$,
[(Headache, yes)] = $\{1, 2, 4, 5\}$,
[(Headache, no)] = $\{3, 6\}$,

$[(\text{Cough, yes})] = \{1, 4, 6\}$,
$[(\text{Cough, no})] = \{2, 3, 5\}$.

The indiscernibility relation $IND(B)$ is known when all elementary blocks of IND(B) are known. Such elementary blocks of B are intersections of the corresponding attribute-value pairs, i.e., for any case $x \in U$,

$$[x]_B = \cap\{[(a, v)] \mid a \in B, \rho(x, a) = v\}.$$

We will illustrate the idea how to compute elementary sets of B for Table 1 and $B = A$.

$[1]_A = [4]_A = \{1, 3, 4\} \cap \{1, 2, 4, 5\} \cap \{1, 4, 6\} = \{1, 4\}$,
$[2]_A = \{2\} \cap \{1, 2, 4, 5\} \cap \{2, 3, 5\} = \{2\}$,
$[3]_A = \{1, 3, 4\} \cap \{3, 6\} \cap \{2, 3, 5\} = \{3\}$,
$[5]_A = \{5, 6\} \cap \{1, 2, 4, 5\} \cap \{2, 3, 5\} = \{5\}$,
$[6]_A = \{5, 6\} \cap \{3, 6\} \cap \{1, 4, 6\} = \{6\}$,

For completely specified decision tables lower and upper approximations are defined using the indiscernibility relation. Any finite union of elementary sets, associated with B, will be called a *B-definable set*. Let X be any subset of the set U of all cases. The set X is called a *concept* and is usually defined as the set of all cases defined by a specific value of the decision. In general, X is not a B-definable set. However, set X may be approximated by two B-definable sets, the first one is called a *B-lower approximation* of X, denoted by $\underline{B}X$ and defined as follows

$$\cup\{[x]_B \mid x \in U, [x]_B \subseteq X\},$$

The second set is called a *B-upper approximation* of X, denoted by $\overline{B}X$ and defined as follows

$$\cup\{[x]_B \mid x \in U, [x]_B \cap X \neq \emptyset).$$

Data set presented in Table 1 is consistent (the lower approximation is equal to the upper approximation for every concept), hence the certain rule set and the possible rule set are identical. Rules in the LERS format (every rule is equipped with three numbers, the total number of attribute-value pairs on the left-hand side of the rule, the total number of examples correctly classified by the rule during training, and the total number of training cases matching the left-hand side of the rule) [6] are:

2, 2, 2
(Temperature, high) & (Headache, yes) -> (Flu, yes)
1, 1, 1
(Temperature, very_high) -> (Flu, yes)
1, 2, 2
(Temperature, normal) -> (Flu, no)
1, 2, 2
(Headache, no) -> (Flu, no)

Note that the above rules were induced by the MLEM2 (Modified Learning from Examples Module, version 2) option of the LERS (Learning from Examples based on Rough Sets) data mining system [2,5,6].

3   Blocks of Attribute-Value Pairs—Incomplete Data

For the rest of the paper we will assume that all decision values are specified, i.e., they are not missing. Additionally, we will assume that lost values will be denoted by "?", "do not care" conditions by "*", and attribute-concept values by "−". Additionally, we will assume that for each case at least one attribute value is specified.

Table 2 is Table 1 with eight attribute values missing. All of these missing attribute values are *lost*.

Table 2. An incomplete decision table (all missing attribute values are *lost* values)

		Attributes		Decision
Case	Temperature	Headache	Cough	Flu
1	high	?	yes	yes
2	?	yes	?	yes
3	?	no	?	no
4	high	?	yes	yes
5	?	yes	no	no
6	normal	no	?	no

For incomplete decision tables, a block of an attribute-value pair must be modified in the following way:

- If for an attribute a there exists a case x such that $\rho(x, a) = ?$, i.e., the corresponding value is lost, then the case x should not be included in any blocks $[(a, v)]$ for all values v of attribute a,
- If for an attribute a there exists a case x such that the corresponding value is a "do not care" condition, i.e., $\rho(x, a) = *$, then the case x should be included in blocks $[(a, v)]$ for all specified values v of attribute a.
- If for an attribute a there exists a case x such that the corresponding value is an attribute-concept value, i.e., $\rho(x, a) = -$, then the corresponding case x should be included in blocks $[(a, v)]$ for all specified values $v \in V(x, a)$ of attribute a, where

$$V(x, a) = \{\rho(y, a) \mid \rho(y, a) \ is \ specified, \ y \in U, \ \rho(y, d) = \rho(x, d)\}.$$

Thus, for Table 2,

[(Temperature, high)] = {1, 4},
[(Temperature, normal)] = {6},
[(Headache, yes)] = {2, 5},
[(Headache, no)] = {3, 6},
[(Cough, yes)] = {1, 4},
[(Cough, no)] = {5}.

For incomplete data sets the idea of the elementary block is extended to a *characteristic set*. For a case $x \in U$ the *characteristic set* $K_B(x)$ is defined as the intersection of the sets $K(x, a)$, for all $a \in B$, where the set $K(x, a)$ is defined in the following way:

- If $\rho(x, a)$ is specified, then $K(x, a)$ is the block $[(a, \rho(x, a)]$ of attribute a and its value $\rho(x, a)$,
- If $\rho(x, a) = ?$ or $\rho(x, a) = *$ then the set $K(x, a) = U$,
- If $\rho(x, a) = -$, then the corresponding set $K(x, a)$ is equal to the union of all blocks of attribute-value pairs (a, v), where $v \in V(x, a)$ if $V(x, a)$ is nonempty. If $V(x, a)$ is empty, $K(x, a) = U$.

Thus, for Table 2

$K_A(1) = \{1, 4\} \cap U \cap \{1, 4\} = \{1, 4\},$
$K_A(2) = U \cap \{2, 5\} \cap U = \{2, 5\},$
$K_A(3) = U \cap \{3, 6\} \cap U = \{3, 6\},$
$K_A(4) = \{1, 4\} \cap U \cap \{1, 4\} = \{1, 4\},$
$K_A(5) = U \cap \{2, 5\} \cap \{5\} = \{5\},$
$K_A(6) = \{6\} \cap \{3, 6\} \cap U = \{6\},$

Characteristic set $K_B(x)$ may be interpreted as the set of cases that are indistinguishable from x using all attributes from B and using a given interpretation of missing attribute values. Thus, $K_A(x)$ is the set of all cases that cannot be distinguished from x using all attributes. In [24] $K_A(x)$ was called a successor neighborhood of x, see also [16,19,24,25].

Obviously, when a data set is complete, for given $B \subseteq A$, all characteristic sets $K_B(x)$ are identical with elementary blocks $[x]_B$.

The characteristic relation $R(B)$ is a relation on U defined for $x, y \in U$ as follows

$$(x, y) \in R(B) \text{ if and only if } y \in K_B(x).$$

The characteristic relation $R(B)$ is reflexive but—in general—does not need to be symmetric or transitive. Also, the characteristic relation $R(B)$ is known if we know characteristic sets $K_B(x)$ for all $x \in U$. In our example, $R(A) = \{(1, 1), (1, 4), (2, 2), (2, 5), (3, 3), (3, 6), (4, 1), (4, 5), (5, 5), (6, 6)\}$. The most convenient way to define the characteristic relation is through the characteristic sets.

For decision tables, in which all missing attribute values are lost, a special characteristic relation was defined in [21], see also, e.g., [20,22].

For incompletely specified decision tables lower and upper approximations may be defined in a few different ways. First, the definition of definability should

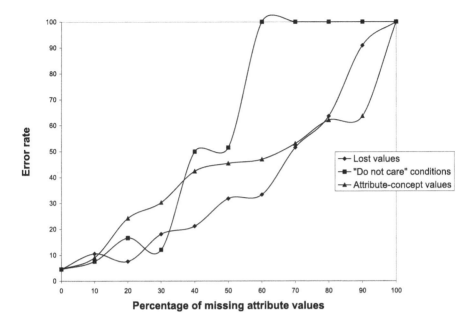

Fig. 1. Bankruptcy data—certain rule sets

be modified. A union of some intersections of attribute-value pair blocks, in any such intersection all attributes should be different and attributes are members of B, will be called B-*locally definable* sets. A union of characteristic sets $K_B(x)$, where $x \in X \subseteq U$ will be called a B-*globally definable* set. Any set X that is B-globally definable is B-locally definable, the converse is not true. In this paper we quote three different definitions of lower and upper approximations [7,8,9].

Let X be a concept, let B be a subset of the set A of all attributes, and let $R(B)$ be the characteristic relation of the incomplete decision table with characteristic sets $K(x)$, where $x \in U$. Our first definition uses a similar idea as in the previous articles on incompletely specified decision tables [14,15,20,21,22], i.e., lower and upper approximations are sets of singletons from the universe U satisfying some properties. Thus, lower and upper approximations are defined by constructing both sets from singletons. We will call these approximations *singleton*. Namely, a singleton B-lower approximation of X is defined as follows:

$$\underline{B}X = \{x \in U \mid K_B(x) \subseteq X\}.$$

A singleton B-upper approximation of X is

$$\overline{B}X = \{x \in U \mid K_B(x) \cap X \neq \emptyset\}.$$

In our example of the decision table presented in Table 2 let us say that $B = A$. Then the singleton A-lower and A-upper approximations of the two concepts: $\{1, 2, 4\}$ and $\{3, 5, 6\}$ are:

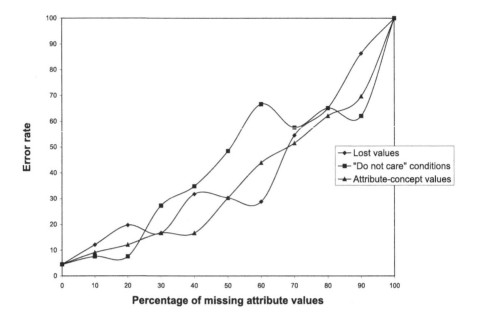

Fig. 2. Bankruptcy data—possible rule sets

$$\underline{A}\{1, 2, 4\} = \{1, 4\},$$

$$\underline{A}\{3, 5, 6\} = \{3, 5, 6\},$$

$$\overline{A}\{1, 2, 4\} = \{1, 2, 4\},$$

$$\overline{A}\{3, 5, 6\} = \{2, 3, 5, 6\}.$$

Note that the set $\overline{A}\{1, 2, 4\}$ is not even A-locally definable. Hence, as it was previously argued in [7,8,9], singleton approximations should not be used for rule induction. Obviously, if a set is not B-locally definable then it cannot be expressed by rule sets using attributes from B.

We may define lower and upper approximations for incomplete decision tables by using characteristic sets instead of elementary sets. There are two ways to do this. Using the first way, a *subset* B-lower approximation of X is defined as follows:

$$\underline{B}X = \cup\{K_B(x) \mid x \in U, K_B(x) \subseteq X\}.$$

A *subset* B-upper approximation of X is

$$\overline{B}X = \cup\{K_B(x) \mid x \in U, K_B(x) \cap X \neq \emptyset\}.$$

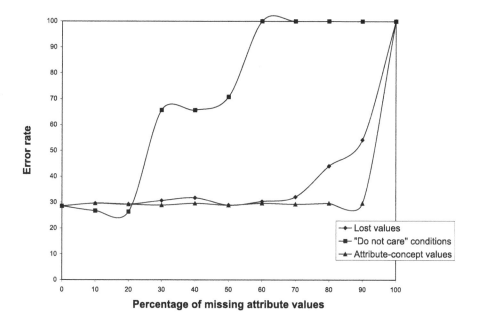

Fig. 3. Breast cancer (Slovenia) data—certain rule sets

Since any characteristic relation $R(B)$ is reflexive, for any concept X, singleton B-lower and B-upper approximations of X are subsets of the subset B-lower and B-upper approximations of X, respectively [9]. For the same decision table, presented in Table 2, the subset A-lower and A-upper approximations are

$$\underline{A}\{1, 2, 4\} = \{1, 4\},$$

$$\underline{A}\{3, 5, 6\} = \{3, 5, 6\},$$

$$\overline{A}\{1, 2, 4\} = \{1, 2, 4, 5\},$$

$$\overline{A}\{3, 5, 6\} = \{2, 3, 5, 6\}.$$

The second possibility is to modify the subset definition of lower and upper approximation by replacing the universe U from the subset definition by a concept X. A *concept* B-lower approximation of the concept X is defined as follows:

$$\underline{B}X = \cup\{K_B(x) \mid x \in X, K_B(x) \subseteq X\}.$$

Obviously, the subset B-lower approximation of X is the same set as the concept B-lower approximation of X [7]. A concept B-upper approximation of the concept X is defined as follows:

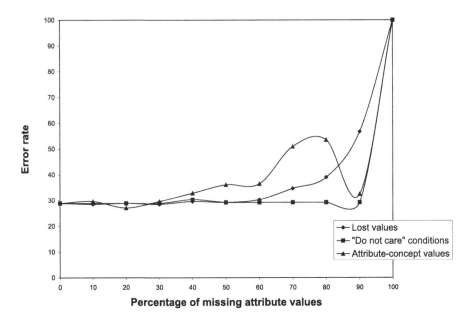

Fig. 4. Breast cancer (Slovenia) data—possible rule sets

$$\overline{B}X = \cup\{K_B(x) \mid x \in X, K_B(x) \cap X \neq \emptyset\} = \cup\{K_B(x) \mid x \in X\}.$$

The concept B-upper approximation of X is a subset of the subset B-upper approximation of X [7]. For the decision table presented in Table 2, the concept A-lower and A-upper approximations are

$$\underline{A}\{1, 2, 4\} = \{1, 4\},$$

$$\underline{A}\{3, 5, 6\} = \{3, 5, 6\},$$

$$\overline{A}\{1, 2, 4\} = \{1, 2, 4, 5\},$$

$$\overline{A}\{3, 5, 6\} = \{2, 3, 5, 6\}.$$

Note that for complete decision tables, all three definitions of lower approximations, singleton, subset and concept, coalesce to the same definition. Also, for complete decision tables, all three definitions of upper approximations coalesce to the same definition. This is not true for incomplete decision tables, as our example shows.

For Table 2, certain rules [3], induced from the *concept* lower approximations are

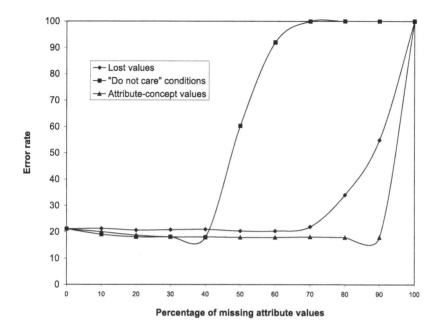

Fig. 5. Breast cancer (Wisconsin) data—certain rule sets

1, 2, 2
(Temperature, high) -> (Flu, yes)
1, 2, 2
(Headache, no) -> (Flu, no)
1, 1, 1
(Cough, no) -> (Flu, no)

and possible rules [3], induced from the *concept* upper approximations, are

1, 2, 2
(Temperature, high) -> (Flu, yes)
1, 2, 2
(Headache, yes) -> (Flu, yes)
1, 2, 2
(Headache, no) -> (Flu, no)
1, 1, 1
(Cough, no) -> (Flu, no)

Table 3 shows a modification of Table 2, where all *lost* values are replaced by "do not care" conditions. For decision tables where all missing attribute values are "do not care" conditions a special characteristic relation was defined in [14], see also, e.g., [15]. Blocks of attribute-value pairs are

[(Temperature, high)] = {1, 2, 3, 4, 5},
[(Temperature, normal)] = {2, 3, 5, 6},

Table 3. An incomplete decision table (all missing attribute values are *lost* values)

	Attributes			Decision
Case	Temperature	Headache	Cough	Flu
1	high	*	yes	yes
2	*	yes	*	yes
3	*	no	*	no
4	high	*	yes	yes
5	*	yes	no	no
6	normal	no	*	no

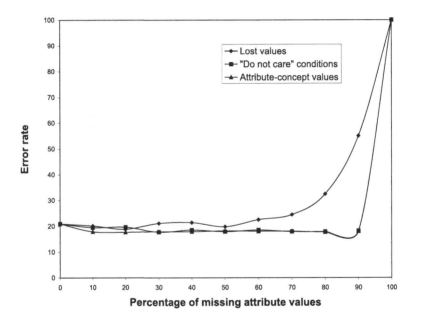

Fig. 6. Breast cancer (Wisconsin) data—possible rule sets

[(Headache, yes)] = {1, 2, 4, 5},
[(Headache, no)] = {1, 3, 4, 6},
[(Cough, yes)] = {1, 2, 3, 4, 6},
[(Cough, no)] = {2, 3, 5, 6}.

Characteristic sets are

$K_A(1) = \{1,2,3,4,5\} \cap U \cap \{1,2,3,4,6\} = \{1,2,3,4\},$
$K_A(2) = U \cap \{1,2,4,5\} \cap U = \{1,2,4,5\},$
$K_A(3) = U \cap \{1,3,4,6\} \cap U = \{1,3,4,6\},$

$K_A(4) = \{1, 2, 3, 4, 5\} \cap U \cap \{1, 2, 3, 4, 6\} = \{1, 2, 3, 4\},$
$K_A(5) = U \cap \{1, 2, 4, 5\} \cap \{2, 3, 5, 6\} = \{2, 5\},$
$K_A(6) = \{2, 3, 5, 6\} \cap \{1, 3, 4, 6\} \cap U = \{3, 6\},$

For the decision table presented in Table 3, the concept A-lower and A-upper approximations are

$$\underline{A}\{1, 2, 4\} = \emptyset,$$

$$\underline{A}\{3, 5, 6\} = \{3, 6\},$$

$$\overline{A}\{1, 2, 4\} = \{1, 2, 3, 4, 5\},$$

$$\overline{A}\{3, 5, 6\} = U.$$

In our example, the concept A-lower approximation of $\{1, 2, 4\}$ is the empty set. With large percentage of missing attribute values interpreted as "do not care" conditions, empty lower approximations cause large increases of error rates during ten-fold cross validation.

For Table 3, the only certain rule, induced from the *concept* lower approximation, is

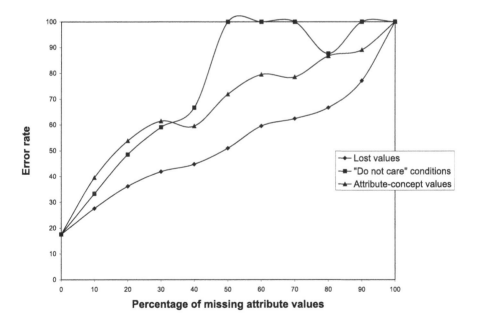

Fig. 7. Image segmentation data—certain rule sets

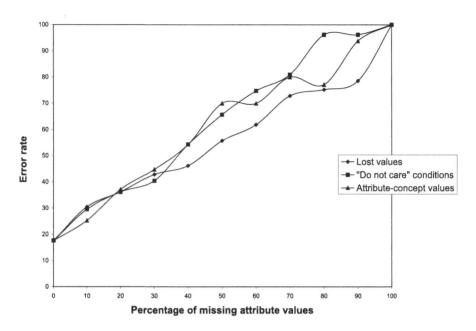

Fig. 8. Image segmentation data—possible rule sets

2, 2, 2
(Temperature, normal) & (Headache, no) -> (Flu, no)

and possible rules, induced from the *concept* upper approximations, are

1, 3, 5
(Temperature, high) -> (Flu, yes)
1, 2, 5
(Temperature, high) -> (Flu, no)
1, 3, 4
(Temperature, normal) -> (Flu, no)

Table 4 is another modification of Table 2, where all *lost* values are replaced by *attribute-concept* values. Blocks of attribute-value pairs are

[(Temperature, high)] = {1, 2, 4},
[(Temperature, normal)] = {3, 5, 6},
[(Headache, yes)] = {1, 2, 4, 5},
[(Headache, no)] = {3, 6},
[(Cough, yes)] = {1, 2, 4},
[(Cough, no)] = {3, 5, 6}.

Characteristic sets are

$K_A(1) = \{1,2,4\} \cap \{1,2,4,5\} \cap \{1,2,4\} = \{1,2,4\}$,
$K_A(2) = \{1,2,4\} \cap \{1,2,4,5\} \cap \{1,2,4\} = \{1,2,4\}$,
$K_A(3) = \{3,5,6\} \cap \{3,6\} \cap \{3,5,6\} = \{3,6\}$,

Table 4. An incomplete decision table (all missing attribute values are *lost* values)

	Attributes			Decision
Case	Temperature	Headache	Cough	Flu
1	high	–	yes	yes
2	–	yes	–	yes
3	–	no	–	no
4	high	–	yes	yes
5	–	yes	no	no
6	normal	no	–	no

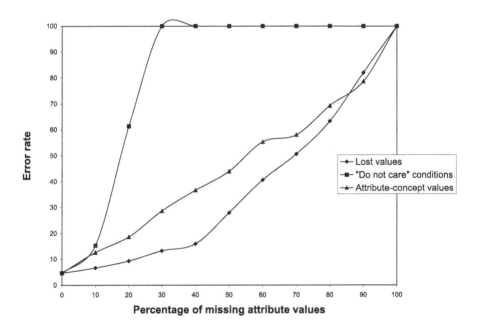

Fig. 9. Iris data—certain rule sets

$$K_A(4) = \{1, 2, 4\} \cap \{1, 2, 4, 5\} \cap \{1, 2, 4\} = \{1, 2, 4\},$$
$$K_A(5) = \{3, 5, 6\} \cap \{1, 2, 4, 5\} \cap \{3, 5, 6\} = \{5\},$$
$$K_A(6) = \{3, 5, 6\} \cap \{3, 6\} \cap \{3, 5, 6\} = \{3, 6\},$$

For the decision table presented in Table 4, the concept A-lower and A-upper approximations are

$$\underline{A}\{1, 2, 4\} = \{1, 2, 4\},$$

$$\underline{A}\{3, 5, 6\} = \{3, 5, 6\},$$

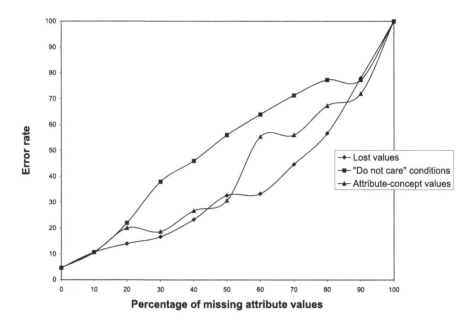

Fig. 10. Iris data—possible rule sets

$$\overline{A}\{1,2,4\} = \{1,2,4\},$$

$$\overline{A}\{3,5,6\} = \{3,5,6\}.$$

For Table 4, certain rules, induced from the *concept* lower approximations, are identical with possible rules, induced from *concept* upper approximations

1, 3, 3
(Temperature, high) -> (Flu, yes)
1, 3, 3
(Temperature, normal) -> (Flu, no)

4   Experiments

For our experiments six typical data sets were used, see Table 5. These data sets were complete (all attribute values were completely specified), with the exception of *breast cancer (Slovenia)* data set, which originally contained 11 cases (out of 286) with missing attribute values. These 11 cases were removed.

In two data sets: *bankruptcy* and *iris* all attributes were numerical. These data sets were processed as numerical (i.e., discretization was done during rule induction by MLEM2). The *image segmentation* data set was converted into symbolic using a discretization method based on agglomerative cluster analysis (this method was described, e.g., in [1]).

Table 5. Data sets used for experiments

| Data set | | Number of | |
	cases	attributes	concepts
Bankruptcy	66	5	2
Breast cancer (Slovenia)	277	9	2
Breast cancer (Wisconsin)	625	9	9
Image segmentation	210	19	7
Iris	150	4	3
Lymphography	148	18	4

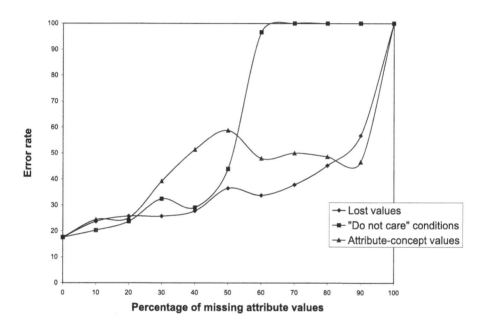

Fig. 11. Lymphography data—certain rule sets

To each data set we conducted a series of three experiments, adding incrementally (with 10% increment) missing attribute values of three different types. Thus, we started each series of experiments with no missing attribute values, then we added 10% of missing attribute values of given type, then we added additional 10% of missing attribute values of the same type, etc., until reaching a level of 100% missing attribute values. For each data set and a specific type of missing attribute values ten additional data sets were created.

Furthermore, for each data set with some percentage of missing attribute values, experiments were conducted separately for certain and possible rule sets, using *concept* lower and upper approximations, respectively. Ten-fold cross

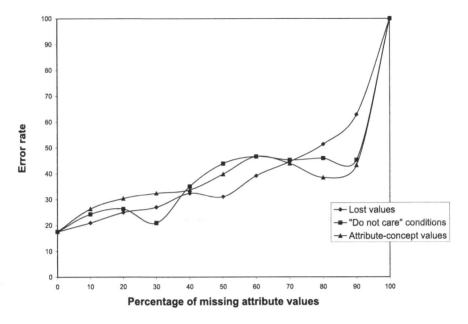

Fig. 12. Lymphography data—possible rule sets

validation was used to compute an error rate. Rule sets were induced by the MLEM2 option of the LERS data mining system [2,5,6]. Results of our experiments are presented in Figures 1–12. In all 12 figures, *lost values, "do not care" conditions*, and *attribute-concept values* denote percentage of error rate for experiments with missing attribute values interpreted as *lost* values, "do not care" conditions, and *attribute-concept* values, respectively.

5   Conclusions

During all series of experiments the error rate was affected by large variance. Moreover, for some data sets (e.g., *breast cancer (Wisconsin)*), adding a small amount of missing attribute values resulted in a decreased error rate. Most likely, in these data sets, attributes affected by missing attribute values were not important. In effect, the induced rule sets were more general and better.

It is clear that inducing certain rule sets while using a "do not care" condition approach to missing attribute values was the worst approach. This was caused by the fact that lower approximations of concepts, with large number of missing attribute values, were empty.

Another surprising conclusion is that for some data sets (*breast cancer (Slovenia)* and *breast cancer (Wisconsin)*) adding a large number of missing attribute values does not affect the error rate seriously—the error rate was almost the same for larger and larger number of missing attribute values.

Overall, it seems that the interpretation of missing attribute values as *lost* is the best approach among our three types of missing attribute value interpretations. Taking into account a large variance, the difference between error rates for certain and possible rule sets is negligible.

References

1. Chmielewski, M.R. and Grzymala-Busse, J.W.: Global discretization of continuous attributes as preprocessing for machine learning. *Int. Journal of Approximate Reasoning* **15** (1996) 319–331.
2. Chan, C.C. and Grzymala-Busse, J.W.: On the attribute redundancy and the learning programs ID3, PRISM, and LEM2. Department of Computer Science, University of Kansas, TR-91-14, December 1991, 20 pp.
3. Grzymala-Busse, J.W.: Knowledge acquisition under uncertainty—A rough set approach. *Journal of Intelligent & Robotic Systems* **1** (1988), 3–16.
4. Grzymala-Busse, J.W.: On the unknown attribute values in learning from examples. Proc. of the ISMIS-91, 6th International Symposium on Methodologies for Intelligent Systems, Charlotte, North Carolina, October 16–19, 1991. Lecture Notes in Artificial Intelligence, vol. 542, Springer-Verlag, Berlin, Heidelberg, New York (1991) 368–377.
5. Grzymala-Busse, J.W.: LERS—A system for learning from examples based on rough sets. In Intelligent Decision Support. Handbook of Applications and Advances of the Rough Sets Theory, ed. by R. Slowinski, Kluwer Academic Publishers, Dordrecht, Boston, London (1992) 3–18.
6. Grzymala-Busse., J.W.: MLEM2: A new algorithm for rule induction from imperfect data. Proceedings of the 9th International Conference on Information Processing and Management of Uncertainty in Knowledge-Based Systems, IPMU 2002, July 1–5, Annecy, France, 243–250.
7. Grzymala-Busse, J.W.: Rough set strategies to data with missing attribute values. Workshop Notes, Foundations and New Directions of Data Mining, the 3-rd International Conference on Data Mining, Melbourne, FL, USA, November 19–22, 2003, 56–63.
8. Grzymala-Busse, J.W.: Data with missing attribute values: Generalization of idiscernibility relation and rule induction. *Transactions on Rough Sets*, Lecture Notes in Computer Science Journal Subline, Springer-Verlag, vol. **1** (2004) 78–95.
9. Grzymala-Busse, J.W.: Characteristic relations for incomplete data: A generalization of the indiscernibility relation. Proc. of the RSCTC'2004, the Fourth International Conference on Rough Sets and Current Trends in Computing, Uppsala, Sweden, June 1–5, 2004. Lecture Notes in Artificial Intelligence 3066, Springer-Verlag 2004, 244–253.
10. Grzymala-Busse, J.W.: Three approaches to missing attribute valuesA rough set perspective. Proceedings of the Workshop on Foundation of Data Mining, associated with the Fourth IEEE International Conference on Data Mining, Brighton, UK, November 1–4, 2004, 55–62.
11. Grzymala-Busse, J.W. and Hu, M.: A comparison of several approaches to missing attribute values in data mining. Proceedings of the Second International Conference on Rough Sets and Current Trends in Computing RSCTC'2000, Banff, Canada, October 16–19, 2000, 340–347.

12. Grzymala-Busse, J.W. and Wang A.Y.: Modified algorithms LEM1 and LEM2 for rule induction from data with missing attribute values. Proc. of the Fifth International Workshop on Rough Sets and Soft Computing (RSSC'97) at the Third Joint Conference on Information Sciences (JCIS'97), Research Triangle Park, NC, March 2–5, 1997, 69–72.
13. Hong, T.P., Tseng L.H. and Chien, B.C.: Learning coverage rules from incomplete data based on rough sets. Proc. of the IEEE International Conference on Systems, Man and Cybernetics, Hague, the Netherlands, October 10–13, 2004, 3226–3231.
14. Kryszkiewicz, M.: Rough set approach to incomplete information systems. Proc. of the Second Annual Joint Conference on Information Sciences, Wrightsville Beach, NC, September 28–October 1, 1995, 194–197.
15. Kryszkiewicz, M.: Rules in incomplete information systems. *Information Sciences* **113** (1999) 271–292.
    and knowledge base systems. Fourth International Symposium on Methodologies of Intelligent Systems (Poster Sessions), Charlotte, North Carolina, October 12–14, 1989, 75–86.
    Tucson, Arizona, December 4–8, 1989, 286–293.
16. Lin, T.Y.: Topological and fuzzy rough sets. In *Intelligent Decision Support. Handbook of Applications and Advances of the Rough Sets Theory*, ed. by R. Slowinski, Kluwer Academic Publishers, Dordrecht, Boston, London (1992) 287–304.
17. Pawlak, Z.: Rough Sets. *International Journal of Computer and Information Sciences* **11** (1982) 341–356.
18. Pawlak, Z.: *Rough Sets. Theoretical Aspects of Reasoning about Data*. Kluwer Academic Publishers, Dordrecht, Boston, London (1991).
19. Slowinski, R. and Vanderpooten, D.: A generalized definition of rough approximations based on similarity. *IEEE Transactions on Knowledge and Data Engineering* **12** (2000) 331–336.
20. Stefanowski, J.: *Algorithms of Decision Rule Induction in Data Mining*. Poznan University of Technology Press, Poznan, Poland (2001).
21. Stefanowski, J. and Tsoukias, A.: On the extension of rough sets under incomplete information. Proc. of the 7th International Workshop on New Directions in Rough Sets, Data Mining, and Granular-Soft Computing, RSFDGrC'1999, Ube, Yamaguchi, Japan, November 8–10, 1999, 73–81.
22. Stefanowski, J. and Tsoukias, A.: Incomplete information tables and rough classification. *Computational Intelligence* **17** (2001) 545–566.
23. Wang, G.: Extension of rough set under incomplete information systems. Proc. of the IEEE International Conference on Fuzzy Systems (FUZZ_IEEE'2002), vol. 2, Honolulu, HI, May 12–17, 2002, 1098–1103.
24. Yao, Y.Y.: Relational interpretations of neighborhood operators and rough set approximation operators. *Information Sciences* **111** (1998) 239–259.
25. Yao, Y.Y. and Lin, T.Y.: Generalization of rough sets using modal logics. *Intelligent Automation and Soft Computing* **2** (1996) 103–119.

Pawlak's Landscaping with Rough Sets

Mihir K. Chakraborty[*]

Department of Pure Mathematics, University of Calcutta
35 Ballygunge Circular Road, Kolkata 700019, India
mihirc99@vsnl.com

Abstract. This paper reviews, rather non-technically, Pawlak's approach to vagueness through rough sets and looks for a foundation of rough sets in an early work of Obtułowicz. An extension of Obtułowicz's proposal is suggested that in turn, hints at a unified approach to rough sets and fuzzy sets.

1 Introduction

The concluding decades of the past century have added several outstanding, significant and elegant contributions to human knowledge of which Rough Set Theory is one. Zdzisław Pawlak, a Professor of Computer Science from Poland, first proposed this theory in 1982 through his publication entitled 'Rough Sets' [20]. Surprisingly, this is again a contribution to humanity from one belonging to the field of computer science – during the same period, the same community gifted several other elegant creations, like Fuzzy Set Theory by Lotfi Zadeh in 1965. It is also interesting to note that both the theories address basically the same issue, viz. 'vagueness' and this fact is not merely a coincidence.

'Vagueness' had been an outstanding issue. Great minds of the antiquity, both of the East and the West delved into the notion exhibited in various forms. (Theseuses' ship [43,17], the Sorites [43,13], or the tetra-lemma (Catuskoti) [35,32], for example). Following Enlightenment, with the rise of modern rationality, embodied in the methods of physical sciences, more specifically physics, 'vagueness' had been gradually pushed aside to the fringes like the indigenous population in Australia or in America and other places. Use of imprecise terms were not only marginalized, but virtually banished from all serious discourses as expressed by the rationalist, humanist Bertrand Russell in the lines (in, Our Knowledge of External World as a Field of Scientific Method in Philosophy) (cf. Hao Wang [42]):

"The study of logic becomes the central study in philosophy: it gives the method of research in philosophy, just as mathematics gives the method in physics;"

---

[*] I would like to express my sincere thanks to Smita Sirker, Dept. of Philosophy, Jadavpur University, for kindly reading the first draft and making valuable comments, particularly on the main philosophy of this article viz. the relationship between the existence of an object in a concept and indiscernibility.

J.F. Peters et al. (Eds.): Transactions on Rough Sets VI, LNCS 4374, pp. 51–63, 2007.
© Springer-Verlag Berlin Heidelberg 2007

and further from his article 'Vagueness' (Australian Journal of Philosophy)[37]:

"Logical words, like the rest, when used by human beings, share the vagueness of all other words. There is however, less vagueness about logical words than about the words of daily life."

However, in spite of the great tide of modern rationality all over the world, a sense of understanding that there exists an essential role of vagueness in human knowledge system as well as life, was not totally wiped out from Eastern thoughts. Ironically, the most advanced technology – computer science, the most recent gift of modernity, has ushered the study of vagueness spectacularly from the disrespectful margin straightaway to centre-stage.

That 'vagueness' in general is different from 'probability' has somewhat been accepted nowadays after the long, fierce debates that took place during the years immediately following the advent of fuzzy set theory in 1965. So Pawlak did not have to fight that battle. Yet he had to utter this warning which is an excellent distinctive criterion viz. "Vagueness is the property of sets... whereas uncertainty is the property of an element"[28]. Uncertainty leads to probabilistic studies. It is often said of course, that vagueness is uncertainty too but not of probabilistic kind. However, right from the beginning Pawlak wanted to point at the distinction between Rough Set theory and Fuzzy Set theory. In the introduction to his short communication [23] he declares "we compare this concept with that of the fuzzy set and we show that these two concepts are different." Different in what sense? Early Pawlak (during the 80s) was firm in his belief that Rough Set is properly addressing vagueness since it talks about 'boundaries' of a set and the property 'rough' is ascribed to a set. On the other hand, although the qualifier 'fuzzy' has been ascribed to sets too, in reality the theory deals with degree of membership of an object in a 'set' and hence is dealing with some kind of uncertainty of belongingness of objects. So according to the above quoted norm, fuzzy set theory is not addressing vagueness proper. However, in later Pawlak, perhaps a change in opinion is observed as reflected in the following categorical remark   "Both fuzzy and rough set theory represent two different approaches to vagueness. Fuzzy set theory addresses gradualness of knowledge, expressed by the fuzzy membership - whereas rough set theory addresses granularity of knowledge expressed by indiscernibility relation" [22].

We shall discuss the role of indiscernibility to some length in the foundations of fuzzy set theory as well as rough set theory and thus in vagueness. But it needs to be mentioned that the relationship between the two theories had quite naturally been a favourite topic of study in those turbulent decades. For example, Wygralak in the 1985 BUSEFAL Conference presented a paper [44] in which he established that basic operations on rough sets (i.e. union and intersection) can be expressed as some special operations as their membership functions. Pawlak already talked about the distinction and the irreducibility of rough sets to fuzzy sets. This point needs a little clarification since rough sets approximations are not distributive with respect to all set theoretic operations. Pawlak in [23] checked for the natural candidate for representation of rough sets by the 3-valued membership function

$$X(x) = \begin{cases} 1 & \text{if } x \text{ belongs to the lower approximation} \\ .5 & \text{if } x \text{ belongs to the boundary region} \\ 0 & \text{otherwise.} \end{cases}$$

With such functions representing rough sets X and Y, the membership function of $X \cup Y$ can not be represented by $max(X(x), Y(x))$ as is done in fuzzy set theory, since it has to coincide with the function

$$X \cup Y(x) = \begin{cases} 1 & \text{if } x \text{ belongs to the lower approximation of } X \cup Y \\ .5 & \text{if } x \text{ belongs to the boundary region} \\ 0 & \text{otherwise,} \end{cases}$$

but which fails.

Similarly, $X \cap Y$ can not be represented by the function $min(X(x), Y(x))$. However if the membership function is modified as below the desired result is obtained. Wygralak proposed to define $\sqcup$ and $\sqcap$ of two rough sets X, Y, by

$$X \sqcup Y(x) = \begin{cases} min(1, X(x) + Y(x)) & \text{if } X(x) = Y(x) = .5 \text{ and } [x] \subseteq X \cup Y \\ max(X(x), Y(x)) & \text{otherwise.} \end{cases}$$

and

$$X \sqcap Y(x) = \begin{cases} max(0, X(x) + Y(x) - 1) & \text{if } X(x) = Y(x) = .5 \text{ and } [x] \cap X \cap Y \neq \emptyset \\ min(X(x), Y(x)) & \text{otherwise.} \end{cases}$$

$X \sqcup Y$ is roughly equal to $X \cup Y$ and
$X \sqcap Y$ is roughly equal to $X \cap Y$.

In spite of this claim of Wygralak, one may doubt about the acceptability or otherwise of such functions as operators for conjunction and disjunction because they might miss some important properties - but it requires detailed investigations to make a final comment.

In Moscow Conference, 1988, Dubois and Prade argue that "fuzzy sets and rough sets aim to different purposes and that it is more natural to try to combine the two models of uncertainty (vagueness for fuzzy sets and coarseness for rough sets) in order to get a more accurate account of imperfect information." They proposed interesting mathematical constructs known as rough-fuzzy sets and fuzzy-rough sets [9].

In the opinion of the present author, there is an essential indiscernibility underlying in all kinds of vagueness - indiscernibility giving rise to both granularity and gradualness. But indiscernibles may be of various types though one can probe into some essential features of this elusive notion. These investigations had also been a favourite topic in the 80s and 90s [1,34,41,12,14,10].

2   Indiscernibilities

One major difference in the approaches to indiscernibility lies in assuming it as a 'yes'/'no' type crisp concept or a graded concept. In the first approach, objects

x and y are either indiscernible or not - this approach is adopted in Rough Set theory by Pawlak. Several generalizations, of course, have taken place. The general feature of these is that starting from some knowledge or data, different components (or clusters) are computed and rough sets are construed out of these components. We shall, however, restrict this study to Pawlak-rough sets only. It would be interesting to investigate if the present 'unified' approach can be extended to generalized rough sets as well. The second approach, as pursued in Fuzzy Set theory, presumes that x and y may be indiscernible to a degree - that is, it is a graded notion. Another very important difference rests in assuming or not assuming or weakening the transitivity property of indiscernibility. In the first case we get standard equivalence relation on the universe of discourse, as is the base of Pawlak-rough sets. The second case generates tolerance relation (reflexive-symmetric) as base – rough set theory on this base is also pursued. In the third case, we have fuzzy transitivity, viz.

$$Ind(x,y) \ \& \ Ind(y,z) \leq Ind(x,z),$$

where $Ind(x,y)$ represents the indiscernibility degree between x and y, and $\&$ is an algebraic operation (a t-norm, perhaps) on a suitable truth set. This graded relation without being reduced to tolerance, relaxes the notion of hard transitivity and elegantly, takes care of the gradualness aspect by using an interactive conjunction operator as follows.

Let $x_1, x_2, x_3, x_4, \ldots$ be a sequence of objects such that $Ind(x_i, x_{i+1}) = .5$, for all i. Now let us take the product ($\times$) as the operator for $\&$. Since $Ind(x_1, x_2) \ \& \ Ind(x_2, x_3) \leq Ind(x_1, x_3)$, we get $Ind(x_1, x_3) \geq .5 \times .5 = .25$. If the least value .25 is taken then

$$Ind(x_1, x_4) \text{ may be taken as } Ind(x_1, x_3) \ \& \ Ind(x_3, x_4) = .25 \times .5 = .125.$$

Thus indiscernibility degree gradually diminishes. It means that the indiscernibility between x_1 and x_4 is less than that between x_1 and x_3, and this is further less than the indiscernibility between x_1 and x_2 – a feature quite intuitively acceptable. Symmetry is naturally expected of indiscernibility. In the fuzzy case it means that $Ind(x,y) = Ind(y,x)$. We shall discuss about reflexivity property later. Before that let us examine the relationship between indiscernibility relation and a concept.

A concept induces an indiscernibility relation in a universe of discourse. For example, the concept A gives rise to Ind_A given by

$$Ind_A(x,y) = \begin{cases} 1 & \text{if } x,y \in A \text{ or } x,y \in A^c \\ 0 & \text{otherwise.} \end{cases}$$

Concepts A, B, and C similarly give rise to the relation

$$Ind_{A,B,C}(x,y) = \begin{cases} 1 & \text{if } x \in X \text{ if and only if } y \in X, \ X = A, B, C \\ 0 & \text{otherwise.} \end{cases}$$

An instance of this latter case is depicted in the following diagram with fifteen objects, x_1 to x_{15}.

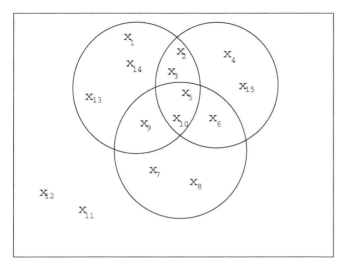

We could also show $Ind_{A,B,C}$ by constructing the following approximation space or information system.

	A	B	C
x_1	Y	N	N
x_2	Y	Y	N
x_3	Y	Y	N
x_4	N	Y	N
x_5	Y	Y	Y
x_6	N	Y	Y
x_7	N	N	Y
x_8	N	N	Y
x_9	Y	N	Y
x_{10}	Y	Y	Y
x_{11}	N	N	N
x_{12}	N	N	N
x_{13}	Y	N	N
x_{14}	Y	N	N
x_{15}	N	Y	N

Y stands for 'yes, belongs to' and N for 'no, does not belong to'. So $Ind_{A,B,C}(x,y)$ $= 1$, if and only if the rows corresponding to x and y are identical.

Thus a set of concepts generates a partition of the universe and hence equivalently gives an approximation space.

Can we retrieve A (in the first example) or A, B, C (in the second example) from the indiscernibility relations Ind_A or $Ind_{A,B,C}$? The two values 0 and 1 of the indiscernibility relation are not sufficient for this purpose. Only from the information about indiscernibility one cannot separate A from its complement A^c since $Ind_A(x,x) = 1$ whether $x \in A$ or $x \in A^c$. So, let us take a third value 2 and stipulate.

$$Ind_A^*(x,x) = \begin{cases} 2 & \text{if } x \in A \\ 1 & \text{if } x \in A^c \end{cases}$$

and

$$Ind_A^*(x,y) = \begin{cases} 1 & \text{if } x \neq y,\ x,y \in A \text{ or } x,y \in A^c \\ 0 & \text{if } x \neq y,\ x \in A \text{ and } y \in A^c, \text{ or } x \in A^c \text{ and } y \in A. \end{cases}$$

It is obvious that $Ind_A(x,y) = 1$ if and only if $Ind_A^*(x,y) \geq 1$ and $Ind_A(x,y) = 0$ if and only if $Ind_A^*(x,y) = 0$.

The two functions Ind_A and Ind_A^* coincide on (x,y), $x \neq y$, but while the first one cannot make a distinction between A and its complement, Ind_A^* can do this.

Similarly, in order to retrieve $A \cup B \cup C$ in the second case, we need the definition

$$Ind_{A,B,C}^*(x,x) = \begin{cases} 2 & \text{if } x \in A \cup B \cup C \\ 1 & \text{if } x \notin A \cup B \cup C \end{cases}$$

and

$$Ind_{A,B,C}^*(x,y) = \begin{cases} 1 & \text{if } x \neq y \text{ and the rows of } x \text{ and } y \text{ in the information} \\ & \text{system coincide} \\ 0 & \text{otherwise.} \end{cases}$$

One point to be observed here is that although while $Ind_X^*(x,y) \geq 1$ implies that x and y are indiscernible in terms of the basic concepts, $Ind_X^*(x,x)$, i.e. the indiscernibility of x with itself is not of the same category for all x. Also to be noted that this representation may be applied to any subset (not only $A \cup B \cup C$) of U obtained by unions of intersections of A, B and C, in other words unions of the indiscernibility classes or blocks in U determined by them. The procedure, however, is to define $Ind_{A,B,C}^*(x,x)$, so that only specified unions of blocks are obtained in the backward process.

We shall summarize and axiomatize such properties of Ind_X^* later. At this stage we only raise an issue. Could we not think that degree of belongingness of an object to a concept is determined by and is the same as the degree of indiscernibility of the object with itself relative to the same concept? Looking from this angle, indiscernibility becomes more a primitive notion and plays a key role in the process of learning as well as of categorization. A category is created depending on the similarity of its members. Based on the degree of indiscernibility of an object, its belongingness to a set/category/class is decided. x belongs to A to the extent $2 = Ind_A^*(x,x)$, means x is within A, x belongs to A to the extent $1 = Ind_A^*(x,x)$ means x is in A^c. That we are interested in A and not in its complement is also represented in the assignment of a lower degree of indiscernibility to elements of the complement with themselves. The idea shall play a key role in what follows and is a basis for enquiries into vagueness. Even in the case of crisp concepts, we have noticed, while retrieving A from Ind_A^* the elements of A have grade 2, and elements of A^c have grade 1, and a third value 0 is also needed to construct the partition.

If, however, the concepts are vague, that is, admit borderline cases, it is necessary to introduce a fourth value and the indiscernibility generated may be defined as above with certain additional conditions. These ideas were published long back in Obtułowicz's paper [18] based on a firm category-theoretic basis proposed by Higgs [12], but probably escaped attention of the researchers in this field. We re-present from his work the main definitions and representation theorems below with little notational changes to fit in the present context.

Let U be a universe and $L(4)$ be the ordered set $\{0 \le 1 \le 2 \le 3\}$ which is a complete distributive lattice (or complete Heyting algebra).

Let $Ind^* : U \times U \to L(4)$ be an indiscernibility relation that satisfies conditions.

$H_1 : Ind^*(x, y) = Ind^*(y, x)$ (Symmetry)
$H_2 : Ind^*(x, y) \wedge Ind^*(y, z) \le Ind^*(x, z)$ (Transitivity)

and the following roughness conditions

$R_1 : 1 \le Ind^*(x, x)$ for $x \in U$
$R_2 : $ if $2 \le Ind^*(x, y)$, then $x = y$
$R_3 : $ if $Ind^*(x, y) = 1$, then $Ind^*(x, x) = Ind^*(y, y)$
$R_4 : $ if $Ind^*(x, x) = 2$, then there exists y such that $Ind^*(x, y) = 1$

The significance of roughness conditions shall be clear from Proposition 2 below. The following two propositions establish that (U, Ind^*) is a representation of any Pawlak-rough set in U.

Proposition 1. *Let (U, Ind^*) be given. Then the relation R defined by xRy if and only if $Ind^*(x, y) \ge 1$ is an equivalence relation, and the pair (I, B) defined by $I = \{x : Ind^*(x, x) = 3\}$, $B = \{x : Ind^*(x, x) = 2\}$ constitute the interior and boundary of a rough set in (U, R).*

Proposition 2. *Let (U, R) be an approximation space in which (I, B), the interior and boundary pair determines a rough set. Consider the mapping $Ind^* : U \times U \to L(4)$ given by*

$$Ind^*(x, x) = \begin{cases} 3 & \text{if } x \in I \\ 2 & \text{if } x \in B \\ 1 & \text{if } x \in U \setminus (I \cup B) \end{cases}$$

and

$$Ind^*(x, y) = \begin{cases} 1 & \text{if } x \ne y, \ xRy \text{ holds} \\ 0 & \text{if } x \ne y, \ xRy \text{ does not hold.} \end{cases}$$

Then Ind^* satisfies the conditions $H_1, H_2, R_1, R_2, R_3, R_4$.

The following important feature is also observed in the two constructions by Propositions 1 and 2.

$$(U, Ind^*) \xrightarrow{Prop1} (U, R, I, B) \xrightarrow{Prop2} (U, Ind^*) \xrightarrow{Prop1} (U, R, I, B).$$

Thus an indiscernibility satisfying the roughness conditions gives rise to a rough set in an approximation space and vice versa.

One can see that the above representation does not amount to the 3-valued semantics where an object falls under a concept with three grades viz. 3 (when it is in the definite region), 2 (when it is in the boundary), and 1 (outside the boundary). We have noticed difficulties with such representation (by Pawlak or Wygralak) in the beginning of this article. The membership function of Obtułowicz also says that

$Ind^*(x, x)$ & $Ind^*(x, y) \leq Ind^*(y, y)$, i.e.

the degree of belongingness of x to a concept & the degree of indiscernibility of x with $y \leq$ the degree of belongingness of y to the concept.

This criterion is the so-called 'saturatedness' condition that has been elaborately discussed in [5]. In Leibniz's terms, this is the version of the doctrine of *Identity of Indiscernibles*, viz. if an object x has a property P and an object y is indiscernible from x, then y has the property P. In the fuzzy context, a more general condition is taken viz.

$\alpha(x)$ & $Ind(x, y) \leq \alpha(y)$,

where $\alpha(x)$ denotes the degree of belongingness of x to the fuzzy set α.

The only addition here is the conceptual indulgence to the assumption that belongingness degree of x to a fuzzy set is the same as $Ind^*(x, x)$.

It is also significant to notice that the fuzzy set theoretic operators 'max' and 'min' are now applicable to obtain the union and intersection.

A summary of what has been said so far is the following:

- The underlying indiscernibility relation for any vague concept in U is a relation Ind^* satisfying the conditions $H_1, H_2, R_1, R_2, R_3, R_4$.
- Such a relation, which is a particular kind of fuzzy equivalence relation determines uniquely a rough set in the approximation space (U, R) where R is virtually the underlying indiscernibility and conditions R_1, R_2, R_3, R_4 determine the interior and boundary of the rough set.
- Conversely, any rough set in (U, R) given by the interior and boundary can be generated by an indiscernibility relations satisfying $H_1, H_2, R_1, R_2, R_3, R_4$.
- This representation is one-to-one.
- In the special case when the boundary region is empty, the condition R_4 is dropped. We need a three-element complete Heyting algebra and Propositions 1 and 2 may be written accordingly. Earlier examples with the concept A, and concepts A, B, C are instances of such representation.

One interesting extension of Obtułowicz's representation of rough sets suggests itself, but has never been taken up.

Formally, one can add more categories other than 0, 1, 2 and 3. For instance, let us take one more viz. 4. All the roughness conditions remain the same except that the condition R_4 shall now be read as

R_4': for each x, $Ind(x, x) = 2$ or 3 implies that there exists y such that $Ind(x, y) = 1$.

The rough set is now extended to (U, R, I, B_1, B_2), where there are two layers in the boundary, elements of I have grade 4, while those of B_1 and B_2 have grades 3 and 2 respectively. The representation theorems now take the following shapes.

Proposition 3. *Let $Ind^* : U \times U \to L(5) (\equiv \{0 \leq 1 \leq 2 \leq 3 \leq 4\}$ be an indiscernibility relation satisfying conditions $H_1, H_2, R_1, R_2, R_3, R_4'$. Then R defined by xRy if and only if $Ind^*(x, y) \geq 1$ is an equivalence relation, and the triple (I, B_1, B_2) defined by $I = \{x : Ind^*(x, x) = 4\}$, $B_1 = \{x : Ind^*(x, x) = 3\}$, $B_2 = \{x : Ind^*(x, x) = 2\}$ constitute the interior and the two layers of boundaries of a rough set in (U, R).*

Proposition 4. *Let (U, R) be an approximation space in which (I, B_1, B_2) determines a rough set. Then the mapping $Ind^* : U \times U \to L(5)$ given by*

$$Ind^*(x, x) = \begin{cases} 4 & \text{if } x \in I \\ 3 & \text{if } x \in B_1 \\ 2 & \text{if } x \in B_2 \\ 1 & \text{if } x \in U \setminus (I \cup B_1 \cup B_2) \end{cases}$$

and

$$Ind^*(x, y) = \begin{cases} 1 & \text{if } x \neq y, \, xRy \text{ holds} \\ 0 & \text{if } x \neq y, \, xRy \text{ does not hold.} \end{cases}$$

Then Ind^* satisfies the conditions $H_1, H_2, R_1, R_2, R_3, R_4'$.

Layers of the boundary may be enhanced arbitrarily, but finitely. The interpretation of these layers is zones of objects of gradually weaker possibilities of falling under the concept that render gradualness along with granularity. One can also see the possibility of interpreting membership values under rough membership function [22] as values of the lattice (after normalization, of course). Elements of a block with lower rough membership may be placed in the weaker layer of the boundary. Element of beyond-possible zone should be given the value 1 instead of 0 which should be retained to denote the discernibility of x, y, $x \neq y$ and to determine the partition of the Universe. In finite situations there should not be any difficulty. The definitions of union intersection and complementation may be suitably defined by using max, min and reversal of grades.

There shall be a departure in this approach from that of Pawlak. 'Roughness' should no longer be considered as an adjective to an ordinary set in an approximation space but as a pair (I, B) of unions of blocks of the space, I being called the interior and B the boundary. This approach is equivalent to that taken by us in [2]. It is also in conformity with the philosophy of rough sets viz. "we 'see' elements of the universe through available informations" and hence "some elements may be 'seen' identical" and "this is to mean that if we see a set through informations, only, the approximations (lower and upper) can be observed." And further "a vague property determines not only a single set of elements falling under the property but a family of sets which can be identified with this property up to indiscernibility." All the above lines within quotation marks are from Pawlak's writings glued together. So the present approach, though a departure, draws the support from his own feelings too.

3   Conclusions

Like fuzzy sets, rough sets also have wide usages for example, in artificial intelligence, cognitive sciences, knowledge discovery from database, machine learning, expert system, inductive reasoning and pattern recognition. As in the case of any successful theory, it has to look back at one point of time. "The rough set theory has reached such a state that some kind of summary of its theoretical foundation is a must" [28] - this realization of Pawlak is quite justifiable. As it appears from the preceding discussions, the approach of Obtułowicz may serve as quite a reasonable foundation. This approach also suggests a kind of unification of fuzzy sets and rough sets. Starting with some sort of indiscernibility in the Universe which is at the base of any vague concept and which arises out of data (concrete or subjective), objects of the Universe are categorized. In such a categorization, some tokens (not necessarily numbers) with varying algebraic structures play a role. A mathematical entity emerges representing the vague concept. In this representation, the following philosophy is adopted:

"The degree of existence of an object in a concept is the degree to which the object is indiscernible with itself relative to the underlying the concept."

The mathematical entity is sometimes a fuzzy set and sometimes a rough set. Divergence occurs because of the nature of the indiscernibility (crisp or fuzzy) and the choice of categories (how many?) and their structures. Incorporation of layers in the boundary as proposed in the paper for the first time brings rough sets closer to fuzzy sets.

We think that this approach could help in erasing Pawlak's persistent feeling of a sort of 'supremacy' of classical set theory over fuzzy set theory or rough set theory. This feeling is expressed in statements like "fuzzy set involves more advanced mathematical concepts real numbers and functions - whereas in classical set theory the notion of set is used as a fundamental notion of whole mathematics and is used to derive any other mathematical concept e.g. numbers and functions. Consequently, fuzzy set theory cannot replace classical set theory, because, in fact, the theory is needed to define fuzzy sets." [21] Again, "in a manner similar to fuzzy set theory, rough set theory is not an alternative to classical set theory but it is embedded in it." [22] Pawlak's concern about the foundational problems of classical set theory and interest in the alternatives like, multisets, multi-fuzzy sets, Blizard sets, general sets, Mereology (Lesniewski), Alternative set theory (Vopenka), Penumbral set theory (Apostoli and Kanda) are well known [21,25]. In the proposal of Obtułowicz, what would, in fact, be needed at the beginning is a Universe, a collection of tokens with some structures including order (in particular the numbers) and the notion of indiscernibility which needs only an understanding of 'pair' and 'correspondence'. These may constitute a nice, intuitively acceptable beginning. If one casts an oblique eye, we can take refuge to categorical foundation (Higgs [12], Banerjee and Chakraborty [3], and others). At this point we would like to draw the attention of readers to a paper of Goguen published in 1974. He claims "Ideally, we would like a foundation for fuzzy sets which justifies the intuitive identification of fuzzy sets with (inexact) concepts, and in which the familiar set operations are uniquely and inevitably determined. These desires are clearly though not explicitly expressed in Zadeh (1965), *and they are all satisfied by the system given in this paper.*" He used as paradigm Lawvere's "relatively unknown axiomatization of sets in the language of category theory". In one of our papers [3] Higgs' and Goguen's categories are discussed and compared. But category-theoretic approach is still not popular among practitioners. If toiled and fuzzy as well as rough sets are found soundly based on categorical grounds, the gains appear to be enormous - reinstatement of 'vagueness' within the discourses of mathematics, logic and thus by sciences. A recent work by Skowron [39] once again brings to focus the importance of the study of vagueness and the role of rough set theoretic methods in such studies. He introduces two operations viz. inductive extension and granulation of approximation spaces and emphasizes on "important consequences of the paper for research on approximation of vague concepts and reasoning about them in the framework of adoptive learning." He thinks that "this (adoptive learning) requires developing a new approach to vague concepts going beyond the traditional rough or fuzzy approaches." This paper extends the notion of approximation space by incorporating rough inclusion or graded inclusion. It would be an interesting project to investigate if Obtułowicz's proposal may be extended to this generalized context also.

We engage not only into crisp (two-valued) talks about vagueness, but into multi-valued talks too (theory of graded consequence [6,7]) or we also talk 'roughly' (theory of rough consequence [5]). The underlying motivation to define graded consequence or rough consequence is to allow room for vagueness in the metalogical concepts like consistency, consequence, tautologihood, completeness, etc. This latter notion viz. rough consequence has its origin in Pawlak's insightful work on rough truth [24] where he states "the rough (approximate) truth and falsity represent our partial knowledge about the world and with the increase of our knowledge the roughly true (or false) formulas tend to be more true (or false) and approach the truth and falsity closer and closer." One is bound to recall Zadeh when he claims that the notion of truth itself is fuzzy. It would not be out of place to mention that the first recorded works on rough logics are by Orłowska and Pawlak [19] and Rasiowa and Skowron [36]. With the advent of graded and rough consequences, the scenario of mathematics should change, in that there may be some mathematical predicates, truth of sentences relative to which may be partial and derivations involving which may not be of full strength.

Professor Pawlak was an artist. His favourite example of vagueness was a 'beautiful' painting [22]. He was fond of landscaping. His work on rough sets is also a beautiful landscape-installation - to which we offer this humble bouquet of ours that might develop roots striving to become an integral part of this scenario.

References

1. M. Banerjee. *A Categorial Approach to the Algebra and Logic of the Indiscernible.* Ph.D Thesis, University of Calcutta, 1993.
2. M. Banerjee and M.K. Chakraborty. Rough sets through algebraic logic. *Fundamenta Informaticae*, 28(3,4): 211–221, 1996.

3. M. Banerjee and M.K. Chakraborty. Foundations of vagueness: a category-theoretic approach. *Electronic Notes in Theoretical Computer Science*, 82(4), 2003.
4. M.K. Chakraborty and M. Banerjee. Rough consequence. *Bulletin of the Polish Academy of Sciences (Mathematics)*, 41(4):299–304, 1993.
5. M.K. Chakraborty and E. Orłowska. Substitutivity principles in some theories of uncertainty. *Fundamenta Informaticae*, 32:107–120, 1997.
6. M.K. Chakraborty and S. Basu. Graded consequence and some metalogical notions generalized. *Fundamenta Informaticae*, 32:299–311, 1997.
7. M.K. Chakraborty and S. Basu. Approximate reasoning methods in vagueness: graded and rough consequences. *ICS Research Report*, 29, Warsaw University of Technology, 1995.
8. S. Demri and E. Orłowska. (eds.) *Incomplete Information: Structure, Inference, Complexity*. Monographs in Theoretical Computer Science, Springer-Verlag, Heidelberg, 2002.
9. D. Dubois and H. Prade. Rough fuzzy sets and fuzzy rough sets. In *Proc. International Conference on Fuzzy Sets in Informatics, Moscow*. 1988, 20–23.
10. M. Eytan. Fuzzy sets: a topos-logical point of view. *Fuzzy Sets and Systems*, 5:47–67, 1981.
11. J. Goguen. Concept representation in natural and artificial languages: axioms, extensions and applications for fuzzy sets. *International Journal for Man-Machine Studies*, 6:513–561, 1975.
12. D. Higgs. A categorical approach to Boolean-valued set theory. Preprint, 1973.
13. D. Hyde. From heaps and gaps to heaps of gluts. *Mind*, 106:440–460, 1997.
14. J. Jacas. On the generators of T-indistinguishability operator. *Stochastica*, XIII: 49–63, 1988.
15. R. Keefe. *Theories of Vagueness*, Cambridge Studies in Philosophy, Cambridge, UK, 2000.
16. R. Keefe and P. Smith. (eds.) *Vagueness: A Reader*, MIT Press, Massachusetts, MA, 1997.
17. F. Keikeben. *http://members.aol.com/kiekeben/theseus.html*, 2000.
18. A. Obtułowicz. Rough sets and Heyting algebra valued sets. *Bulletin of the Polish Academy of Sciences (Mathematics)*, 13(9-10):667–671, 1987.
19. E. Orłowska and Z. Pawlak. Representation of non-deterministic information. *Theoretical Computer Science*, 29:27–39, 1984.
20. Z. Pawlak. Rough sets. *International Journal of Computer and Information Sciences*, 11:341–356, 1982.
21. Z. Pawlak. Some issues on rough sets. *Transactions of Rough sets I*, 1–58, 1998.
22. Z. Pawlak. A treatise on rough sets. *Transactions on Rough sets IV*, 1–17, 2005.
23. Z. Pawlak. Rough sets and fuzzy sets. *Fuzzy Sets and Systems* 17:99–102, 1985.
24. Z. Pawlak. Rough logic. *Bulletin of the Polish Academy of Sciences(Technical Sciences)*, 35(5-6):253–258, 1987.
25. Z. Pawlak. Hard and soft sets. *ICS Research Report*, 10/94, Warsaw University of Technology, 1994.
26. Z. Pawlak. Vagueness – a rough set view. *LNCS 1261*, Springer, 1997, 106–117.
27. Z. Pawlak. Vagueness and uncertainty: a rough set perspective. *Computational Intelligence: An International Journal*, 11:217–232, 1995.
28. Z. Pawlak. Rough sets, present state and further prospects. *ICS Research Report*, 15/19, Warsaw University of Technology, 1995.
29. Z. Pawlak and A. Skowron. Rudiments of rough sets. *Information Sciences*, to appear.

30. Z. Pawlak and A. Skowron. Rough sets: some extensions. *Information Sciences*, to appear.
31. Z. Pawlak and A. Skowron. Rough sets and Boolean reasoning, *Information Sciences*, to appear.
32. G. Priest and R. Routley. First historical introduction: a preliminary history of paraconsistent and dialethic approaches. In Priest, Routley and Normann, editors, *Paraconsistent Logic, Essays on the Inconsistent.* Philosophia Verlag, Munchen-Hamden-Wien, 1989, 1–75.
33. L. Polkowski. *Rough Sets: Mathematical Foundations.* Advances in Soft Computing, Physica Verlag, Hiedelberg, 2002.
34. A. Pultr, Fuzziness and fuzzy equality. In H.J. Skala, S. Termini and E. Trillas, editors, *Aspects of Vagueness.* D. Reidel Publishing Co., Dordrecht, Holland, 1984, 119–135.
35. P.T. Raju. The principle of four-coloured negation in Indian philosophy. *Review of Metaphysics* 7:694-713, 1953.
36. H. Rasiowa and A. Skowron. Rough concepts logic in computation theory. In A. Skowron, editor, *LNCS 208*, Springer, 1985, 288–297.
37. B. Russell. Vagueness. *Australian Journal of Philosophy*, 1:84–92, 1923.
38. A. Skowron. The relationship between the rough set theory and evidence theory. *Bulletin of the Polish Academy of Sciences (Technical Sciences)*, 37(1-2):87–90, 1989.
39. A. Skowron. Rough sets and vague concepts. *Fundamenta Informaticae*, 64: 417–431, 2005.
40. A. Skowron and J.W. Grzymała-Busse. From rough set theory to evidence theory. In Yager, Fedrizzi and Kacprzyk, editors, *Advances in the Dempster-Shafer Theory of Evidence.* John Wiley & Sons, New York, 1994, 193–236.
41. E. Trillas and L. Valverde. An inquiry into indistinguishability operators. In H.J. Skala, S. Termini and E. Trillas, editors, *Aspects of Vagueness.* D. Reidel Publishing Co., Dordrecht, Holland, 1984, 231–256.
42. H. Wang. *Beyond Analytic Philosophy.* MIT Press, Cambridge, 1986.
43. D. Wiggins. *Sameness and Substance.* Oxford Blackwell, 1980, 92–94.
44. M. Wygralak. Some remarks on rough and fuzzy sets. *BUSEFAL 21*, 1985, 43–49.
45. L.A. Zadeh. Fuzzy Sets. *Information and Control*, 8:338–353, 1965.

A Comparison of Pawlak's and Skowron–Stepaniuk's Approximation of Concepts

Anna Gomolińska*

University of Białystok, Department of Mathematics,
Akademicka 2, 15267 Białystok, Poland
anna.gom@math.uwb.edu.pl

Abstract. In this article, we compare mappings of Pawlak's lower and upper approximations of concepts with those proposed by Skowron and Stepaniuk. It is known that both approaches coincide for the standard rough inclusion, so we consider the case of an arbitrary rough inclusion function. Even if the approximation space investigated is based on an arbitrary non-empty binary relation, the lower approximation mappings are equal in both approaches. Nevertheless, the upper approximation mappings are different in general.

In view of many generalizations and extensions of rough set theory some kind of unification of the basic theory seems to be badly needed.

(Z. Pawlak [1], p. 10)

1   Introduction

Nowadays, the Pawlak rough approximation of concepts [2,3,4,5] has become a classical research topic. Lower and upper rough approximations have been investigated by many researchers in the rough set community, to mention [6,7,8] by way of example. Skowron and Stepaniuk's proposal regarding rough approximation of concepts [9,10] is well-known, yet less popular among researchers as a subject of study. A possible reason may be the fact that both approaches coincide for the standard rough inclusion, whereas this very function is the most known among rough inclusions.

The aim of this paper is to study and to compare both Pawlak's and Skowron–Stepaniuk's approaches to approximation of concepts in the rough-set framework. To this end, we relax the usual assumptions about the approximation space considered. We start with an approximation space, understood as a structure $M = (U, \varrho, \kappa)$, where U (the universe) is a non-empty set of objects, ϱ is a non-empty binary relation on U, and κ is a mapping on the set of pairs of sets of objects called a rough inclusion function. Step by step, we consider spaces

* The research was supported by the grant 3T11C00226 from the Ministry of Science of the Republic of Poland.

J.F. Peters et al. (Eds.): Transactions on Rough Sets VI, LNCS 4374, pp. 64–82, 2007.
© Springer-Verlag Berlin Heidelberg 2007

based on serial relations, reflexive relations, transitive relations, symmetric (and in particular, tolerance) relations, and equivalence relations. By definition, κ is assumed to satisfy two postulates only which are in accordance with the axioms of rough mereology. However, we also investigate cases, where κ fulfills additional conditions.

In the paper, we examine two pairs of approximation mappings in line with Pawlak's approach and two pairs of approximation mappings in line with Skowron–Stepaniuk's proposal. Each pair consists of a lower approximation mapping and an upper approximation mapping. Both in the Pawlak case as well as in the Skowron–Stepaniuk case, approximation mappings constituting one pair are viewed as basic, whereas mappings forming the remaining pair are "definable" versions of the basic mappings. As regarding the results, the basic lower approximation mappings are equal in both approaches, and similarly for their "definable" variants. Basic upper approximation mappings (and similarly for their "definable" counterparts) are different in general and may be compared only under special conditions put on κ. Therefore, we mainly try to compare these mappings indirectly via their properties. Apart from that, we aim at a uniform presentation of facts about lower and upper approximation mappings. Some of the facts are new, others are only recalled. We try to answer such questions as: What are the results of application of a given approximation mapping to the empty set and to the whole universe? What are the relationships among various forms of approximation? For instance, how is the lower approximation related to the upper one? Moreover, how are approximations of a concept related to the concept itself? Are the approximation mappings under investigation monotone, and if it is the case, what are the consequences? Last but not least, what a mapping may be obtained by various compositions of the approximation mappings? As we shall see, the mappings investigated can lack some essential properties attributed to an approximation mapping if the relation ϱ, underlying a given approximation space M, is not reflexive. For example, a lower approximation of a concept may not be included in that concept. However, slightly abusing the terminology, we shall use the names 'lower approximation' and 'upper approximation' for the sake of uniformity.

Basic terminology and notation is introduced in Sect. 2. In Sect. 3, we present the notion of a rough approximation space and the mappings of Pawlak's as well as Skowron–Stepaniuk's lower and upper rough approximation of concepts. In Sect. 4, properties of these mappings are studied in the case of an approximation space based on an arbitrary non-empty binary relation ϱ. In Sect. 5, we examine the approximation mappings for special cases of approximation spaces, where ϱ and/or its converse relation ϱ^{-1} are serial and where ϱ is, in turn, a reflexive relation, a transitive relation, a symmetric relation, and an equivalence relation. The results are summarized briefly in the last section.

2   Preliminaries

Let X, Y be any sets. Throughout the paper, the power set of X, the cardinality of X, the Cartesian product $X \times X$, the identity mapping on X, and the set of all mappings $f : X \mapsto Y$ will be denoted by $\wp X$, $\#X$, X^2, id_X, and Y^X, respectively.

Consider (partially) ordered sets $(X, \leq)$ and $(Y, \preceq)$. A mapping $f : X \mapsto Y$ is referred to as *monotone*, written $f \in \mathrm{MON}$, if for any $x, y \in X$, $x \leq y$ implies $fx \preceq fy$. The operation of composition of relations will be denoted by $\circ$. In the case of mappings, the composition of $f : X \mapsto Y$ with $g : Y \mapsto Z$ is a mapping $g \circ f : X \mapsto Z$ such that for any $x \in X$, $(g \circ f)x = g(fx)$. For any sets X, Y, define a relation $\sqsubseteq$ on $(\wp Y)^{\wp X}$ and operations $\sqcap, \sqcup$ on $((\wp Y)^{\wp X})^2$ such that for any mappings $f, g : \wp X \mapsto \wp Y$ and any $Z \subseteq X$,

$$f \sqsubseteq g \overset{\text{def}}{\Leftrightarrow} \forall Z \subseteq X. fZ \subseteq gZ,$$
$$(f \sqcap g)Z \overset{\text{def}}{=} fZ \cap gZ \ \& \ (f \sqcup g)Z \overset{\text{def}}{=} fZ \cup gZ. \tag{1}$$

By assumption, $\circ$ will take the precedence of the operations just defined, whereas the logical connectives of conjunction and disjunction will take the precedence of implication and double implication.

Proposition 1. *For any mappings $f, g, h : \wp X \mapsto \wp Y$, we have:*

> (a) $f \sqsubseteq f$
> (b) $f \sqsubseteq g \ \& \ g \sqsubseteq f \Rightarrow f = g$
> (c) $f \sqsubseteq g \ \& \ g \sqsubseteq h \Rightarrow f \sqsubseteq h$
> (d) $f \sqsubseteq g \Rightarrow f \circ h \sqsubseteq g \circ h$
> (e) $h \in \mathrm{MON} \ \& \ f \sqsubseteq g \Rightarrow h \circ f \sqsubseteq h \circ g$
> (f) $f \sqcup g \sqsubseteq h \Leftrightarrow f \sqsubseteq h \ \& \ g \sqsubseteq h$
> (g) $f \sqsubseteq g \sqcap h \Leftrightarrow f \sqsubseteq g \ \& \ f \sqsubseteq h$

The proof is easy and, hence, omitted. Let us only note that $\sqsubseteq$ is a partial ordering on $(\wp Y)^{\wp X}$ in virtue of (a)–(c).

3   Rough Approximation Spaces

The notion of a rough approximation space was obtained by Prof. Pawlak in the early 80's of the 20th century as one of the results of investigations on approximation of vague concepts in information systems [2,3,4,5]. This basic notion was next refined and generalized in several directions (see, e.g., [11,12,13,14,15]), yet we shall only focus upon the extension proposed by Skowron and Stepaniuk in [9,10], and elaborated in a series of research articles [16,17,18].

In [19,20][1], Polkowski and Skowron introduced and characterized axiomatically the formal notion of a rough inclusion. Although this notion is unnecessary when discussing Pawlak's classical approach, it is fundamental for Skowron–Stepaniuk's one. Consider a non-empty set U of entities called objects. In com-

---

[1] See also more recent papers.

pliance with rough mereology, by a *rough inclusion function* (RIF for short) upon U we understand any mapping $\kappa : (\wp U)^2 \mapsto [0,1]$ satisfying rif_1 and rif_2 below:

$$\mathrm{rif}_1(\kappa) \overset{\mathrm{def}}{\Leftrightarrow} \forall X, Y.(\kappa(X,Y) = 1 \Leftrightarrow X \subseteq Y)$$

$$\mathrm{rif}_2(\kappa) \overset{\mathrm{def}}{\Leftrightarrow} \forall X, Y, Z.(Y \subseteq Z \Rightarrow \kappa(X,Y) \leq \kappa(X,Z))$$

RIFs are intended as mappings measuring the degrees of inclusion of sets of objects in sets of objects. Apart from the above postulates, one may consider other conditions, for instance,

$$\mathrm{rif}_3(\kappa) \overset{\mathrm{def}}{\Leftrightarrow} \forall X \neq \emptyset.\kappa(X,\emptyset) = 0,$$

$$\mathrm{rif}_4(\kappa) \overset{\mathrm{def}}{\Leftrightarrow} \forall X \neq \emptyset.\forall Y.(\kappa(X,Y) = 0 \Rightarrow X \cap Y = \emptyset),$$

$$\mathrm{rif}_{4*}(\kappa) \overset{\mathrm{def}}{\Leftrightarrow} \forall X \neq \emptyset.\forall Y.(X \cap Y = \emptyset \Rightarrow \kappa(X,Y) = 0),$$

$$\mathrm{rif}_5(\kappa) \overset{\mathrm{def}}{\Leftrightarrow} \forall X \neq \emptyset.\forall Y.(\kappa(X,Y) = 0 \Leftrightarrow X \cap Y = \emptyset).$$

One can easily see that $\mathrm{rif}_{4*}(\kappa)$ implies $\mathrm{rif}_3(\kappa)$, whereas $\mathrm{rif}_5(\kappa)$ if and only if $\mathrm{rif}_4(\kappa)$ and $\mathrm{rif}_{4*}(\kappa)$. The most famous RIF is the *standard* one, defined for the finite universe and denoted by $\kappa^{\mathcal{L}}$ here, which goes back to Łukasiewicz [21] and is based on the frequency count. $\kappa^{\mathcal{L}}$ is given by

$$\kappa^{\mathcal{L}}(X,Y) = \begin{cases} \frac{\#(X \cap Y)}{\#X} & \text{if } X \neq \emptyset \\ 1 & \text{otherwise} \end{cases} \tag{2}$$

and fulfills not only $\mathrm{rif}_1, \mathrm{rif}_2$ but also rif_5 and some other conditions.

By a *rough approximation space* we mean a triple $M = (U, \varrho, \kappa)$, where U — the universe of M — is a non-empty set of objects as earlier, ϱ is a non-empty binary relation on U, and κ is a RIF upon U. Objects will be denoted by u with sub/superscripts if needed. Sets of objects of U are viewed as *concepts* of M. With every object u, there are associated two basic concepts: the image and the co-image of $\{u\}$, $\varrho^{\rightarrow}\{u\}$ and $\varrho^{\leftarrow}\{u\}$, defined along the standard lines and called *elementary granules of information*[2] drawn to u. It is worth recalling that $\varrho^{\rightarrow}\{u\} = \varrho^{-1\leftarrow}\{u\}$ and $\varrho^{\leftarrow}\{u\} = \varrho^{-1\rightarrow}\{u\}$.

Let us note that ϱ induces mappings $\Gamma_\varrho, \Gamma_\varrho^* : U \mapsto \wp U$, called *uncertainty* mappings in line with Skowron–Stepaniuk's approach, such that for every object $u \in U$,

$$\Gamma_\varrho u = \varrho^{\leftarrow}\{u\} \quad \& \quad \Gamma_\varrho^* u = \varrho^{\rightarrow}\{u\}. \tag{3}$$

Thus, elementary granules of information are simply values of Γ, Γ^*. Clearly, $\Gamma_\varrho^* = \Gamma_{\varrho^{-1}}$, and $u' \in \Gamma_\varrho^* u$ if and only if $(u, u') \in \varrho$, i.e., if and only if $u \in \Gamma_\varrho u'$. Moreover, $\Gamma_\varrho^* = \Gamma_\varrho$ if ϱ is symmetric. On the other hand, every mapping $\Gamma : U \mapsto \wp U$ induces a relation ϱ_Γ on U such that for any objects $u, u' \in U$,

$$(u, u') \in \varrho_\Gamma \Leftrightarrow u \in \Gamma u'. \tag{4}$$

---

[2] The term 'information granule' was proposed by Zadeh [22] to denote a clump of objects drawn together on the basis of indiscernibility, similarity or functionality.

Since $\Gamma_{\varrho_\Gamma} = \Gamma$ and $\varrho_{\Gamma_\varrho} = \varrho$, structures (U, ϱ, κ) and $(U, \Gamma_\varrho, \kappa)$ are interdefinable, and similarly for (U, Γ, κ) and $(U, \varrho_\Gamma, \kappa)$.

In the classical Pawlak approach, a rough approximation space is a pair (U, ϱ), where U is a finite non-empty set and ϱ is an equivalence relation understood as a relation of indiscernibility of objects. Then, elementary granules of information are equivalence classes, i.e. sets of objects indiscernible from one another. Clearly, a natural augmentation of (U, ϱ) with a RIF κ, results in an approximation space (U, ϱ, κ) in line with our approach.

Keeping with the recent state-of-art, one can say that Skowron–Stepaniuk's approximation spaces, introduced in [9,10], are of the form $N = (U, \Gamma_\$, \kappa_\$)$, where $\Gamma_\$$ is an uncertainty mapping such that for every object u, $u \in \Gamma_\$ u$, $\kappa_\$$ is a RIF, and $\$$ is a list of tuning parameters to obtain a satisfactory quality of approximation of concepts. For the latter, such spaces are called *parameterized* approximation spaces as well. Henceforth, the parameters $\$$ will be dropped for simplicity. One can easily see that N is based on a reflexive relation[3], e.g. $\varrho_{\Gamma_\$}$. Due to our earlier observations on interdefinability of approximation spaces and the corresponding structures based on uncertainty mappings, and slightly abusing the original terminology, we shall think of Skowron–Stepaniuk's approximation spaces as structures of the form (U, ϱ, κ), where ϱ is a reflexive relation on U.

In the sequel, a concept X is referred to as ϱ-*definable* (resp., ϱ^{-1}-*definable*) if it is a set-theoretical union of elementary granules of the form $\Gamma_\varrho u$ (resp., $\Gamma_\varrho^* u$). Henceforth, references to ϱ will be omitted whenever possible. For instance, we shall write Γ and Γ^* instead of Γ_ϱ and Γ_ϱ^*, respectively. Where ϱ is symmetric, both forms of definability coincide, so we may simply speak of definable or undefinable concepts. The main idea underlying the Pawlak rough approximation of concepts is that even if a concept is not definable in a given space, it can be approximated from the inside and the outside by definable concepts. In this way, the *Pawlak lower* and *upper rough approximation* mappings, $\text{low}^\cup, \text{upp}^\cup \in (\wp U)^{\wp U}$, respectively, are obtained such that for any concept X,

$$\text{low}^\cup X \overset{\text{def}}{=} \bigcup\{\Gamma u \mid \Gamma u \subseteq X\} \ \& \ \text{upp}^\cup X \overset{\text{def}}{=} \bigcup\{\Gamma u \mid \Gamma u \cap X \neq \emptyset\}. \quad (5)$$

The lower approximation of X, $\text{low}^\cup X$, is the largest ϱ-definable concept included in X, whereas the upper approximation of X, $\text{upp}^\cup X$, is the least ϱ-definable concept containing X provided that ϱ is serial. The difference

$$\text{bnd}^\cup X \overset{\text{def}}{=} \text{upp}^\cup X - \text{low}^\cup X \quad (6)$$

is called the *boundary region* of X. When this region is empty, X is referred to as *exact*; otherwise it is *rough*. In Pawlak's approximation spaces, it turns out that a concept is exact if and only if it is definable. Apart from $\text{low}^\cup, \text{upp}^\cup$, we shall also refer to the mappings $\text{low}, \text{upp} \in (\wp U)^{\wp U}$ given below as the *Pawlak lower* and *upper rough approximation* mappings, respectively:

---

[3] Primarily, parameterized approximation spaces were based on reflexive and symmetric (i.e., tolerance) relations.

$$\text{low}X \stackrel{\text{def}}{=} \{u \mid \Gamma u \subseteq X\} \ \& \ \text{upp}X \stackrel{\text{def}}{=} \{u \mid \Gamma u \cap X \neq \emptyset\} \tag{7}$$

As a matter of fact, $\text{low}^{\cup} = \text{low}$ and $\text{upp}^{\cup} = \text{upp}$ for the Pawlak approximation spaces. In a general case, however, low, upp differ from their ϱ-definable versions $\text{low}^{\cup}, \text{upp}^{\cup}$, respectively. The lower approximation of X, $\text{low}X$, may be viewed as the set consisting of all objects u which surely belong to X since their elementary granules Γu are included in X. On the other hand, the upper approximation of X, $\text{upp}X$, may be perceived as the set consisting of all objects u which possibly belong to X since their elementary granules Γu overlap with X.

The *Skowron–Stepaniuk lower* and *upper rough approximation* mappings, low^{S}, $\text{upp}^{\text{S}} \in (\wp U)^{\wp U}$, respectively, are defined by the following conditions, for any concept X,

$$\text{low}^{\text{S}}X \stackrel{\text{def}}{=} \{u \mid \kappa(\Gamma u, X) = 1\} \ \& \ \text{upp}^{\text{S}}X \stackrel{\text{def}}{=} \{u \mid \kappa(\Gamma u, X) > 0\}. \tag{8}$$

That is, the lower approximation of X, $\text{low}^{\text{S}}X$, consists of all objects u that their elementary granules Γu are included in X to the highest degree 1. On the other hand, the upper approximation of X, $\text{upp}^{\text{S}}X$, is the set of all objects u that their elementary granules Γu are included in X to some positive degree. The *boundary region* of X is defined as the set

$$\text{bnd}^{\text{S}}X \stackrel{\text{def}}{=} \{u \mid 0 < \kappa(\Gamma u, X) < 1\}. \tag{9}$$

Mappings $\text{low}^{\text{SU}}, \text{upp}^{\text{SU}}$, being ϱ-definable versions of $\text{low}^{\text{S}}, \text{upp}^{\text{S}}$, are also referred to as the Skowron–Stepaniuk lower and upper rough approximation mappings, respectively. They are given by the following equalities:

$$\text{low}^{\text{SU}}X \stackrel{\text{def}}{=} \bigcup \{\Gamma u \mid \kappa(\Gamma u, X) = 1\}$$
$$\text{upp}^{\text{SU}}X \stackrel{\text{def}}{=} \bigcup \{\Gamma u \mid \kappa(\Gamma u, X) > 0\} \tag{10}$$

Obviously, we can repeat the construction of approximation mappings for ϱ^{-1} what can be useful if ϱ is not symmetric. As a result, mappings $\text{low}^{*}, \text{upp}^{*}, \text{low}^{\text{S}*}$, and $\text{upp}^{\text{S}*}$ may be derived (as well as their ϱ^{-1}-definable versions which will not be presented here), where for any concept X,

$$\text{low}^{*}X \stackrel{\text{def}}{=} \{u \mid \Gamma^{*}u \subseteq X\} \ \& \ \text{upp}^{*}X \stackrel{\text{def}}{=} \{u \mid \Gamma^{*}u \cap X \neq \emptyset\},$$
$$\text{low}^{\text{S}*}X \stackrel{\text{def}}{=} \{u \mid \kappa(\Gamma^{*}u, X) = 1\} \ \& \ \text{upp}^{\text{S}*}X \stackrel{\text{def}}{=} \{u \mid \kappa(\Gamma^{*}u, X) > 0\}. \tag{11}$$

The mapping upp^{*} is particularly important for our purposes. It turns out that for any concept X, $\bigcup\{\Gamma u \mid u \in X\} = \text{upp}^{*}X$. Mappings low, upp, low^{S}, and upp^{S} will be viewed as basic. Thus, for every basic mapping f,

$$f^{\cup} = \text{upp}^{*} \circ f. \tag{12}$$

Example 1. For the sake of illustration of the approximation mappings, consider a set $U = \{3, \ldots, 10\}$ consisting of 8 objects denoted by $3, \ldots, 10$, and a binary relation ϱ on U generating the uncertainty mappings[4] Γ, Γ^* shown in Table 1. ϱ is reflexive, so it is a similarity relation (but even not a tolerance relation). Let κ be any RIF as earlier and κ_1, κ_2 be such RIFs that for any concepts X, Y where $X \neq \emptyset$,

$$\kappa_1(X, Y) = 0 \Leftrightarrow X \cap \mathrm{upp}^* Y = \emptyset,$$
$$\kappa_2(X, Y) = 0 \Leftrightarrow \mathrm{upp} X \cap \mathrm{upp}^* Y = \emptyset. \tag{13}$$

Both κ_1 and κ_2 satisfy rif$_4$ since $\kappa_2(X, Y) = 0$ implies $\kappa_1(X, Y) = 0$, and the latter implies $X \cap Y = \emptyset$. Indeed, $X \subseteq \mathrm{upp} X$ and $Y \subseteq \mathrm{upp}^* Y$ due to reflexivity of ϱ. It is easy to see that $\kappa_1(X, Y) > 0$ if and only if there exist $u \in X$ and $u' \in Y$ such that $(u, u') \in \varrho$. Furthermore, $\kappa_2(X, Y) > 0$ if and only if there exist $u \in X$ and $u' \in Y$ such that $(u, u') \in \varrho \circ \varrho$. Let upp_1^S and upp_2^S denote the Skowron–Stepaniuk upper approximation mappings based on κ_1 and κ_2, respectively, i.e., for any concept X and $i = 1, 2$,

$$\mathrm{upp}_i^S X = \{u \mid \kappa_i(\Gamma u, X) > 0\}. \tag{14}$$

That is, $u \in \mathrm{upp}_1^S X$ if and only if $\Gamma u \cap \mathrm{upp}^* X \neq \emptyset$, and $u \in \mathrm{upp}_2^S X$ if and only if $(\mathrm{upp} \circ \Gamma) u \cap \mathrm{upp}^* X \neq \emptyset$. Values of $\mathrm{upp} \circ \Gamma$ are given in Table 1. One can show that

$$\mathrm{upp} \sqsubseteq \mathrm{upp}_1^S \sqsubseteq \mathrm{upp}_2^S, \tag{15}$$

yet the converse inclusions may not hold in general. To see this, consider $X = \{3, 4\}$. Note that $\mathrm{low} X = \{4\}$, $\mathrm{low}^* X = \emptyset$, $\mathrm{upp} X = \{3, 4, 5\}$, $\mathrm{upp}^* X = \{3, 4, 6\}$, $\mathrm{upp}_1^S X = \{3, 4, 5, 6\}$, and $\mathrm{upp}_2^S X = \{3, 4, 5, 6, 10\}$.

Table 1. Values of Γ, Γ^*, and $\mathrm{upp} \circ \Gamma$

u	Γu	$\Gamma^* u$	$\mathrm{upp}(\Gamma u)$
3	{3,4,6}	{3,4,5}	{3,4,5,6}
4	{3,4}	{3,4,5}	{3,4,5}
5	{3,4,5,6}	{5}	{3,4,5,6}
6	{6,10}	{3,5,6}	{3,5,6,10}
7	{7,8,9}	{7,8}	{7,8,9,10}
8	{7,8}	{7,8,9}	{7,8,9}
9	{8,9}	{7,9,10}	{7,8,9,10}
10	{9,10}	{6,10}	{6,7,9,10}

4   Properties of Approximation Mappings

We first investigate properties of the basic mappings low, upp, low^S, and upp^S, where ϱ is an arbitrary non-empty relation on U. Henceforth, f will denote low, upp or upp^S.

---

[4] We drop references to ϱ for simplicity.

Proposition 2. *Let X, Y be any concepts. Then, we have:*

(a) $\text{low} = \text{low}^S \sqsubseteq \text{upp}^S$

(b) $\text{rif}_4(\kappa) \Rightarrow \text{upp} \sqsubseteq \text{upp}^S$

(c) $\text{upp}\emptyset = \emptyset$ & $\text{low}U = \text{upp}^S U = U$

(d) $\{u \mid \Gamma u = \emptyset\} \subseteq \text{low}X = U - \text{upp}(U - X)$

(e) $X \subseteq Y \Rightarrow fX \subseteq fY$

(f) $f(X \cap Y) \subseteq fX \cap fY \subseteq fX \cup fY \subseteq f(X \cup Y)$

Proof. We prove (b), (d), and (e) only. To this end, consider any concepts X, Y and any object u.

For (b) assume that $\text{rif}_4(\kappa)$ holds. Consider $u \in \text{upp}X$. By the definition of upp, (b1) $\Gamma u \cap X \neq \emptyset$. Hence, $\Gamma u \neq \emptyset$. As a consequence, $\kappa(\Gamma u, X) > 0$ by the assumption and (b1). By the definition of upp^S, $u \in \text{upp}^S X$. Thus, $\text{upp}X \subseteq \text{upp}^S X$. Immediately, $\text{upp} \sqsubseteq \text{upp}^S$ by the definition of $\sqsubseteq$.

In case (d), first suppose that $\Gamma u = \emptyset$. Hence, $\kappa(\Gamma u, X) = 1$ in virtue of $\text{rif}_1(\kappa)$, i.e., $u \in \text{low}X$ by the definition of low. In the sequel, $u \in \text{low}X$ if and only if (by the definition of low) $\Gamma u \subseteq X$ if and only if $\Gamma u \cap (U - X) = \emptyset$ if and only if (by the definition of upp) $u \notin \text{upp}(U - X)$ if and only if $u \in U - \text{upp}(U - X)$.

For (e) assume (e1) $X \subseteq Y$. First, let $f = \text{low}$ and suppose that $u \in \text{low}X$. By the definition of low, $\Gamma u \subseteq X$. Hence, $\Gamma u \subseteq Y$ by (e1). Again by the definition of low, $u \in \text{low}Y$. Thus, $\text{low}X \subseteq \text{low}Y$. Next, let $f = \text{upp}$ and $u \in \text{upp}X$. By the definition of upp, $\Gamma u \cap X \neq \emptyset$. Hence, $\Gamma u \cap Y \neq \emptyset$ by (e1). By the definition of upp, $u \in \text{upp}Y$. Thus, $\text{upp}X \subseteq \text{upp}Y$. Finally, where $f = \text{upp}^S$, assume that $u \in \text{upp}^S X$. By the definition of upp^S, (e2) $\kappa(\Gamma u, X) > 0$. In virtue of (e1) and $\text{rif}_2(\kappa)$, $\kappa(\Gamma u, X) \leq \kappa(\Gamma u, Y)$. Hence, $\kappa(\Gamma u, Y) > 0$ by (e2). By the definition of upp^S, $u \in \text{upp}^S Y$. Thus, $\text{upp}^S X \subseteq \text{upp}^S Y$. □

Let us comment upon the properties. In virtue of property (a), Pawlak's and Skowron–Stepaniuk's approaches coincide as regarding the lower approximation. In the latter approach, the lower approximation of a concept is always included in the upper one. The Pawlak upper approximation is, in general, incomparable with the Skowron–Stepaniuk upper approximation unless some additional assumptions like $\text{rif}_4(\kappa)$ are made. Due to (b), the Pawlak upper approximation of a concept is included in the Skowron–Stepaniuk one if $\text{rif}_4(\kappa)$ is assumed. As we shall see later on, $\text{rif}_{4*}(\kappa)$ will guarantee that the Skowron–Stepaniuk upper approximation of a concept is included in the Pawlak one provided that ϱ^{-1} is serial (see Proposition 6g). By (c), the Pawlak upper approximation of the empty set is empty as well. On the other hand, both the lower approximation of the universe as well as the Skowron–Stepaniuk upper approximation of the universe are equal to the whole universe. According to (d), the Pawlak lower and upper approximations are dual to each other. Moreover, all objects u with empty elementary granules Γu belong to the lower approximation of any concept. By (e), the lower and upper approximation mappings are monotone. Property (f), being a direct consequence of (e), may be strengthened for the Pawlak approximation mappings as follows:

$$\text{low}(X \cap Y) = \text{low}X \cap \text{low}Y \quad \& \quad \text{upp}(X \cup Y) = \text{upp}X \cup \text{upp}Y \qquad (16)$$

We show the 1st property. "$\subseteq$" holds by Proposition 2f. For "$\supseteq$" assume that $u \in \text{low}X \cap \text{low}Y$. Hence, $u \in \text{low}X$ and $u \in \text{low}Y$. By the definition of low, $\Gamma u \subseteq X$ and $\Gamma u \subseteq Y$. Thus, $\Gamma u \subseteq X \cap Y$. By the definition of low, $u \in \text{low}(X \cap Y)$.

Below, we collect several facts about compositions of the approximation mappings examined.

Proposition 3. *In cases (d)–(f), assume* $\text{rif}_4(\kappa)$. *The following may be obtained:*

(a) $\text{low} \circ \text{low} \sqsubseteq \text{upp}^S \circ \text{low} \sqsubseteq \text{upp}^S \circ \text{upp}^S$

(b) $\text{low} \circ \text{low} \sqsubseteq \text{low} \circ \text{upp}^S \sqsubseteq \text{upp}^S \circ \text{upp}^S$

(c) $\text{upp} \circ \text{low} \sqsubseteq \text{upp} \circ \text{upp}^S \quad \& \quad \text{low} \circ \text{upp} \sqsubseteq \text{upp}^S \circ \text{upp}$

(d) $\text{upp} \circ \text{low} \sqsubseteq \text{upp}^S \circ \text{low} \quad \& \quad \text{low} \circ \text{upp} \sqsubseteq \text{low} \circ \text{upp}^S$

(e) $\text{upp} \circ \text{upp} \sqsubseteq \text{upp}^S \circ \text{upp} \sqsubseteq \text{upp}^S \circ \text{upp}^S$

(f) $\text{upp} \circ \text{upp} \sqsubseteq \text{upp} \circ \text{upp}^S \sqsubseteq \text{upp}^S \circ \text{upp}^S$

Proof. We prove (a), (e) only. In case (a), $\text{low} \circ \text{low} \sqsubseteq \text{upp}^S \circ \text{low}$ by Proposition 2a and Proposition 1d. Next, $\text{upp}^S \circ \text{low} \sqsubseteq \text{upp}^S \circ \text{upp}^S$ by Proposition 2a, monotonicity of upp^S, and Proposition 1e.

In case (e), assume $\text{rif}_4(\kappa)$. By Proposition 2b, $\text{upp} \sqsubseteq \text{upp}^S$. Hence, $\text{upp} \circ \text{upp} \sqsubseteq \text{upp}^S \circ \text{upp}$ by Proposition 1d, whereas $\text{upp}^S \circ \text{upp} \sqsubseteq \text{upp}^S \circ \text{upp}^S$ by monotonicity of upp^S and Proposition 1e. □

Clearly, properties analogous to Proposition 2, Proposition 3, and (16) hold for the $*$-versions of the basic mappings[5]. Now, we can formulate several properties of the ϱ-definable versions of low, low^S, upp, and upp^S.

Proposition 4. *For any concepts* X, Y, *we can prove that:*

(a) $\text{low}^{\cup} = \text{low}^{S\cup} \sqsubseteq \text{upp}^{S\cup} \sqcap \text{upp}^{\cup} \sqcap \text{id}_{\varrho U}$

(b) $\text{rif}_4(\kappa) \Rightarrow \text{upp}^{\cup} \sqsubseteq \text{upp}^{S\cup}$

(c) $\text{upp}^{\cup} \emptyset = \emptyset$

(d) $X \subseteq Y \Rightarrow f^{\cup}X \subseteq f^{\cup}Y$

(e) $f^{\cup}(X \cap Y) \subseteq f^{\cup}X \cap f^{\cup}Y \subseteq f^{\cup}X \cup f^{\cup}Y \subseteq f^{\cup}(X \cup Y)$

Proof. We prove (a) only. To this end, let X be any concept and u be any object. First, $\text{low}^{\cup} = \text{low}^{S\cup}$ directly by (12) and Proposition 2a. In virtue of

---

[5] To safe space, we do not formulate them explicitly. When referring to them, we shall attach $*$ to the name of a property as a superscript. For instance, we shall refer to $\forall X, Y.(X \subseteq Y \Rightarrow f^*X \subseteq f^*Y)$, being the counterpart of Proposition 2e, as Proposition 2e*.

Proposition 1g, it remains to show that (a1) $\mathrm{low}^{\cup} \sqsubseteq \mathrm{upp}^{S\cup}$, (a2) $\mathrm{low}^{\cup} \sqsubseteq \mathrm{upp}^{\cup}$, and (a3) $\mathrm{low}^{\cup} \sqsubseteq \mathrm{id}_{\wp U}$. (a1) holds by Proposition 2a and Proposition 2e* for $f = \mathrm{upp}^*$. For (a2) it suffices to prove $\mathrm{low}^{\cup}X \subseteq \mathrm{upp}^{\cup}X$. To this end, assume $u \in \mathrm{low}^{\cup}X$. By (12), $u \in \mathrm{upp}^*(\mathrm{low}X)$, i.e., there is u' such that (a4) $u \in \Gamma u'$ and $u' \in \mathrm{low}X$. Hence, $\Gamma u' \subseteq X$ by the definition of low. Note that $\Gamma u' \cap X \neq \emptyset$ since $\Gamma u' \neq \emptyset$ in virtue of (a4). Hence, $u' \in \mathrm{upp}X$ by the definition of upp. As a consequence, $u \in \mathrm{upp}^*(\mathrm{upp}X) = \mathrm{upp}^{\cup}X$ due to (a4) and (12). For (a3) we prove that $\mathrm{low}^{\cup}X \subseteq X$. To this end, let $u \in \mathrm{low}^{\cup}X$. By arguments as earlier, there is u' such that $u \in \Gamma u' \subseteq X$. Immediately, $u \in X$.   □

In the case of the ϱ-definable versions of approximation mappings, the Pawlak lower approximation and the Skowron–Stepaniuk lower approximation are equal. The novelty is that the lower approximation of a concept is included not only in the Skowron–Stepaniuk upper approximation of that concept but also in the concept itself and in the Pawlak upper approximation of that concept. Obviously, (b) is a counterpart of Proposition 2b, whereas (c) corresponds to the 1st part of Proposition 2c. The remaining two properties are counterparts of Proposition 2e and Proposition 2f, respectively. For example, (d) says that the mappings of lower and upper approximations are monotone. Additionally, we can derive the counterpart of (16) for the Pawlak upper approximation:

$$\mathrm{upp}^{\cup}(X \cup Y) = \mathrm{upp}^{\cup}X \cup \mathrm{upp}^{\cup}Y. \tag{17}$$

Several observations upon compositions of the ϱ-definable versions of the basic approximation mappings are presented below.

Proposition 5. *In cases (e)–(g), assume* $\mathrm{rif}_4(\kappa)$. *The following can be derived:*

(a) $\mathrm{low}^{\cup} \circ \mathrm{low}^{\cup} = \mathrm{low}^{\cup} \sqsubseteq \mathrm{upp}^{\cup} \circ \mathrm{low}^{\cup} \sqsubseteq \mathrm{low}^{\cup} \circ \mathrm{upp}^{\cup} = \mathrm{upp}^{\cup} \sqsubseteq \mathrm{upp}^{\cup} \circ \mathrm{upp}^{\cup}$

(b) $\mathrm{low}^{\cup} \circ \mathrm{low}^{\cup} \sqsubseteq \mathrm{upp}^{S\cup} \circ \mathrm{low}^{\cup} \sqsubseteq \mathrm{low}^{\cup} \circ \mathrm{upp}^{S\cup} = \mathrm{upp}^{S\cup} \sqsubseteq \mathrm{upp}^{S\cup} \circ \mathrm{upp}^{S\cup}$

(c) $(\mathrm{upp}^{\cup} \circ \mathrm{low}^{\cup}) \sqcup \mathrm{upp}^{S\cup} \sqsubseteq \mathrm{upp}^{\cup} \circ \mathrm{upp}^{S\cup}$

(d) $(\mathrm{upp}^{S\cup} \circ \mathrm{low}^{\cup}) \sqcup \mathrm{upp}^{\cup} \sqsubseteq \mathrm{upp}^{S\cup} \circ \mathrm{upp}^{\cup}$

(e) $\mathrm{upp}^{\cup} \sqsubseteq \mathrm{upp}^{\cup} \circ \mathrm{upp}^{S\cup}$ & $\mathrm{upp}^{\cup} \circ \mathrm{low}^{\cup} \sqsubseteq \mathrm{upp}^{S\cup} \circ \mathrm{low}^{\cup}$

(f) $\mathrm{upp}^{\cup} \circ \mathrm{upp}^{\cup} \sqsubseteq \mathrm{upp}^{S\cup} \circ \mathrm{upp}^{\cup} \sqsubseteq \mathrm{upp}^{S\cup} \circ \mathrm{upp}^{S\cup}$

(g) $\mathrm{upp}^{\cup} \circ \mathrm{upp}^{\cup} \sqsubseteq \mathrm{upp}^{\cup} \circ \mathrm{upp}^{S\cup} \sqsubseteq \mathrm{upp}^{S\cup} \circ \mathrm{upp}^{S\cup}$

Proof. We prove (a), (b) only. In case (a), first note that $\mathrm{low}^{\cup} \circ \mathrm{low}^{\cup} \sqsubseteq \mathrm{low}^{\cup}$, $\mathrm{low}^{\cup} \circ \mathrm{low}^{\cup} \sqsubseteq \mathrm{upp}^{\cup} \circ \mathrm{low}^{\cup}$, $\mathrm{low}^{\cup} \circ \mathrm{upp}^{\cup} \sqsubseteq \mathrm{upp}^{\cup} \circ \mathrm{upp}^{\cup}$, and $\mathrm{low}^{\cup} \circ \mathrm{upp}^{\cup} \sqsubseteq \mathrm{upp}^{\cup}$ by Proposition 4a and Proposition 1d. Subsequently, $\mathrm{upp}^{\cup} \circ \mathrm{low}^{\cup} \sqsubseteq \mathrm{upp}^{\cup}$ holds by Proposition 4a, monotonicity of $\mathrm{upp}^{\cup}$, and Proposition 1e.

Now, we prove that $\mathrm{low}^{\cup} \sqsubseteq \mathrm{low}^{\cup} \circ \mathrm{low}^{\cup}$. To this end, it suffices to show $\mathrm{low} \sqsubseteq \mathrm{low} \circ \mathrm{low}^{\cup}$. Then, by monotonicity of upp^* and Proposition 1e, we obtain $\mathrm{upp}^* \circ \mathrm{low} \sqsubseteq \mathrm{upp}^* \circ (\mathrm{low} \circ \mathrm{low}^{\cup}) = (\mathrm{upp}^* \circ \mathrm{low}) \circ \mathrm{low}^{\cup}$ which finally results in $\mathrm{low}^{\cup} \sqsubseteq \mathrm{low}^{\cup} \circ \mathrm{low}^{\cup}$ by (12). Thus, consider a concept X and an object u such that $u \in \mathrm{low}X$. Hence, (a1) $\Gamma u \subseteq X$ by the definition of low. Then, for every $u' \in \Gamma u$, $u' \in \mathrm{low}^{\cup}X$ by (a1) and the definition of $\mathrm{low}^{\cup}$. In other words, $\Gamma u \subseteq \mathrm{low}^{\cup}X$.

By the definition of low, $u \in \text{low}(\text{low}^{\cup}X) = (\text{low} \circ \text{low}^{\cup})X$. In this way, we have proved that $\text{low}X \subseteq (\text{low} \circ \text{low}^{\cup})X$. Hence, immediately, $\text{low} \sqsubseteq \text{low} \circ \text{low}^{\cup}$ by the definition of $\sqsubseteq$.

Finally, we show that $\text{upp}^{\cup} \sqsubseteq \text{low}^{\cup} \circ \text{upp}^{\cup}$. As in the preceding case, it suffices to prove $\text{upp} \sqsubseteq \text{low} \circ \text{upp}^{\cup}$ and, then, to apply Proposition 2e*, Proposition 1e, and (12). Thus, consider a concept X and an object u such that $u \in \text{upp}X$. Immediately, (a2) $\Gamma u \cap X \neq \emptyset$ by the definition of upp. As earlier, for every $u' \in \Gamma u$, $u' \in \text{upp}^{\cup}X$ by (a2) and the definition of $\text{upp}^{\cup}$. As a consequence, $\Gamma u \subseteq \text{upp}^{\cup}X$. Hence, $u \in \text{low}(\text{upp}^{\cup}X) = (\text{low} \circ \text{upp}^{\cup})X$ by the definition of low. We have shown that $\text{upp}X \subseteq (\text{low} \circ \text{upp}^{\cup})X$. Finally, $\text{upp} \sqsubseteq \text{low} \circ \text{upp}^{\cup}$ by the definition of $\sqsubseteq$.

In case (b), note that $\text{low}^{\cup} \circ \text{low}^{\cup} \sqsubseteq \text{upp}^{S\cup} \circ \text{low}^{\cup}$, $\text{low}^{\cup} \circ \text{upp}^{S\cup} \sqsubseteq \text{upp}^{S\cup} \circ \text{upp}^{S\cup}$, and $\text{low}^{\cup} \circ \text{upp}^{S\cup} \sqsubseteq \text{upp}^{S\cup}$ follow from Proposition 4a and Proposition 1d. Moreover, $\text{upp}^{S\cup} \circ \text{low}^{\cup} \sqsubseteq \text{upp}^{S\cup}$ by Proposition 4a, monotonicity of $\text{upp}^{S\cup}$, and Proposition 1e. It remains to show that $\text{upp}^{S\cup} \sqsubseteq \text{low}^{\cup} \circ \text{upp}^{S\cup}$. It suffices to prove $\text{upp}^{S} \sqsubseteq \text{low} \circ \text{upp}^{S\cup}$ and, then, to apply Proposition 2e*, Proposition 1e, and (12). To this end, consider a concept X and an object u such that $u \in \text{upp}^{S}X$. Hence, (b1) $\kappa(\Gamma u, X) > 0$ by the definition of upp^{S}. Then, for every $u' \in \Gamma u$, $u' \in \text{upp}^{S\cup}X$ by (b1) and the definition of $\text{upp}^{S\cup}$. Thus, $\Gamma u \subseteq \text{upp}^{S\cup}X$. Hence, $u \in \text{low}(\text{upp}^{S\cup}X) = (\text{low} \circ \text{upp}^{S\cup})X$ by the definition of low. That is, we have proved $\text{upp}^{S}X \subseteq (\text{low} \circ \text{upp}^{S\cup})X$. Immediately, $\text{upp}^{S} \sqsubseteq \text{low} \circ \text{upp}^{S\cup}$ by the definition of $\sqsubseteq$. $\qquad\square$

In comparison to Proposition 3, more and stronger relationships may be noted than in the basic case. It is due to Proposition 4a and to the fact that if an object u belongs to the ϱ-definable lower or upper approximation of a concept (in either of the senses considered), then its elementary granule $\Gamma^{*}u$ is non-empty. By way of example, the composition of the lower approximation mapping with itself equals to the lower approximation mapping, whereas the compositions of the upper approximation mappings with the lower approximation mapping equal to the former mappings in virtue of (a), (b). Such results cannot be obtained for the basic approximation mappings without extra assumptions about ϱ.

5   Properties of Approximation Mappings II

In this section, we present and discuss properties of approximation mappings for special cases of approximation spaces. In detail, we consider approximation spaces which, in turn, are based on serial relations, reflexive relations, transitive relations, symmetric relations (and, in particular, tolerance relations), and — last but not least — equivalence relations[6].

---

[6] A technical remark can be handy here. Except for a few cases, in a given subsection, we only present these properties which can be derived for the kind of approximation space investigated, yet were not obtained under weaker assumptions. For instance, when discussing approximation spaces based on reflexive relations, we do not recall the properties obtained for spaces based on serial relations and, the more, for arbitrary spaces.

5.1   The Case of Serial Relations

In this section, we discuss two independent cases, viz., the case where ϱ^{-1} is serial and the case where ϱ is serial. Let $\mathrm{SER}(U)$ denote the set of all serial relations on U.

First, let $\varrho^{-1} \in \mathrm{SER}(U)$. That is, for every $u \in U$, there exists $u' \in U$ such that $(u', u) \in \varrho$ (in other words, $\varrho^{\leftarrow}\{u\} \neq \emptyset$ or $\Gamma u \neq \emptyset$).

Proposition 6. *Let X be any concept. Then, we have:*

(a) $\mathrm{upp}U = U$ & $\mathrm{low}\emptyset = \mathrm{low}^{\cup}\emptyset = \emptyset$

(b) $\mathrm{rif}_3(\kappa) \Rightarrow \mathrm{upp}^S\emptyset = \mathrm{upp}^{S\cup}\emptyset = \emptyset$

(c) $\mathrm{low} \sqsubseteq \mathrm{upp}$

(d) $\mathrm{low} \circ \mathrm{low} \sqsubseteq \mathrm{upp} \circ \mathrm{low} \sqsubseteq \mathrm{upp} \circ \mathrm{upp}$

(e) $\mathrm{low} \circ \mathrm{low} \sqsubseteq \mathrm{low} \circ \mathrm{upp} \sqsubseteq \mathrm{upp} \circ \mathrm{upp}$

(f) $\mathrm{upp}^S \circ \mathrm{low} \sqsubseteq \mathrm{upp}^S \circ \mathrm{upp}$ & $\mathrm{low} \circ \mathrm{upp}^S \sqsubseteq \mathrm{upp} \circ \mathrm{upp}^S$

(g) $\mathrm{rif}_{4*}(\kappa) \Rightarrow \mathrm{upp}^S \sqsubseteq \mathrm{upp}$ & $\mathrm{upp}^{S\cup} \sqsubseteq \mathrm{upp}^{\cup}$

(h) $\mathrm{rif}_5(\kappa) \Rightarrow \mathrm{upp}^S = \mathrm{upp}$ & $\mathrm{upp}^{S\cup} = \mathrm{upp}^{\cup}$

Proof. We prove (a), (b) only. Consider any object u.

For (a) note that $\Gamma u \cap U \neq \emptyset$ by seriality of ϱ^{-1}. Hence, $u \in \mathrm{upp}U$ by the definition of upp. Thus, $U \subseteq \mathrm{upp}U$ and, finally, $\mathrm{upp}U = U$. Moreover, it can never be $\Gamma u \subseteq \emptyset$. Immediately, $\mathrm{low}\emptyset = \emptyset$ by the definition of low. Hence, $\mathrm{low}^{\cup}\emptyset = (\mathrm{upp}^* \circ \mathrm{low})\emptyset = \mathrm{upp}^*(\mathrm{low}\emptyset) = \mathrm{upp}^*\emptyset = \emptyset$ by (12) and Proposition 2c*.

For (b) assume that κ satisfies rif_3. By seriality of ϱ^{-1}, $\Gamma u \neq \emptyset$. Then, (b1) $\kappa(\Gamma u, \emptyset) = 0$ in virtue of $\mathrm{rif}_3(\kappa)$. Hence, (b2) $\mathrm{upp}^S\emptyset = \emptyset$ by the definition of upp^S. Next, $\mathrm{upp}^{S\cup}\emptyset = (\mathrm{upp}^* \circ \mathrm{upp}^S)\emptyset = \mathrm{upp}^*(\mathrm{upp}^S\emptyset) = \mathrm{upp}^*\emptyset = \emptyset$ by (b2), (12), and Proposition 2c*. □

Let us briefly comment upon the results. Seriality of ϱ^{-1} guarantees that elementary granules of the form Γu are non-empty. Thanks to that, the Pawlak upper approximation of the universe is the universe itself and the lower approximations of the empty set are empty by (a). Moreover, in virtue of (b), if rif_3 is satisfied by κ, then the Skowron–Stepaniuk upper approximations of the empty set are empty. By (c), the lower approximation of a concept is included in the Pawlak upper approximation of that concept. The next three properties, being consequences of (c), augment Proposition 3 by new facts on compositions of approximation mappings. If rif_{4*} is satisfied by κ, then the Skowron–Stepaniuk upper approximations of a concept are included in the corresponding Pawlak upper approximations of that concept by (g). Moreover, if $\mathrm{rif}_5(\kappa)$ holds, then the Pawlak upper approximations and the Skowron–Stepaniuk upper approximations coincide due to (h).

We now consider the case, where $\varrho \in \mathrm{SER}(U)$. Then, for every $u \in U$, there exists $u' \in U$ such that $(u, u') \in \varrho$ (in other words, $\varrho^{\rightarrow}\{u\} \neq \emptyset$ or $\Gamma^* u \neq \emptyset$). First, observe that properties analogous to Proposition 6 can be obtained for the *-versions of the basic approximation mappings. In particular, $\mathrm{upp}^*U = U$. Furthermore, we can prove the following properties.

Proposition 7. *Let X be any concept. Then, it holds:*

(a) $\mathrm{upp}^{SU}U = U$ & $(\varrho^{-1} \in \mathrm{SER}(U) \Rightarrow \mathrm{upp}^U U = U)$

(b) $\mathrm{upp}^U X \cup \mathrm{low}^U(U - X) = U$

(c) $\mathrm{id}_{\wp U} \sqsubseteq \mathrm{upp}^U$ & $(\mathrm{rif}_4(\kappa) \Rightarrow \mathrm{id}_{\wp U} \sqsubseteq \mathrm{upp}^{SU})$

(d) $\mathrm{upp}^{SU} \sqsubseteq \mathrm{upp}^{SU} \circ \mathrm{upp}^U$

Proof. We show (a), (b) only. To this end, consider any concept X.

For the 1st part of (a) note that $\mathrm{upp}^{SU}U = (\mathrm{upp}^* \circ \mathrm{upp}^S)U = \mathrm{upp}^*(\mathrm{upp}^S U) = \mathrm{upp}^* U = U$ by (12), seriality of ϱ, Proposition 2c, and Proposition 6a*. Now, assume additionally that ϱ^{-1} is serial. Hence, $\mathrm{upp}^U U = (\mathrm{upp}^* \circ \mathrm{upp})U = \mathrm{upp}^*(\mathrm{upp}U) = \mathrm{upp}^* U = U$ by (12), Proposition 6a, seriality of ϱ, and Proposition 6a*.

In case (b), $\mathrm{upp}^U X \cup \mathrm{low}^U(U - X) = (\mathrm{upp}^* \circ \mathrm{upp})X \cup (\mathrm{upp}^* \circ \mathrm{low})(U - X) = \mathrm{upp}^*(\mathrm{upp}X) \cup \mathrm{upp}^*(\mathrm{low}(U - X)) = \mathrm{upp}^*(\mathrm{upp}X \cup \mathrm{low}(U - X)) = \mathrm{upp}^* U = U$ by (12), (16*), Proposition 2d, seriality of ϱ, and Proposition 6a*.   □

Most of the properties above strongly depend on Proposition 6a* (a counterpart of Proposition 6a) which is a consequence of seriality of ϱ. By (a), the Skowron–Stepaniuk upper approximation of the universe is the whole universe, whereas the Pawlak upper approximation of the universe is the universe if both ϱ, ϱ^{-1} are serial. According to (b), every object belongs to the Pawlak upper approximation of a concept and/or to the lower approximation of the complement of that concept. (c) states that every concept is included in its Pawlak upper approximation, and similarly for the Skowron–Stepaniuk upper approximation if κ satisfies rif_4. Finally, the Skowron–Stepaniuk upper approximation of a concept is included in the Skowron–Stepaniuk upper approximation of the Pawlak upper approximation of that concept due to (d).

5.2   The Case of Reflexive Relations

Assume that ϱ is reflexive. Then, for every $u \in U$, $(u, u) \in \varrho$. Immediately, $(u, u) \in \varrho^{-1}$, so ϱ^{-1} is reflexive as well. Thus, $u \in \Gamma u \cap \Gamma^* u$. Clearly, every reflexive relation is serial as well. In the context of approximation spaces, reflexive relations are referred to as *similarity* relations. The set of all reflexive relations on U will be denoted by $\mathrm{RF}(U)$.

Proposition 8. *We can prove that:*

(a) $\mathrm{low} \sqsubseteq \mathrm{id}_{\wp U} \sqsubseteq \mathrm{upp}$ & $(\mathrm{rif}_4(\kappa) \Rightarrow \mathrm{id}_{\wp U} \sqsubseteq \mathrm{upp}^S)$

(b) $f \sqsubseteq f^U$

(c) $\mathrm{low} \circ \mathrm{low} \sqsubseteq \mathrm{low} \sqsubseteq \mathrm{upp} \circ \mathrm{low} \sqsubseteq \mathrm{upp} \sqsubseteq \mathrm{upp} \circ \mathrm{upp}$

(d) $\mathrm{low} \sqsubseteq \mathrm{low} \circ \mathrm{upp} \sqsubseteq \mathrm{upp}$

(e) $\mathrm{upp}^S \circ \mathrm{low} \sqcup \mathrm{low} \circ \mathrm{upp}^S \sqsubseteq \mathrm{upp}^S \sqsubseteq \mathrm{upp}^S \circ \mathrm{upp} \sqcap \mathrm{upp} \circ \mathrm{upp}^S$

Proof. We prove (e) only. First, $\text{upp}^S \circ \text{low} \sqsubseteq \text{upp}^S$, $\text{upp}^S \sqsubseteq \text{upp}^S \circ \text{upp}$ by (a), monotonicity of upp^S, and Proposition 1e. Next, $\text{low} \circ \text{upp}^S \sqsubseteq \text{upp}^S$, $\text{upp}^S \sqsubseteq \text{upp} \circ \text{upp}^S$ by (a) and Proposition 1d.   □

If ϱ is reflexive, then Proposition 6c may be strengthened to the property (a) above. In detail, the lower approximation of a concept is included in that concept and, on the other hand, every concept is included in its Pawlak upper approximation. The same holds for the Skowron–Stepaniuk approximation if $\text{rif}_4(\kappa)$ is satisfied. By (b), the lower (resp., upper) approximation of a concept is included in the ϱ-definable version of the lower (upper) approximation of that concept. Moreover, the list of properties of compositions of approximation mappings can be extended with several new dependencies given by (c)–(e).

5.3   The Case of Transitive Relations

Now, suppose that ϱ is transitive, i.e., for every $u, u', u'' \in U$, $(u, u') \in \varrho$ and $(u', u'') \in \varrho$ imply $(u, u'') \in \varrho$. Either both ϱ, ϱ^{-1} are transitive or both of them are not transitive. We denote the set of all transitive relations on U by $\text{TR}(U)$.

Proposition 9. *The following dependencies can be proved:*

(a) $\text{low}^\cup \sqsubseteq \text{low}$

(b) $\text{low} \sqsubseteq \text{low} \circ \text{low}$ & $\text{upp} \circ \text{upp} \sqsubseteq \text{upp}$

(c) $\varrho \in \text{RF}(U) \Rightarrow \text{low} \circ \text{low} = \text{low} = \text{low}^\cup$ & $\text{upp} \circ \text{upp} = \text{upp}$

Proof. We prove (a), (b) only. Consider any concept X and any object u.

In case (a), we show that $\text{low}^\cup X \subseteq \text{low}X$ which results in $\text{low}^\cup \sqsubseteq \text{low}$ by the definition of $\sqsubseteq$. To this end, assume that $u \in \text{low}^\cup X$. By the definition of $\text{low}^\cup$, there is u' such that $u \in \Gamma u'$, i.e., (a1) $(u, u') \in \varrho$, and (a2) $u' \in \text{low}X$. Next, (a3) $\Gamma u' \subseteq X$ by (a2) and the definition of low. Consider any $u'' \in \Gamma u$ (i.e., $(u'', u) \in \varrho$). Hence, $(u'', u') \in \varrho$ in virtue of (a1) and transitivity of ϱ. In other words, $u'' \in \Gamma u'$. Hence, immediately, $u'' \in X$ by (a3). Thus, $\Gamma u \subseteq X$. Hence, $u \in \text{low}X$ by the definition of low.

For the 1st part of (b) it suffices to prove that $\text{low}X \subseteq (\text{low} \circ \text{low})X$, i.e., $\text{low}X \subseteq \text{low}(\text{low}X)$, and to apply the definition of $\sqsubseteq$. To this end, assume that $u \in \text{low}X$. By the definition of low, (b1) $\Gamma u \subseteq X$. We need to prove that $u \in \text{low}(\text{low}X)$, i.e., $\Gamma u \subseteq \text{low}X$ in virtue of the definition of low. Thus, consider any $u' \in \Gamma u$ (i.e., (b2) $(u', u) \in \varrho$). It remains to show that $u' \in \text{low}X$, i.e., $\Gamma u' \subseteq X$. Let $u'' \in \Gamma u'$, i.e., $(u'', u') \in \varrho$. Hence, $(u'', u) \in \varrho$ by (b2) and transitivity of ϱ. That is, $u'' \in \Gamma u$. In virtue of (b1), $u'' \in X$ which ends the proof of this part of (b).

For the remaining part of (b) it suffices to show that $(\text{upp} \circ \text{upp})X \subseteq \text{upp}X$, i.e., $\text{upp}(\text{upp}X) \subseteq \text{upp}X$, and to apply the definition of $\sqsubseteq$. To this end, let $u \in \text{upp}(\text{upp}X)$. By the definition of upp, $\Gamma u \cap \text{upp}X \neq \emptyset$. Hence, there is (b3) $u' \in \Gamma u$ such that $u' \in \text{upp}X$. By the definition of upp, $\Gamma u' \cap X \neq \emptyset$. Hence, there is (b4) $u'' \in \Gamma u'$ such that (b5) $u'' \in X$. Note that $(u', u) \in \varrho$ and $(u'', u') \in \varrho$ by (b3) and (b4), respectively. Hence, by transitivity of ϱ,

$(u'', u) \in \varrho$ as well. That is, $u'' \in \Gamma u$. As a consequence, $\Gamma u \cap X \neq \emptyset$ by (b5). By the definition of upp, $u \in \text{upp}X$.   □

Some comments can be useful. Due to transitivity of ϱ, the ϱ-definable version of the lower approximation of a concept is included in the lower approximation of that concept by (a). Moreover, the lower approximation of a concept is included in the lower approximation of the lower approximation of that concept, whereas the Pawlak upper approximation of the same form of the upper approximation of a concept is included in the Pawlak upper approximation of that concept by (b). Unfortunately, a similar result does not seem to hold for the Skowron–Stepaniuk upper approximation. In the sequel, assuming reflexivity and transitivity of ϱ, the number of different compositions of approximation mappings may substantially be reduced thanks to (c) since, then, both versions of the lower approximation are equal to the composition of the lower approximation with itself, and the Pawlak upper approximation is equal to the composition of the very form of the upper approximation with itself.

5.4   The Case of Symmetric Relations

In this section, we examine the case, where ϱ is symmetric. Then, for every $u, u' \in U$, $(u, u') \in \varrho$ implies $(u', u) \in \varrho$. Immediately, $\varrho = \varrho^{-1}$ and $\Gamma = \Gamma^*$. Relations which are both reflexive and symmetric are called *tolerance* relations. Obviously, every tolerance relation is also a similarity relation[7]. The sets of all symmetric relations and all tolerance relations on U will be denoted by $\text{SYM}(U)$ and $\text{TL}(U)$, respectively.

Proposition 10. *The following properties hold:*

> (a) $f = f^* \ \& \ f^{\cup} = \text{upp} \circ f$
>
> (b) $\text{upp} \circ \text{low} \sqsubseteq \text{low} \circ \text{upp}$
>
> (c) $\varrho \in \text{TR}(U) \ \Rightarrow \ \text{upp} \circ \text{low} \sqsubseteq \text{low} \ \& \ \text{upp}^{\cup} \sqsubseteq \text{upp} \sqsubseteq \text{low} \circ \text{upp}$
>
> $\ \& \ \text{upp}^{\cup} \circ f^{\cup} \sqsubseteq f^{\cup}$

Proof. We show (b), (c) only. Consider any concept X and any object u.

In case (b), assume that $u \in \text{upp}(\text{low}X)$. By the definition of upp, $\Gamma u \cap \text{low}X \neq \emptyset$. Hence, there is u' such that (b1) $u' \in \Gamma u$ and $u' \in \text{low}X$. From the latter, (b2) $\Gamma u' \subseteq X$ by the definition of low. In virtue of (b1), $(u', u) \in \varrho$. Hence, by symmetry of ϱ, $(u, u') \in \varrho$. That is, $u \in \Gamma u'$. Hence, (b3) $u \in X$ due to (b2). We need to show that $u \in \text{low}(\text{upp}X)$, i.e., $\Gamma u \subseteq \text{upp}X$ by the definition of low. To this end, consider $u'' \in \Gamma u$ (i.e., $(u'', u) \in \varrho$). By symmetry of ϱ, $(u, u'') \in \varrho$ as well, i.e., $u \in \Gamma u''$. Hence, $\Gamma u'' \cap X \neq \emptyset$ by (b3). By the definition of upp, $u'' \in \text{upp}X$ as required. Thus, we have proved that $\text{upp}(\text{low}X) \subseteq \text{low}(\text{upp}X)$, i.e., $(\text{upp} \circ \text{low})X \subseteq (\text{low} \circ \text{upp})X$. Immediately, $\text{upp} \circ \text{low} \sqsubseteq \text{low} \circ \text{upp}$ by the definition of $\sqsubseteq$.

---

[7] Tolerance relations will not be a subject to a separate study in this article. To list their properties, it suffices to merge the proposition below with the facts presented in Sect. 5.2.

In case (c), assume additionally that ϱ is transitive. First, $\mathrm{upp} \circ \mathrm{low} \sqsubseteq \mathrm{low}$ by the assumption, the 2nd part of (a), and Proposition 9a. Next, $\mathrm{upp}^{\cup} \sqsubseteq \mathrm{upp}$ by the assumption, the 2nd part of (a), and Proposition 9b. In the sequel, we show that $\mathrm{upp}X \subseteq (\mathrm{low} \circ \mathrm{upp})X$, i.e., $\mathrm{upp}X \subseteq \mathrm{low}(\mathrm{upp}X)$. Then, $\mathrm{upp} \sqsubseteq \mathrm{low} \circ \mathrm{upp}$ by the definition of $\sqsubseteq$. To this end, suppose that $u \in \mathrm{upp}X$. By the definition of upp, $\Gamma u \cap X \neq \emptyset$. Hence, there is u' such that (c1) $u' \in X$ and $u' \in \Gamma u$ (i.e., (c2) $(u', u) \in \varrho$). Consider any $u'' \in \Gamma u$ (i.e., $(u'', u) \in \varrho$). By symmetry of ϱ, $(u, u'') \in \varrho$. Hence, $(u', u'') \in \varrho$ by (c2) and transitivity of ϱ. In other words, $u' \in \Gamma u''$. Hence, $\Gamma u'' \cap X \neq \emptyset$ in virtue of (c1). By the definition of upp, $u'' \in \mathrm{upp}X$. Thus, we have proved that $\Gamma u \subseteq \mathrm{upp}X$, i.e., $u \in \mathrm{low}(\mathrm{upp}X)$ by the definition of low. For the remaining part of (c) note that $\mathrm{upp}^{\cup} \circ f^{\cup} = (\mathrm{upp} \circ \mathrm{upp}) \circ (\mathrm{upp} \circ f) \sqsubseteq \mathrm{upp} \circ (\mathrm{upp} \circ f) = (\mathrm{upp} \circ \mathrm{upp}) \circ f \sqsubseteq \mathrm{upp} \circ f = f^{\cup}$ by (a), transitivity of ϱ, Proposition 9b, and Proposition 1d. □

Thus, whenever symmetry of ϱ is assumed, the ∗-versions of the basic approximation mappings coincide with the very mappings in virtue of (a). As a consequence, the ϱ-definable versions of the basic mappings are compositions of these mappings with the Pawlak upper approximation mapping. Due to (b), the Pawlak upper approximation of the lower approximation of a concept is included in the lower approximation of the Pawlak upper approximation of that concept. In virtue of (c), if ϱ is both symmetric and transitive, then — among others — the Pawlak upper approximation of the lower approximation of a concept is included in the lower approximation of that concept, the ϱ-definable version of the Pawlak upper approximation of a concept is included in the Pawlak upper approximation of that concept, and the latter one is included in the lower approximation of the Pawlak upper approximation of the very concept.

5.5   The Case of Equivalence Relations

Finally, we consider the case, where ϱ is an equivalence relation. In the context of Pawlak's information systems and approximation spaces, equivalence relations on the set of objects are understood as *indiscernibility* relations. By definition, an equivalence relation is simultaneously reflexive, symmetric, and transitive, i.e., it is a transitive tolerance relation. Let $\mathrm{EQ}(U)$ denote the set of all equivalence relations on U. Thus, $\mathrm{EQ}(U) = \mathrm{TL}(U) \cap \mathrm{TR}(U)$. Note that $\mathrm{EQ}(U) = \mathrm{SER}(U) \cap \mathrm{SYM}(U) \cap \mathrm{TR}(U)$ as well. In the sequel, Γu is called an *equivalence class* of u and it may be denoted by $[u]_\varrho$ (or simply, by $[u]$ if ϱ is understood) along the standard lines. Note that

$$\forall u, u' \in U.(u' \in \Gamma u \Leftrightarrow \Gamma u' = \Gamma u). \tag{18}$$

Proposition 11. *It holds that:*

(a) $f^{\cup} = f$

(b) $\mathrm{low} = \mathrm{low} \circ \mathrm{low} = \mathrm{upp} \circ \mathrm{low} \sqsubseteq \mathrm{low} \circ \mathrm{upp} = \mathrm{upp} \circ \mathrm{upp} = \mathrm{upp}$

(c) $\mathrm{low} \sqsubseteq \mathrm{upp}^{\mathrm{S}} \circ \mathrm{low} \sqsubseteq \mathrm{low} \circ \mathrm{upp}^{\mathrm{S}} = \mathrm{upp}^{\mathrm{S}} = \mathrm{upp} \circ \mathrm{upp}^{\mathrm{S}} \sqsubseteq \mathrm{upp}^{\mathrm{S}} \circ \mathrm{upp}$

Proof. We show (a) only. Note that "$\sqsupseteq$" holds by reflexivity of ϱ and Proposition 8b. In the sequel, $\text{low}^{\cup} \sqsubseteq \text{low}$ by transitivity of ϱ and Proposition 9a. Moreover, $\text{upp}^{\cup} \sqsubseteq \text{upp}$ by symmetry and transitivity of ϱ, and Proposition 10c. Now, consider any concept X and any object u. To prove $\text{upp}^{S\cup} \sqsubseteq \text{upp}^{S}$, it suffices to show that $\text{upp}^{S\cup}X \subseteq \text{upp}^{S}X$. Thus, suppose that $u \in \text{upp}^{S\cup}X$. Hence, there is u' such that (a1) $u \in \Gamma u'$ and $u' \in \text{upp}^{S}X$ by the definition of $\text{upp}^{S\cup}$. It follows from the latter that $\kappa(\Gamma u', X) > 0$ by the definition of upp^{S}. By (18) and (a1), $\kappa(\Gamma u, X) > 0$. Finally, $u \in \text{upp}^{S}X$ by the definition of upp^{S}. $\square$

Thus, whenever ϱ is an equivalence relation, the ϱ-definable versions of the basic approximation mappings coincide with the mappings by (a), respectively. As regarding the reduction in the number of different compositions of the mappings considered, both the composition of the lower approximation mapping with itself and the composition of the lower approximation mapping with the Pawlak upper approximation mapping are equal (and hence, may be reduced) to the lower approximation mapping. Furthermore, both the composition of the Pawlak upper approximation mapping with the lower approximation mapping and the composition of the Pawlak upper approximation mapping with itself are equal to the Pawlak upper approximation mapping thanks to (b). Moreover by (c), both the composition of the Skowron–Stepaniuk upper approximation mapping with the lower approximation mapping and the composition of the Skowron–Stepaniuk upper approximation mapping with the Pawlak upper approximation mapping are equal to the Skowron–Stepaniuk upper approximation mapping. Finally, the lower approximation of a concept is included in the Skowron–Stepaniuk upper approximation of the lower approximation of that concept, the latter set is included in the Skowron–Stepaniuk upper approximation of the concept, and the very upper approximation of a concept is included in the Skowron–Stepaniuk upper approximation of the Pawlak upper approximation of that concept.

Last but not least, observe that the upper approximation mappings upp_{1}^{S}, upp_{2}^{S} investigated in Example 1 are equal to upp if ϱ is an equivalence relation. Indeed, $\text{upp} \circ \Gamma = \Gamma$ then, and for any object u and any concept X, $u \in \text{upp}_{2}^{S}X$ if and only if $\Gamma u \cap \text{upp}^{*}X \neq \emptyset$ if and only if $\Gamma u \cap \text{upp}X \neq \emptyset$ if and only if $u \in \text{upp}(\text{upp}X)$ if and only if $u \in \text{upp}X$.

6   Summary

In this article, we studied and compared Pawlak's rough approximation of concepts with Skowron–Stepaniuk's approach within a general framework of approximation spaces of the form (U, ϱ, κ), where U is a non-empty set of objects, ϱ is a non-empty binary relation on U, and κ is a RIF satisfying rif$_1$ and rif$_2$. The lower approximation mappings are the same in both approaches unlike the upper approximation ones[8]. The latter mappings cannot be compared directly without additional assumptions made about κ. For the sake of illustration, we

---

[8] The fact that Pawlak's and Skowron–Stepaniuk's upper approximations coincide for the standard RIF was known earlier.

considered two special cases of κ and two corresponding Skowron–Stepaniuk upper approximation mappings in Example 1. In general, these two kinds of mappings are different from each other and from Pawlak's upper approximation mapping. However — as turned out — the three cases coincide if ϱ is an equivalence relation.

In the paper presented, we compared Pawlak's and Skowron–Stepaniuk's upper approximation mappings indirectly by investigation of their properties. It is difficult to say which mappings are totally better. While Pawlak's approximation seems to be more suitable in some aspects, it is the Skowron–Stepaniuk approach which seems to provide more interesting results in other cases. As a side-effect, we have obtained a fairly exhaustive list of basic mathematical properties of the mappings investigated, to be used in the future research and applications.

References

1. Pawlak, Z.: Rough set elements. In Polkowski, L., Skowron, A., eds.: Rough Sets in Knowledge Discovery 1. Volume 18 of Studies in Fuzziness and Soft Computing. Physica-Verlag, Heidelberg (1998) 10–30
2. Pawlak, Z.: Information systems – Theoretical foundations. Information Systems **6** (1981) 205–218
3. Pawlak, Z.: Rough sets. Computer and Information Sciences **11** (1982) 341–356
4. Pawlak, Z.: Information Systems. Theoretical Foundations (in Polish). Wydawnictwo Naukowo-Techniczne, Warsaw (1983)
5. Pawlak, Z.: Rough Sets. Theoretical Aspects of Reasoning About Data. Kluwer, Dordrecht (1991)
6. Pomykała, J.A.: Approximation operations in approximation space. Bull. Polish Acad. Sci. Math. **35** (1987) 653–662
7. Wybraniec-Skardowska, U.: On a generalization of approximation space. Bull. Polish Acad. Sci. Math. **37** (1989) 51–62
8. Żakowski, W.: Approximations in the space (U, Π). Demonstratio Mathematica **16** (1983) 761–769
9. Skowron, A., Stepaniuk, J.: Generalized approximation spaces. In: Proc. 3rd Int. Workshop Rough Sets and Soft Computing, San Jose, USA, 1994, November 10-12. (1994) 156–163
10. Skowron, A., Stepaniuk, J.: Tolerance approximation spaces. Fundamenta Informaticae **27** (1996) 245–253
11. Gomolińska, A.: Variable-precision compatibility spaces. Electronical Notices in Theoretical Computer Science **82** (2003) 1–12
    http://www.elsevier.nl/locate/entcs/volume82.html.
12. Słowiński, R., Vanderpooten, D.: Similarity relation as a basis for rough approximations. In Wang, P.P., ed.: Advances in Machine Intelligence and Soft Computing. Volume 4. Duke University Press (1997) 17–33
13. Yao, Y.Y., Wong, S.K.M., Lin, T.Y.: A review of rough set models. In Lin, T.Y., Cercone, N., eds.: Rough Sets and Data Mining: Analysis of Imprecise Data. Kluwer, Boston London Dordrecht (1997) 47–75
14. Ziarko, W.: Variable precision rough set model. J. Computer and System Sciences **46** (1993) 39–59
15. Ziarko, W.: Probabilistic decision tables in the variable precision rough set model. J. Comput. Intelligence **17** (2001) 593–603

16. Peters, J.F.: Approximation space for intelligent system design patterns. Engineering Applications of Artificial Intelligence **17** (2004) 1–8

17. Skowron, A.: Approximation spaces in rough neurocomputing. In Inuiguchi, M., Hirano, S., Tsumoto, S., eds.: Rough Set Theory and Granular Computing. Volume 125 of Studies in Fuzziness and Soft Computing. Springer-Verlag, Berlin Heidelberg (2003) 13–22

18. Skowron, A., Swiniarski, R., Synak, P.: Approximation spaces and information granulation. Transactions on Rough Sets III: Journal Subline to Lecture Notes in Computer Science **3400** (2005) 175–189

19. Polkowski, L., Skowron, A.: Rough mereology. Lecture Notes in Artificial Intelligence **869** (1994) 85–94

20. Polkowski, L., Skowron, A.: Rough mereology: A new paradigm for approximate reasoning. Int. J. Approximated Reasoning **15** (1996) 333–365

21. Łukasiewicz, J.: Die logischen Grundlagen der Wahrscheinlichkeitsrechnung. In Borkowski, L., ed.: Jan Łukasiewicz – Selected Works. North Holland, Polish Scientific Publ., Amsterdam London, Warsaw (1970) 16–63 First published in Kraków in 1913.

22. Zadeh, L.A.: Outline of a new approach to the analysis of complex system and decision processes. IEEE Trans. on Systems, Man, and Cybernetics **3** (1973) 28–44

Data Preparation for Data Mining in Medical Data Sets

Grzegorz Ilczuk[1] and Alicja Wakulicz-Deja[2]

[1] Siemens AG Medical Solutions,
Allee am Roethelheimpark 2, 91052 Erlangen, Germany
Grzegorz.Ilczuk@ilczuk.com
[2] Institut of Informatics University of Silesia,
Bedzinska 39, 41-200 Sosnowiec, Poland
wakulicz@us.edu.pl

Abstract. Data preparation is a very important but also a time consuming part of a Data Mining process. In this paper we describe a hierarchical method of text classification based on regular expressions. We use the presented method in our data mining system during a pre-processing stage to transform Latin free-text medical reports into a decision table. Such decision tables are used as an input for rough sets based rule induction subsystem. In this study we also compare accuracy and scalability of our method with a standard approach based on dictionary phrases.

Keywords: rough sets, data preparation, regular expression.

1 Introduction

Preparation of data takes about 60% of a time needed for the whole Data Mining process and it is also defined by Pyle as the most important part of a Data Exploration Process which leads to success [1]. This estimation is also valid in case of our Data Exploration system, where the entry stage of data processing is a key element for the further analysis. The mentioned Data Exploration system will be used in medicine (especially in cardiology) as a complete solution suitable for improving medical care and clinical work flow through revealing new patterns and relations among data. Functional blocks of the system are:

- Import subsystem-responsible for importing data from medical information systems into our storage subsystem
- Data recognition subsystem-during this stage we use algorithms and methods described in this paper to transform the raw data to a form suitable for further Data Exploration
- Data preprocessing-based on the statistical analysis of the transformed information noise and redundant data are removed [9]
- Feature selection-this stage utilizes a few attribute reduction methods such as CFS (Correlation-based Feature Selection), Quickreduct and conjunction of these methods to select an optimal set of attributes for a further analysis

J.F. Peters et al. (Eds.): Transactions on Rough Sets VI, LNCS 4374, pp. 83–93, 2007.
© Springer-Verlag Berlin Heidelberg 2007

- Rule induction subsystem based on Rough Set Theory [6,7,11]. Early research on this area was described in [8,12]
- Visualization of the knowledge discovery in a form easily understandable by humans for validating and extending of the collected knowledge [13]

Most medical information useful for Data Mining is still written in form of free-text Latin reports. These reports are mostly used to extend a lapidary diagnosis written with statistical ICD-10 codes. There are some challenges to solve during analyzing such reports such as: different descriptions for the same disease, non-standard abbreviations, misspelled words and a floating structure of such reports. In our first solution of these problems we had used a phrase dictionary to map information from a report to an attribute. The main disadvantage of this approach was a lack of scalability and a difficult maintenance. These facts leaded us to develop a different approach. The proposed method and results achieved with it are presented in this study. Described in this paper technique is used in our Data Exploration system as a preprocessing step, which prepares data for rule induction. For generation of decision rules an own implementation of MLEM2 algorithm is used.

2   Rough Sets: Basic Notions and Medical Appliance

Developed by Pawlak and presented in 1982 Rough Sets theory is a mathematical approach to handle imprecision and uncertainty [4]. The main goal of rough set analysis is to synthesize approximation of concepts from the acquired data. Some basic definitions are presented below.

Information system [4] is a pair $\mathbf{A} = (U, A)$ where U is a non-empty, finite set called the universe and A is a non-empty, finite set of *attributes*, i.e. $a : U \rightarrow V_a$ for $a \in A$, where V_a is called the *value set* of attribute a. Elements of U are called *objects*.

The special case of information systems called *decision system* is defined as $\mathbf{A} = (U, A \cup \{d\})$, where $d \notin A$ is a distinguished attribute called *decision* and elements of A are called *conditions*.

A *decision rule* is defined as $r = (a_{i1} = v_1) \wedge \ldots \wedge (a_{im} = v_m) \Rightarrow (d = k)$ where $1 \leq i_1 < \ldots < i_m \leq |A|, v_i \in V_{ai}$. We say an object matches a rule if its attributes satisfy all *atomic formulas* $(a_{ij} = v_j)$ of the rule. A rule is called *minimal consistent* with $\mathbf{A}$ when any decision rule r' created from r by removing one of atomic formula of r is not consistent with $\mathbf{A}$.

In our Data Exploration system we use a modified version of LEM2 algoritm - MLEM2 to generate decision rules. LEM2 (Learning from Examples Module, version 2) algorithm was firstly presented in [14,15] and then implemented in [16]. LEM2 induces a rule set by exploring the space of blocks of attribute-value pairs to generate a local covering. Afterwards the found local covering is converted into the rule set. Following definitions must be quoted prior to define a local covering [11].

For a variable (attribute or decision) x and its value v, a block $[(x, v)]$ of a variable-value pair (x, v) is the set of all cases for which variable x has value v.

Let B be a nonempty lower or upper approximation of a concept represented by a decision-value pair (d, w). Set B *depends* on a set T of attribute-value pairs (a, v) if and only if

$$\emptyset \neq [T] = \bigcap_{(a,v) \in T} [(a, v)] \subseteq B. \qquad (1)$$

Set T is a *minimal complex* of B if and only if B depends on T and no proper subset T' of T exists such that B depends on T'. Let $\mathbf{T}$ be a nonempty collection of nonempty sets of attribute-value pairs. Then $\mathbf{T}$ is a *local covering* of B if and only if the following conditions are satisfied:

- each member T of $\mathbf{T}$ is a minimal complex of B,
- $\bigcup_{T \in \mathbf{T}} [T] = B$, and
- $\mathbf{T}$ is minimal, i.e., $\mathbf{T}$ has the smallest possible number of members.

Modified LEM2 (MLEM2) proposed by Grzymala-Busse in [11] in compare to LEM2 allows inducing rules from data containing numerical attributes without a need of a separate discretization step. Our implementation of MLEM2 algorithm induces decision rules from both *lower approximation* (certain rules) and *upper approximation* (possible rules). This technique allows us reasoning from "real" data, which contains uncertain, noisy and redundant information. Decision rules are used in our system to present the extracted knowledge from medical data. This approach in medical domain has several advantages over other data mining techniques:

- Decision rules are easy to understand and verify
- Decision rules can be easily validated with existing knowledge
- Gathered decision rules can be modified and extended with a new knowledge
- If decision rules are used for classification it is easy to explain the choice
- Simple structure of decision rules allows several ways of visualization

These advantages lead to a rapid growth of interest in appliance of rough set theory in medical domain. Many interesting case studies reported a successful appliance of rough set software systems. Some of them were:

- Treatment of duodental ulcer by HSV described by Slowinski in [21,23]
- Multistage analysis in progressive encephalopathy presented by Paszek in [20,22]
- Preterm birth prediction researched by Grzymala-Busse [19]
- Analysis of medical databases (headaches, CVD) [24]
- Acute abdominal pain in childhood (MET system applied in Children's Hospital of Eastern Ontario) [18]
- Cardiac Tests analysis [25]

More successful studies, not only from medical domain, are described in [17].

3   Regular Expression

The origins of, belonging to automata and formal language theory, regular expressions lie in 1940, when McCulloch and Pitts described a nervous system as a neurons in a small automata [3]. These models were then described by Kleene and Kozen using *regular expression* (regular set) notation [2].

Regular expressions consist of constants and operators that denote sets of strings and operations over these sets, respectively. Given a finite alphabet Σ the following constants are defined:

- *empty set* $\emptyset$ denoting the set $\emptyset$
- *empty string* ϵ denoting the set $\{\epsilon\}$
- *literal character* α in Σ denoting the set $\{\alpha\}$

Following operations are defined:

- *concatenation RS* denoting the set $\{\alpha\beta | \alpha$ in R and β in $S\}$. For example {"ab", "c"}{"d", "ef"} = {"abd", "abef", "cd", "cef"}.
- *alternation R|S* denoting the set union of R and S.
- *Kleene star R** denoting the smallest superset of R that contains ϵ and is closed under string concatenation. This is the set of all strings that can be made by concatenating zero or more strings in R. For example, {"ab","c"}* = $\{\epsilon, "ab", "c", "abab", "abc", "cab", "cc", "ababab", ...\}$.

To avoid brackets it is assumed that the Kleene star has the highest priority, then concatenation and then set union. If there is no ambiguity then brackets may be omitted. For example, $(ab)c$ is written as abc and $a|(b(c*))$ can be written as $a|bc*$.

4   Methods

In our research of analyzing medical data we would like to extend and complement information collected from clinical information systems in form of ICD-10 codes with additional information stored in free-text descriptions. A typical example of a such description is shown below:

```
Status post implantationem pacemakeri VVI (1981, 1997) ppt.
diss. A-V gr. III.Exhaustio pacemakeri. Morbus ischaemicus
cordis. Insufficientia coronaria chronica CCS I.
Myocardiopathia ischaemica in stadio comp. circulatoriae.
Fibrillatio atriorum continua. Pacemaker dependent.
```

A method suitable for our needs should therefore fulfill the following requirements:

- it shall recognize misspelled and abbreviated words
- it shall interpretable whole sentences

– it shall provide a back tracing so that an expert can always validate an assigned mapping
– all mappings must done based 100% on information from an original text
– it shall be easily maintainable and extendible

With these requirements in mind we have developed a method which bases on a fixed number of user defined records each containing following three attributes: a level value (shown at figure 1 as 'LEVEL'), a mask coded using regular expressions for searching a phrase of text ('FIND TEXT') and a string of text which will be used for replacing if the searched phrase is found ('REPLACE FOUND TEXT'). Defined records are sorted incrementally based on their level value, so that, when the algorithm starts a group of records having the lowest value can be firstly selected. During the next step for each record from the group the algorithm tries to replace a found text with a specified string. When all records are processed then a next group of records with a next higher level value is selected and the process of searching/replacing text repeats. This algorithm ends when the last record from the group with the highest level value is processed. A simplified, but based on a real implementation, example is shown at figure 1.

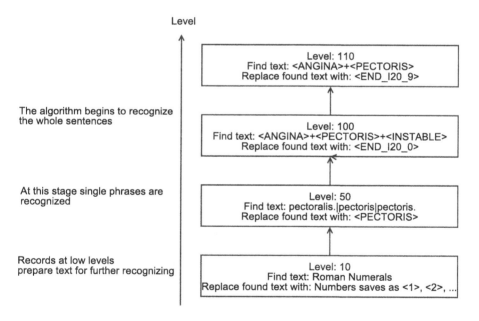

Fig. 1. Sample of records used by the algorithm

From this figure it can be seen that the lowest level value is 10, so that the algorithm begins to select a group of records having this level value. In our case it is only one record which replace all found Roman numbers with the following schema '<'+number+'>' for example a number 'II' will be replaced to <2> and

'IV' results in <4>. After this processing a next higher level value is selected (50) together with a group of records having the same value of their level attribute. In the shown example there are only one such record. But it is possible and common that, there are a lot of records which have the same level value and thus can be processed in parallel. It is important to note that a definition of a mask used for searching (field 'FIND TEXT') contains not only a correct version of a phrase but also several misspelled combinations stored using the alternation operation of regular expressions for example (pectoralis|pectoris|...). If a phase is found then it will be replaced with a specified replace string independently if it was written correctly or incorrectly. This allows to correct a simple type errors and focus on sentence analysis. As an example records with their level value 100 and 110 can be used. These two records search for a combination of symbolic replacements previously replaced by records with the level value 50, so that these two records can correctly assign an '<END_I20_0> code not only to a properly written 'angina pectoris' diagnose but also to a whole bunch of misspelled combinations of these two words as for example 'angin pectoralis'.

The described algorithm has following advantages:

- it allows filtering of redundant and noisy information at entry processing stage
- it correctly recognizes misspelled diagnoses
- the process of interpreting whole sentences is simplified because only connections between symbolic phrases must be analyzed and not all possible combinations which can be found in an input text
- it is possible to stop the algorithm at stage and analyze or eventually correct the replacing process
- in our implementation we only use already found in input text combinations of words what decreases a possibility of false positive recognitions

In the next section we will compare the recognition accuracy and scalability of the described algorithm with our previous dictionary based algorithm.

5   Dataset Preparation and Experimental Environment

Data used in our research was obtained from the Cardiology Department of Silesian Medical Academy in Katowice - the leading Electrocardiology Department in Poland specializing in hospitalization of severe heart diseases. For our experiments we took a data set of 4000 patients hospitalized in this Department between 2003 and 2005. This data were imported into a PostgreSQL database and then divided in eight groups (G-C1, ..., G-C8), where G-C1 contained first 500 records from the database and each next group had 500 more records then the previous group, so that the last group G-C8 contained all 4000 records. Each record in a group contained a single free-text report which was then analyzed of the presence of one of the following diseases:

- Essential (primary) hypertension - if found mapped to I10 code
- Old myocardial infarction - if found mapped to I25-2 code

- Atrioventricular block, first degree - if found mapped to I44-0 code
- Atrioventricular block, second degree - if found mapped to I44-1 code
- Atrioventricular block, complete - if found mapped to I44-2 code
- Sick sinus syndrome - if found mapped to I49-5 code

We implemented the presented algorithm in Java version 1.5. and used the Java implementation of regular expressions from the 'java.util.regex.Pattern' class.

6   Results

Results presented the table 1 show an absolute number of cases recognized by the described in this paper method within each of the tested group. These results are additionally compared with the dictionary method and this comparison is shown as a number in brackets, where a positive number means a number of cases additionally recognized by the method based on regular expressions. Visualization of these numbers is shown at figure 2, where it can be seen that the proposed method recognized more cases then the dictionary method but with a different, depending on a selected disease, characteristic. For hypertension and old myocardial infarction a number of additionally recognized cases is rather low what can be attributable to the fact, that the most diagnosis variants are already covered by the dictionary method. Recognition of atrioventricular block poses a bigger challenge, so that a difference in a number of recognized cases for all three types of this disease oscillates between 20-40% additional cases identified by the proposed method. The most spectacular results were achieved for recognizing Sick sinus syndrome what can be assignable with a huge number of possible combinations used to specify this diagnosis. These combinations were better covered by regular expressions and a difference to the dictionary method was almost 42%.

It can be also seen, that a number of identified cases, shown at figure 3, increased for all tested diseases almost linearly. This satisfactory result shows a good ability of the presented method to recognize new records with a relatively small number of definitions (500 regular expressions compared to more then 4800 dictionary phrases).

We had also randomly selected a set of 100 records and with a help from domain experts from the Cardiology Department manually identified them for three diseases. These numbers were then compared with a results achieved by both the regular expression and the dictionary method. This comparison is shown in the table 2. From this table it can be seen that for a relatively small group of records the method based on regular expression recognized all hypertension and atrioventricular block (first degree) cases. Of course it will be only a matter of additional time effort needed to extend the recognition accuracy of the dictionary method but this is exactly the advantage of the proposed algorithm, which with a significant smaller number of records presents better scalability and in case of new data also a better update ability.

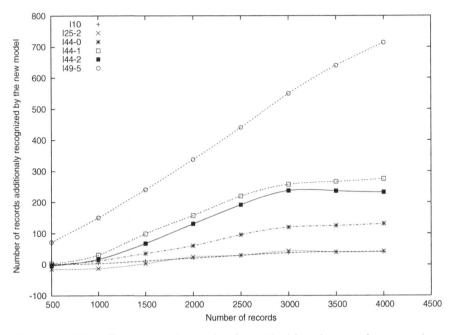

Fig. 2. Additionally recognized cases by the method based on regular expressions

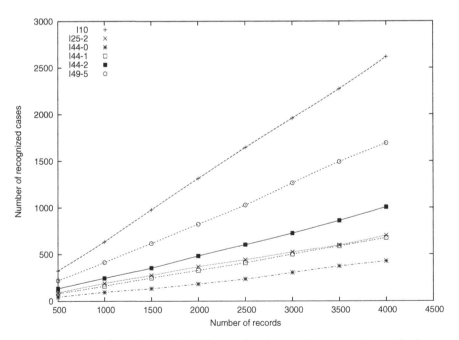

Fig. 3. Number of recognized diseases by the regular expression method

Table 1. Number of recognized cases by the proposed method

Group	Number of records	I10	I25-2	I44-0	I44-1	I44-2	I49-5
G-C1	500	325 (1)	91 (-15)	45 (0)	82 (3)	135 (-4)	220 (72)
G-C2	1000	636 (3)	190 (-13)	95 (12)	161 (31)	245 (17)	414 (150)
G-C3	1500	978 (12)	276 (3)	134 (36)	248 (99)	354 (68)	618 (241)
G-C4	2000	1314 (21)	368 (25)	183 (61)	329 (158)	483 (131)	824 (338)
G-C5	2500	1645 (30)	442 (30)	236 (96)	410 (220)	604 (192)	1029 (441)
G-C6	3000	1959 (38)	524 (44)	305 (120)	502 (258)	728 (238)	1265 (550)
G-C7	3500	2275 (41)	600 (40)	373 (125)	590 (267)	863 (237)	1493 (640)
G-C8	4000	2616 (41)	704 (43)	427 (131)	678 (276)	1007 (233)	1693 (714)

Table 2. Recognition accuracy comparison between methods

Disease	Number of cases	Regular exp. found cases	Regular exp. accuracy [%]	Dictionary found cases	Dictionary accuracy [%]
I10	61	61	100.0	61	100.0
I44-0	11	11	100.0	9	81.9
I44-1	15	14	93.3	10	66.7

7 Conclusions

In this paper we presented an algorithm for recognition of free-text Latin medical reports which is based on hierarchically organized records. These records use regular expressions to find a specified phrase in an input text and replace it with a user defined text. The hierarchically organized records convert an input text step by step replacing firstly simple words into symbolic phrases then these symbolic phrases into more complicated expressions and at the end the whole sentences are mapped to user defined codes. Such codes can be then easily used to construct a decision table used by next data mining algorithms.

Our experiments shown that the presented method achieves better recognition accuracy then the method based on fixed dictionary phrases and this result can be achieved with a significant smaller number of records used for definition. This small number of easily modifiable and very flexible records is truly an advantage of the described method.

Our idea to reduce the complexity of recognizing Latin diagnosis through defining a short parts of the whole sentence using regular expressions and then to join hierarchically such pieces of information together allowed us to cover with a finite, small number of records a huge number of possible combinations. This advantage and the fact that the presented method fulfill all the specified requirements it is used in our data exploration system during a preprocessing stage for processing not only Latin free-text reports but also laboratory, electrocardiogram (ECG) and cardiovascular ultrasound descriptions.

Acknowledgements

We would like to thank Rafal Mlynarski from the Cardiology Department of Silesian Medical Academy in Katowice, Poland for providing us the data and giving us feedbacks.

References

1. Pyle, D.: Data preparation for data mining. Morgan Kaufmann, San Francisco (1999)
2. Kozen, D.: On Kleene Algebras and Closed Semirings. In: Mathematical Foundations of Computer Science, Banská Bystrica (1990) 26–47
3. McCulloch, W. and Pitts, W.: A logical calculus of the ideas immanent in nervous activity. In: Bulletin of Mathematical Biophysics, (1943) 115–133
4. Pawlak, Z.: Rough sets. International Journal of Computer and Information Science **11** (1982) 341–356
5. Sipser, M.: Introduction to the Theory of Computation. Course Technology, (2006)
6. Pawlak, Z.: Knowledge and Uncertainty: A Rough Set Approach. SOFTEKS Workshop on Incompleteness and Uncertainty in Information Systems (1993) 34–42
7. Pawlak, Z. and Grzymala-Busse, J. W. and Slowinski, R. and Ziarko, W.: Rough Sets. Commun. ACM **38** (1995) 88–95
8. Ilczuk, G. and Wakulicz-Deja, A.: Rough Sets Approach to Medical Diagnosis System. In: AWIC 2005, Lodz (2005) 204–210
9. Ilczuk, G. and Wakulicz-Deja, A.: Attribute Selection and Rule Generation Techniques for Medical Diagnosis Systems. In: RSFDGrC 2005, Regina (2005) 352–361
10. Wakulicz-Deja, A. and Paszek, P.: Applying Rough Set Theory to Multi Stage Medical Diagnosing. Fundam. Inform. **54** (2003) 387–408
11. Grzymala-Busse, J. W.: MLEM2 - Discretization During Rule Induction. In: IIS 2003, Zakopane (2003) 499–508
12. Ilczuk, G. and Mlynarski, R. and Wakulicz-Deja, A. and Drzewiecka, A. and Kargul, W.: Rough Sets Techniques for Medical Diagnosis Systems. In: Computers in Cardiology 2005, Lyon (2005) 837–840
13. Mlynarski, R. and Ilczuk, G. and Wakulicz-Deja, A. and Kargul, W.: Automated Decision Support and Guideline Verification in Clinical Practice. In: Computers in Cardiology 2005, Lyon (2005) 375–378
14. Chan, C. C. and Grzymala-Busse, J. W.: On the two local inductive algorithms: PRISM and LEM2. Foundations of Computing and Decision Sciences **19** (1994) 185–203
15. Chan, C. C. and Grzymala-Busse, J. W.: On the attribute redundancy and the learning programs ID3, PRISM, and LEM2.Department of Computer Science, University of Kansas,TR-91-14, (1991)
16. Grzymala-Busse, J. W.: A new version of the rule induction system LERS. Fundam. Inform. **31** (1997) 27–39

17. Komorowski, H. J. and Pawlak, Z. and Polkowski, L. T. and Skowron, A.: Rough Sets: A Tutorial. Springer-Verlag, Singapore (1999)
18. Farion, K. and Michalowski, W. and Slowinski, R. and Wilk, S. and Rubin, S.: Rough Set Methodology in Clinical Practice: Controlled Hospital Trial of the MET System. Rough Sets and Current Trends in Computing. **3066** (2004) 805–814
19. Grzymala-Busse, J. W. and Goodwin, L. K.: Predicting pre-term birth risk using machine learning from data with missing values. Bulletin of the International Rough Set Society (IRSS). **1** (1997) 17–21
20. Paszek, P. and Wakulicz-Deja, A.: The Application of Support Diagnose in Mitochondrial Encephalomyopathies. Rough Sets and Current Trends in Computing. **2475** (2002) 586–593
21. Pawlak, Z. and Slowinski, K. and Slowinski, R.: Rough Classification of Patients After Highly Selective Vagotomy for Duodenal Ulcer. International Journal of Man-Machine Studies. **24** (1986) 413–433
22. Tsumoto, S. and Wakulicz-Deja, A. and Boryczka, M. and Paszek, P.: Discretization of continuous attributes on decision system in mitochondrial encephalomyopathies. Proceedings of the First International Conference on Rough Sets and Current Trends in Computing. **1424** (1998) 483–490
23. Slowinski, K. and Slowinski, R. and Stefanowski, J.: Rough sets approach to analysis of data from peritoneal lavage in acute pancreatitis. Medical Informatics. **13** (1988) 143–159
24. Tsumoto, S. and Tanaka, H.: Induction of Disease Description based on Rough Sets. 1st Online Workshop on Soft Computing. (1996) 19–30
25. Komorowski, H.J. and Øhrn, A.: Modelling prognostic power of cardiac tests using rough sets. Artificial Intelligence in Medicine. **15** (1999) 167–191

A Wistech Paradigm for Intelligent Systems

Andrzej Jankowski[1,2] and Andrzej Skowron[3]

[1] Institute of Decision Processes Support
[2] AdgaM Solutions Sp. z o.o.
Wąwozowa 9 lok. 64, 02-796 Warsaw, Poland
andrzejj@adgam.com.pl
[3] Institute of Mathematics,
Warsaw University
Banacha 2, 02-097 Warsaw, Poland
skowron@mimuw.edu.pl

*If controversies were to arise, there would be no
more need of disputation between two philosophers
than between two accountants. For it would suffice to
take their pencils in their hands, and say to each other:
'Let us calculate'.*
     – Gottfried Wilhelm Leibniz,
       Dissertio de Arte Combinatoria (Leipzig, 1666).

*... Languages are the best mirror of the human mind,
and that a precise analysis of the signification of words
would tell us more than anything else about the operations
of the understanding.*
     – Gottfried Wilhelm Leibniz,
       New Essays on Human Understanding (1705)
       Translated and edited by
       Peter Remnant and Jonathan Bennett
       Cambridge: Cambridge UP, 1982

Abstract. The problem considered in this article is how does one go about discovering and designing intelligent systems. The solution to this problem is considered in the context of what is known as wisdom technology (wistech), an important computing and reasoning paradigm for intelligent systems. A rough-granular approach to wistech is proposed for developing one of its possible foundations. The proposed approach is, in a sense, the result of the evolution of computation models developed in the Rasiowa–Pawlak school. We also present a long-term program for implementation of what is known as a wisdom engine. The program is defined in the framework of cooperation of many Research & Development (R & D) institutions and is based on a wistech network (WN) organization.

Keywords: wisdom technology, adaptive rough-granular computing, rough sets, wisdom engine, open innovation, wisdom network.

J.F. Peters et al. (Eds.): Transactions on Rough Sets VI, LNCS 4374, pp. 94–132, 2007.
© Springer-Verlag Berlin Heidelberg 2007

1   Introduction

Huge technological changes occurred during the second half of the 20th century affecting every one of us. These changes affect practically all objects manufactured by man such as spoons, clothing, books, and space rockets. There are many indications that we are currently witnessing the onset of an era of radical changes. These radical changes depend on the further advancement of technology to acquire, represent, store, process, discover, communicate and learn wisdom. In this paper, we call this technology *wisdom technology* (or wistech, for short). The term *wisdom* commonly means "judging rightly" [50]. This common notion can be refined. By *wisdom*, we understand an adaptive ability to make judgements correctly to a satisfactory degree (in particular, correct decisions) having in mind real-life constraints.

One of the basic objectives of the paper is to indicate the potential directions for the design and implementation of wistech computation models. An important aspect of wistech is that the complexity and uncertainty of real-life constraints mean that in practise we must reconcile ourselves to the fact that our judgements are based on non-crisp concepts and also do not take into account all the knowledge accumulated and available to us. This is why consequences of our judgements are usually imperfect. But as a consolation, we also learn to improve the quality of our judgements via observation and analysis of our experience during interaction with the environment. Satisfactory decision-making levels can be achieved as a result of improved judgements.

The intuitive nature of wisdom understood in this way can be expressed metaphorically as shown in (1).

$$wisdom = KSN + AJ + IP, \tag{1}$$

where *KSN, AJ, IP* denote *knowledge sources network, adaptive judgement,* and *interactive processes,* respectively. The combination of the technologies represented in (1) offers an intuitive starting point for a variety of approaches to designing and implementing computational models for wistech. In this paper, (1) is called the *wisdom equation.* There are many ways to build wistech computational models. In this paper, the focus is on an adaptive rough-granular approach.

The issues discussed in this article are relevant for the current research directions (see, e.g., [16,15,31,38,51,90,108] and the literature cited in these articles).

This paper is organized as follows.

2   Wisdom Technology

This section briefly introduces the wistech paradigm.

2.1   What Do We Mean by Wistech?

On the one hand, the idea expressed by (1) (the wisdom equation paradigm) is a step in the direction of a new philosophy for the use of computing machines in our daily life, referred to as ubiquitous computing (see [66]). This paradigm is strongly connected with various applications of autonomic computing [64]. On the other hand, it should be emphasized that the idea of integrating many basic AI concepts (e.g., interaction, knowledge, network, adaptation, assessment, pattern recognition, learning, network, simulation of behavior in an uncertain environment, planning and problem solving) is as old as the history of AI itself. Many examples of such an approach adopted by researchers in the middle of the 20th century can be found in [27]. This research was intensively continued in the second half of the 20th century. For example, the abstracts of thousands of interesting reports from the years 1954 -1985 can be found in [91,92].

This paper contains the conclusions of the authors' experiences during numerous practical projects implementing wistech technologies in specific applications, e.g., fraud detection (MERIX – a prototype system for Bank of America), dialogue based search engine (EXCAVIO – intelligent search engine), UAV control (WITAS project), Intelligent marketing (data mining and optimization system for Ford Motor Company, General Motors), robotics, EVOLUTIONARY CHECKERS (adaptive checker R&D program at the University of North Carolina at Charlotte) and many other applications. These experiences are summarized by the authors in the metaphoric wisdom equation (1). This equation can also be illustrated using the following diagram presented in Figure 1.

In Figure 1 the term 'data' is understood as a stream of symbols without any interpretation of their meaning.

From the perspective of the metaphor expressed in the wisdom equation (1), wistech can be perceived as the integration of three technologies (corresponding to three components in the wisdom equation (1)). At the current stage two of them seem to be conceptually relatively clear, namely

1. *knowledge sources network* – by knowledge we traditionally understand every organized set of information along with the inference rules; in this context one can easily imagine the following examples illustrating the concept of knowledge sources network:
   - representation of states of reality perceived by our senses (or observed by the "receptors" of another observer) are integrated as a whole in our minds in a network of sources of knowledge and then stored in some part of our additional memory,
   - a network of knowledge levels represented by agents in some multi-agent system and the level of knowledge about the environment registered by means of receptors;
2. *interactive processes* – interaction understood as a sequence of stimuli and reactions over time; examples are:
   - the dialogue of two people,
   - a sequence of actions and reactions between an unmanned aircraft and the environment in which the flight takes place, or
   - a sequence of movements during some multi-player game.

Far more difficult conceptually seems to be the concept of adaptive judgement distinguishing wisdom from the general concept of problem solving. Intuitions behind this concept can be expressed as follows:

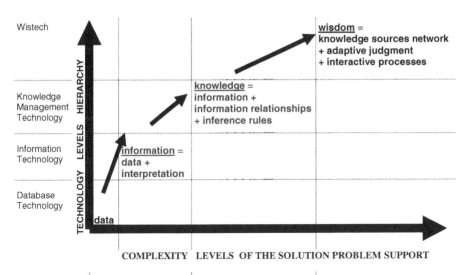

Fig. 1. Wisdom equation context

1. *adaptive judgement* – understood here as arriving at decisions resulting from the evaluation of patterns observed in sample objects. This form of judgement is made possible by mechanisms in a metalanguage (meta-reasoning) which on the basis of selection of available sources of knowledge and on the basis of understanding of history of interactive processes and their current status enable us to perform the following activities under real-life constraints:

   – identification and judgement of importance (for future judgement) of sample phenomena, available for observation, in the surrounding environment;

- planning current priorities for actions to be taken (in particular, on the basis of understanding of history of interactive processes and their current status) toward making optimal judgements;
- selection of fragments of ordered knowledge (hierarchies of information and judgement strategies) satisfactory for making a decision at the planned time (a decision here is understood as a commencing interaction with the environment or as selecting the future course to make judgements);
- prediction of important consequences of the planned interaction of processes;
- adaptive learning and, in particular, reaching conclusions deduced from patterns observed in sample objects leading to adaptive improvement in the adaptive judgement process.

One of the main barriers hindering an acceleration in the development of wistech applications lies in developing satisfactory computation models implementing the functioning of "adaptive judgement". This difficulty primarily consists in overcoming the complexity of the process of integrating the local assimilation and processing of changing non-crisp and incomplete concepts necessary to make correct judgements. In other words, we are only able to model tested phenomena using local (subjective) models and interactions between them. In practical applications, usually, we are not able to give global models of analyzed phenomena (see, e.g., [110,62,64,45,25,21]). However, we can only approximate global models by integrating the various incomplete perspectives of problem perception. One of the potential computation models for "adaptive judgement" might be the *rough-granular approach*.

2.2   Main Differences Between Wisdom and Inference Engine

In natural language, the concept of wisdom is used in various semantic contexts. In particular, it is frequently semantically associated with such concepts as inference, reasoning, deduction, problem solving, judging rightly as a result of pattern recognition, common sense reasoning, reasoning by analogy, and others. As a consequence this semantic proximity may lead to misunderstandings. For example, one could begin to wonder what the difference is between the widely known and applied concept in AI of "inference engine" and the concept of "wisdom engine" defined in this paper? In order to avoid this type of misunderstanding it is worth explaining the basic difference between the understanding of wisdom and such concepts as inference, reasoning, deduction and others.

Above all, let as start with explaining how we understand the difference between problem solving and wisdom. The widespread concept of problem solving is described as some slight modification of this notion defined in the context of solving mathematical problems by George Pólya in [84]. The concept of problem solving is understood in [84] as the following set of activities:

1. *First, you have to understand the problem.*
2. *After understanding, then make a plan.*

3. *Carry out the plan.*
4. *Look back on your work. How could it be better?*

An attempt at explaining the concept of wisdom can be taken using the concept of *problem solving* in the following manner: wisdom is the ability to identify important problems, search for sufficiently correct solutions to them, having in mind real life, available knowledge sources, personal experience, constraints, etc. Having in mind this understanding of wisdom we get at once the first important difference. Namely, in the problem solving process we do not have the following important wisdom factor in the above sequence (1-4) of activities:

0. Learning to recognize patterns that identify important problems and problem solution constraints.

Certainly, this is not the only difference. Therefore, one can illustrate the general difference between the concept of problem solving and wisdom as the difference between the concept of flying in an artificially controlled environment (e.g., using a flying simulator and problem solving procedures) and the concept of flying Boeing 767 aeroplane in real-life dangerous environment (wisdom in a particular domain).

One can therefore think that wisdom is very similar to *the ability of problem solving in a particular domain of application*, which in the context of the world of computing machines is frequently understood as an *inference engine*. The commonly accepted definition of the concept of inference engine can be found for example in Wikipedia (http://en.wikipedia.org/wiki/Inference_engine). It refers to understanding of "problem solving" in the spirit of the book [84]. It reads as follows:

An inference engine is a computer program that tries to derive answers from a knowledge base. It is the "brain" that expert systems use to reason about the information in the knowledge base, for the ultimate purpose of formulating new conclusions.

An inference engine has three main elements. They are:

1. An interpreter. The interpreter executes the chosen agenda items by applying the corresponding base rules.
2. A scheduler. The scheduler maintains control over the agenda by estimating the effects of applying inference rules in light of item priorities or other criteria on the agenda.
3. A consistency enforcer. The consistency enforcer attempts to maintain a consistent representation of the emerging solution.

In other words, the concept of inference engine relates to generating strategies for the inference planning from potentially varied sources of knowledge which are in interaction together. So this concept is conceptually related to the following two elements of the wisdom equation:

1. knowledge sources network,
2. interactive processes.

However, it should be remembered that wisdom in our understanding is not only some general concept of *inference*. The basic characteristic of wisdom, distinguishing this concept from the general understanding of inference, is *adaptive ability to make correct judgements having in mind real-life constraints*. The significant characteristic differentiating wisdom from the general understanding of such concepts as problem solving or inference engine is adaptive judgement.

In analogy to what we did in the case of *problem solving*, we can now attempt to explain the concept of wisdom based on the notion of an *inference engine* in the following manner: Wisdom is an inference engine interacting with a real-life environment, which is able to identify important problems and to find for them sufficiently correct solutions having in mind real-life constraints, available knowledge sources and personal experience. In this case, one can also illustrate the difference between the concept of inference engine and the concept of wisdom using the metaphor of flying a plane.

One could ask the question of which is the more general concept: wisdom or problem solving? Wisdom is a concept carrying a certain additional structure of adaptive judgement which in a continuously improving manner assists us in identifying the most important problem to resolve in a given set of constraints and what an acceptable compromise between the quality of the solution and the possibility of achieving a better solution is. Therefore, the question of what the more general concept is closely resembles the question from mathematics: What is the more general concept in mathematics: the concept of a field (problem solving), or the concept of the vector space over a field (wisdom understood as problem solving + adaptive judgement)? The vector space is a richer mathematical structure due to the action on vectors. Analogously to wisdom it is a richer process (it includes adaptive judgement - a kind of meta-judgement that encompasses recognition of patterns common to a set of sample objects that leads to judgements relating to problem solving). On the other hand, research into single-dimensional space can be treated as the research of fields. In this sense, the concept of vector space over a field is more general than the concept of a field.

2.3 Why Does Wistech Seem to Be One of the Most Important Future Technologies?

Nobody today doubts that technologies based on computing machines are among the most important technology groups of the 20th century, and, to a considerable degree, have been instrumental in the progress of other technologies. Analyzing the stages in the development of computing machines, one can quite clearly distinguish the following three stages in their development in the 20th century:

1. Database Technology (gathering and processing of transaction data).
2. Information Technology (understood as adding to the database technology the ability to automate analysis, processing and visualization of information).
3. Knowledge Management Technology (understood as systems supporting organization of large data sets and the automatic support for knowledge processing and discovery (see, e.g., [59,18])).

The three stages of development in computing machine technology show us the trends for the further development in applications of these technologies. These trends can be easily imagined using the further advancement of complexity of information processing (Shannon Dimension) and advancement of complexity of dialogue intelligence (Turing Dimension), viz.,

- *Shannon Dimension* level of information processing complexity (representation, search, use);
- *Turing Dimension* the complexity of queries that a machine is capable of understanding and answering correctly. One of the objectives of AI is for computing machines to reach the point in Turing Dimension that is well-known Turing Test (see [114]).

In this framework, the development trends in the application of computing machines technology can be illustrated in Figure 2.

Technology	Additional attributes	Shannon Dimensions	Turing Dimensions
Database Technology	data is the most basic level	How to represent information?	SQL
Information Technology	information = data + interpretation	Where to find information?	Who? What? When? Where? How much?
Knowledge Management Technology	knowledge = information + information relationships + inference rules	How to use information?	How? Why? What if?

Fig. 2. Computing machines technology

Immediately from the beginning of the new millennium one can see more and more clearly the following new application of computing machine technology, viz., wisdom technology (wistech) which put simply can be presented in table (see Figure 3, being an extension of the table presented in Figure 2).

In other words, the trends in the development of technology of computing machines can be presented using the so-called DIKW hierarchy (i.e., Data, Information, Knowledge, Wisdom). Intuitively speaking, each level of the DIKW hierarchy adds certain attributes over and above the previous one. The hierarchy is presented graphically in Figure 4.

Technology	Additional attributes	Shannon Dimensions	Turing Dimensions
Wisdom Technology (Wistech)	Wisdom equation, i.e. wisdom = knowledge sources network + adaptive judgment + interactive processes	*Learn when to use information* *Learn how to get important information*	*How to make correct judgements (in. particular correct decisions) heaving in mind real-life constraints?*

Fig. 3. Computing machine technology (continued)

DIKW hierarchy can be traced back to the well-known poem by T. S. Eliot, "The Rock", written in 1932. He wrote:

> *Where is the life we have lost in living?*
> *Where is the wisdom we have lost in knowledge?*
> *Where is the knowledge we have lost in information?*

It is a truism to state that the effects of any activity depend to a decisive degree on the wisdom of the decisions taken, both in the start, during the implementation, improvement and completion of the activity. The main objective of wistech is to automate support for the process leading to wise actions. These activities cover all areas of man's activities, from the economy, through medicine, education, research, development, etc.

In this context, one can clearly see how important role may have the wistech development in the future. The following comment from G.W. Leibniz on the idea to automate the processing of concepts representing thoughts should not surprise us either:

No one else, I believe, has noticed this, because if they had ... they would have dropped everything in order to deal with it; because there is nothing greater that man could do.

2.4   A General Approach to Wistech Computation Model

In order to create general Wistech computational models let us start an analysis of the concept of adaptive judgement.

For better familiarization of adaptive judgement we shall use the visualization of processes based on the IDEFO standard. Put simply, this means of visualisation is described in the diagram presented in Figure 5.

An intrinsic part of the concept of judgement is relating it to the entity implementing the judgement. Intuitively this can be a person, animal, machine, abstract agent, society of agents, etc. In general, we shall call the entity making

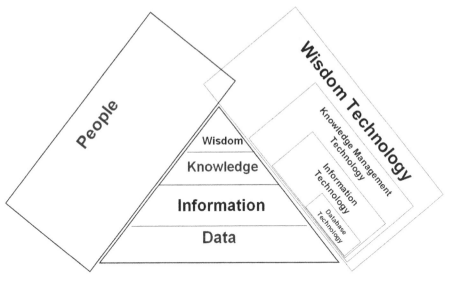

Fig. 4. DIKW hierarchy

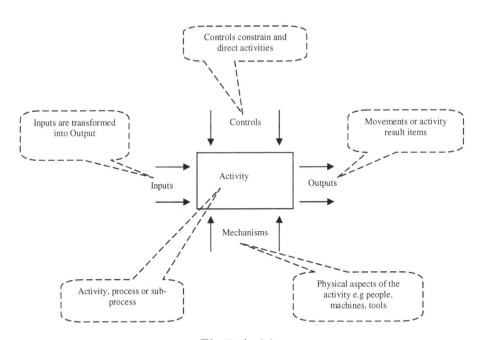

Fig. 5. Activity

a judgement the *judge*. We shall also assume that knowledge sources network is divided into external sources, i.e., sources of knowledge that are also available to other judges, internal sources, which are only available to the specific judge in question.

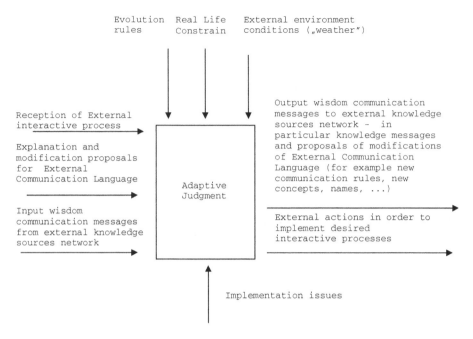

Fig. 6. The first level of the model

The first level of the model is presented in Figure 6. Of course, successive levels of the model are more complex. Its details may depend on the assumed paradigms for the implementation of adaptive judgement. However, these details should include such elements as:

1. Learning of the *External Communication Language* understood as a language based on concepts used to communicate and process knowledge with a network of external sources of knowledge;
2. Learning of the *Internal Communication Language* understood as a hierarchy of meta-languages based on concepts used to process and improve External Communication Language and a language based on concepts used to communicate and process knowledge with a network of internal sources of knowledge;
3. Receiving in memory signals from signal receptors and interactive processes and expressing their significance in the *External Communication Language and the Internal Communication Language*;
4. Planning the current priorities for internal actions (mainly related to the processing of wisdom) on the basis of an assessment in relation to the hierarchy of values controlling the adaptive judgement process;
5. Selection of fragments of ordered knowledge (hierarchies of information and judgement strategies) sufficient to take a decision at the planned time (a decision here is understood as commencing interaction with the environment or selecting the future course to resolve the problem);

6. Output wisdom communication messages to external knowledge sources network, in particular, knowledge messages and proposals of modifications of the External Communication Language (e.g., new communication rules, new concepts, names);
7. External actions in order to implement the desired interactive processes.

All elements occurring in the above list are very complex and important but the following two problems are particularly important for adaptive judgement computational models:

1. *Concept learning and integration* - this is the problem of computational models for implementation of learning concepts important for the representation, processing and communicating of wisdom and, in particular, this relates to learning of concepts improving the quality of approximation of the integration of incomplete local perceptions of a problem (arising during local assimilation and processing of vague and incomplete concepts (see, *e.g.*, [78,79])).
2. *Judge hierarchy of habit habits controls* - this is the problem of computational models for implementation of process of the functioning of a hierarchy of habit controls by a judge controlling the judgement process in an adaptive way.

Now, we sketch the idea of a framework for solution of the problem of implementation of judge hierarchy of habit controls. In this paper, we treat a concept of *habit* as an elementary and repeatable part of behavioral pattern. In this context, the meaning of elementary should be considered by comparison to the required reasoning (knowledge usage) complexity necessary for the behavioral pattern implementation. In other words, by a habit we mean any regularly repeated behavioral pattern that requires little or no reasoning effort (knowledge usage). In general, any behavioral pattern could be treated as a sequence of habits and other activities which use knowledge intensively. Among such activities those leading to new habits are especially important. We assume that such habit processing is controlled by so-called *habit controls* which support the following aspects of adaptive judgement process for a considered situation by a *judge*:

1. *Continuous habit prioritization* to be used in a particular situation after identification of habits. This is a prioritization from the point of view of the following three criteria:
   - The predicted consequences of the phenomena observed in a considered situation;
   - Knowledge available to a judge;
   - The actual plans of a judge's action.
2. *Knowledge prioritization* is used if we do not identify any habit to be used in a considered situation, then we have to make prioritization of pieces of available knowledge which could be used to choose the best habit or for a construction of a new habit for the considered situation.
3. *Habit control assessment* for continuous improvement of adaptive judgement process and for construction of new habits and habit controls.

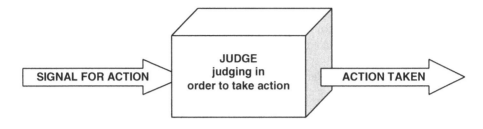

Fig. 7. Judge diagram

As it can be seen from the above considerations, one of the key components of wistech, judge hierarchy of habit control, is essential for optimal decision making and is closely correlated with the knowledge held and interactions with the environment. Judge hierarchy also means the desire of the judge to satisfy his/her needs in interactions with his/her environment. Put very simply, the judge receives and sends out signals according to the diagram presented in Figure 7.

The interior of the box is the place for the judge to process signals and to take an action. By the judge environment adaptation we understand the interaction of the following two adaptive processes:

1. *adaptation of the environment*, in which the judge *lives* to the *needs and objectives* of the judge so as to best fit the *needs and objectives* of the environment,
2. *adaptation of the internal processes taking place in a judge* in such a way as to best realize his/her *needs and objectives* based on the resources available in the environment.

The judge environment adaptation is the basis for computational models of judge learning. The key part of this is the evolution of judge hierarchy of habit controls. The judge hierarchy of habits controls constitutes a catalyst for evolutionary processes in the environment, and also constitutes an approach to expressing various paradigms of computation models to be used in the machine implementation of this concept. For example, these paradigms can be based on the metaphorically understood principle of Newtonian dynamics (e.g., action = reaction), thermodynamics (e.g., increase in entropy of information), quantum mechanics (the principle of it being impossible to determine *location and speed simultaneously*) and quantum computational models [44], psychology (e.g., based on metaphorical understanding of Maslow's hierarchy of needs; see also [53,80,40]). Particularly worthy of attention in relation to wistech is the metaphoric approach to Maslow's hierarchy of needs in reference to the abstractly understood community of agents. Put simply, this hierarchy looks as in Figure 8. It could be used for direct constructions of computational models of judge hierarchy of habit controls.

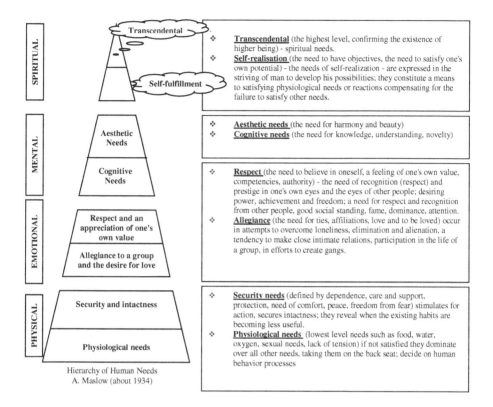

Fig. 8. The Maslow Hierarchy of human needs (about 1934) as an example of judge hierarchy of habit controls

2.5 A Rough-Granular Computing Approach to Wistech Computation Models

In this section, we outline basic ideas for the rough-granular approach to wisdom.

2.5.1 Evolution of Reasoning Computation Models in the Rasiowa–Pawlak School

By the Rasiowa–Pawlak school we mean a continuation of approaches to computational models of approximate reasoning developed by Rasiowa [86], Pawlak [74,87], and their students. In some sense, it is a continuation of ideas initiated by Leibniz, Boole and currently continued in a variety of forms over the world. Of course, the Rasiowa–Pawlak school is also some kind of continuation of the Polish School of Mathematics and Logics. The achievements of this school led to the development of the modern understanding of the basic computational aspects of logic, epistemology, ontology, foundations of mathematics and natural deduction (S. Banach, S. Eilenberg, R. Ingarden, S. Jaśkowski, K. Kuratowski,

S. Leśniewski, A. Lindenbaum, J. Łukasiewicz, S. Mazur, A. Mostowski, H. Rasiowa, R. Sikorski, W. Sierpiński, A. Tarski, S. Ulam, and many others). Two fundamental tools of the Rasiowa–Pawlak school are the following:

- *Computation models of a logical concept (especially of such concepts as deduction or algebraic many-valued models for classical, modal, and constructive mathematics).*

    The Rasiowa–Pawlak approach is based on the method of treating the sets of logically equivalent statements (or formulas) as abstract algebras known as the Lindenbaum–Tarski algebras.
- *Computation models of vague concept.*

    Łukasiewicz originally has proposed to treat uncertainty (or vague concepts) in logic as concepts of many-valued logic. However, software developed for today's computers is based on two-valued Boolean algebra. Therefore it is more practical to treat uncertainty and vagueness using the classical logic concept based on two-valued Boolean algebra. The concept of a rough set introduced by Pawlak [74] and developed in the Rasiowa–Pawlak school is based on the classical two-valued logic and, hence, the rough set approach is important and suitable for the applications mentioned above. The rough set approach intended to deal with uncertainty and vagueness has been developed to deal with uncertainty and vagueness. The rough set approach makes it possible to reason precisely about approximations of vague concepts. These approximations are tentative, subjective, and varying accordingly to changes in the environment [75,76,77,8].

Both the above mentioned fundamental tools can be applied in many contexts. It is interesting to illustrate evolution of the both above fundamental tools from the Rasiowa–Pawlak school perspective (see Figure 9 and Figure 10).

2.5.2   Rough-Granular Computing (RGC)

Solving complex problems by multi-agent systems in distributed environments requires approximate reasoning methods based on new computing paradigms. One such emerging recently computing paradigm is RGC. Computations in RGC are performed on information granules representing often vague, partially specified, and compound concepts delivered by agents engaged in tasks such as knowledge representation, communication with other agents, and reasoning.

We discuss the rough-granular approach for modeling computations in complex adaptive systems and multiagent systems.

Information granules are any objects constructed when modeling of computations, and in approximating compound concepts, and approximate reasoning about these concepts. Information granules are constructed in an optimization process based on the minimal length principle. This process is aiming at constructing approximations of concepts satisfying some (vague and/or uncertain) constraints. Examples of information granules are information systems and decision systems, elementary information granules defined by indiscernibility neighborhoods, families of elementary granules (e.g., partitions and coverings),

Domain & Operators	Natural Numbers Calculus	Algebra of subsets	Boolean Algebra	Logical concepts in Lindenbaum – Tarski algebra	Semantical models of constructive mathematics	Topoi	Wisdom Granular Computing for a given application domain
X < Y	X is smaller than Y	X is a subset of Y	X is smaller than Y in Boolean algebra	Y can be deduced from Y	Logical value of X is smaller than logical value of Y in a Heyting algebra	Morphism from X to Y	Wisdom granule Y is a consequence of wisdom granule X in the domain
0	Zero	Empty set	The smallest element	Falsity	0 in a Heyting algebra	Initial element	The smallest wisdom granule in the domain
1	One	Full set	The biggest element	Truth	1 in a Heyting algebra	Final element	The biggest wisdom granule in the domain
+	Addition	Join of two sets	Maximum	Disjunction	Maximum	Coproduct	Relative coproduct of two wisdom granules
*	Multiplication	Intersection of two sets	Minimum	Conjunction	Minimum	Product	Relative product of two wisdom granules
XY	Exponentiation of X to power Y	Join of (−Y) and X	Join of (−Y) and X	Implication (Y implies X)	Relative pseudo – complementation Y→X in a Heyting algebra	Object corresponding to all morphisms from Y to X	Granule corresponding to all consequences from granule Y to granule X
Mod (X)	Modulo X calculus	Quotient algebra of the filter generated by set X	Quotient Boolean algebra of the filter generated by set X	Lindenbaum – Tarski algebra for a theory generated by a set of axioms X	Models for a theory generated by axioms X	Category of sheaves over X	All consequences from a given granule X
Logical values	True False	True False	True False	Algebra of logical values	Elements of Heyting Algebra	Subobject classifier	Identification of subgranules of granules

ANCIENT	CONTEMPORARY	FUTURE

Fig. 9. Evolution of computational models of logical concepts from the Rasiowa–Pawlak school perspective (the last column is hypothetical for a further research)

relational structures obtained by granulation of objects or classes of relational structures (representing structured objects and their classes), elementary and compound patterns (e.g., clusters of already defined patterns, hierarchical or behavioral patterns, protocols of cooperation), decision rules on different levels, interaction patterns, sets of decision rules, strategies of searching for relevant features, rough inclusions, approximation spaces, fusion operations on information granules, negotiation and conflict resolution strategies, classifiers constructed for compound and vague concepts. We discuss some aspects of rough set-based foundations for information granule calculi and methods for inducing relevant information granule constructions from data and background knowledge.

RGC has been applied for solving complex problems in areas such as identification of objects or behavioral patterns by autonomous systems, web mining, and sensor fusion (see, e.g., [3,5,6,7,8,22,68,69,93,94,95,98,99,100,101,102,105,106]).

2.5.3 Vague Concept Approximation

The RGC methods should make it possible to construct vague concept approximation and to perform approximate reasoning about such concepts. There is a long debate in philosophy on vague concepts [54]. Nowadays, computer scientists are also interested in vague (imprecise) concepts, e.g., many intelligent systems

Evolution of AI models of computing in the Rasiowa – Pawlak School

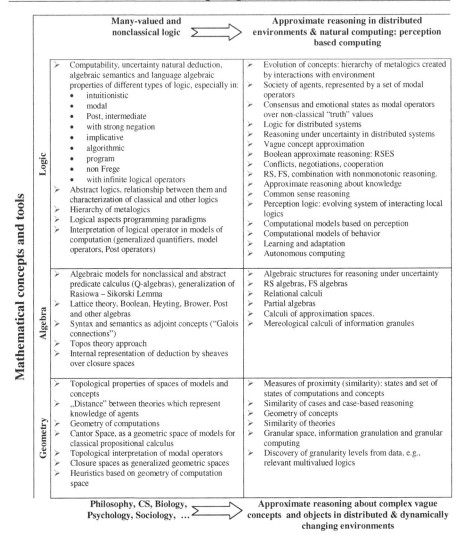

Fig. 10. Evolution of computational models of vagueness from the Rasiowa–Pawlak school perspective

should satisfy some constraints specified by vague concepts. Hence, the problem of vague concept approximation as well as preserving vague dependencies (especially in dynamically changing environments) is important for such systems.

Lotfi Zadeh [120] introduced a very successful approach to vagueness. In this approach, sets are defined by partial membership in contrast to crisp membership used in the classical definition of a set.

Rough set theory [74] expresses vagueness not by means of membership but by employing the boundary region of a set. If the boundary region of a set is empty it means that a particular set is crisp, otherwise the set is rough (inexact). The non-empty boundary region of the set means that our knowledge about the set is not sufficient to define the set precisely. Inductive extensions of approximation spaces and adaptive concept learning yield better understanding of vague concepts [8]. A discussion on vagueness in the context of fuzzy sets and rough sets can be found in [88].

The central role of the rough set approach in RGC comes from the necessity of modeling different interactions between agents. These interactions are operations on information granules. Different agents use different languages to describe information granules. Information granules expressed in one language often cannot be expressed precisely in another language. Hence, there is the need for developing methods which could be used by judges in approximation of (vague) concepts (partially) specified by other judges. Initially, the approximation spaces, which can be treated as information granules, were introduced for decision tables (samples of objects). The assumption was made that the partial information about objects is given by values of attributes and that the approximations of subsets of objects from the universe restricted to sample have been defined on the basis of such information about objects [74]. Starting at least from the early 1990s, many researchers have been using the rough set approach for constructing classification algorithms (classifiers) defined over extensions of samples. This is based on the assumption that available information about concepts is partial.

In recent years, there have been reported methods based on approximation spaces and operations on approximation spaces to develop for approximation of concepts over the extensions of samples (see, e.g., [8,102]). Among the basic operations on extension of samples related to concept approximation are inductive extensions of approximation spaces (see, e.g., [8,102]). Neighborhoods of objects are the basic components of approximation spaces. They are defined by the available information about objects and rough inclusion functions between sets of objects. Observe that searching for relevant (to approximation of concepts) extensions of approximation spaces requires tuning more parameters than in the case of approximation of concepts on samples. The important conclusion is that the inductive extensions defining classification algorithms (classifiers) are defined by arguments "for" and "against" the concepts. Each argument is defined by a tuple consisting of a degree of inclusion of objects into a pattern and a degree of inclusion of the pattern into the concepts. In the case of rule-based classifiers, patterns can be interpreted as the left-hand sides of decision rules. The arguments are discovered from available data and can be treated as the basic information granules used in the concept approximation process. For any new object, it is possible to check the satisfiability of arguments and to select the arguments which are satisfied (at least to a satisfactory degree). Such selected arguments are fused by conflict resolution strategies for obtaining the classification decision.

Searching for relevant approximation spaces in the case of approximations over extensions of samples requires discovery of many parameters and patterns including selection of relevant attributes defining information about objects, discovery of relevant patterns for approximated concepts, selection of measures (similarity or closeness) of objects into the discovered patterns for concepts, the structure and parameters of conflict resolution strategy. This causes infeasibility of the searching process in the case of more compound concepts the searching process becomes infeasible (see, e.g., [12,32,81,116]).

We have proposed to use additional domain knowledge as hints in searching for relevant approximation spaces for compound concepts. This additional knowledge is represented by a concept ontology [3,5,6,7,8] including concepts expressed in natural language and some dependencies between them. We assume that the ontology of concepts has a hierarchical structure. Moreover, we assume that for each concept from the ontology a labeled set of examples of objects is given. The labels reflect the degree of membership of objects relative to the approximated concepts. The aim is to discover relevant conditional attributes for concepts on different levels of the hierarchy. Such attributes can be constructed using the so-called production rules, productions, and approximate reasoning schemes (AR schemes, for short) discovered from data (see, e.g. [3,5,6,7,8,100,101,102,105,106]). Searching for relevant arguments "for" and "against" for more compound concepts can be simplified because it can be organized along the derivations over the ontology using the domain knowledge.

It should be mentioned that the searching process for relevant approximation spaces is driven by some selected quality measures. While in some learning problems such measures can be selected in a relatively easy way and remain unchanged during learning, in other learning processes they can only be approximated on the basis of a partial information about such measures received, e.g., as the result of interaction with the environment. This applies to, e.g., adaptive learning. We present an example illustrating the complexity of the searching process for relevant approximation spaces in different tasks of adaptive learning [21]. In our recent projects, we develop methods for adaptation of observation to an agent's scheme, incremental learning, reinforcement learning, and adaptive planning.

Our discussion is presented within the framework of MAS. The main conclusion is that the approximation of concepts in adaptive learning requires new advanced methods for modeling of computations based on information granules. Among them are those which, in particular, will make it possible to approximate the quality measures together with approximation of concepts.

In adaptive learning, the approximation of concepts is constructed gradually and the tentative approximations change dynamically in the learning process where we try to achieve the approximation of the desired quality. In particular, this changes boundary conditions during the learning process in which we attempt to find the relevant approximation of the boundary regions of vague concepts. This is consistent with the requirement of the higher-order vagueness [54] stating that the borderline cases of vague concepts are not crisp sets. This paper is a continuation of our research (see, e.g., [3,5,6,7,8,68,69], [93,94,95,98,99], [100,101,102,105,106]) on approximation spaces and vague concept approximation processes conducted for several years.

In this paper, we concentrate on some issues of RGC relevant to adaptive processes [8,21,62,64]. In particular, we present some basic schemes relevant for adaptive concept learning. The aim is to illustrate the complexity of spaces in which RGC should enable us to construct relevant information granules and to develop searching methods for many compound kinds of relevant information granules. These information granules should make it possible to construct high quality approximation of concepts and to reason efficiently about performed computations. In such granular computations many compound information granules are involved.

Let us consider an example of the measure of approximation quality. When searching for relevant approximation of the compound concepts, methods for constructing appropriate measures are necessary. At a given step of the learning process, only partial information about such a measure is available. On the basis of such information we construct approximation of the measure and we use it for inducing approximation spaces relevant for concept approximation. However, at the next stages of the learning process, it may happen that after receiving new information from the environment, it will be necessary to reconstruct the approximation of the quality measure, and in this way we obtain a new "driving force" in searching for relevant approximation spaces during the learning process. Adaptive learning strategies create information granules of higher level. Evolutionary techniques for modeling computations on such information granules to synthesize relevant learning strategies (toward achieving the given goals) are critical for many applications. In the next two sections, we outline some basic concepts related to agents and their interactions. The agents are called judges to emphasize that among tasks they perform are judgements. We discuss two more examples.

2.5.4   Basic Concepts Relative to Judges

In this section, we discuss basic components of agents (called judges here) which perform their tasks. In particular, they interact with other agents in the environment, approximate vague concepts and have the ability of reasoning about concepts. Each judge can also be treated as an information granule. The basic concepts relative to the judge J are the following:

- *Information granules accessible by J.* N is the set of information granules accessible by J (e.g., neighborhoods of objects, sets defined by the left-hand sides of decision rules, sets defined by classifiers together with the classifiers, granules accessible by J through interaction with other judges, e.g., representing other sources of knowledge). Information granules are all constructive objects definable (accessible, generated) by the judge J which are used by J for representing knowledge, approximate reasoning, and interaction with other judges and/or the environment.

- *Goals (Targets) of J.* There are some goals (targets) by the judge J to be achieved, e.g., preservation of some constraints with some priorities or achievement of a state with a given property. G denotes the granule representing goals for J. In particular, constraints and targets are defined by means of information granules.
- *Environment of J.* ENV_J denotes the set of all judges interacting with J (directly or indirectly).
- *Information Function of J.* $Inf : States(ENV_J) \longrightarrow N$ is the information function about states of the environment from the set

$$States(ENV_J)$$

perceived by J. By N_{Inf} we denote the set $Inf(States(ENV_J))$.
- *Judgemental Strategies of J.* S is the set of judgemental strategies of J. Some examples of judgemental strategies are listed below. First let us introduce some notation. $\models^+_{deg}, \models^-_{deg}$ are binary relations in $N \times N$, called the rough inclusions of J, with the following intended meaning: $u \models^+_{deg} u'$ if and only if the granule u matches the granule u' to a degree at least deg; $u \models^-_{deg} u'$ if and only if the granule u matches the granule u' to a degree at most deg. For simplicity of reasoning we assume that $deg \in [0,1]$. Assume that a set D of information granules (e.g., the set of decision classes for a rule-based classifier) is given. Let t_+, t_- be two thresholds from the interval $[0,1]$ and let

$$N^+(u) = \{u' \in D : \exists deg > t_+ (u \models^+_{deg} u')\}$$

for $u \in N_{Inf}$. Granule u votes "for" granules from $N^+(u)$ (relative to t_+) (see [46]). Let us assume

$$N^-(u) = \{u' \in D : \exists deg < t_- (u \models^-_{deg} u')\}$$

for $u \in N_{Inf}$. Then granule u votes "against" granules from $N^-(u)$ (relative to t_-). We assume that B is a distinguished set of information granules called behavioral patterns of J (e.g., decisions, actions, plans [34,115]) and $Lab : D \longrightarrow B$ is the (partial) labeling function assigning the behavioral patterns to (some) information granules from D.

- S is one of the judgemental strategies of J making it possible to select a particular behavioral pattern as a reaction to the perceived information about the environment. In particular, S uses granules from $Lab(N^+(u))$ and $Lab(N^-(u))$, where $u = Inf(x)$ and x is the current state of the environment, and the labeling of these sets of granules by behavioral patterns. Observe that the strategy S should resolve conflicts arising due to the fact that information granules should satisfy some constraints. For example, some information granules cannot be matched by one information granule to a degree higher than a given threshold t_+.
- *Quality strategy of J.* Q is the quality strategy of J for estimation of the closeness (similarity) between granules. The closeness estimation is

based on arguments "for" and "against" the satisfiability of the compound concept of "closeness" represented by Q. In this judgement J uses relevant granules from available granules representing knowledge accessible for J, often distributed among other judges, as well as the relationships between granules represented by matching degrees.

- *Adaptation strategy of J. Adap* is the adaptation strategy transforming a tuple

$$(N, G, Inf, B, Lab, \models^+_{deg}, \models^-_{deg}, S, Q)$$

into a new such tuple. Observe that judgements performed by J during adaptation can, in particular, lead to construction of new granules (e.g., through cooperation with other judges [2]), changing some strategies such as the matching strategy, the labeling strategy, the selection strategy for relevant behavioral patterns, and the strategy for estimation of closeness of granules. *Adap* can also be changed, e.g., by tuning some of its parameters.

2.5.5 Basic Cycle of Judge

Each judge realizes some goals using behavioral patterns. The basic cycle of each judge J is the following:

1. *Step 1: Initialization.*
   $$(N, G, Inf, B, Lab, \models^+_{deg}, \models^-_{deg}, S, Q) :=$$
   $$(N_0, G_0, Inf_0, B_0, Lab_0, \models^+_{deg,0}, \models^-_{deg,0}, S_0, Q_0).$$
2. *Step 2: Perception granule construction by J representing the current state.*

   $$u := Inf(x);$$

   where u is the granule representing perception by J of the current environment state x.
3. *Step 3: J selects the relevant granules from $N^+(u)$, $N^-(u)$ and performs judgements to select (construct) the relevant behavior b toward achieving the current goal (target).* During selection of b the judge J is also predicting the information $Inf_{pred}(b, x)$ returned from ENV_J as a reaction to the behavior b applied to the current state x of ENV_J. This is realized by another special judgemental strategy of J. By applying S to $Lab(N^+(u))$ and $Lab(N^-(u))$ J searches for a relevant behavior b.
4. *Step 4: Estimation of the closeness.*
   The judge J uses the quality measure Q for estimation of the closeness (similarity) of $Inf_{pred}(b, x)$ and $Inf_{real}(b, x)$ by

   $$Q(Inf_{pred}(b, x), Inf_{real}(b, x)),$$

   where $Inf_{real}(b, x)$ is information about the real reaction of the environment in state x to the behavior b.

5. *Step 5: J uses a special judgemental strategy in testing whether the closeness is satisfactory.*

   If the closeness is satisfactory, then J continues from *Step*2; otherwise J goes to the next step.
6. Step 6: Adaptation step.
   $$(N, G, Inf, B, Lab, \models^+_{deg}, \models^-_{deg}, S, Q) :=$$
   $$Adapt(N, G, Inf, B, Lab, \models^+_{deg}, \models^-_{deg}, S, Q).$$
7. Step 7: Go to Step 2.

All constructive objects involved in computations realized by means of the above judgement schemes are information granules.

2.5.6   Remark on Task Solving by Systems of Judges

The above examples illustrate the complexity and richness of the information granule spaces we deal with when modeling adaptive processes and reasoning about such processes. Systems of judges solve tasks by searching in the information granule spaces for information granules satisfying the task specification to a satisfactory degree (not necessarily exactly), i.e., matching information granules representing the task specification to a satisfactory degree. The requirement of "matching to a degree" used instead of "matching exactly" often makes searching for solutions feasible in information granule spaces [122].

In a number of papers (see, e.g., [99,105,106]), we have developed methods for construction of information granules (satisfying a given specification to a satisfactory degree) by means of operations on information systems called constrained sums. In particular, this approach proved to be general enough for modeling compound spatio-temporal information granules (e.g., information granules representing processes or behavioral patterns specified by vague concepts) and interactions between them.

3   Wistech Network (WN)

In this section, we discuss shortly the organization of cooperation for the projects based on wistech.

3.1   What We Mean by Wistech Network

The huge complexity of the problem of designing effective wistech computation models means that wistech progress significantly depends on forming effective and systematic cooperation between the numerous interdisciplinary teams verifying the Wistech calculation models developed in practical experiments. Moreover, in order to make a really essential progress in wistech it is important to involve the best possible specialists for making it possible to combine in wistech based projects knowledge of such areas as: psychology, sociology, ethics and domain dependent knowledge, e.g., neuroscience, medicine, economics, security, law, robotics, telecommunications, banking. This research, like all other research, requires a significant effort in other fundamental sciences, such as logic, epistemology, ontology, mathematics, computer science, philosophy and others. Of course such activity is very expensive. Moreover, in general, research of this type does not translate directly into economic results. No private company can afford to implement such extensive research by itself. It is also unlikely that there would be any significant commitment by government agencies in the coordination and development of research on such a wide scale. Unfortunately, current attempts at extending the international coordination of such type of research are not effective.

A dilemma therefore arises whether to develop wistech within the framework of expensive and highly risky closed research programs, or to support open programs in which the costs and risk are spread among many entities? It is our opinion that both directions are equally important and the key to the success is an environment for creating and developing harmony mechanisms between open and closed research (see [19]). In [19], among others, the contrasting principles of closed and open innovation are clarified (see Figure 11).

At the current stage of building an environment for creating and developing harmony mechanisms between open and closed research it is very important to develop a powerful framework for effective Open Innovation Wistech R&D network. The current stage of development in wistech above all requires the development of coordinated interdisciplinary basic research with a well-coordinated and easily accessible environment for experiments. Such activities are not possible in hermetically sealed companies, which are paralyzed by security procedures and guided by the criterion of rapid economic return. This is also why it is proposed to start up mechanisms for the systematized and relatively coordinated cooperation of centers interested in developing Wistech under a Wistech Network (WN) cooperating with one another in accordance with jointly perfected open principles based on Open Innovation Principles. It is worth stressing that organizations preferring Closed Innovation Principles may also draw great benefits from active participation in WN. This participation gives the possibility of testing solutions that have little chance of giving rapid market results, and also in the case of the appearance of such opportunities they can be translated into economic results in accordance with the principles accepted. At the same time, in the case of basic research, which in general does not translate directly into market effects, the understanding of progress in basic research gives greater opportunities for developing new market applications of one's own. A further great benefit of active participation in WN should be the possibility of comparing the various paradigms for building calculation models for Wistech. The times have long since gone when people believed that there is only one perfect paradigm in AI. Hybrid solutions adapted to the specific nature of the sphere of application dominate in applications. Hybrid applications themselves also use a variety of construction paradigms in a platform for integrating various approaches. Similarly we are also assuming that the WN environment would be represented in

Contrasting Principles of Closed and Open Innovation	
Closed Innovation Principles	Open Innovation Principles
The smart people in our field work for us.	Not all smart people work for us. We need to work with smart people inside *and* outside of our company.
To profit from R&D, we must discover it, develop it, and ship it ourselves.	External R&D can create significant value; internal R&D in needed to claim some portion of that value.
If we discover it ourselves, we will get it to the market first.	We don't have to originate the research to profit from it.
The company that gets an innovation to the market first will win.	Building a better business model is better than getting to the market first.
If we create the most and the best ideas in the industry, we will win.	If we make the best use of internal and external ideas, we will win.
We should control our intellectual properties (IP), so that our competitors don't profit from our ideas.	We should profit from others' use of our IP, and we should buy others' IP whenever it advances our own business model.

Fig. 11. Contrasting principles of closed and open innovation [19]

the form of a sub-network with various paradigms for the construction of an integration platform. WN would provide the data and criteria to assess the results of experiments used for the assessment of various paradigms. In the remainder of this work we present, among others, a proposal to start up a sub-network based on a paradigm for the integration of various technologies based on an *adaptive rough-granular computing approach* (RGC).

3.2   A Potential Example Scenario of WN Establishment

3.2.1   WN Long-Term Vision and Role
The basic objectives of WN are supporting open innovation and the development of wistech and its applications through:

1. creating new paradigms and trends in Wistech and its applications,
2. creating a platform (e.g. intranet, symposia, training programs, e-learning, etc.) for communication and the exchange of knowledge and experience on the practical applications and achievements of basic research,
3. preparing educational and research programs,
4. starting up projects for specific practical applications, as well as for basic research,

5. establishing the conditions and criteria used to compare the quality of various approaches to Wistech (especially having in mind applications in medicine, economy, agriculture, energy and forex market),
6. popularization of Wistech.

3.2.2 WN Organization and Financial Support

We assume that participation in WN is absolutely voluntary in nature, and WN itself also does not assume any additional financial fees or obligatory participation in conferences. The organization is open in nature and any person or organization can take part in it. The form of organization is based on communities cooperating together, which jointly use and develop open software (see, e.g., http://www.opensource.org/).

At the same time we assume that at some stage WN may take part in commercial projects. The project participants will mutually agree upon the principles for cooperation in every such case. It is expected that in the long-term some products or components created by WN will function according to the principles of open software (e.g. similar to the principles of http://www.opensource.org/). We continue to assume the organization of working groups in the network which would deal with jointly agreed packets of problems and projects.

It is expected in our exemplary scenario that WN will develop in accordance with the stages for development of a mature organization modeled on the ideas of Carnegie Mellon Capability Maturity Model (http://www.sei.cmu.edu/cmm/). This model consists of the six stages presented in Figure 12 and Figure 13.

The basic assumption to WN is the realization of projects financed by WN participants who cover the costs and risk of their own activities in the network. It is also assumed that in WN there will be several specialist centers which will coordinate the activities in individual areas (competency centers), e.g. the multi-agent approach, the rough mereology approach. The coordination work of these centers would be financed from voluntary financial contributions from participants of the group in question. It follows from this that the intensity and quality of work in a given group will to a large degree depend on the level of financial support from participants in the group.

4 Wisdom Engine

We discuss some exemplary projects proposed as pilot projects in development of wistech.

4.1 Wisdom Engine Concept

By wisdom engine we understand a machine system which implements the concept of wisdom. In other words, the basic functions of the wisdom engine would be acquiring, processing, discovering, learning and communicating wisdom. One of the main first objectives of WN can be to create an open international R&D environment for the design and implementation of the concept of *universal domain-independent wisdom engine*. A universal wisdom engine implementation should

Name of stage	Organization	Content
ESTABLISHMENT	Starting up the first projects in the network and defining the principles for the cooperation of the first group of participants who confirm their participation in WN. Starting up the first forms of communication.	Developing the initial catalogue of paradigms for approaches to development of wistech (e.g., multi-agents, evolution, symbolic processing, neural nets, statistics, **adaptive rough granular approach**, formal concepts, ontology engineering, information semiotics, cognitive and epistemological approach, etc., and their combinations).
INITIAL	Developing a common language to describe the concepts relating to starting up, implementing and closing projects in WN.	The preliminary allocation of categorized paradigms for approaches to wistech to their respective competency centers. Allocating a paradigm to a competency center, e.g. multi-agent approach, **adaptive rough granular approach**, etc. This does not mean that at a given competency center only and exclusively this method will be developed. On the contrary, it is assumed that every competency center will develop hybrid solutions combining various approaches. At the same time, a competency center will particularly strongly develop aspects relating to the paradigms allocated to this center.

Fig. 12. Six stages of the Carnegie Mellon Capability Maturity Model

be independent of any specific application domain. At the same time, functionality of the universal wisdom engine should enable the configuration and tuning of modules for it in the form of a series of products dependent on specific application domains such as, e.g., medicine, economics, stock market, forex market, security, law, tourism, telecommunications, banking, job market. In particular universal wisdom engine should be able to learn domain knowledge by reading, discussing with experts and gathering wisdom from experience. Of course, the design and implementation of a universal wisdom engine is an extremely difficult task and probably unrealistic today in a short term. First of all, we have to do some experiments with several application domains and several different paradigms for wistech implementation. Based on an analysis of the results of such experiments we can create a more general wistech ontology which should provide a better formal framework for the implementation of a universal wisdom engine.

REPEATABLE	Establishing the principles for selecting good practices specific for the implementation of a project in wistech, designed to repeat the successes of projects realized in similar conditions and to avoid failures. Establishing the list of first conditions and criteria used to compare the quality of various approaches to wistech.	Establishing the mutually tied objectives to achieve at the individual competency centers in order to verify the effectiveness and possibilities of developing various approaches.
DEFINED	Putting in writing and the effective implementation of a list of joint standards for organization and management of projects specific to wistech, that will be binding for the WTN community.	Starting up the first projects realized in the common standards by a variety of centers within the WN
MEASURABLE	Enhancing the standards arising at the previous stage to include sets of measurable indices used to verify and optimize the benefits to costs of wistech projects.	Starting up mechanisms for competitiveness between communities working on various approaches to wistech in the network.
CONTINUOUS IMPROVEMENT	Enhancing the standards and indices defined at the MEASURABLE stage to set out in writing and effectively implement procedures for continuously improving the functioning of WN.	Developing the optimum methods for harmonious co-operation between WN and commercial companies.

Fig. 13. Six stages of the Carnegie Mellon Capability Maturity Model (continued)

Thus, it is assumed that in parallel with the work on a universal concept of a wisdom engine, work would also be conducted on utilizing the wisdom engine in selected areas of application, e.g., medicine, economics, stock market, forex market, security, law, tourism, telecommunications, banking, or job market. The long-term vision is as follows: "wisdom engineers" will receive the task to create the configuration for the wisdom engine for applications in a specific field of life, and then, after having carried out the necessary analytical and design work, to configure the wisdom engine and to enter the necessary initial data. The wisdom engine should have properties for self-growth and adaptation to changing conditions of its environment, as well as advances in wisdom in the fields of application. This is why one should strongly emphasize the planned property of automatic adaptation of the system – a feature not taken into account in the construction of the numerous systems in the past that were intended to perform similar tasks. A classic example here is the long-standing MYCIN project implemented by Stanford University.

The implementation of the idea expressed by the wisdom equation is very difficult and it would be unreasonable to expect its full implementation in a short period of time. We assume that the creativity cycle for the first product prototypes implementing this concept would take several years of intensive work with cooperation of product managers, scientists, engineers, programmers and domain experts. On the other hand, it is not desirable to implement such long projects without any clear interim effects. This is why we assume that the wisdom engine implementation project would go through several phases. For example, initially we assume they will go through five phases in the implementation of the wisdom engine. We propose a route, to achieving the target wisdom engine products through continuously improving intermediary products that meet successive expansions in functionality. The five phases are called as follows:

1. Summary,
2. Spider,
3. Conceptual Clustering and Integration,
4. Wisdom Extraction,
5. Wisdom Assistant.

The effect of each of these phases will be a prototype product that after acceptance would be interesting for the WN community.

Stated in simple terms the functional effects of the individual phases would be as presented in Figure 14.

4.2   Examples of Wisdom Engine Domain-Dependent Product Lines

The above five phases (i.e., Summary, Spider, Conceptual Clustering and Integration, Wisdom Extraction, and Wisdom Assistant) should be applied to several directions for potential product lines which would be developed in the WN. Of course, it can theoretically be any product relating to applications in robotics, unmanned aircraft, space rockets, etc. However, if we wish to have as many people as possible cooperating in the WN, then the product lines must be chosen so that experimenting with them does not prove expensive. On the other hand, these product lines must be sufficiently attractive so as to interest as many people as possible. We propose that these product lines relate to applications in such areas as medicine, economics, the stock market, forex market, security, law, tourism, telecommunications, banking, job market and others.

The list of products that could be expanded in accordance with the above scheme is potentially unlimited. The proposals for the descriptions of specific

Phase	Summary	Spider	Conceptual Clustering and Integration	Adaptive Wisdom Extraction	Adaptive Wisdom Assistant
Key functions	key concept extraction	document searching related to key concept and indexing documents	document clustering based on key concept and Integration	extracting data structures, information structures, knowledge structures and adaptive wisdom structures from documents, in particular generating thesauruses, conceptual hierarchies and constraints to acceptable solutions	user query / answering processing in order to support users in solving their problems as effectively as possible

Fig. 14. Functional effects of the individual phases

Phase/Product	Summary	Spider	Conceptual Clustering	Wisdom Extraction	Wisdom Assistant
Document Manager	Document Summary	Document Spider	Document Conceptual Clustering	Document adaptive wisdom Extraction	Document adaptive wisdom Assistant
Job Market	Job Market Summary	Job Market Spider	Job Market Conceptual Clustering	Job Market adaptive wisdom Extraction	Job Market adaptive wisdom Assistant
Brand Monitoring	Brand Monitoring Summary	Brand Monitoring Spider	Brand Monitoring Conceptual Clustering	Brand Monitoring adaptive wisdom Extraction	Brand Monitoring adaptive wisdom Assistant
World Communication	World Communication Summary	World Communication Spider	World Communication Conceptual Clustering	World Communication adaptive wisdom Extraction	World Communication adaptive wisdom Assistant
World Forex	World Forex Summary	World Forex Spider	World Forex Conceptual Clustering	World Forex adaptive wisdom Extraction	World Forex adaptive wisdom Assistant
World Stock Market	World Stock Market Summary	World Stock Market Spider	World Stock Market Conceptual Clustering	World Stock Market adaptive wisdom Extraction	World Stock Market adaptive wisdom Assistant
World Tourist	World Tourist Summary	World Tourist Spider	World Tourist Conceptual Clustering	World Tourist adaptive wisdom Extraction	World Tourist adaptive wisdom Assistant
Physician	Physician Summary	Physician Spider	Physician Conceptual Clustering	Physician adaptive wisdom Extraction	Physician adaptive wisdom Assistant
Lawyer	Lawyer Summary	Lawyer Spider	Lawyer Conceptual Clustering	Lawyer adaptive wisdom Extraction	Lawyer adaptive wisdom Assistant
Economy Monitoring	Economy Monitoring Summary	Economy Monitoring Spider	Economy Monitoring Conceptual Clustering	Economy Monitoring adaptive wisdom Extraction	Economy Monitoring adaptive wisdom Assistant

Fig. 15. Proposed products

products, included in the later part of this report, should be treated as flexible and primarily constitute material for discussion, and not a final decision. On the other hand, the list of products described is not entirely accidental in nature.

Product / Phase	Summary	Spider	Conceptual Clustering	Adaptive Wisdom Extraction	Adaptive Wisdom Assistant
Document Manager	automatic summarizing of a document and groups of documents, the contents of which are not connected with any specific field	automatic searching and downloading of any documents	conceptual clustering of documents on any subject	extracting data structures, information structures, knowledge structures and adaptive wisdom structures from documents, in particular generating thesauruses, conceptual hierarchies and constraints to acceptable solutions in any domain	general user query / answering processing in order to support users in solving their problems as effectively as possible
Job Market	automatic summarizing of a document and groups of documents relating to job market, carried out from the perspective of the following groups of users: potential employers and potential employees	automatic searching and downloading of documents relating to job market, carried out with particular emphasis on the needs of the following groups of users: potential employers and potential employees	conceptual clustering of documents relating to job market with particular emphasis on the specific nature of queries submitted by the following types of users: potential employers and potential employees	extracting data structures, information structures, knowledge structures and adaptive wisdom structures from documents, in particular generating thesauruses, conceptual hierarchies and constraints to acceptable solutions in job market domain	job market related to user query / answering processing in order to support users in solving their problems as effectively as possible

Fig. 16. Functionality of individual products

This is because they form a certain logical continuity, connected both with the degree of difficulty in successive products and current preferences resulting from the previous experiences of the human resources that would be engaged to carry out the work on individual products. The initial selection of product lines is as follows:

- Document Manager,
- Job Market,
- Brand Monitoring,
- World Communication,
- World Forex,
- World Stock Market,
- World Tourist,
- Physician,
- Lawyer,
- Economy Monitoring.

Product / Phase	Summary	Spider	Conceptual Clustering	Adaptive Wisdom Extraction	Adaptive Wisdom Assistant
Brand Monitoring	automatic summarizing of a document and groups of documents relating to brand, carried out from the perspective of the following groups of users: brand owners, detectives looking for frauds, buyers	automatic searching and downloading of documents relating to brand, carried out with particular emphasis on the needs of the following groups of users: brand owners, detectives looking for frauds, buyers	conceptual clustering of documents relating to brand with particular emphasis on the specific nature of queries submitted by the following types of users: brand owners, detectives looking for frauds, buyers	extracting data structures, information structures, knowledge structures and adaptive wisdom structures from documents, in particular generating thesauruses, conceptual hierarchies and constraints to acceptable solutions in brand monitoring domain	brand monitoring related to user query / answering processing in order to support users in solving their problems as effectively as possible
World Communication	automatic summarizing of a document and groups of documents relating to communication, carried out from the perspective of people looking for optimal connections	automatic searching and downloading of documents relating to communication carried out with particular emphasis on the needs of the following groups of users: people looking for optimal connections	conceptual clustering of documents relating to communication with particular emphasis on the specific nature of queries submitted by the following types of users: people looking for optimal connections	extracting data structures, information structures, knowledge structures and adaptive wisdom structures from documents, in particular generating thesauruses, conceptual hierarchies and constraints to acceptable solutions in communication domain	communication related to user query / answering processing in order to support users in solving their problems as effectively as possible

Fig. 17. Functionality of individual products (continued)

This initial selection for the product list generates several dozen of products that would be the effect of work on the individual phases of implementing each of the products, i.e., Summary, Spider, Conceptual Clustering, Wisdom Extraction, Wisdom Assistant. We present the proposed products in Figure 15.

The scope of the program described in this paper should be considered as dynamic and more as a basis for further discussion than a final version of the specific definitions of the projects. This is why the innovative ideas presented and the vision for their implementation do not contain any detailed cost benefits analysis. It will only be possible to specify revenues, costs and cash flow forecasts with any accuracy after the planned scope of work and the role of the WN has stabilized. As there are as yet no final decisions on the scope of operations or role of the WN, this means that at the current stage it is impossible to precisely estimate the planned requirements for human resources. This is why in this document we present only the general human resources requirements and a description of the general mechanisms for acquiring these resources to implement WN.

Each of these products would have their own individual functionality which would result from adapting the wisdom engine to the specific characteristics of their specialist fields. Figure 16 and Figure 17 show the functionality of the individual products.

5 Conclusions

We have discussed the main features of wistech and its importance for further progress in the development of intelligent systems. The proposed approach is based on Rough Granular Computing (RGC).

One of the central problems of science today is to develop methods for approximation of compound vague concepts and approximate reasoning about them [32,81].

Today, we do not have yet satisfactory tools for discovery of relevant patterns for approximation of compound concepts directly from sample objects. However, we have developed methods for compound concept approximation using sample objects and domain knowledge acquired from experts (this is the approach pioneered by Zdzisław Pawlak in [73]). The performed experiments based on approximation of concept ontology (see, e.g., [3,5,6,7,8,22,68,69,70], [78,79,93,94,95,98,99], [100,101,102,105,106]) showed that domain knowledge enables to discover relevant patterns in sample objects for compound concept approximation. Our approach to compound concept approximation and approximate reasoning about compound concepts is based on the rough-granular approach.

One of the RGC challenges is to develop approximate reasoning techniques for reasoning about dynamics of distributed systems of judges. These techniques should be based on systems of evolving local perception logics rather than on a global logic [94,95]. Approximate reasoning about global behavior of judges' system is infeasible without methods for approximation of compound vague concepts and approximate reasoning about them. One can observe here an analogy to phenomena related to the emergent patters in complex adaptive systems [21].

Let us observe that judges can be organized into a hierarchical structure, i.e., one judge can represent a coalition of judges in interaction with other agents existing in the environment [2,56,62]. Such judges representing coalitions play an important role in hierarchical reasoning about behavior of judges' populations. Strategies for coalition formation and cooperation [2,62,64] are of critical importance in designing systems of judges with dynamics satisfying to a satisfactory degree a given specification. Developing strategies for discovery of information granules representing relevant (for the given specification) coalitions and cooperation protocols is another challenge for RGC.

RGC will become more and more important for analysis and synthesis of the discussed compound adaptive processes. The impact of RGC on real-life applications will be determined by techniques based on the rough-granular approach to modeling of relevant computations on compound information granules and methods for approximate reasoning about complex adaptive processes over such information granules. RGC techniques for modeling of complex processes will also have impact on the development of new non-conventional computation models.

Acknowledgments

The research of Andrzej Jankowski was supported by Institute of Decision Process Support. The research of Andrzej Skowron has been supported by the grant 3 T11C 002 26 from Ministry of Scientific Research and Information Technology of the Republic of Poland.

Many thanks to Professors James Peters and Anna Gomolińska for their incisive comments and for suggesting many helpful ways to improve this article.

References

1. G. Antoniou, F. van Harmelen, A Semantic Web Primer (Cooperative Information Systems) The MIT Press, 2004.
2. R. Axelrod, The Complexity of Cooperation, Princeton, NJ: Princeton University Press, 1997.
3. A. Bargiela, W. Pedrycz, Granular Computing: An Introduction, Dordrecht: Kluwer Academic Publishers, 2003.
4. J. Barwise, J. Seligman, Information Flow: The Logic of Distributed Systems, Cambridge University Press, 1997.
5. J. G. Bazan, J. F. Peters, and A. Skowron, Behavioral pattern identification through rough set modelling, in [71], pp. 688–697.
6. J. Bazan, A. Skowron, On-line elimination of non-relevant parts of complex objects in behavioral pattern identification, in [111], pp. 720–725.
7. J. Bazan, A. Skowron, Classifiers based on approximate reasoning schemes, in Monitoring, Security, and Rescue Tasks in Multiagent Systems (MSRAS 2004), B. Dunin-Keplicz, A. Jankowski, A. Skowron, and M. Szczuka, Eds., Advances in Soft Computing, pp. 191-202, Heidelberg: Springer, 2005.
8. J. Bazan, A. Skowron, R. Swiniarski, Rough sets and vague concept approximation: From sample approximation to adaptive learning, Transactions on Rough Sets V: Journal Subline, Lecture Notes in Computer Science, vol. 3100, pp. 39-63, Heidelberg: Springer, 2006.
9. R. Baeza-Yates, B. Ribeiro-Neto, Modern Information Retrieval, Addison Wesley, 1999.
10. M. W. Berry, Survey of Text Mining : Clustering, Classification, and Retrieval, Springer, 2003.
11. R. Brachman, H. Levesque, Knowledge Representation and Reasoning, Morgan Kaufmann, 2004.
12. L. Breiman, Statistical modeling: The two Cultures, Statistical Science 16(3) (2001) 199–231.

13. S. Brin, L. Page, The Anatomy of a Large-Scale Hypertextual Web Search Engine, Stanford University, 1998.
14. C. Carpineto, G. Romano, Concept Data Analysis: Theory and Applications, John Wiley & Sons, 2004.
15. N. L. Cassimatis, A cognitive substrate for achieving human-level intelligence, AI Magazine 27(2) (2006) 45-56.
16. N. L. Cassimatis, E. T. Mueller, P. H. Winston, Achieving human-level intelligence through integrated systems and research, AI Magazine 27(2) (2006) 12-14.
17. S. Chakrabarti, Mining the Web: Analysis of Hypertext and Semi Structured Data, The Morgan Kaufmann Series in Data Management Systems, Morgan Kaufmann, 2002.
18. Mu-Y. Chen, An-P. Chen, Knowledge management performance evaluation: a decade review from 1995 to 2004, Journal of Information Science 32 (1) 2006 17-38.
19. H. W. Chesbrough, Open Innovation: The New Imperative for Creating and Profiting from Technology, Cambridge MA: Harvard Business School Publishing, 2003.
20. J. Coleman, Introducing Speech and Language Processing (Cambridge Introductions to Language and Linguistics), Cambridge University Press, 2005.
21. A. Desai, Adaptive complex enterprises, Communications ACM 48(5) (2005) 32-35.
22. P. Doherty, W. Łukaszewicz, A. Skowron, and A. Szałas, Knowledge Engineering: A Rough Set Approach, Studies in Fuzziness and Soft Computing, vol. 202, Heidelberg: Springer, 2006.
23. R. Dornfest, T. Calishain, Wistech Network Hacks O'Reilly Media, Inc.; 2004.
24. R. Duda, P. Hart, and R. Stork, Pattern Classification, New York, NY: John Wiley & Sons, 2002.
25. B. Dunin-Kęplicz, A. Jankowski, A. Skowron, M. Szczuka, Monitoring, Security, and Rescue Tasks in Multiagent Systems (MSRAS'2004), Series in Soft Computing, Heidelberg: Springer, 2005.
26. A. E. Eiben, J. E. Smith, Introduction to Evolutionary Computing, Natural Computing Series, Springer, 2003.
27. E. Feigenbaum, J. Feldman (Eds.), Computers and Thought, New York: McGraw Hill, 1963.
28. S. Feldman, Why Search is Not Enough (white paper), IDC, 2003.
29. S. Feldman, Enterprise Search Technology: Information Disasters and the High Cost of Not Finding Information (Special IDC Report), IDC, 2004.
30. D. Fensel, Ontologies: A Silver Bullet for Knowledge Management and Electronic Commerce, Springer, 2003.
31. K. D. Forbus, T. R. Hinrisch, Companion cognitive systems: A step toward human-level AI, AI Magazine 27(2) (2006) 83-95.
32. M. Gell-Mann, The Quark and the Jaguar, NY: Freeman and Co., 1994.
33. J. H. Friedman, T. Hastie, R. Tibshirani, The Elements of Statistical Learning: Data Mining, Inference, and Prediction, Heidelberg: Springer, 2001.
34. M. Ghallab, D. Nau, and P. Traverso, Automated Planning: Theory and Practice, CA: Morgan Kaufmann, 2004.
35. S. Ghemawat, H. Gobioff, Shun-Tak Leung, The Wistech Network File System, Wistech Network, 2005.
36. D. E. Goldberg, Genetic Algorithms in Search, Optimization, and Machine Learning, Addison-Wesley Professional, 1989.

37. A. Gomez-Perez, O. Corcho, M. Fernandez-Lopez, Ontological Engineering with examples from the areas of Knowledge Management, e-Commerce and the Semantic Web (Advanced Information and Knowledge Processing), Springer, 2004.
38. R. Granger, Engines of the brain: The computational instruction set of human cognition, AI Magazine 27(2) (2006) 15-31.
39. S. Grimes, The Developing Text Mining Market, A white paper prepared for Text Mining Summit 2005, Boston, June 7-8 2005, Alta Plana Corporation, 2005.
40. A. OHagan, C. E. Buck, A. Daneshkhah, J. R. Eiser, P. H. Garthwaite, D. J. Jenkinson, J. E. Oakley, T. Rakow, Uncertain Judgements: Eliciting Expert Probabilities, Wiley, New York, 2006.
41. J. Heaton, Programming Spiders, Bots, and Aggregators in Java, Sybex, 2002.
42. K. Hemenway, T. Calishain, Spidering Hacks O'Reilly Media, Inc.; 2003.
43. M. Henzinger, S. Lawrence, Extracting knowledge from the World Wide Web, Wistech Network, 2004.
44. M. Hirvensalo, Quantum Computing, Springer-Verlag, Heidelberg 2001.
45. M. N. Huhns, M. P. Singh, Readings in Agents, Morgan Kaufmann, 1998.
46. P. Jackson, I. Moulinier, Natural Language Processing for Online Applications: Text Retrieval, Extraction, and Categorization (Natural Language Processing, 5), John Benjamins Publishing Co, 2002.
47. Z. Janiszewski, On needs of mathematics in Poland (O potrzebach matematyki w Polsce) (in Polish), In: Nauka Polska. Jej Potrzeby, Organizacja i Rozwój, Warszawa, 1918; see also reprint in Wiadomości Matematyczne VII (1963) 3-8.
48. A. Jankowski, An alternative characterization of elementary logic, Bull. Acad. Pol. Sci., Ser. Math. Astr. Phys.XXX (1-2) (1982) 9-13.
49. A. Jankowski, Galois structures, Studia Logica 44(2) (1985) 109-124.
50. S. Johnson, Dictionary of the English Language in Which the Words are Deduced from Their Originals, and Illustrated in their Different Significations by Examples from the Best Writers, 2 Volumes. London: F.C. and J. Rivington, 1816.
51. R. M. Jones, R. E. Wray, Comparative analysis of frameworks for knowledge-intensive intelligent agents, AI Magazine 27(2) (2006) 57-70.
52. D. Jurafsky, J. H. Martin, Speech and Language Processing: An Introduction to Natural Language Processing, Computational Linguistics and Speech Recognition, Prentice Hall, 2000.
53. D. Kahneman, P. Slovic, A. Tversky, A. (Eds.). Judgement under Uncertainty: Heuristics and Biases, Cambridge University Press, New York, 1982.
54. R. Keefe, Theories of Vagueness, Cambridge, UK: Cambridge Studies in Philosophy, 2000.
55. W. Kloesgen, J. Żytkow, Handbook of Knowledge Discovery and Data Mining, New York: Oxford University Press, 2002.
56. S. Kraus, Strategic Negotiations in Multiagent Environments, Massachusetts: The MIT Press, 2001.
57. J. Lambek, P. J. Scott, Introduction to Higher-Order Categorical Logic (Cambridge Studies in Advanced Mathematics 7), Cambridge University Press, 1986.
58. M. Lamb, Build Your Own Army of Web Bots Within 24 Hours (Army of Web Bots Series) Authorhouse, 2003.
59. P. Langley, H. A. Simon, G. L. Bradshaw, J. M. Zytkow, Scientific Discovery: Computational Explorations of the Creative Processes, MIT Press, 1987.
60. P. Langley, Cognitive architectures and general intelligent systems, AI Magazine 27(2) (2006) 33-44.
61. G. W. Leibniz, New Essays on Human Understanding, Cambridge UP, 1982.

62. J. Liu, Autonomous Agents and Multi-Agent Systems: Explorations in Learning, Self-Organization and Adaptive Computation, Singapore: World Scientific Publishing, 2001.
63. J. Liu, L. K. Daneshmend, Spatial Reasoning and Planning: Geometry, Mechanism, and Motion, Springer, 2003, Hardcover.
64. J. Liu, X. Jin, K. Ch. Tsui, Autonomy Oriented Computing: From Problem Solving to Complex Systems Modeling, Heidelberg: Kluwer Academic Publisher/Springer, 2005.
65. S. MacLane, I. Moerdijk, Sheaves in Geometry and Logic: A First Introduction to Topos Theory (Universitext), Springer, 1994.
66. A. Madhavapeddy, N. Ludlam, Ubiquitious Computing needs to catch up with Ubiquitous Media, University of Cambridge Computer Laboratory, Interceptor Communications Ltd., 2005.
67. C. D. Manning, H. Schütze, H., Foundations of Statistical Natural Language Processing, The MIT Press, 1999.
68. S. H. Nguyen, J. Bazan, A. Skowron, and H. S. Nguyen, Layered learning for concept synthesis, Transactions on Rough Sets I: Journal Subline, Lecture Notes in Computer Science, vol. 3100, pp. 187-208, Heidelberg: Springer, 2004.
69. T. T. Nguyen, Eliciting domain knowledge in handwritten digit recognition, in [71] 762-767.
70. S. K. Pal, L. Polkowski, and A. Skowron (Eds.), Rough-Neural Computing: Techniques for Computing with Words, Cognitive Technologies, Heidelberg: Springer-Verlag, 2004.
71. S. K. Pal, S. Bandoyopadhay, and S. Biswas (Eds.), Proceedings of the First International Conference on Pattern Recognition and Machine Intelligence (PReMI'05), December 18-22, 2005, Indian Statistical Institute, Kolkata, Lecture Notes in Computer Science vol. 3776, Heidelberg: Springer, 2005.
72. T. B. Passin, Explorer's Guide to the Semantic Web Mining Publications, 2004.
73. Pawlak, Z.: Classification of objects by means of attributes, Research Report PAS 429, Institute of Computer Science, Polish Academy of Sciences, ISSN 138-0648, January (1981).
74. Z. Pawlak, Rough Sets: Theoretical Aspects of Reasoning about Data, System Theory, Knowledge Engineering and Problem Solving 9, Dordrecht: Kluwer Academic Publishers, 1991.
75. Z. Pawlak, A. Skowron, Rudiments of rough sets. Information Sciences. An International Journal. 177(1) (2007) 3-27.
76. Z. Pawlak, A. Skowron, Rough sets: Some extensions. Information Sciences. An International Journal. 177(1) (2007) 28-40.
77. Z. Pawlak, A. Skowron, Rough sets and Boolean reasoning. Information Sciences. An International Journal. 177(1) (2007) 41-73.
78. J. F.Peters, Rough ethology: Toward a biologically-inspired study of collective behavior in intelligent systems with approximation spaces. Transactions on Rough Sets III: Journal Subline, Lecture Notes in Computer Science, vol. 3400, pp. 153-174, Heidelberg: Springer, 2005.
79. J. F. Peters, C. Henry, C., Reinforcement learning with approximation spaces. Fundamenta Informaticae 71(2-3) (2006) 323-349.
80. S. Plous, The Psychology of Judgement and Decision Making, McGraw-Hill, New York, 1993.
81. T. Poggio, S. Smale, The mathematics of learning: Dealing with data, Notices of the AMS 50(5) (2003) 537-544.

82. L. Polkowski, A. Skowron, Rough mereology: A new paradigm for approximate reasoning, International Journal of Approximate Reasoning 15 (1996) 333-365.
83. L. Polkowski, S. Tsumoto, T. Y. Lin (Eds.), Rough Set Methods and Applications: New Developments in Knowledge Discovery in Information Systems, Studies in Fuzziness and Soft Computing vol. 56, Physica-Verlag Heidelberg, 2000.
84. G. Pólya, How to Solve It, 2nd ed., Princeton University Press, 1957; see also http://en.wikipedia.org/wiki/How_to_Solve_It.
85. H. Rasiowa, W. Marek, On reaching consensus by groups of intelligent agents, In: Z. W. Ras (Ed.), Methodologies for Intelligent Systems, North-Holland, Amsterdam, 1989, 234-243.
86. H. Rasiowa, Algebraic Models of Logics, Warsaw University, 2001.
87. H. Rasiowa, R. Sikorski, The Mathematics of Metamathematics, Monografie Matematyczne vol. 41, PWN Warsaw, 1963.
88. S. Read, Thinking about Logic. An Introduction to the Philosophy of Logic, Oxford, New York: Oxford University Press, 1995.
89. P. Saint-Dizier, E. Viegas, B. Boguraev, S. Bird, D. Hindle, M. Kay, D. McDonald, H. Uszkoreit, Y. Wilks, Computational Lexical Semantics (Studies in Natural Language Processing), Cambridge University Press, 2005.
90. C. Schlenoff, J. Albus, E. Messina, A. J. Barbera, R. Madhavan, S. Balakirsky, Using 4D/RCS to address AI knowledge integration, AI Magazine 27(2) (2006) 71-81.
91. Scientific Datalink, The Scientific DataLink index to artificial intelligence research, 1954-1984, Scientific DataLink, 1985.
92. Scientific Datalink, The Scientific DataLink index to artificial intelligence research, 1985 Supplement, Scientific DataLink, 1985.
93. A. Skowron, Approximate reasoning in distributed environments, in N. Zhong, J. Liu (Eds.), Intelligent Technologies for Information Analysis, Heidelberg: Springer, pp. 433-474.
94. A. Skowron, Perception logic in intelligent systems (keynote talk), In: S. Blair et al (Eds.), Proceedings of the 8th Joint Conference on Information Sciences (JCIS 2005), July 21-26, 2005, Salt Lake City, Utah, USA, X-CD Technologies: A Conference & Management Company, ISBN 0-9707890-3-3, 15 Coldwater Road, Toronto, Ontario, M3B 1Y8, 2005, pp. 1-5.
95. A. Skowron, Rough sets in perception-based computing (keynote talk), in [71], pp. 21-29.
96. A. Skowron, R. Agrawal, M. Luck, T. Yamaguchi, O. Morizet-Mahoudeaux, J. Liu, N. Zhong (Eds.), Proceedings of the 2005 IEEE/WIC/ACM International Conference on WEB Intelligence, Compiegne, France, September 19-22, 2005, IEEE Computer Society Press, Los Alamitos, CA, 2005, pp. 1-819.
97. A. Skowron, J.-P. Barthes, L. Jain, R. Sun, P. Morizet-Mahoudeaux, J. Liu, N. Zhong (Eds.), Proceedings of the 2005 IEEE/WIC/ACM International Conference on Intelligent Agent Technology, Compiegne, France, September 19-22, 2005, IEEE Computer Society Press, Los Alamitos, CA, 2005, pp. 1-766.
98. A. Skowron, J. Stepaniuk, Tolerance approximation spaces, Fundamenta Informaticae 27 (1996) 245-253.
99. A. Skowron, J. Stepaniuk, Information granules: Towards foundations of granular computing, International Journal of Intelligent Systems 16(1) (2001) 57-86.
100. A. Skowron, J. Stepaniuk, Information granules and rough-neural computing, in [70], pp. 43-84.
101. A. Skowron, P. Synak, Complex patterns, Fundamenta Informaticae 60(1-4) (2004) 351-366.

102. A. Skowron, R. Swiniarski, and P. Synak, Approximation spaces and information granulation, Transactions on Rough Sets III: Journal Subline, Lecture Notes in Computer Science, vol. 3400, pp. 175-189, Heidelberg: Springer, 2005.
103. J. F. Sowa, Knowledge Representation: Logical, Philosophical, and Computational Foundations, Course Technology, 1999.
104. S. Staab, R. Studer, Handbook on Ontologies, in International Handbooks on Information Systems, Heidelberg: Springer 2004.
105. J. Stepaniuk, J. Bazan, and A. Skowron, Modelling complex patterns by information systems, Fundamenta Informaticae 67 (1-3) (2005) 203-217.
106. J. Stepaniuk, A. Skowron, J. Peters, and R. Swiniarski, Calculi of approximation spaces, Fundamenta Informaticae 72 (1-3) (2006) 363-378.
107. P. Stone, Layered Learning in Multi-Agent Systems: A Winning Approach to Robotic Soccer, Cambridge, MA: The MIT Press, 2000.
108. W. Swartout, J. Gratch, R. W. Hill, E. Hovy, S. Marsella, J. Rickel, D. Traum, Towards virtual humans, AI Magazine 27(2) (2006) 96-108.
109. R. Sun (Ed.), Cognition and Multi-Agent Interaction. From Cognitive Modeling to Social Simulation. New York, NY: Cambridge University Press, 2006.
110. K. Sycara, Multiagent systems, in AI Magazine, Summer 1998, 79-92.
111. D. Ślęzak, J. T. Yao, J. F. Peters, W. Ziarko, and X. Hu (Eds.), Proceedings of the 10th International Conference on Rough Sets, Fuzzy Sets, Data Mining, and Granular Computing (RSFDGrC'2005), Regina, Canada, August 31-September 3, 2005, Part II, Lecture Notes in Artificial Intelligence, vol. 3642, Heidelberg: Springer, 2005.
112. A. S. Troelstra, H. Schwichtenberg, Basic Proof Theory, Cambridge University Press, 2000.
113. A. S. Troelstra, D. Van Dalen, Constructivism in Mathematics: An Introduction, Studies in Logic and the Foundations of Mathematics Vol. 1 & 2, Elsevier Science Publishing Company, 1988.
114. A. Turing, Computing machinery and intelligence, Mind LIX(236) (October 1950) 433-460.
115. W. Van Wezel, R. Jorna, and A. Meystel, Planning in Intelligent Systems: Aspects, Motivations, and Methods. Hoboken, New Jersey: John Wiley & Sons, 2006.
116. V. Vapnik, Statistical Learning Theory, New York: John Wiley & Sons, 1998.
117. S. Weiss, N. Indurkhya, T. Zhang, F. Damerau, Text Mining: Predictive Methods for Analyzing Unstructured Information, Springer, 2004.
118. I. H. Witten, E. Frank, Data Mining: Practical Machine Learning Tools and Techniques, Second Edition, Morgan Kaufmann Series in Data Management Systems, Morgan Kaufmann, 2005.
119. I. H. Witten, A. Moffat, T. C. Bell, Managing Gigabytes: Compressing and Indexing Documents and Images, The Morgan Kaufmann Series in Multimedia and Information Systems, Morgan Kaufmann, 1999.
120. L. A. Zadeh, Fuzzy sets, Information and Control 8 (1965) 333-353.
121. L. A. Zadeh, From computing with numbers to computing with words - from manipulation of measurements to manipulation of perceptions, IEEE Transactions on Circuits and Systems - I: Fundamental Theory and Applications 45(1) (1999) 105-119.
122. L. A. Zadeh, A new direction in AI: Toward a computational theory of perceptions, AI Magazine 22 (1) (2001) 73-84.
123. W. Ziarko, Variable precision rough set model, Journal of Computer and System Sciences 46 (1993) 39-59.

The Domain of Acoustics Seen from the Rough Sets Perspective

Bozena Kostek

Multimedia Systems Department, Gdansk University of Technology
and
Excellence Center Communication Process: Hearing and Speech,
PROKSIM, Warsaw, Poland
bozenka@sound.eti.pg.gda.pl

Abstract. This research study presents rough set-based decision systems applications to the acoustical domain. Two areas are reviewed for this purpose, namely music information classification and retrieval and noise control. The main aim of this paper is to show results of both measurements of the acoustic climate and a survey on noise threat, conducted in schools and students' music clubs. The measurements of the acoustic climate employ multimedia noise monitoring system engineered at the Multimedia Systems Department of the Gdansk University of Technology. Physiological effects of noise exposure are measured using pure tone audiometry and otoacoustic emission tests. All data are gathered in decision tables in order to explore the significance of attributes related to hearing loss occurence and subjective factors that attribute to the noise annoyance. Future direction of experiments are shortly outlined in Summary.

1 Opening Thoughts

Before introducing the particular topic of research presented in the paper, I would like to share a few thoughts. This Section is devoted to some personal aspects of the research carried out by the author for many years. It concerns the fascination of the rough set methodology and the philosophy that lies behind it, and also (or rather in the first place) the fascination of the rough set method creator, Professor Zdzislaw Pawlak [34,37]. His personality stands out very clearly amongst other researchers. It happened that his plenary talk I've listened to on the occasion of the 2nd International Conference on Rough Sets in Banff guided me toward new interests, namely the applications of decisions rule-based systems which are formidably fitted for uncertainty so often found in acoustics and its analysis. From this time on, we have met many times on various occasions, and I was always inspired by his presentations that led me into new directions and horizons. Professor Pawlak was a mentor to me and I benefited greatly because he was very kind to write Foreword for my two books showing his interest in the research carried out by me and my colleagues. These books perhaps would not happen without his wise patronage.

J.F. Peters et al. (Eds.): Transactions on Rough Sets VI, LNCS 4374, pp. 133–151, 2007.
© Springer-Verlag Berlin Heidelberg 2007

This is very valuable to me and I will be always grateful to him. Altogether within the rough set society, a clear interest appeared to pursue the rough set-acoustic applications [7,21,22], especially the domain of music evoked many research studies [2,3,8,9,12,13,16,17,18,21,22,23,24,25,29,30,31,40,41,42,43]. For some years now, many researches have published in the joint area of rough sets and acoustic/music, thus some of these names being recalled in References. Lately, also Chinese and Korean contributors to this domain appeared. As a result of the interest in this area a new domain of applications emerged, which focuses on interests such as musical instrument recognition based on timbre descriptors, musical phrase classification based on its parameters or contour, melody classification (e.g. query-by-humming systems), rhythm retrieval (different approaches), high-level-based music retrieval such as looking for emotions in music or differences in expressiveness, music search based on listeners' preferences, and others. One may also find research studies which try to correlate low-level descriptor analysis to high-level human perception.

The semantic description is becoming a basis of the next web generation, i.e., the Semantic Web. Several important concepts have been introduced recently by the researchers associated with the rough set community with regard to semantic data processing including techniques for computing with words [20,33]. Moreover, Zdzislaw Pawlak in his papers [35,36] promoted his new mathematical model of flow networks which can be applied to mining knowledge in databases. Such topics are reflected also in papers that followed Prof. Pawlak's original idea on flow graphs [9,26].

Studies performed on the verge of two domains: soft computing (and particularly rough sets) and acoustics enabled the author to apply for many research grants and many of these projects have been successfully awarded. Once again, the current research is also a good example of the need for employing decision systems to the area which at first glance seems far away from the soft computing interests.

2   Introduction

This paper deals with a particular topic which is noise threat-related. As indicated in numerous reports, noise threats occur very frequently nowadays. Occupational exposure limits (OELs) for noise are typically given as the maximum duration of exposure permitted for various noise levels. Environmental noise regulations usually specify a maximum outdoor level of 60 to 65 dB(A), while occupational safety organizations recommend that the maximum exposure to noise is 40 hours per week at 85 to 90 dB(A). For every additional 3 dB(A), the maximum exposure time is reduced by a factor of 2, e.g. 20 hours per week at 88 dB(A). Sometimes, a factor of 2 per additional 5 dB(A) is used. However, these occupational regulations are recognized by the health literature as inadequate to protect against hearing loss and other health effects, especially for sensitive individuals, adverse subjective effects might be expected to appear earlier than for others [4,5,38].

The background of this study is the fact that younger and younger people experience a noticeable loss in hearing. In previous decades a procedure was established within the audiology field, that a group of young males of the age between 18-21 could constitute a reference group for hearing measurements. However, during the last decade numerous studies have shown that this statement is no longer valid. Also, hearing characteristics of students measured during psychoacoustic laboratory sessions at the Multimedia Systems Department have shown that students of this age typically have a threshold shift at 6 kHz. On average, this accounts to 20 dB HL (hearing loss), which is for this age rather unexpected taking into account that students that have any history of ear illnesses were excluded from the experiments. That is why the starting point is to look for the causes of loss in hearing in younger groups of population.

The study aimed at showing results of a survey on noise threat which was conducted in schools and students' music clubs. Noise has an enormous impact on health and life quality of human beings. Noise pollution in Poland is greater than in others UE countries, moreover recently it has been reported to be on the increase [14]. Taking into account the European 2002/49/WE directive related to the control and assessment of environmental noise, monitoring of these threats becomes a necessity [38]. That is why a thorough study on many aspects of noise was envisioned and is carried out for some time at the Multimedia Systems Department [6,10,27,28].

First of all, measurements of the acoustic climate that employed telemetry stations for continuous noise monitoring engineered at the Multimedia Systems Department were conducted. Also, physiological effects of noise were measured among pupils and students. Hearing tests were performed twice, before and after the exposure to noise. For this purpose a so-called distortion product otoacoustic emission method (DPOAE) was utilized. As derived from numerous studies, otoacoustic emission is treated as an early indicator of the occurrence of hearing loss for which reason this method was chosen. The obtained results of noise measurements revealed that an unfavorable noise climate was found in examined schools and music clubs. This was also confirmed by the results of a subjective examination. For the latter purpose students and pupils filled in a questionnaire expressing their feelings as to noise presence and its annoyance. The noise dose analysis based on average time spent by pupils in schools was also calculated. It revealed that noise in schools did not constitute a risk to the pupils' hearing system, however, it may be considered as an essential source of annoyance. On the other hand, noise in music clubs surpassed all permitted noise limits, thus could be treated as dangerous to hearing. Hearing tests revealed changes in the cochlea activity of examined students, also the Tinnitus (ringing in the ear) effect was experienced temporarily. In addition, noise annoyance and noise threat criteria and analysis were proposed and verified based on the acquired and analyzed data.

All factors recognized in the study constitute the basis of two types of decision tables that were created. The first one consists of the attributes derived from the measurements and calculation of the noise dose, also the presence or absence of the Tinnitus (ringing in the ear) effect is included in this table. The second decision table gathers data from the survey on noise annoyance. The conditional attributes are subject-driven in this case. Examples of the questions included in this survey are shown in the following sections. The paper aims at showing that a complex and thorough study may lead to better understanding of noise threats and the correlation between the measurement data and the survey responses concerning noise annoyance. Another aim is to show that current regulations are not adequate to predict PTS (Permanent Threshold Shift) early. Finally the data dependency is analyzed, in the reduced database, to find the minimal subset of attributes called reduct. The analysis of collected data is done by employing the rough set decision system.

3   Multimedia Noise Monitoring System

The MNMS (Multimedia Noise Monitoring System), developed at the Multimedia Systems Department of the Gdansk University of Technology enables to proceed with the environmental noise measurements in cities on an unparalleled scale now. In general, the MNMS consists of a central database which serves as a repository of measurement results, and numerous equipment tools which execute noise meter functions. One of the proposed devices is a mobile noise monitoring station. The station realizes all measuring functions typical for a sound level meter. It also includes special solutions for long-term measurements and introduces a new type of noise indicators. The application of wireless data transmission technology enables to send data to the server and to remotely control the performance of the station. Since this subject was already published [6], thus its main features are only outlined above.

4   Noise and Hearing Measurements

The noise measurement results, obtained by means of the MNMS, are presented below. The measurements were done in selected schools, musical clubs, and during a musical band rehearsals. Participation in music bands concerts and staying in students' clubs are a common way of entertainment amongst students. This is why the investigation was carried out also in these locations. The acquired data were utilized to perform the noise dose analysis. This is done to determine the noise exposure of a person staying in the considered places. In selected cases (i.e. schools and musical clubs), the noise dose analysis was expanded by the assessment of hearing. To achieve this, a so-called distortion product otoacoustic emission (DPOAE) measurement and pure tone audiometry were applied. Hearing was examined twice. First, directly before the exposure to noise of a given type, and then immediately after. The performed analysis combined the obtained noise and hearing measurement results.

Hearing examinations employed the DPOAE method using GSI 60 DPOAE system. The following parameters of the stimuli were used during tests: L1 equals 65 dB, L2 equals 55 dB, f2/f1 = 1.2, DP frequency (geometric mean): 1062, 1312, 1562, 1812, 2187, 2625, 3062, 3687, 4375, 5187, 6187, 7375 Hz. A DP signal level and a noise floor for every stimuli were registered. The test result was accepted if the difference between the evoked otoacoustic emission signals and the noise floor was not less than 10 dB. For pure tone audiometry only selected frequencies were examined: 1000, 1500, 2000, 3000, 4000, 6000, 8000 Hz. The stimuli for each frequency were presented starting from the minimal loudness. The reason of such a selection of parameters was because the noise impact on the hearing system is the strongest for middle and high frequencies. The test was carried out in rooms specially adapted for this purpose. Some measurements performed in schools were interfered with sounds coming from adjoining rooms.

Typical response to stimuli is shown in Figure 1. The DP otoacoustic response was found at 3187 Hz, and because the difference between the noise level and DP signal is larger than 10 dB, thus this measurement is accepted.

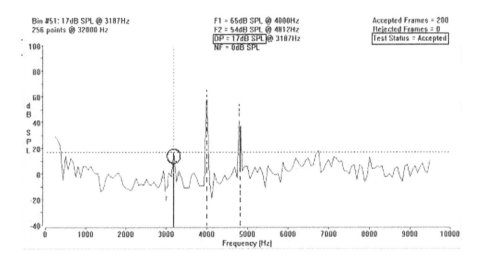

Fig. 1. Example of the DPOAE analysis

The following noise parameters L_{AFmin}, L_{Aeq}, L_{Amax} (see the Equation and definitions below) were measured independently over broadband and in one-third octave bands. A histogram of time history of L_{AF} instantaneous levels was also calculated. A measuring microphone was located 1.9 m above the floor level for every measurement. For all measuring series, a place where people gather most often was selected. This was to determine correctly a real noise dose to which they were exposed.

$$L_{eq} = 10 \log \frac{1}{N} \sum_{i=1}^{N} 10^{0.1 \cdot L_{AdBi}} \tag{1}$$

where:

L_{eq}– A-weighted equivalent continuous noise level,
N –number of L_{AdBi} values,
L_{AdBi} – A-weighted instantaneous sound levels,

On the other hand, L_{AFmin}, L_{AFmax} denote the lowest and highest A-weighted sound levels for fast time weighting that occurred during the measurement.

4.1 Noise Dose Analysis

The evaluation of both the occupational noise exposure and the risk of developing a permanent hearing loss that may result from the noise exposure are shown in Fig. 2.

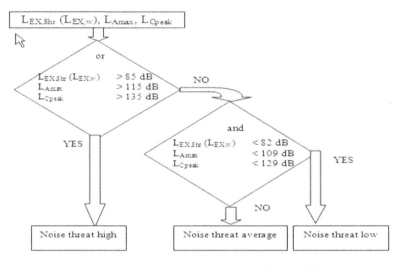

Fig. 2. Evaluation of the occupational noise exposure ($L_{EX,8h}$(D) – daily noise exposure level, L_{Amax} – maximum sound level in working conditions, L_{Cpeak} – peak sound level)

The presented evaluation of the occupational noise exposure and the risk of developing a permanent hearing loss is based on noise regulations [4]. The regulations recommend a limit for occupational noise exposure of 85 decibels, A-weighted, as an 8-hour time-weighted average [85 dBA as an 8-hr TWA]). This enables to evaluate whether occupational noise may cause hearing loss, and (or) whether personal a hearing protector (i.e. earmuffs, formable earplugs, earplugs, etc.) is required. The A filter is a weighting curve which approximates equal loudness perception characteristics of the human hearing for pure tones with reference to 40 dB SPL at 1 KHz. It is worth noticing that this curve was established for pure tones, and a potential noise is typically broadband, which means that A-weighting may not provide the best estimate of potential noise-induced hearing loss. The formal definition of the noise dose defines the dose as the amount of actual exposure relative to the amount of allowable exposure, and for which 100% and above represents exposures that are hazardous. The noise dose is calculated according to the following formula:

$$D = [C_1/T_1 + C_2/T_2 + ... + C_n/T_n]100\% \tag{2}$$

where D is a dose in the allowable percent, C_n refers to the total time of exposure at a specified noise level, and T_n denotes the exposure time at which noise for this level becomes hazardous. This definition allows to calculate a TWA (*Time-weighted average*) from a Noise Dose:

$$TWA = 90 + 16.61log(D/100) \tag{3}$$

where TWA is the 8-hour time-weighted average noise exposure, and D denotes the dose.

Unfortunately, the definition given above is arguable since other formulas of the TWA are also known.

The above regulations are recalled before the noise dose calculated for this study is presented, because it is to show that the norms may not provide the best estimate of the potential occurrence of the permanent hearing loss. One may find also such conclusions in many research studies [1,15,11,32,39].

Time of the noise exposure for the presented activities is much longer in real conditions than in a time-controlled experiment. A simple survey that included questions about how long pupils/students stay in clubs, play or listen to loud music, stay in school, etc. was also carried out. On the basis of the answers, an average time of the exposure for different type of activities was specified. The total time of the noise exposure in schools, clubs and rehearsing musicians' is respectively equal to 3600 s, 13500 s, and 5400 s. Based on the assumption that in the indicated places the noise climate is the same, it is possible to obtain the noise dose for people staying in these places. The noise dose for school amounts to not more than 26%, for rehearsing in a musical band to 673%, for club No.1 - 506% and for club No. 2 - 1191% of the daily dose.

4.2   Noise Investigation Results

The obtained noise measurement results are presented in Table 1. Noise investigation was performed in three different schools. They differed from each other in the age of the pupils. The youngest pupils attended a primary school (school No. 1). The second school was for children between the age of 13 and 15. The third school was a high school attended by the youth aged from 16 to 19. The biggest noise occurred at the primary school. This is because small children are the main source of noise in schools. They behave extremely vigorously at this age. This entailed a very high noise level. In this school, additional source of noise was loud music played from loudspeakers. In high school No. 3 the noise was produced by loud conversations. It should also be mentioned that in all investigated schools there was no sufficient absorption materials covering walls and ceilings, which fact further increased sound level. The fourth measuring series was done during a rehearsal of a small students' music band. The band consisted of a drummer, a bass player and a keyboard player. This measurement revealed high dynamics of noise level. This is because the musicians often paused and

Table 1. Noise measurement results. Time of noise exposure is expressed in seconds, noise levels in dB and noise dose in per cent of allowed daily noise dose.

Measurem. No.	Exposure	L_{AFmin}	L_{Aeq}	L_{AFmax}	Exposure Time	Noise Dose
1	School No. 1	67.4	89.0	105.5	600	5.2
2	School No. 2	67.2	85.5	106.8	900	3.5
3	School No. 3	72.0	83.6	97.4	600	1.5
4	**Music band**	**52.5**	**100.5**	**114.4**	**4058**	**506.1**
5	Club No. 1	76.2	95.3	108.2	4529	169.9
6	**Club No. 2**	**68.9**	**99.0**	**114.2**	**5330**	**470.0**

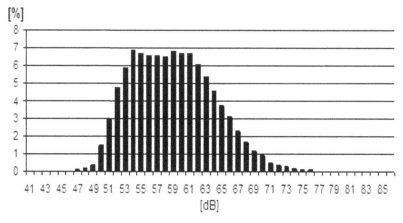

Fig. 3. L_{AF} histogram measured during lessons

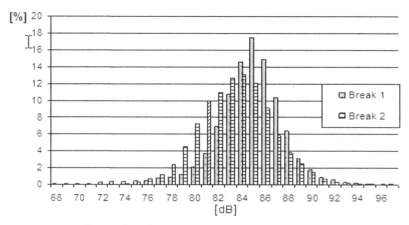

Fig. 4. L_{AF} histogram measured during breaks

consulted each. The 5th and 6th measurement series were carried out in two students' clubs. They differed in the type of music. In the first club a pop music was dominating, while in the second only rock was played. The results of the noise dose analysis for the measured time exposures are presented in Table 1. Figures 3 and 4 show histograms measured during lessons and breaks in schools.

The breaks were 15 minutes long. The results of noise measurements during two breaks were similar (see Fig. 4). L_{eq} for the breaks was approx. 85dB. Sounds louder than 90 dB appeared at 3-4% of measured time span, however a noise level of 122 dB was also registered. On the other hand, noise during lessons turned out to be less annoying, L_{eq} equaled to 62 dB. In all examined places, a very high noise level was observed. Staying for a long time in places such as music clubs, discotheques, etc. where noise level reaches the values shown in Table 1, can be hazardous to hearing (i.e. may produce a permanent threshold shift).

4.3   Hearing Measurement Results

Several dozens of persons took part in the presented hearing tests. The total number of people examined for different types of exposure is presented in Table 2. Table 2 also includes the results of average changes of a hearing threshold (pure tone audiometry), and the results of the DPOAE tests. The average changes of a hearing threshold after the exposure to noise for individual frequencies obtained for pure tone audiometry are presented in Table 3. Two different aspects were taken into consideration while analyzing the DPOAE results. First, the number

Table 2. Results of hearing testing (in [%])

Measurem. No.	Pure tone audiometry				DPOAE test results			
	No. of persons	Decrease of threshold	Increase of threshold	No change	No. of persons	+Pass	-Pass	No change
1	-	-	-	-	10	11.0	13.6	75.4
2	-	-	-	-	5	10.0	19.2	70.8
3	-	-	-	-	5	3.3	12.5	84.2
4	9	21.4	**37.3**	41.3	9	5.1	11.6	83.3
5	10	14.3	**62.9**	22.8	11	3.4	10.6	86.0
6	12	12.5	**75.0**	12.5	12	4.5	**20.5**	75.0

Table 3. The average changes of hearing threshold for pure tone audiometry (in [dB])

Type of noise exposure		1000	1500	2000	3000	4000	6000	8000
Music band	L	2.8	0.0	2.2	2.2	**3.3**	1.7	1.1
	R	0.0	-1.1	1.1	0.6	2.2	3.3	0.5
	AVG	1.4	-0.6	1.7	1.4	2.8	2.5	0.8
Club No. 1	L	2.5	6.5	7.5	10.5	11.5	5	1
	R	-1.5	3	7	10.5	10.5	5.5	3.5
	AVG	0.5	4.75	7.25	10.5	11.0	5.25	2.25
Club No. 2	L	2.9	6.7	8.8	10.4	**15.9**	12.1	-1.7
	R	3.8	6.7	10.0	12.1	12.9	10.8	0.8
	AVG	3.3	6.7	9.4	11.3	**14.4**	11.5	-0.4

Table 4. The average changes of DP signal level for particular types of exposures (in [%])

Measurement No.	1	2	3	4	5	6
Increase	30.3	27.5	36.7	25.0	27.3	16.3
Decrease	28.1	30.0	34.2	**44.4**	**45.4**	**53.5**
No change	41.6	42.5	29.1	30.6	27.3	30.2

of "passed" and "failed" tests for the DPOAE examination were determined. The result of the first examination served as reference. The symbol "+Pass" indicates that a pupil failed the first examination and passed the second one. The symbol "–Pass" signifies a reverse situation (a test passed in the first examination and failed after the exposure to noise). The results are presented in Table 2, in the "DPOAE test results" column. The second kind of analysis determined how the DP signal level changed under the influence of the exposure to noise. The results of this analysis are presented in Table 4.

As seen from the tables, in most cases the hearing threshold was increased. It means that almost every person had a TTS after noise exposure. Also some people reported the perception of the Tinnitus effect as well. The most significant TTS occurred for 4000 Hz. The data obtained from DPOAE tests confirm negative after-effects of the noise exposure analysis acquired by means of pure tone audiometry.

5   Psychophysiological Noise Dosimeter

Methods of the estimation of noise-induced hearing loss presented in this paper are based mainly on the equal energy hypothesis [5]. This approach focuses on an assessment of the quantity of energy which affects the hearing system. The time characteristics of noise is neglected, and the main emphasis is placed on the

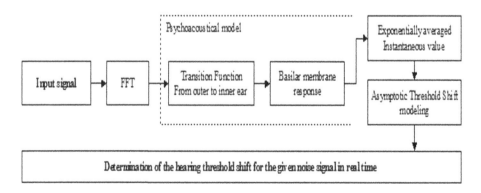

Fig. 5. General scheme of the psychophysiological noise dosimeter

assessment of the equivalent sound level value. However, in many cases this approach could be insufficient. The analysis of numerous literature data, including testing of the exposure to noise from different sources, provides a knowledge that time characteristics and noise spectrum have an essential significance in generating loss in hearing [1,11,15]. Taking these data into consideration, a method of estimating the risk of hearing impairment has been proposed by J. Kotus, the author's Ph.D. student. The proposed model is based on a modified Johnston's psychoacoustical model [19], which provides a global distribution of a basilar membrane deflection within critical bands. In Figure 5, a scheme of the psychophysiological noise dosimeter is presented.

The model is based on the analysis of a basilar membrane's answer to noise in critical bands. First, the power spectrum of noise is determined. Afterwards, it is corrected by taking into account the transition from the outer to the inner ear. Subsequently, particular spectrum coefficients are grouped into critical bands, according to the bark scale. Then the noise level in particular critical bands is calculated. The result let us assess the extent to which the basilar membrane is stimulated. Its answer is determined by the multiplication of the instate stimulation level value by the characteristics of hearing filters for the particular critical band. A basilar membrane displacement value obtained in this way is exponentially averaged. This action reflects the inertia of the processes occurring in the inner ear. The obtained averaged values are used for the assessment of the asymptotic hearing threshold shift. Finally, these values are subjected to exponential averaging, that reflects a process of the hearing threshold shift. Therefore this model enables to assess TTS in critical bands and the recovery time of a hearing threshold to its initial value. The model enables to determine the hearing threshold shift for a given noise during the exposure.

The initial simulation for two selected noise exposures was done (series 4 and 6). A theoretical time series of noise was created based on the histogram of the L_{AF} levels. The obtained TTS levels for a particular measurement series amounted to: TTS = 13.5 dB (measurement No. 4), and TTS = 18.5 dB (measurement No. 6). A comparison of the noise exposure for series 4 and 6 has shown that the time characteristic of noise influences the occurrence and level of the TTS. The results obtained through the use of the presented model, confirmed a greater TTS for the 6th exposition (club No. 2). Overall, this confirms harmful effects of noise.

6   Survey on Noise Annoyance

An objective noise measurement was extended by a subjective measurement by means of a dedicated survey. The survey consisted of three parts. The first part involved getting information such as age, sex, class, school. The second part included questions about noise in places of residence and exposure to noise related to musical preferences. The last part concentrated on noise climates in schools in typical circumstances (lessons, breaks, etc.). In the following Figures (Figs. 6-10) some sample results are shown.

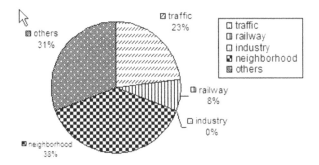

Fig. 6. Main type of noise sources in the neighborhood during day and night

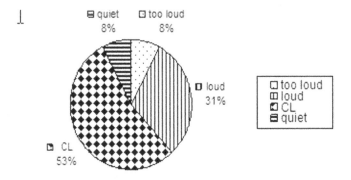

Fig. 7. Subjective evaluation of noise annoyance during day (CL – comfortable)

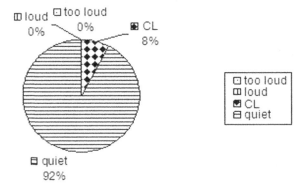

Fig. 8. Subjective evaluation of noise annoyance during night

Main type of noise sources in the neighborhood

As seen in Figure 6 the main sources of noise during day and night are communication and the neighborhood. They constitute 60% of all noise sources. On the other hand, industrial noise does not present a threat to the place of living.

Subjective evaluation of noise annoyance

Overall, 8 % of asked pupils evaluated their place of living as quiet, 40% said that the neighborhood is either too loud or loud. On the other hand, during the night most places were evaluated as quiet, apart from special events occurring during the night. It should be mentioned that 70% of our responders live in towns or in close proximity to towns.

Subjective evaluation of noise annoyance in schools

Almost 70% of pupils evaluate noise as too loud during breaks, and the remaining 30% as loud. As to noise evaluation during lessons pupils differ in their opinions.

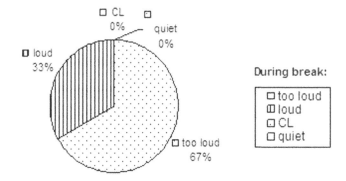

Fig. 9. Subjective evaluation of noise annoyance during breaks

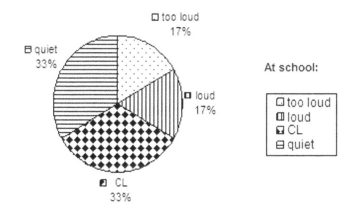

Fig. 10. Subjective evaluation of noise annoyance during lessons

Evaluation of noise sources related to personal way of living

In addition, pupils were asked how often they listen to music and how loud this music is, and also how often they use personal music players and listen to music via headphones. It occurs that on average younger groups of pupils do not often listen to loud music, contrarily to older pupils and students. The latter group also attends musical events very frequently. As a result, one may notice problems with speech understanding when noise is present, which in students' subjective opinions occurred in 50% of situations.

7 Data Analysis Employing Rough Sest

Regulations and gathered data enabled to define attributes that may provide ways to assess the risk of developing PTS (*Permanent Threshold Shift*) and noise annoyance.

Conditional attributes from the decision table (Table 5) are derived from objective measurements and noise dose calculation. One of the attributes driven more subjectively, namely the absence and presence of the Tinnitus effect after the exposure to noise, is also taken into consideration. As mentioned before, otoacoustic emission is treated as an early indicator of hearing loss, that is why the DPOAE analysis should be included in the decision table. In a way the decision table follows results included in Tables 1 and 2.

Another decision table (Table 6) includes answers gathered at schools based on a questionnaire form published in the Internet (MNMS service [7]). The decision attribute are in this case noise annoyance/nuisance or noise threat. A legal definition of a nuisance says that this is the noise that offends or upsets the receiver because it is occurring at the wrong time, in the wrong place, or is of a character that annoys due to excessive tonal components or impulses.

When looking at the table, one may expect that sensitivity to noise is more frequent among children and increases with age. This could be explained by the fact that small children get easily tired in noisy conditions, and on the other hand, people active professionally prefer some rest after work, the same as older people. This means that the age of the survey respondents' should be taken into consideration. Also, it is quite obvious that females are less noisy than males while entertaining. When analyzing other attributes from Table 5 the relation between them and the decision attribute is not always that clear, and since this is a subjectively-driven questionnaire many contradictions may occur in the acquired data. A good example of such a confusing attribute is '*hearing*'. Hearing impairments of the inner ear are characterized by the so-called recruitment phenomenon, thus people with the inner-ear hearing loss may suffer more in noisy conditions. This means that they do not hear soft sounds, and perceive loud sounds as too loud. This phenomenon is due to an abnormally rapid rise of the loudness sensation with only a small increase in sound stimulus intensity.

In this early study, the survey did not include questions related to for example social class, since it was directed at specific groups (pupils attending schools and students). The observations and earlier researches on noise annoyance indicate that these factors should also be taken into account.

All these factors constitute huge data, especially as the survey will be progressively filled in by the Internet respondents. This enables to consolidate data on noise annoyance on a very large scale. The only way to analyze subjectively-driven data is to use a soft computing approach, and particularly the rough set method which seems to be most suitable for this problem. The rough set-based analysis of the decision tables enables to derive rules in which attributes and their values support a claim that noise dose analysis based on average daily exposure is not only insufficient to assess noise annoyance, but also to predict the risk of permanent threshold shift early enough. There are a lot of data in which we see noise events recalled by the survey respondents even if they live in a quiet neighborhood with no evident source of noise during the day and night. This means that they evaluate the noise annoyance or nuisance as high because of noise events that happen sporadically and were of short duration. In such a case, factors derived from the noise dose analysis will not exceed any norm or regulations, but still these noise events may be highly stressful and harmful.

Table 5. Decision table based on measurements and calculation according to noise regulations (denotation same as before, in addition: Ts denotes absence or presence of the Tinnitus effect)

No.	L_{AFmin}	L_{eq}	L_{AFmax}	Exposure Time	...	D	DPOAE	TTS	Ts	**PTS**
1	67.4	89	105.5	600		5.2	...	YES	YES	NO
2	52.5	100.5	114.4	4058		506.1	...	YES	YES	**YES**
...	...	...	...	...	...	...	...	...	...	...
n	...	...	...	...	...	...	...	...	...	...

Table 6. Decision table based on survey (denotations: StoN – sensitivity to noise, H – hearing, SC – social class, WP/S – type of work performed/school, SR/D – Survey results/neighborhood day, SR/N – Survey results/neighborhood night, SR/NS – Survey results/noise source, NE/AdNS – Noise events/additional noise sources, NA/N – noise annoyance/nuissance)

No.	Age	Sex	StoN	H	SC	WP/S	SR/D	SSR/N	SR/NS	...	NE/AdNS	NA/N
1	41-50	M	high	good	high	educat.	quiet	quiet	no	...	dogs	**HIGH**
2	10-15	F	high	good	NO	high school	noisy	quiet	road traffic	...	ambul. siren	**HIGH**
...	...	...	...	...	...	...	...	...	...	...	...	...
n	...	...	...	...	...	...	...	...	...	...	...	...

A rule prototype derived from the rough set-based analysis is presented below:

$$(attribute\_1) = (val\_1) \wedge ... \wedge (attribute\_k) = (val\_k) => (PTS_{YES/NO}) \quad (4)$$

It should be mentioned that some of the numerical attributes require quantization, however in the case of the noise-related attributes the quantization should be based on norms and standards. This means a process of replacing the original values of the input data with for example the number of an interval to which a selected parameter value belongs should be performed taking into account specific constraints. For example, it is well-known that values of L_{AFmax}, L_{eq}, Exposure

Time exceeding the limits for which these values start to be harmful constitute the cut-points. It is obvious that after rule derivation not all attributes are to be retained and some of the rules are not longer valid. Most important attributes are to be found in reducts. Preliminary experiments show that amongst the most significant attributes resulting from measurements one may find: L_{AFmax}, L_{eq}, Exposure Time and TTS, though some of these parameters are interrelated, and are eliminated in the rough set-based analysis, reducing the number of the correlated attributes. On the other hand, looking into the Decision Table based on survey results, one may find that such attributes as: age of a person, specific type of noise events, duration of these events, neighborhood noise background are present in the evaluation of noise annoyance/nuissance.

Having the rough set analysis performed and rules derived based on both approaches (objective and subjective), this may constitute a scientific basis of an advertising campaign against noise pollution and its adverse consequences, which in addition may contribute to better regulations on noise. This is why noise monitoring stations should be installed in all agglomerations larger than 250,000 inhabitants in the coming year, thus both measurement quantities shown in a form of noise maps and subjective opinions would be available for the same place, in future. However, it is worth noticing that without changing our habits, regulations may still be ineffective as seen from the survey.

8   Summary

To sum up, on the basis of the investigations, it was confirmed that noise climates in schools is adverse to pupils' and teachers' health. The main reasons of the high noise level in schools are: the behavior of pupils, loudspeaker systems and low sound absorption of the classrooms and corridors. The data analysis of the hearing measurements at schools does not confirm negative influence of noise on the hearing system. Especially because the time of exposure to noise is too short to produce measurable changes in the activity of the inner ear.

Noise measured during of the students' music band rehearsals and in clubs reaches very high levels and exceeds all related norms. Measurements of the hearing characteristics of people working or entertaining in these places confirmed harmful effects of noise. A significant TTS and the reduction of the DP level were observed. A comparison of the noise exposure for series 4 and 6 has shown that the time characteristic of noise influences the occurrence of the TTS. The results obtained through the use of the presented model, confirmed a greater TTS for the 6th exposition (club No. 2). In addition, the Tinnitus effect was perceived by some students.

The data presented are very complex, interrelated, and in some cases contradictory, thus for an adequate analysis they require the use of a decision system. For this particular problem, it is thought that the rough set method is the most suitable solution for the analysis, since it allows to derive rules/reduct capable of identifying the most significant attributes. It is then possible to determine the weights that should be taken into consideration when constructing a feature vector.

On the basis of the survey, it may be said that even if norms are not exceeded, noise can be still perceived as annoying, stressful and in consequence harmful. The analysis of the acquired data has shown that all measurement quantities should first be included as conditional attributes in the constructed decision tables along with subjective factors. At the moment, two types of decision tables have been proposed. The first one provides indications as to the significance of some attributes in relation to the PTS occurrence (decision attribute). The second gathers results of the survey, and relates subjective factors to the noise annoyance. The results of this analysis may provide the basis for an adequate campaign against noise pollution and lead to better regulations.

Acknowledgment. This work was supported by the Polish Ministry of Science and Education within the projects No. 3T11E02829 and No. R0201001. The author wishes to acknowledge her Ph.D. student J. Kotus for his valuable input to the presented research.

References

1. Borg, E., Engstrom, B.: Noise level, inner hair cell damage, audiometric features and equal-energy hypothesis, J Acoust. Soc. Am. 86 (5) (1989) 1776–1782
2. Budzynska, L., Jelonek, J., Lukasik, E., Slowinski, R.,: Supporting Experts in Ranking Generic Audio and Visual Objects, Proc. IEEE Workshop "Signal Processing'2004", Poznan (2004) 81–86.
3. Budzynska, L., Jelonek, J., Lukasik, E., Susmaga, R., Slowinski R.: Multistimulus ranking versus pairwise comparison in assessing quality of musical instruments sounds, 118 AES Convention Paper, 6482, Barcelona (2005)
4. Criteria for a recommended standard, Occupational Noise Exposure, U.S. Department of Health and Human Services (1998)
5. http://www.cdc.gov/niosh/98-126.html (CRITERIA FOR A RECOMMENDED STANDARD)
6. Czyzewski, A., Kostek, B., Skarzynski, H.: Intelligent System for Environmental Noise Monitoring, in Monitoring, Security, and Rescue Techniques in Multiagent Systems, Series: Advances in Soft Computing, Dunin-Keplicz, B.; Jankowski, A.; Skowron, A.; Szczuka, M.(eds.), chapter, 397–410 , XII, Springer Verlag, Heidelberg, New Yorkc(2005)
7. Czyzewski, A., Kostek, B., Skarzynski, H.: IT applications for the remote testing of communication senses" chapter in INFORMATION TECHNOLOGY SOLUTIONS FOR HEALTH CARE, Spinger-Verlag (2006)
8. Czyzewski, A., Szczerba M., Kostek B.: Musical Phrase Representation and Recognition by Means of Neural Networks and Rough Sets, Rough Set Theory and Applications (RSTA), vol. 1, 259-284, Advances in Rough Sets, Subseries of Springer-Verlag Lecture Notes in Computer Sciences, LNCS 3100, Transactions on Rough Sets, Grzymala-Busse, J.W., Kostek, B., Swiniarski, R.W., Szczuka M. (eds.) (2004)
9. Czyzewski A., Kostek B.: Musical Metadata Retrieval with Flow Graphs, in Rough Sets and Current Trends in Computing, RSCTC, Uppsala, Sweden, Lecture Notes in Atificial Intelligence, LNAI 3066, Springer Verlag, Berlin, Heidelberg, New York (2004) 691–698

10. Czyzewski, A., Kotus, J., Kostek, B.,: Comparing Noise Levels and Audiometric Testing Results Employing IT Based Diagnostic Systems, The 33rd International Congress and Exposition on Noise Control Engineering INTERNOISE'2004, August 22-24, Prague (2004)

11. Dunn, D.E., Davis, R.R., Merry, C.J., Franks, J.R.: Hearing loss in the chinchilla from impact and continuous noise exposure, J Acoust. Soc. Am. 90 (4) (1991) 1979–1985

12. Dziubinski, M., Dalka, P., Kostek, B.: Estimation of Musical Sound Separation Algorithm Effectiveness Employing Neural Networks, J. Intelligent Information Systems, Special Issue on Intelligent Multimedia Applications, 24, 2(2005) 133–157

13. Dziubinski, M., Kostek, B.: Octave Error Immune and Instantaneous Pitch Detection Algorithm, J. of New Music Research, vol. 34, 292-273, Sept. 2005.

14. Engel, Z.W., Sadowski J., et al.: Noise protection in Poland in European Legislation, The Committee on Acoustics of the Polish Academy of Science & CIOP-PIB, Warsaw, (2005) (in Polish)

15. Henderson, D., Hamernik, R.P.: Impulse noise: Critical review, J Acoust. Soc. Am. 80(2) (1986) 569–584

16. Hippe, M.P.: Towards the Classification of Musical Works: A Rough Set Approach Third International Conference, RSCTC 2002, Malvern, PA, USA, October 14-16, 2002. Proceedings Editors: J.J. Alpigini, J.F. Peters, A. Skowron, N. Zhong (eds.) (2002) 546-553

17. Jelonek, J., Lukasik, E., Naganowski, A., Slowinski, R.: Inferring Decision Rules from Jurys' Ranking of Competing Violins, Proc. Stockholm Music Acoustic Conference, KTH, Stockholm (2003) 75–78

18. Jelonek, J., Lukasik, E., Naganowski, A., Slowinski, R.: Inducing jury's preferences in terms of acoustic features of violin sounds, Lecture Notes in Computer Science, LNCS 3070, Springer (2004) 492–497

19. Johnston, J.D.: Transform Coding of Audio Signals Using Perceptual Noise Criteria. IEEE Journal on Selected Areas in Communications, vol. 6(2) (1988) 314–323

20. Komorowski, J, Pawlak, Z, Polkowski, L, Skowron, A. Rough Sets: A Tuto-rial. In: Pal SK, Skowron A (eds) Rough Fuzzy Hybridization: A New Trend in Decision-Making. Springer-Verlag (1998), 3–98

21. Kostek, B.: Soft Computing in Acoustics, Applications of Neural Networks, Fuzzy Logic and Rough Sets to Musical Acoustics, Physica Verlag, Heidelberg, New York (1999)

22. Kostek, B.: Perception-Based Data Processing in Acoustics. Applications to Music Information Retrieval and Psychophysiology of Hearing, Springer Verlag, Series on Cognitive Technologies, Berlin, Heidelberg, New York (2005)

23. Kostek, B.: Musical Instrument Classification and Duet Analysis Employing Music Information Retrieval Techniques, Proc. of the IEEE, 92, 4 (2004) 712–729

24. Kostek, B.: Intelligent Multimedia Applications - Scanning the Issue, J. Intelligent Information Systems, Special Issue on Intelligent Multimedia Applications, 24, 2 (2005) 95–97 (Guest Editor)

25. Kostek, B., Wojcik, J.: Machine Learning System for Estimating the Rhythmic Salience of Sounds, International J. of Knowledge-based and Intelligent Engineering Systems, 9 (2005), 1–10

26. Kostek, B., Czyzewski, A.: Processing of Musical Metadata Employing Pawlak's Flow Graphs, Rough Set Theory and Applications (RSTA), vol. 1, 285–305, Advances in Rough Sets, Subseries of Springer-Verlag Lecture Notes in Computer Sciences, LNCS 3100, Transactions on Rough Sets, Grzymala-Busse, J.W., Kostek, B., Swiniarski, R.W., Szczuka, M., (eds.) (2004)
27. Kotus, J., Kostek, B.: Investigation of Noise Threats and Their Impact on Hearing in Selected Schools, OSA' 2006, Archives of Acoustics (2006) (in print).
28. Kotus, J.: Evaluation of Noise Threats and Their Impact on Hearing by Employing Teleinformatic Systems, (Kostek, B., supervisor) (2007) (in preparation).
29. Lukasik, E.: AMATI-Multimedia Database of Violin Sounds. In: Proc Stockholm Music Acoustics Conference, KTH Stockholm (2003a) 79–82
30. Lukasik, E.: Timbre Dissimilarity of Violins: Specific Case of Musical Instruments Identification. Digital Media Processing for Multimedia Interactive Services, World Scientific, Singapore (2003b) 324–327
31. Lukasik, E., Susmaga, R.: Unsupervised Machine Learning Methods in Timbral Violin Characteristics Visualization. In: Proc Stockholm Music Acoustics Conference, KTH Stockholm (2003) 83–86
32. Melnick, W.: Human temporary threshold shift (TTS) and damage risk, J Acoust. Soc. Am. 90(1) (1991) 147–154
33. Pal, S.K., Polkowski, L., Skowron, A. Rough-Neural Computing. Techniques for Computing with Words. Springer Verlag, Berlin Heidelberg New York (2004)
34. Pawlak, Z.: Rough Sets. International J Computer and Information Sciences (1982)
35. Pawlak, Z.: Probability, Truth and Flow Graph. Electronic Notes in Theoretical Computer Science 82, International Workshop on Rough Sets in Knowledge Discovery and Soft Computing, Satellite event of ETAPS 2003, Elsevier, Warsaw (2003)
36. Pawlak, Z.: Elementary Rough Set Granules: Towards a Rough Set Processor. In: Pal SK, Polkowski L, Skowron A (eds) Rough-Neural Computing. Techniques for Computing with Words. Springer Verlag, Berlin Heidelberg New York, 5–13(2004)
37. Pawlak, Z.: A Treatise on Rough Sets. Transactions on Rough Sets IV, Peters, J.F., Skowron, A. (Eds) 1–17 (2005)
38. Polish Standard PN-N-01307 (1994), Permissible sound level values in work-places and general requirements concerning taking measurements (in Polish).
39. Seixas, N., et al.: Alternative Metrics for Noise Exposure Among Construction Workers, Ann Occup Hyg. 49 (2005) 493–502
40. A. Wieczorkowska, P. Synak, R. Lewis, Z. W. Ras, Creating Reliable Database for Experiments on Extracting Emotions from Music. In: M. A. Klopotek, S. Wierzchon, K. Trojanowski (eds.), Intelligent Information Processing and Web Mining, Proceedings of the International IIS: IIPWM'05 Conference, Gdansk, Poland Advances in Soft Computing, Springer (2005), 395-402
41. Wieczorkowska, A., Synak, P., Lewis, R., Ras, Z.W.: Extracting Emotions from Music Data, in: M.-S. Hacid, Murray, N.V., Ras Z.W., Tsumoto, S. (eds.), Foundations of Intelligent Systems, 15th International Symposium, ISMIS 2005, Saratoga Springs, NY, USA, 2005, Proceedings; LNAI 3488, Springer, 456-465
42. Wieczorkowska, A., Ras, Z.W.: Do We Need Automatic Indexing of Musical Instruments?, in: Warsaw IMTCI, International Workshop on Intelligent Media Technology for Communicative Intelligence, Warsaw, Poland, September 13–14, Proceedings, PJIIT - Publishing House (2004), 43–38
43. Wieczorkowska, A.: Towards Extracting Emotions from Music. In: Warsaw IMTCI, International Workshop on Intelligent Media Technology for Communicative Intelligence, Warsaw, Poland, September 13–14, Proceedings, PJIIT - Publishing House (2004) 181–183

Rule Evaluations, Attributes, and Rough Sets: Extension and a Case Study

Jiye Li[1], Puntip Pattaraintakorn[2], and Nick Cercone[3]

[1] David R. Cheriton School of Computer Science, University of Waterloo
200 University Avenue West, Waterloo, Ontario, Canada N2L 3G1
j27li@uwaterloo.ca
[2] Department of Mathematics and Computer Science, Faculty of Science
King Mongkut's Institute of Technology Ladkrabang, Thailand
kppuntip@kmitl.ac.th
[3] Faculty of Science and Engineering, York University
4700 Keele Street, North York, Ontario, Canada M3J 1P3
ncercone@yorku.ca

Abstract. Manually evaluating important and interesting rules generated from data is generally infeasible due to the large number of rules extracted. Different approaches such as rule interestingness measures and rule quality measures have been proposed and explored previously to extract interesting and high quality association rules and classification rules. Rough sets theory was originally presented as an approach to approximate concepts under uncertainty. In this paper, we explore rough sets based rule evaluation approaches in knowledge discovery. We demonstrate rule evaluation approaches through a real-world geriatric care data set from Dalhousie Medical School. Rough set based rule evaluation approaches can be used in a straightforward way to rank the importance of the rules. One interesting system developed along these lies in HYRIS (HYbrid Rough sets Intelligent System). We introduce HYRIS through a case study on survival analysis using the geriatric care data set.

1 Introduction

The general models of knowledge discovery in databases (KDD) contains processes including data preprocessing, knowledge discovery algorithms, rule generations and evaluations. Rule evaluation is a significant process in KDD. How to automatically extract important, representative rules to the human beings instead of selecting those useful rules manually are the main problems. Specific difficulties make the research of rule evaluation very challenging.

One of the difficulties is that real-world large data sets normally contain missing attribute values. They may come from the collecting process, or redundant scientific tests, change of the experimental design, privacy concerns, ethnic issues, unknown data and so on. Discarding all the data containing the missing attribute values cannot fully preserve the characteristics of the original data, and

J.F. Peters et al. (Eds.): Transactions on Rough Sets VI, LNCS 4374, pp. 152–171, 2007.
© Springer-Verlag Berlin Heidelberg 2007

wastes part of the data collecting effort. Knowledge generated from missing data may not fully represent the original data set, thus the discovery may not be as sufficient. Understanding and utilizing of original context and background knowledge to assign the missing values seem to be an optimal approach for handling missing attribute values. In reality, it is difficult to know the original meaning for missing data from certain application domains. Another difficulty is that huge amount of rules are generated during the knowledge discovery process, and it is infeasible for humans to manually select useful and interesting knowledge from such rule sets.

Rough sets theory, originally proposed in the 1980's by Pawlak [1], was presented as an approach to approximate concepts under uncertainty. The theory has been widely used for attribute selection, data reduction, rule discovery and many knowledge discovery applications in the areas such as data mining, machine learning and medical diagnoses.

We are interested in tackling difficult problems in knowledge discovery from a rough sets perspective. In this paper, we introduce how rough sets based rule evaluations are utilized in knowledge discovery systems. Three representative approaches based on rough sets theory are introduced. The first approach is to provide a rank of how important is each rule by rule importance measure (RIM) [2]. The second approach is to extract representative rules by considering rules as condition attributes in a decision table [3]. The third approach is applied to data containing missing values. This approach provides a prediction for all the missing values using frequent itemsets as a knowledge base. Rules generated from the complete data sets contain more useful information. The third approach can be used at the data preprocessing process, combining with the first or second approach at the rule evaluation process to enhance extracting more important rules. It can also be used alone as preprocessing of missing attribute values. An interesting system based on this rule-enhanced knowledge discovery system, HYRIS (HYbrid Rough sets Intelligent System) [4], is developed. Case studies on using HYRIS on survival analysis are further demonstrated.

We address particular problems from real-world data sets, using recent missing attribute value techniques and rule evaluations based on rough sets theory to facilitate the tasks of knowledge discovery. The rule discovery algorithm focuses on association rule algorithms, although it can be classification algorithm, decision tree algorithm and other rule discovery algorithms from data mining and machine learning. We demonstrate the rule evaluation approaches using a real-world geriatric care medical data set.

We discuss related work on rough sets theory, current knowledge discovery system based on rough sets, and rule evaluations in Section 2. Section 3 presents three rough sets based rule evaluations methods. We show experiments on the geriatric care data set in Section 4. Section 5 contains a case study of HYRIS system developed based on the proposed approaches, and experiments on survival analysis are demonstrated. Section 6 gives the concluding remarks.

2   Related Work

We introduce related work to this paper including rough sets theory, knowledge discovery systems based on rough sets theory and existing rule evaluations approaches.

2.1   Rough Sets Theory

Rough sets theory, proposed in the 1980's by Pawlak [1], has been used for attribute selection, rule discovery and many knowledge discovery applications in the areas such as data mining, machine learning and medical diagnoses.

We briefly introduce rough sets theory [1] as follows. U is the set of objects we are interested in, where $U \neq \phi$.

Definition 1. *Equivalence Relation.* *Let R be an equivalence relation over U, then the family of all equivalence classes of R is represented by U/R. $[x]_R$ means a category in R containing an element $x \in U$. Suppose $P \subseteq R$, and $P \neq \phi$, $IND(P)$ is an equivalence relation over U. For any $x \in U$, the equivalence class of x of the relation $IND(P)$ is denoted as $[x]_P$.*

Definition 2. *Lower Approximation and Upper Approximation.* *X is a subset of U, R is an equivalence relation, the lower approximation of X and the upper approximation of X is defined as:*

$$\underline{R}X = \cup\{x \in U | [x]_R \subseteq X\} \tag{1}$$

$$\overline{R}X = \cup\{x \in U | [x]_R \cap X \neq \phi\} \tag{2}$$

respectively.

From the original definitions [1], reduct and core are defined as follows. R is an equivalence relation and let $S \in R$. We say, S is dispensable in R, if $IND(R) = IND(R - \{S\})$; S is indispensable in R if $IND(R) \neq IND(R - \{S\})$. We say R is independent if each $S \in R$ is indispensable in R.

Definition 3. *Reduct.* *Q is a reduct of P if Q is independent, $Q \subseteq P$, and $IND(Q) = IND(P)$.*

An equivalence relation over a knowledge base can have many reducts.

Definition 4. *Core.* *The intersection of all the reducts of an equivalence relation P is defined to be the Core, where*

$$Core(P) = \cap All\ Reducts\ of\ P.$$

Reduct and core are among the most important concepts in this theory. A reduct contains a subset of condition attributes that are sufficient enough to represent the whole data set. The reducts can be used in attribute selection process. There may exist more than one reduct for each decision table. Finding all the reduct sets for a data set is NP-hard [5]. Approximation algorithms are used to obtain reduct sets [6]. The intersection of all the possible reducts is called the *core*.

The core is contained in all the reduct sets, and it is the essential of the whole data. Any reduct generated from the original data set cannot exclude the core attributes.

Reduct Generations. There are several reduct generation approaches, such as ROSETTA [7], RSES [8], ROSE2 [9], QuickReduct algorithm [10] and Hu et al. [11]'s reduct generation combining the relational algebra with the traditional rough sets theory. ROSETTA rough set system GUI version 1.4.41 [7] provides Genetic reducer, Johnson reducer, Holte1R reducer, Manual reducer, Dynamic reducer, RSES Exhaustive reducer and so on. Genetic reducer is an approximation algorithm based on genetic algorithm for multiple reducts generation. Johnson reducer generates only a single reduct with minimum length. In this research, we use both genetic and Johnson's reduct generations to develop rule evaluations approaches.

Core Generation. Hu et al. [11] introduced a core generation algorithm based on rough sets theory and efficient database operations, without generating reducts. The algorithm is shown in Algorithm 1, where C is the set of condition attributes, and D is the set of decision attributes. $Card$ denotes the count operation in databases, and Π denotes the projection operation in databases.

Algorithm 1. Hu's Core Generating Algorithm [11]

---

input : Decision table $T(C, D)$, C is the condition attributes set; D is the decision attribute set.
output: $Core$, Core attributes set.

$Core \leftarrow \phi$;
for *each condition attribute $A \in C$* **do**
  if $Card(\Pi(C - A + D)) \neq Card(\Pi(C - A))$ **then**
   | $Core = Core \cup A$;
  end
end
return $Core$;

This algorithm is developed to consider the effect of each condition attribute on the decision attribute. The intuition is that, if the core attribute is removed from the decision table, the rest of the attributes will bring different information to the decision making. Theoretical proof of this algorithm is provided in [11]. The algorithm takes advantage of efficient database operations such as count and projection. This algorithm requires no inconsistency in the data set.

2.2 Rough Sets Based KDD Systems

We briefly survey current rough sets based knowledge discovery systems. We discuss the individual functions of each system based on general characteristics, such as the input data sets, the preprocessing tasks, the related rough sets tasks, the rule generations and so on.

1. **ROSETTA.** ROSETTA [7] software is a general purpose rough set toolkit for analyzing the tabular data, and is freely distributed. The downloadable versions for both the Windows and Linux operating systems are available. The software supports the complete data mining process, from data pre-processing, including processing incomplete data, data discretization, generating reduct sets which contain essential attributes for the given data set, to classification, rule generation, and cross validation evaluation. Some discretization and reducts generation packages are from RSES library [8].

2. **RSES2.2.** RSES [8] stands for Rough Set Exploration System. There are downloadable versions for both the Windows and Linux operating systems. It is still maintained and being developed. The system supports data pre-processing, handling incomplete data, discretization, data decomposition into parts that share the same properties, reducts generation, classification, and cross validations and so on.

3. **ROSE2.** ROSE [9] stands for Rough Sets Data Explorer. This software is designed to process data with large boundary regions. The software supports data preprocessing, data discretization, handling missing values, core and reducts generation, classifications and rule generation, as well as evaluations. This software provides not only the classical rough set model, but also the variable precision model, which is not provided by [7] and [8].

4. **LERS.** LERS [12] stands for Learning from Examples based on Rough Sets. It is not publicly available. The system was designed especially to process missing values of attributes and inconsistency in the data set. Certain rules and possible rules are both extracted based on the lower and upper approximations.

In addition to the rough sets based systems mentioned above, there are other available knowledge discovery systems based on the methodologies of rough sets such as GROBIAN [13] and DBROUGH [14].

2.3   Current Research on Rule Evaluations

Rule generation often brings a large amount of rules to analyze. However, only part of these rules are distinct, useful and interesting. How to select only useful, interesting rules among all the available rules to help people understand the knowledge in the data effectively has drawn the attention of many researchers. Research on designing effective measures to evaluate rules comes from statistic, machine learning, data mining and other fields. These measures fall into two categories of evaluation measures.

Rule Interestingness Measures. One category of evaluating rules is to rank the rules by rule interestingness measures. Rules with higher interestingness measures are considered more interesting. The rule interestingness measures, originated from a variety of sources, have been widely used to extract interesting rules. Different applications may have different interestingness measures emphasizing on different aspect of the applications. Hilderman provided an extensive survey on the current interestingness measures [15] for different data mining tasks. For example, *support* and *confidence* are the most common interestingness measures to evaluate the association rules.

Not all the interestingness measures generate the same rank of interestingness for the same set of rules. Depending on different application purpose, appropriate rule interestingness measures should be selected to extract proper rules. More than one measure can be applied together to evaluate and explain the rules. Tan et. al. [16] evaluate twenty one measures in their comparative experiments and suggest different usage domains for these measures. They provide several properties of the interestingness measures so that one can choose a proper measure for certain applications. Their experiments also imply that not all the variables perform equally good at capturing the dependencies among the variables. Furthermore, there is no measure that can perform constantly better than the others in all application domains. Different measure is designed towards different domains.

Rule Quality Measures. The concept of rule quality measures was first proposed by Bruha [17]. The motivation for exploring this measure is that decision rules are different with different predicting abilities, different degrees to which people trust the rules and so on. Measures evaluating these different characteristics should be used to help people understand and use the rules more effectively. These measures have been known as rule quality measures.

The rule quality measures are often applied in the post-pruning step during the rule extraction procedure [18]. For example, some measures are used to evaluate whether the rules overfit the data. When removing an attribute-value pair, the quality measure does not decrease in value, this pair is considered to be redundant and will be pruned. As one of the applications, rule generation system uses rule quality measures to determine the stopping criteria for the rule generations and extract high quality rules. In [19] twelve different rule quality measures were studied and compared through the ELEM2 [18] system on their classification accuracies. The measures include empirical measures, statistical measures and measures from information theory.

3 Rule Evaluation on Knowledge Discovery

In this section, we first examine a current rough set knowledge discovery system, and suggest the importance of rule evaluations. We propose rule evaluation approaches and their functions in knowledge discovery systems.

3.1 Analyzing RSES – Rough Set Exploration System

We take the RSES [8] system as an example system, and study in more detail of the role of rule evaluations. We show that current systems are limited with regard to rule evaluation, and we emphasize the importance of rule evaluation in current knowledge discovery systems.

RSES (Rough Set Exploration System) is a well developed knowledge discovery system focusing on data analysis and classification tasks, which is currently under development. Figure 1 shows a use of the system on a heart disease data set for classification rule generation.

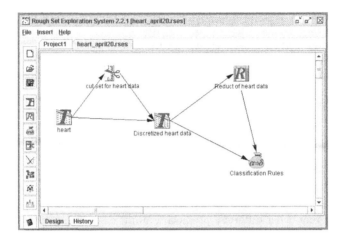

Fig. 1. Using Rough Set Exploration System on heart data

The data input to RSES is in the form of decision table $T = (C, D)$, where C is the condition attribute set and D is the decision attribute set. Preprocessing is conducted once the data is imported to the system, during which stage the missing attribute values are handled and discretization is performed if necessary as well. Reducts are then generated, classification rules based on the reducts are extracted.

RSES provides four approaches on processing missing attribute values, such as removing data records with missing values, assigning the most common values of the missing attribute within the same decision class and without the same decision class, and considering missing attribute values as a special value of the attribute [8]. These approaches are used during the data preprocessing stage in the system. Although these approaches are fast and can be directly applied in the data, they lack the ability of preserving the semantic meanings of the original data set. Missing values may be assigned, however, the filled values may not be able to fully represent what is missing in the data.

RSES provides rule postprocessing, which are "rule filter", "rule shorten" and "rule generalize". "Rule filter" removes from the rule set rules that do not satisfy certain support. "Rule shorten" shortens the length of the rules according to certain parameters [8]. "Rule generalization" generalizes rules according to a system provided parameter on the precision level. Although these rule postprocessing approaches provide an easier presentation of all the rule sets, these approaches do not provide ways to evaluate which rules are more interesting, and which rules have higher quality. These functions cannot provide a rank of rules according to a rule's significance to the users.

3.2   Enhanced Knowledge Discovery System Based on Rough Sets

We present a rough set based knowledge discovery system, as shown in Figure 2.

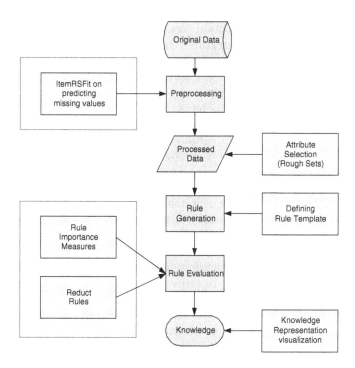

Fig. 2. The Knowledge Discovery Based on Rough Sets Theory

In this general purpose knowledge discovery system, data from different application domains are first imported into the system. Preprocessings including missing attribute values processing, discretization, are conducted in this stage. After the data is preprocessed attribute selections are conducted. Depending on the output, different attribute selection approaches can be applied here. Rule generation algorithms extract rules. After the rule sets are obtained, the important postprocessing - rule evaluations are performed in this stage. Rules are finally represented, possibly visualized in a certain format, as knowledge to the end users.

We introduce three approaches integrated into this general purpose KDD system as shown in Figure 2. The first approach *ItemRSFit* is used in the data preprocessing stage. The second approach, *rule importance measure* is used to rank rules during the rule evaluation process. The third approach of extracting *reduct rules* is also used during the rule evaluation stage. We will elaborate these approaches in the following.

I. Predicting missing attribute values based on Frequent Itemset. ItemRSFit approach is a recently developed approach on predicting missing attribute values based on association rules algorithm and rough sets theory. It has been shown on both large scale real world data set and UCI machine learning data sets on the improved prediction accuracies.

ItemRSFit approach is an integration of two other approaches from association rule algorithm and rough sets theory. Priori to the association rule generation, frequent itemsets are generated based on the item-item relations from the large data set according to a certain *support*. Thus the frequent itemsets of a data set represent strong correlations between different items, and the itemsets represent probabilities for one or more items existing together in the current transaction. When considering a certain data set as a transaction data set, the implications from frequent itemsets can be used to find to which attribute value the missing attribute is strongly connected. Thus the frequent itemset can be used for predicting the missing values. We call this approach "itemset-approach" for prediction. The larger the frequent itemsets used for the prediction, the more information from the data set itself will be available for prediction, hence the higher the accuracy will be obtained. However, generating frequent itemset for large data set is time-consuming. Although itemsets with higher support need less computation time, they restrict item-item relationships, therefore not all the missing values can be predicted. In order to balance the tradeoff between computation time and the percentage of the applicable prediction, another approach must be taken into consideration.

A reduct contains a subset of condition attributes that are sufficient enough to represent the whole data set. The intersection of all the possible reduct is the core. Therefore the attributes contained in the reduct or core are more important and representative than the rest of the attributes. Thus by examining only attributes within the same core or reduct to find the similar attribute value pairs for the data instance containing the missing attribute values, we can assign the most relevant value for the missing attribute. Since this method only considers a subset of the data set, which is either the core or the reduct, the prediction is quite fast. This approach "RSFit" is recently proposed in [20], and it is an alternative approach designed for fast prediction. It can be used to predict missing attributes that cannot be predicted by the frequent itemset.

We integrate the prediction based on frequent itemset and RSFit approach into a new approach **ItemRSFit** to predict missing attribute values. Frequent itemsets are used to predict missing values first, and RSFit approach is used to predict the rest of the missing values that cannot be predicted by the frequent itemsets. This integrated approach can predict missing values from the data itself, therefore less noise is brought into the original data. The details on the ItemRSFit approach is presented in [21].

Properly processed data can improve the quality of the generated knowledge. Therefore the ItemRSFit approach is used in this system at the preprocessing stage. It helps to preserve the qualities of the original input data to this system, thus facilitate the rule evaluation process.

II. Rule Importance Measures. Rule importance measure [2] is developed to provide a diverse rank of how important the association rules are, although this approach can also be applied to rules generated by other rule discovery algorithms.

Association rules algorithm can be applied on this transaction data set to generate rules, which have condition attributes on the antecedent part and decision attributes on the consequent part of the rules. Rules generated from different reduct sets can contain different representative information. If only one reduct set is being considered to generate rules, other important information might be omitted. Using multiple reducts, some rules will be generated more frequently than other rules. We consider the rules that are generated more frequently more important.

The *Rule Importance* is defined to be important by the following definition.

Definition 5. *If a rule is generated more frequently across different rule sets, we say this rule is* more important *than rules generated less frequently across those same rule sets.*

Rule importance measure is defined as follows,

Definition 6

$$Rule\ Importance\ Measure = \frac{\begin{array}{c}Number\ of\ times\ a\ rule\ appears\ in\ all\\ the\ generated\ rules\ from\ the\ reduct\ sets\end{array}}{Number\ of\ reduct\ sets}.$$

The definition of the rule importance measure can be elaborated by Eq. 3. Let n be the number of reducts generated from the decision table $T(C, D)$. Let $RuleSets$ be the n rule sets generated based on the n reducts. $ruleset_j \in RuleSets$ $(1 \leq j \leq n)$ denotes individual rule sets containing rules generated based on reducts. $rule_i$ $(1 \leq i \leq m)$ denotes the individual rule from $RuleSets$. RIM_i represents the rule importance measure for the individual rule. Thus the rule importance measures can be computed by the following

$$RIM_i = \frac{|\{ruleset_j \in RuleSets | rule_i \in ruleset_j\}|}{n}. \tag{3}$$

The details of how to use rule importance measures can be found in [2]. Rule importance measure can be integrated into the current rough sets based knowledge discovery system to be used during the rule evaluation process. A list of ranked important rules can therefore be presented with their rule importance measures to facilitate the understanding of the extracted knowledge.

III. Extracting Reduct Rules. In [3] a method of discovering and ranking important rules by considering rules as attributes was introduced. The motivation comes from the concept of reduct. A reduct of a decision table contains attributes that can fully represent the original knowledge. If a reduct is given, rules extracted based on this reduct are representative of the original decision table. Can we take advantage of the concept of a reduct to discover important rules?

We construct a new decision table $A_{m \times (n+1)}$, where each record from the original decision table $u_0, u_1, ..., u_{m-1}$ are the rows, and the columns of this new table consists of $Rule_0, Rule_1, ..., Rule_{n-1}$ and the decision attribute. We say a rule can be applied to a record in the decision table if both the antecedent and the consequent of the rule appear together in the record. For each $Rule_j$ ($j \in [0, ..., n-1]$), we assign 1 to cell $A[i, j]$ ($i \in [0, ..., m-1]$) if the rule $Rule_j$ can be applied to the record u_i. We set 0 to $A[i, j]$ otherwise. The decision attribute $A[i, n]$ ($i \in [0, ..., m-1]$) remains the same as the original values of the decision attribute in the original decision table. Eq. 4 shows the conditions for the value assignments of the new decision table.

$$A[i, j] = \begin{cases} 1, & \text{if } j < n \text{ and } Rule_j \text{ can be applied to } u_i \\ 0, & \text{if } j < n \text{ and } Rule_j \text{ cannot be applied to } u_i \\ d_i, & \text{if } j = n \text{ and } d_i \text{ is the corresponding decision attributes for } u_i \end{cases}$$
(4)

where $i \in [0, ..., m-1], j \in [0, ..., n-1]$.

We further define *Reduct Rule Set* and *Core Rule Set*.

Definition 7. *Reduct Rule Set*. *We define a reduct generated from the new decision table A as **Reduct Rule Set**. A Reduct Rule Set contains Reduct Rules.*

The *Reduct Rules* are representative rules that can fully describe the decision attribute.

Definition 8. *Core Rule Set*. *We define the intersection of all the* Reduct Rule Sets *generated from this new decision table A as* Core Rule Set. *A Core Rule Set contains **Core Rules**.*

The *Core Rules* are contained in every *Reduct Rule Set*.

By considering rules as attributes, reducts generated from the new decision table contain all the important attributes, which represent the important rules generated from the original data set; and it excludes the less important attributes. Core attributes from the new decision table A contain the most important attributes, which represent the most important rules.

Other Enhancements. The three approaches discussed in our research have shown to effectively evaluate rules. There are other techniques that can be used along with these approaches in Figure 2. For example, during the rule generation process, properly defined rule templates can not only reduce the computation of rule generations, but it also ensures high quality rules, or interesting rules generated according to the application purposes. Important attributes, such as probe attributes (discussed in Section 5) can be defined in the data preprocessing stage for generating rules containing such attributes for generating expected rules.

Our motivation is, proposing approaches to enhance the current knowledge discovery system, to facilitate the knowledge discovery process on discovering more interesting and higher quality rules.

4   Experiments

We demonstrate, through a series of experiments, that systems improved by the proposed rule evaluation approaches can help humans discover and understand more important rules.

4.1   Specifying Rule Templates

Apriori association rules algorithm is used to generate rules. Because our interest is to make decisions or recommendations based on the condition attributes, we are looking for rules with only decision attributes on the consequent part. Therefore, we specify the following 2 rule templates to extract rules we want as shown by Template 5, and to subsume rules as shown by Template 6.

$$\langle Attribute_1, Attribute_2, \ldots, Attribute_n \rangle \rightarrow \langle DecisionAttribute \rangle \qquad (5)$$

Template 5 specifies only decision attributes can be on the consequent part of a rule, and $Attribute_1$, $Attribute_2$, ..., $Attribute_n$ lead to a decision of $DecisionAttribute$.

We specify the rules to be removed or subsumed using Template 6. For example, given rule

$$\langle Attribute_1, Attribute_2 \rangle \rightarrow \langle DecisionAttribute \rangle \qquad (6)$$

the following rules

$$\langle Attribute_1, Attribute_2, Attribute_3 \rangle \rightarrow \langle DecisionAttribute \rangle \qquad (7)$$

$$\langle Attribute_1, Attribute_2, Attribute_6 \rangle \rightarrow \langle DecisionAttribute \rangle \qquad (8)$$

can be removed because they are subsumed by Template 6. Take the geriatric care data in Table 1 as an example, in the rule set, a rule shown as Eq. 9 exists

$$SeriousChestProblem \rightarrow Death \qquad (9)$$

the following rule is removed because it is subsumed.

$$SeriousChestProblem, TakeMedicineProblem \rightarrow Death \qquad (10)$$

4.2 Geriatric Care Data Set

We perform experiments on a geriatric care data set as shown in Table 1. This data set is an actual data set from Dalhousie University Faculty of Medicine to determine the survival status of a patient giving all the symptoms he or she shows. The data set contains 8,547 patient records with 44 symptoms and their survival status (dead or alive). We use *survival status* as the decision attribute, and the 44 symptoms of a patient as condition attributes, which includes *education level, the eyesight, hearing, be able to walk, be able to manage his/her own meals, live alone, cough, high blood pressure, heart problem, cough, gender, the age of the patient at investigation* and so on.[1] There is no missing value in this data set. There are 12 inconsistent data entries in the medical data set. After removing these instances, the data contains 8,535 records. [2]

Table 1. Geriatric Care Data Set

edulevel	eyesight	hearing	health	trouble	livealone	cough	hbp	heart	stroke	...	sex	livedead
0.6364	0.25	0.50	0.25	0.00	0.00	0.00	0.00	0.00	0.00	...	1	0
0.7273	0.50	0.25	0.25	0.50	0.00	0.00	0.00	0.00	0.00	...	2	0
0.9091	0.25	0.50	0.00	0.00	0.00	0.00	1.00	1.00	0.00	...	1	0
0.5455	0.25	0.25	0.50	0.00	1.00	1.00	0.00	0.00	0.00	...	2	0
0.4545	0.25	0.25	0.25	0.00	1.00	0.00	1.00	0.00	0.00	...	2	0
0.2727	0.00	0.00	0.25	0.50	1.00	0.00	1.00	0.00	0.00	...	2	0
0.0000	0.25	0.25	0.25	0.00	0.00	0.00	0.00	1.00	0.00	...	1	0
0.8182	0.00	0.50	0.00	0.00	0.00	0.00	1.00	0.00	0.00	...	2	0
...	...	...	...	...	...	...	...	...	...	...	...	...

The ItemRSFit approach is implemented by Perl and the experiments are conducted on Sun Fire V880, four 900Mhz UltraSPARC III processors. We use apriori frequent itemset generation [23] to generate frequent 5-itemset. The core generation in RSFit approach is implemented with Perl combining the SQL queries accessing MySQL (version 4.0.12). ROSETTA software [7] is used for reduct generation.

4.3 Experiments on Predicting Missing Attribute Values

In order to show the ItemRSFit approach obtains better prediction accuracy than the existing approach (i.e., RSFit), we perform the experiments on the geriatric care data set by randomly selecting 150 missing values from the original data. We then apply both RSFit approach and ItemRSFit approach on predicting missing values, and compare the accuracy of the prediction. Figure 3 demonstrates the comparison predicting abilities between RSFit and ItemRSFit approaches. We can see from the figure that the smaller the support is, the more accurate the

---

[1] Refer to [22] for details about this data set.
[2] Notice from our previous experiments that core generation algorithm cannot return correct core attributes when the data set contains inconsistent data entries.

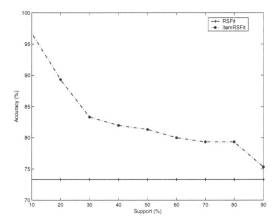

Fig. 3. Accuracy Comparisons for Geriatric Care Data with 150 Missing Attribute Values

prediction of the missing attribute values for the ItemRSFit approach obtains; whereas for the RSFit approach, the accuracy remains the same as the value of support gets smaller; and the accuracy obtained by RSFit is always lower than the ItemRSFit approach. This result demonstrates that frequent itemsets as knowledge base can be effectively applied for predicting missing attribute values.

4.4   Experiments on Rule Importance Measure

In our experiment, we use the genetic algorithm to generate multiple reduct sets with the option of full discernibility. The apriori algorithm [23] for large item sets generation.

The core attributes for this data set are *eartrouble, livealone, heart, high-bloodpressure, eyetrouble, hearing, sex, health, educationlevel, chest, housework, diabetes, dental, studyage.*

Table 2. Reduct Sets for the Geriatric Care Data Set after Preprocessing

No.	Reduct Sets
1	{edulevel,eyesight,hearing,shopping,housewk,health,trouble,livealone, cough,sneeze,hbp,heart,arthriti,eyetroub,eartroub,dental, chest,kidney,diabetes,feet,nerves,skin,studyage,sex}
2	{edulevel,eyesight,hearing,phoneuse,meal,housewk,health,trouble,livealon, cough,sneeze,hbp,heart,arthriti,evetroub,eartroub,dental, chest,bladder,diabetes,feet,nerves,skin,studyage,sex}
...	...
86	{edulevel,eyesight,hearing,shopping,meal,housewk,takemed,health, trouble,livealone,cough,tired,sneeze,hbp,heart,stroke,arthriti, eyetroub,eartroub,dental,chest,stomach,kidney,bladder,diabetes, feet,fracture,studyage,sex}

Table 3. Rule Importance for the Geriatric Care Data

No.	Selected Rules	Rule Importance
0	SeriousHeartProblem → Death	100%
1	SeriousChestProblem → Death	100%
2	SeriousHearingProblem, HavingDiabetes → Death	100%
3	SeriousEarTrouble → Death	100%
4	SeriousEyeTrouble → Death	100%
5	Sex_Female → Death	100%
...	...	...
10	Livealone, HavingDiabetes, NerveProblem → Death	95.35%
...	...	...
216	SeriousHearingProblem, ProblemUsePhone → Death	1.16%
217	TakeMedicineProblem, NerveProblem → Death	1.16%

Table 2 shows selected reduct sets among the 86 reducts generated by ROSETTA. All of these reducts contain the core attributes. For each reduct set, association rules are generated with $support = 30\%, confidence = 80\%$. [3] 218 unique rules are generated over these 86 reducts. These rules as well as their rule importance are shown in Table 3. Among these 218 rules, 87 rules have rule importance of no less than 50% , 8 of which have rule importance of 100%. All the rules with rule importance of 100% contain only core attributes.

4.5   Experiments on Generating Reduct Rules

The new decision table $A_{8535 \times 219}$ is constructed by using the 218 rules [4] as condition attributes, and the original decision attribute as the decision attribute. Note that after reconstructing the decision table, we must check for inconsistency again before generating reduct rules for this table. After removing the inconsistent data records, there are 5709 records left in the new decision table. The core rule set is empty. We use Johnson's reduct generation algorithm on this table $A'_{5709 \times 219}$ and the reduct rule set is $\{Rule_0, Rule_1, Rule_3, Rule_5, Rule_{19}, Rule_{173}\}$. We show these rules in Table 4. From Table 4 we can see that the reduct rule sets contain 6 rules. There are 4 rules judged to be the most important. The rule importance for $Rule_0$, $Rule_1$, $Rule_3$ and $Rule_5$ are all 100%. The $Rule_{19}$ has the importance of 82.56%, which is more important among the 218 rules.

---

[3] Note that the value of support and confidence can be adjusted to generate as many or as few rules as required.

[4] There are 1615 rules generated by apriori algorithm from the original data set with $support = 30\%, confidence = 80\%$, after applying the rule template. We can circumvent problems inherent in considering all 1615 generated rules using the 218 unique rules that are derived from the 86 reducts obtained by ROSETTA's genetic algorithm.

Table 4. Reduct Rules for the Geriatric Care Data

No. in Table 3	Reduct Rules	Rule Importance
0	SeriousHeartProblem → Death	100%
1	SeriousChestProblem → Death	100%
3	SeriousEarTrouble → Death	100%
5	Sex_Female → Death	100%
19	Livealon, OftenSneeze, DentalProblems, HavingDiabetes → Death	82.56%
173	ProblemHandleYourOwnMoney → Death	27.91%

5   A Case Study: Survival Analysis in HYRIS

This section provides a case study to illustrate how a rough sets based KDD system provides a useful mechanism for analyzing and distilling essential attributes and rules from survival data, and evaluates the generated rules in postprocessing for survival analysis.

Use of medical survival data challenges researchers because of the size of data sets and vagaries of their structures. Among prognostic modeling techniques that induce models from medical data, survival analysis warrants special treatment in the type of data required and its modeling. Data required for medical analysis includes demographic, symptoms, laboratory tests and treatment information. Special features for survival data are the events of interest, censoring, follow-up time and survival time specific for each type of disease. Such data demands powerful analytical models for survival analysis. The studies applying rough sets to survival analysis related to our work are [24][25]. They illustrated rough sets contribution to a medical expert system for throat cancer patients successfully. Rough sets and decision trees have been used to study kidney dialysis patients' survival [26].

HYRIS (HYbrid Rough sets Intelligent System) [27][28] is designed specifically to consider survival analysis with several statistical approaches. HYRIS uses the CDispro algorithm from a previous study [4]. HYRIS successively derives dispensable, probe attribute, reduct and probe reduct together with life time table and Kaplan-Meier survival curves [27]. In addition to survival analysis, HYRIS provides a general data analysis and decision rule generation and evaluation [27] as well.

HYRIS Case Study I. HYRIS is able to analyze *censor variable* and *survival time* attributes that are a speciality for survival analysis. Given the survival data set, the system can identify the covariant levels of particular attributes according to rough sets and several statistical approaches. The Kaplan-Meier method, hazard function, hypothesis testing, log-rank, Brewslow, Tarone-Ware tests, p-value and CDispro [4]. incorporate the rough sets framework to generate core, dispensable attributes, probe attribute, reducts, and probe reducts which are the informative attributes. Consequently, the rules are derived and validated with ELEM2 [18].

We demonstrate the utility of HYRIS by investigating a particular problem using both actual and benchmark medical data sets: geriatric data, melanoma data [29], pneumonia data [30] and primary biliary cirrhosis data (PBC) [31]. For the geriatric care data set, time lived in months are used as the survival time target function. Inconsistent records are removed. Data is discretized according to percentile groups. The *age* is used as probe attribute, the reducts and probe reducts are generated successfully. The rules generated for geriatric data care are the decision rules for predicting survival time. Note that in Section 4 the experimental results showing important rules are used to predict the survival status, not the survival time.

Two sample survival prediction rules out of 1,600 rules of geriatric care data set (when considering the probe attribute {*ExperienceDiabetes*}) generated from HYRIS are provided as follows:

Rule 1: *UnHealthy, SevereHearingDamage, NerveProblem, FootProblem, SeriousHeartProblem, DentalDisease, StomachDisease, HighBloodPressure, ExperienceDiabetes → SurvivalTime = 7-18 months.*

Rule 2: *FemalePatient, LowEducationLevel, EyesightProblemLowToSeriousType, HealthProblemFromLowToSeriousType, HearQuiteWell, DoNotHaveDiabetesExperience, EasilyTired, FootProblem, → SurvivalTime = 56-73 months.*

When comparing the accuracy of rules that were generated from original attributes and those generated from reducts, the accuracy of all data sets range between 83.7851%–90.5686%. Rule performance outcomes are improved significantly as reported in [4].

HYRIS Case Study II. HYRIS can accomplish preprocessing, learning and model construction and broaden further to use in rule evaluation and postprocessing. We continue a series of studies in [28]. In this case study, we propose an alternative approach for decision rule learning with rough sets theory in the postprocessing step called *ROSERULE - Rough Sets Rule Reducts Learning Algorithm.* ROSERULE learned and analyzed from the rule set to generate rule reducts which can be used to reduce the number of the rules. Results imply a reduced number of rules that successfully preserve the original classification. The rule numbers of geriatric data set reduced from 1,600 to 1,150, melanoma data set reduced from 16 to 15, pneumonia data set reduced from 606 to 42 and PBC data set reduced from 83 to 72. At the same time, the prediction accuracy is preserved for all data sets.

6   Conclusion

We study the work of rough sets based rule evaluations on knowledge discovery system. We propose solutions to the challenging problems brought by large real world data sets, such as the existence of missing values and analyzing huge amount of generated rules manually. Three rough set based approaches to enhance the current KDD systems on rule evaluations are introduced. The *ItemRS-Fit* approach is used to predict missing attribute values using frequent itemset as a knowledge base. Complete data can be obtained using this approach. The *rule importance measure* provides a ranking of how important is a rule. Finally, the *reduct rules* are extracted using the concept of reduct by considering rules as condition attributes in a decision table. Experimental results on a real world geriatric care data set demonstrate the utilities of applying rough sets based rule evaluations to enhance current KDD systems. A case study of a recent knowledge discovery system shows the applications of approaches which have been incorporated into HYRIS with an emphasis on survival analysis.

Acknowledgements

This research is supported by the Natural Sciences and Engineering Research Council of Canada (NSERC). The research of Puntip Pattraintakorn has also been supported by a grant from the King Mongkut's Institute of Technology Ladkrabang (KMITL) research fund, Thailand. We would also like to thank Arnold Mitnitski from Dalhousie University for providing the geriatric care data set, and the anonymous reviewers for their helpful comments.

References

1. Pawlak, Z.: Rough Sets. In Theoretical Aspects of Reasoning about Data. Kluwer Academic Publishers (1991)
2. Li, J. and Cercone, N.: Introducing A Rule Importance Measure. Transactions on Rough Sets, Springer LNCS, vol **5** (2006)
3. Li, J., Cercone, N.: Discovering and Ranking Important Rules. In Proceedings of IEEE International Conference on Granular Computing, vol **2**, Beijing China 25-27 July (2005) 506–511
4. Pattaraintakorn, P., Cercone, N., Naruedomkul, K.: Hybrid Intelligent Systems: Selecting Attributes for Soft-Computing Analysis. In Proc. of the 29th Annual International Computer Software and Applications Conference (COMPSAC), vol **2** (2005) 319–325
5. Kryszkiewicz, M., Rybinski, H.: Finding Reducts in Composed Information Systems, Rough Sets, Fuzzy Sets Knowldege Discovery. In W.P. Ziarko (Ed.), Proceedings of the International Workshop on Rough Sets, Knowledge Discovery, Heidelberg/Berlin: Springer-Verlag (1994) 261–273
6. Jan Bazan, Hung Son Nguyen, Sinh Hoa Nguyen, Piotr Synak, and Jakub Wroblewski.: Rough set algorithms in classification problems. Rough Set Methods and Applications: New Developments in Knowledge Discovery in Information Systems, volume 56 of Studies in Fuzziness and Soft Computing, pages 49-88. Physica-Verlag, Heidelberg, Germany (2000).
7. Øhrn, A.: Discernibility and Rough Sets in Medicine: Tools and Applications. PhD Thesis, Department of Computer and Information Science, Norwegian University of Science and Technology, Trondheim, Norway. (1999)
8. RSES 2.2 User's Guide. Warsaw University. http://logic.mimuw.edu.pl/~rses/

9. Predki, B., Wilk, Sz.: Rough Set Based Data Exploration Using ROSE System. In: Foundations of Intelligent Systems. Ras, Z. W., Skowron, A., Eds, LNAI **1609**, Springer-Verlag, Berlin (1999) 172-180
10. Chouchoulas, A. and Shen, Q.: Rough Set-Aided Keyword Reduction For Text Categorization. Applied Artificial Intelligence, vol **15** (2001) 843-873
11. Hu, X., Lin, T., Han, J.: A New Rough Sets Model Based on Database Systems. Fundamenta Informaticae **59** no.2-3 (2004) 135-152
12. Freeman, R. L., Grzymala-Busse, J. W., Riffel, L. A., Schroeder, S. R.: Analyzing the Relation Between Heart Rate, Problem Behavior, and Environmental Events Using Data Mining System LERS. In 14th IEEE Symposium on Computer-Based Medical Systems (CBMS'01) (2001)
13. Ivo, D., Gunther, G.: The Rough Set Engine GROBIAN. In Proc. of the 15th IMACS World Congress, vol **4**, Berlin, August (1997)
14. Hu, T., Shan, N., Cercone, N. and Ziarko, W.: DBROUGH: A Rough Set Based Knowledge Discovery System, Proc. of the 8th International Symposium on Methodologies for Intelligent System, LNAI **869**, Spring Verlag (1994) 386-395
15. Hilderman, R. and Hamilton, H.: Knowledge discovery and interestingness measures: A survey. Technical Report 99-04, Department of Computer Science, University of Regina, October (1999)
16. Pang-Ning Tan and Vipin Kumar and Jaideep Srivastava: Selecting the right interestingness measure for association patterns. Processings of SIGKDD. (2002) 32-41
17. Bruha, Ivan: Quality of Decision Rules: Definitions and Classification Schemes for Multiple Rules. In Machine Learning and Statistics, The Interface, Edited by G. Nakh aeizadeh and C. C. Taylor. John Wiley & Sons, Inc. (1997) 107-131
18. An, A. and Cercone, N.: ELEM2: A Learning System for More Accurate Classifications. In: Proceedings of Canadian Conference on AI (1998) 426-441
19. An, A. and Cercone, N.: Rule Quality Measures for Rule Induction Systems: Description and Evaluation. Computational Intelligence. **17-3** (2001) 409-424.
20. Li, J. and Cercone, N.: Assigning Missing Attribute Values Based on Rough Sets Theory. In Proceedings of IEEE Granular Computing, Atlanta, USA. (2006)
21. Li, J. and Cercone, N.: Predicting Missing Attribute Values based on Frequent Itemset and RSFit. Technical Report, CS-2006-13, School of Computer Science, University of Waterloo (2006)
22. Li, J. and Cercone, N.: Empirical Analysis on the Geriatric Care Data Set Using Rough Sets Theory. Technical Report, CS-2005-05, School of Computer Science, University of Waterloo (2005)
23. Borgelt, C.: Efficient Implementations of Apriori and Eclat. Proceedings of the FIMI'03 Workshop on Frequent Itemset Mining Implementations. In: CEUR Workshop Proceedings (2003) 1613-0073 http://CEUR-WS.org/Vol-90/borgelt.pdf
24. Bazan, J., Osmolski, A., Skowron, A., Slezak, D., Szczuka, M., Wroblewski, J.: Rough Set Approach to the Survival Analysis. In Alpigini, J. J., et al. (Eds.): The Third International Conference on Rough Sets and Current Trends in Computing (RSCTC), Proceedings, LNAI **2475**, Springer-Verlag Berlin Heidelberg (2002) 522-529
25. Bazan, J., Skowron, A., Slezak, D., Wroblewski, J.: Searching for the Complex Decision Reducts: The Case Study of the Survival Analysis, LNAI **2871**, Springer-Verlag, Berlin Heidelberg (2003) 160-168
26. A. Kusiak, B. Dixon, S. Shah: Predicting Survival Time for kidney Dialysis Patients: A Data Mining Approach, Computers in Biology and Medicine 35 (2005) 311-327

27. Pattaraintakorn, P., Cercone, N., Naruedomkul, K.: Selecting Attributes for Soft-Computing Analysis in Hybrid Intelligent Systems. In D. Slezak et al. (Eds.): Rough Sets, Fuzzy Sets, Data Mining, and Granular Computing 10th International Conference (RSFDGrC), Proceedings, Part II Series: Lecture Notes in Computer Science, Subseries: LNAI **3642**, Springer-Verlag, Berlin, Heidelberg (2005) 698–708
28. Pattaraintakorn, P., Cercone, N., Naruedomkul, K.: Rule Analysis with Rough Sets Theory, The IEEE International Conference on Granular Computing, Atlanta, USA (2006)
29. Elisa, L.T., John, W.W.: Statistical methods for survival data analysis, 3rd edn. New York: John Wiley and Sons (2003)
30. Klein, J.P., Moeschberger, M.L.: Survival analysis: techniques for censored and truncated data, 2nd edn. Berlin: Springer (2003)
31. Newman, D.J., Hettich, S., Blake, C.L. and Merz, C.J.: UCI Repository of machine learning databases. University of California, Irvine, Department of Information and Computer Seiences (1998) http://www.ics.uci.edu/~mlearn/MLRepository.html

The Impact of Rough Set Research in China: In Commemoration of Professor Zdzisław Pawlak

Qing Liu[1,2] and Hui Sun[1]

[1] Department of Computer Science & Technology
Nanchang Institute of Technology, Nanchang 330099, China
qliu_ncu@yahoo.com.cn
[2] Department of Computer Science & Technology
Nanchang University, Nanchang 330029, China

This article is dedicated to the creative genius Zdzisław Pawlak for his contribution to the theoretical development of science and technology in China. His distinguished discovery of Rough Set Theory is a formal theory which is well suited for uncertainty computing to analyze imprecise, uncertain or incomplete information of data. Inspired by his work scientists and engineers in China has developed many theories and applications in various science and technology fields. For instance, J.H.Dai studied the theories of Rough Algebras and Axiom Problem of Rough 3-Valued Algebras [1, 2]. G.L.Liu studied the Rough Sets over Fuzzy Lattices [3, 4]. D.W.Pei studied the Generalized Model of Fuzzy Rough Sets [5]. W.Z.Wu Studied the On Random Rough Sets [6]. D.Q.Miao studied the Rough Group and Their Properties [7]. These are part of their recent research results related to rough set theory. As a matter of fact, there are still many researchers working in the field of rough sets in China, who have proposed many creative results for last few years. These results are not listed one by one in this short commemorative article. We will try to review all the "Rough Set" researchers and their research results in the appeared next article.

In this article, we present only a recent partial research results of the authors. Based on Rough Logic and Decision Logic defined by Pawlak [8, 9], first author Liu has proposed a rough logic in a given information system [10]. Influenced by the concept of granular language proposed by Skowron [11], the granular logic defined by Polkowski [12], and the work of Lin, Yao in [13, 14], we also have defined a granular logic by applying the semantics of rough logical formulas in a given information system, and have created the deductive systems as well as have discussed many properties in [15, 16]. The proposed granular logic is a set which consists of granular formulas of form $m(F)$, where F is the rough logical formula in the given information system. It is used as individual variable of semantic function symbol m, so, we call it a paradigm of higher order logic. Truth values of granular formula of form $m(F)$ in the logic have two types. One is the function value, which is the meaning of rough logical formula, a subset in U; Another is the truth value of a degree, which is equal to a degree of meaning of the formula to close to universe U of objects. Pawlak introduced the concept of rough truth in 1987, assuming that a formula is roughly true in a

J.F. Peters et al. (Eds.): Transactions on Rough Sets VI, LNCS 4374, pp. 172–175, 2007.
© Springer-Verlag Berlin Heidelberg 2007

given information system if and only if the upper approximation of meaning of the formula is equal to the whole universe. So, our approach extends Pawlak's approach [9]. Skolem clause form, resolution principles, λ-resolution strategies and deductive reasoning of the granular logic are also discussed in the next article. These logic systems should be an extension of Rough Logic proposed by Pawlak in 1987 [9]. The practicability of the higher order logic will offer the new idea for studying classical logic. It could also be a theoretical tool for studying granular computing. Based on reference [23], we further propose to use rough set theory to find out minimal approximate solution set in the approximate solution space of differential equations and functional variation problems in mechanics. This could be a new studying style in rough set applications.

The significance and future development direction of the proposed Rough Sets are described. Any undefinable subset on the universe of a given information system is constructed into precise definable lower and upper approximations via indiscernibility relation. Hence, complex and difficult problems on undefinable sets are resolved or transformed into precise definable lower and upper approximations [8, 17]. This is one of a great contribution of Pawlak's Rough Sets Theory. Successful applications of rough sets in many fields offer a lot of new idea of studying granular computing, which also promote the development of granular computing.

Founder of predicate logic, G. Frege proposed the vague boundary in 1904, that is, how to compute the number of elements on vague boundary [18, 19]. Many mathematicians and computer scientists have made hard efforts on the question.

L. A. Zadeh proposed the Fuzzy Sets (FS) in 1965. He attempted to solve the computability of Frege's vague boundary by Fuzzy Set concept [20]. Unfortunately, Fuzzy Sets are not mechanically computable, that is, the formula of exact describing for the fuzzy concept hasn't been given. Therefore, the number of elements on the vague boundary could not be computed by exact formula. For example, the membership μ in Fuzzy Sets and fuzzy operator λ in operator fuzzy logic [21], could not be computed exactly.

Z.Pawlak proposed the Rough Sets (RS) in 1982 for computing Frege's vague boundary [8, 17], and the number of elements on the vague boundary could be exactly computed by it.

Rough Set Theory is a new tool to deal with incomplete and uncertainty problems. In the field of computer applications nowadays, this theory is no doubt a challenge to other uncertainty theories. Since rough set theory is one of the most important, newest theories and with the rapid development, it is also very important in artificial intelligence and cognitive science. Especially methods,which are based on rough set theory alone or are in combination with other approaches, have been used over a wide range of applications in many areas. More and more people in the region of China are attracted by them.

In this article, we also present state of art of RS in China. Since May, 2001 Professor Pawlak being invited to China made his keynote speech in a conference and gave many invited talks among universities, the research of RS have been rapidly developed in China. Rough Set Theory has been used for important applications in human life, such as, data reduction, approximate classification of data, management systems of business information, computing the average value and the standard deviation in quality test of products of statistics and so on.

We developed successfully the Business Information Management Systems with rough set approach, which is a management for price of houses of some region [22].

We defined an indiscernibility relation on numerical interval $[a, b]$ by using ancient mathematical Golden Cut method and created rough sets by the defined relation. Based on the rough set approach, we developed a "Diagnosis Software of Blood Viscosity Syndrome on Hemorheology", which is used to test Blood Viscosity chroma of patients. The systems has been applied in the clinic for many years. The medicine experts review that the diagnosis software is precursive, creative, scientific and practical [24].

In this article, we present the state of art of primary rough set research results and their applications in practice in China during last few years. Especially, in 2001, Pawlak was invited to China, his keynote speech had made a big influence on the development of science and technology of China.

We would like to thank Pawlak for his fundamental and significant contribution to the development of rough set research in China. We would like to thank the editor-in-chief Professor James F. Peters, Professor Andrej Skowron and Professor Ewa Orlowska for their kindness to let us publish article in this historical event to commemorate the great scientist Zdzisław Pawlak for his contribution to the science and technology world. Thanks are also to the support of Natural Science Fund of China (NSFC-60173054). At last we would like to thank Dr. James Kuodo Huang (who is a IEEE member and have taught in the universities of USA for over 20 years) for his kind suggestions of English in this article. Still we would like to take the whole responsibility for any further errors made in this article.

References

1. Dai,J.H., Axiom Problem of Rough 3-Valued Algebras, The Proceedings of International Forum on Theory of GrC from Rough Set Perspective (IFTGrCRSP2006), Nanchang, China, Journal of Nanchang Institute of Technology, Vol.25, No.2, 2006, 48-51.
2. Dai,J.H., On Rough Algebras, Journal of software 16 (2005),1197-1204.
3. Liu,G.L., Rough Set Theory Over Fuzzy Lattices, The Proceedings of International Forum on Theory of GrC from Rough Set Perspective (IFTGrCRSP2006), Nanchang, China, Journal of Nanchang Institute of Technology, Vol.25, No.2, 2006, 51-51.
4. Liu,G.L., The Topological Structure of Rough Sets over Fuzzy Lattices, 2005 IEEE International Conference on Granualr Computing, Vol.I, Proceedings, Beijing, China, July 25-27,2005, 535-538.
5. Pei,D.W., A Generalized Model of Fuzzy Rough Sets, Int. J. General Systems 34 (5)2005, 603-613.

6. Wu,W.Z., On Random Rough Sets, The Proceedings of International Forum on Theory of GrC from Rough Set Perspective (IFTGrCRSP2006), Nanchang, China, Journal of Nanchang Institute of Technology, Vol.25, No.2, 2006, 66-69.
7. Miao,D.Q., Rough Group, Rough Subgroup and their Properties, LNAI 3641, 10th International Conference, RSFDGrC2005, Regina, Canada, August/September 2005, Proceedings, Part I, 104-113.
8. Pawlak, Z., Rough Sets: Theoretical Aspects of Reasoning about Data, Kluwer Academic Publishers, Dordrecht , 1991.
9. Pawlak, Z., Rough Logic, Bulletin of the Polish Academy of Sciences, Technical Sciences, Vol.35, No.5-6, 1987, 253-259.
10. Liu, Q., Liu, S.H. and Zheng, F., Rough Logic and Applications in Data Reduction, Journal of Software, Vol.12, No.3, March 2001,415-419. (In Chinese).
11. Skowron, A., Toward Intelligent Systems: Calculi of Information Granules, Bulletin of International Rough Set Society, Vol.5, No.1/2, Japan, 2001,9-30.
12. Polkowski,L., A Calculus on Granules from Rough Inclusions in Information Systems, The Proceedings of International Forum on Theory of GrC from Rough Set Perspective (IFTGrCRSP2006), Nanchang, China, Journal of Nanchang Institute of Technology, Vol.25, No.2, 2006, 22-27.
13. Lin,T.Y., Liu,Q., First Order Rough Logic I: Approximate Reasoning via Rough Sets, Fundamenta Informaticae 2-3, 1996, 137-153.
14. Yao,Y.Y., Liu,Q., A Generalized Decision Logic in Interval-Set-Valued Information table, Lecture Notes in AI 1711, Springer-Verlag, Berlin, 1999, 285-294.
15. Liu,Q.,Sun,H., Theoretical Study of Granular Computing, LNAI 4062, The Proceedings of RSKT2006, by Springer, China, July 2006, 93-102.
16. Liu,Q. and Huang,Z.H., G-Logic and Its Resolution Reasoning, Chinese Journal of Computer, Vol.27,No.7, 2004, 865-873. (In Chinese).
17. Pawlak,Z., Rough Sets, Int. J. Inform. Comp. Sci., 11(1982), 341-356.
18. Frege,G., Grundgesetze der Arithmentic, In: Geach and Black (eds.) , Selection from the philosophical Writings of Gotlob Frege, Blackwei, Oxford 1970
19. Pawlak,Z., Rough Sets present State and Further Prospects, The Proceedings of Third International Workshop on Rough Sets and Soft Computing, Nov. 10-12,1994, 72-76.
20. Zadeh,L.A., Fuzzy Sets, Information and Control, No.,8, 1965, 338-353.
21. Liu,X.H., Fuzzy Logic and Fuzzy Reasoning [M], Press. of Jilin University, Jilin, 1989,(In Chinese).
22. Liu,Q.,Sun H., Studying Direction of Granular Computing from Rough Set Perspective, Journal of Nanchang Institute of TechnologyVol. 25, No.3, 2006, 1-5.
23. Sun,H.,Liu,Q., The Research of Rough Sets in Normed Linear Space, LNAI 4259The Proceedings of RSCTC2006, by Springer, Japan, 8-11 Nov., 2006, 91-98.
24. Liu,Q., Jiang,F. and Deng,D.Y., Design and Implement for the Diagnosis Software of Blood Viscosity Syndrome Based on Hemorheology on GrC., Lecture Notes in Artificial Intelligence 2639, Springer-Verlag, 2003,413-420.

A Four-Valued Logic for Rough Set-Like Approximate Reasoning

Jan Małuszyński[1], Andrzej Szałas[1,2], and Aida Vitória[3]

[1] Linköping University, Department of Computer and Information Science
581 83 Linköping, Sweden
janma@ida.liu.se
[2] The University of Economics and Computer Science
Olsztyn, Poland
andsz@ida.liu.se
[3] Dept. of Science and Technology, Linköping University
S 601 74 Norrköping, Sweden
aidvi@itn.liu.se

Abstract. This paper extends the basic rough set formalism introduced by Pawlak [1] to a rule-based knowledge representation language, called Rough Datalog, where rough sets are represented by predicates and described by finite sets of rules. The rules allow us to express background knowledge involving rough concepts and to reason in such a knowledge base. The semantics of the new language is based on a four-valued logic, where in addition to the usual values TRUE and FALSE, we also have the values BOUNDARY, representing uncertainty, and UNKNOWN corresponding to the lack of information. The semantics of our language is based on a truth ordering different from the one used in the well-known Belnap logic [2, 3] and we show why Belnap logic does not properly reflect natural intuitions related to our approach. The declarative semantics and operational semantics of the language are described. Finally, the paper outlines a query language for reasoning about rough concepts.

1 Introduction

The seminal ideas of Pawlak [1, 4, 5, 6] on the treatment of imprecise and incomplete data opened a new area of research, where the notion of rough sets is used in theoretical studies as well as practical applications.

Rough sets are constructed by means of approximations obtained by using elementary sets which partition a universe of considered objects. The assumption as to partitioning of the universe has been relaxed in many papers (see, e.g., [7, 8, 9, 10, 11, 12, 13, 14, 15]), however the Pawlak's idea of approximations has remained the same.

This paper extends the basic rough set formalism to a rule-based language, where rough sets are represented by predicates and are described by finite sets of rules. The rules allow one to express background knowledge concerning rough concepts and to reason in such a knowledge base. The new language is different from that proposed in [14, 15], where the rules described rough sets by combining their regions (lower approximation, upper approximation and boundary region). In contrast to the language

J.F. Peters et al. (Eds.): Transactions on Rough Sets VI, LNCS 4374, pp. 176–190, 2007.
© Springer-Verlag Berlin Heidelberg 2007

described in this paper, the rules expressed in the language presented in [14, 15] refer explicitly to different regions of a rough set.

Lifting the level of description makes the definitions easier to understand, also for the people not familiar with the technicalities of rough sets. The semantics of the new language is based on a four-valued logic, where in addition to the usual values TRUE and FALSE we have the values BOUNDARY representing uncertain/inconsistent information and UNKNOWN corresponding to the lack of information. As discussed in Section 3.2, the well-known four-valued Belnap logic [3, 2] does not properly reflect the natural intuitions related to our approach. We propose instead a slightly different truth ordering and use it, together with the standard knowledge ordering, for defining a declarative semantics of our language.

By using the four-valued logic we propose, we are then able to deal with some important issues.

First of all, we are able to provide a natural semantics for Datalog-like rules where negation can be used freely, both in the bodies and in the heads of rules. This, in previous approaches to various variants of negation, has always been problematic either due to the high computational complexity of queries or to a nonstandard semantics of negation, often leading to counterintuitive results (for an overview of different approaches to negation see, e.g., [16]). Our semantics reflects intuitions of fusing information from various independent sources. If all sources claim that a given fact is true (respectively, false) then we have an agreement and attach TRUE (respectively FALSE) to that fact. If information sources disagree in judgement of a fact, we attach to it the value BOUNDARY. If no source provides an information about a given fact, we then make it UNKNOWN.

Second, we are able to import knowledge systems based on the classical logic without any changes and make them work directly within the rough framework. In such cases these systems would act as single information sources providing answers TRUE, FALSE, when queried about facts. Possible conflicting claims of different systems would then be solved by the same, uniform four-valued approach we propose. This might be useful in combining low level data sources, like classifiers as well as higher level expert systems.

Third, one can import rough set-based systems, or systems supporting approximate reasoning, like for example, those described in [14, 15], or [17, 18]. In the latter three-valued logics are used (identifying BOUNDARY and UNKNOWN).

The paper is structured as follows. First, in Section 2, we recall basic definitions related to rough sets and approximations. Next, in Section 3, we discuss our choice of four-valued logic. In Section 4 we introduce Rough Datalog and provide its semantics. Section 5 outlines a query language and discusses its implementation in logic programming. Finally, Section 6 concludes the paper.

2   Rough Sets

According to Pawlak's definition (see, e.g., [19]), a rough set S over a universe U is characterized by two subsets of U:

Table 1. Test results considered in Example 1

car	station	safe
a	s1	yes
a	s2	no
b	s2	no
c	s1	yes
d	s1	yes

– the set $\underline{S}$, of all objects which can be *certainly* classified as belonging to S, called *the lower approximation* of S, and
– the set $\overline{S}$, of all objects which can be *possibly* classified as belonging to S, called *the upper approximation* of S.

The set difference between the upper approximation and the lower approximation, denoted by $\overline{\underline{S}}$, is called the *boundary* region.

In practice, in order to describe a given reality, one chooses a set of attributes and the elements of the underlying universe are described by tuples of attribute values. Rough sets are then defined by *decision tables* associating membership decisions with attribute values. The decisions are not exclusive: a given tuple of attribute values may be associated with the decision "yes", with the decision "no", with both or with none, if the tuple does not appear.

Example 1. Consider a universe consisting of cars. If a car passed a test then it may be classified as safe (and as not safe, if it failed the test). Tests may be done independently at two test stations. The upper approximation of the rough set of safe cars would then include cars which passed at least one test. The lower approximation of the set would include the cars which passed all tests (and therefore, they did not fail at any test). The boundary region consists of the cars which passed one test and failed at one test. Notice that there are two other categories of cars, namely those which were not tested and those which failed all tests.

As an example consider the situation described in Table 1, where the first column consists of cars, the second column consists of test stations and the third one contains test results. Denote by "Safe" the set of safe cars. Then:

– the upper approximation of Safe consists of cars for which there is a decision "yes", i.e., $\overline{\text{Safe}} = \{a, c, d\}$
– the lower approximation of Safe consists of cars for which all decisions are "yes", i.e., $\underline{\text{Safe}} = \{c, d\}$
– the boundary region of Safe consists of cars for which there are both decisions "yes" and "no", i.e., $\overline{\underline{\text{Safe}}} = \{a\}$.   ◁

A decision table, representing a concept t, may be represented as a finite set of literals of the form $t(y)$ or $\neg t(x)$, where y ranges over the tuples of attribute values associated with the decision "yes" and x ranges over the tuples of attribute values associated with the decision "no".

Example 2. For the Example 1 with the universe of cars $\{a, b, c, d, e\}$ and with two test stations, we may have the decision table, shown in Table 1, encoded as

$$\{\texttt{safe}(a), \neg\texttt{safe}(a), \neg\texttt{safe}(b), \texttt{safe}(c), \texttt{safe}(d)\} \, .$$

Notice that the literal $\texttt{safe}(a)$ indicates that car a has passed a safety test in one of the stations while literal $\neg\texttt{safe}(a)$ states that the same car as failed a safety test in another test station.

In this case the rough set Safe has the approximations

$$\overline{\text{Safe}} = \{a, c, d\} \text{ and } \underline{\text{Safe}} = \{c, d\}.$$

The rough set ¬Safe, describing those cars that have failed some test, has the approximations $\overline{\neg\text{Safe}} = \{a, b\}$ and $\underline{\neg\text{Safe}} = \{b\}$.

Note that it is totally unknown what is the status of car e.   ◁

We notice that a decision table $\mathcal{T}$ of this kind defines two rough sets, T and $\neg T$, with a common boundary region which is the intersection of the upper approximations of both sets, i.e. $\overline{T} \cap \overline{\neg T}$. As rough sets are usually defined by decision tables, we then adopt the following definition (used also in [20, 14, 15]).

Definition 1. *A rough set S over a universe U is a pair $\langle \overline{S}, \overline{\neg S} \rangle$ of subsets of U.*   ◁

Intuitively, the rough set S describes those elements of U having certain property. The set $\overline{S}$ is the upper approximation of S, and consists of the elements of U for which there is an indication of having the given property. On the other hand, the set $\overline{\neg S}$ consists of the elements for which there is an indication of not having the property. In Example 2, $\overline{\text{Safe}} = \{a, c, d\}$ and $\overline{\neg\text{Safe}} = \{a, b\}$.

Remark 1

1. Observe that Definition 1 differs from the understanding of rough sets as defined by Pawlak. In fact, the definition of Pawlak requires the underlying elementary sets used in approximations to be based on equivalence relations, while Definition 1 relaxes this requirement. Such differences are examined and discussed in depth in [12].
2. Since relations are sets of tuples, we further on also use the term *rough relation* to mean a rough set of tuples.   ◁

3 A Four-Valued Logic for Rough Sets

3.1 The Truth Values for Rough Membership

Our objective is to define a logical language for rough set reasoning. The vocabulary of the language includes predicates to be interpreted as rough relations and constants to be used for representing attribute values. Consider an atomic formula of the form $p(t_1, \cdots, t_n)$, where p is a predicate, denoting a rough set P, and $t_1, \ldots, t_n$ (with $n > 0$) are constants. We now want to define the truth value represented by an atom $p(t_1, \cdots, t_n)$. Let $v = \langle t_1, \ldots, t_n \rangle$ and "$-$" denote the set difference operation. Then, the following cases are possible:

- $v \in \overline{P} - \overline{\neg P}$: intuitively, we only have evidence that the element of the universe described by the attributes v has property P. Thus, the truth value of $p(v)$ is defined to be TRUE.
- $v \in \overline{\neg P} - \overline{P}$: intuitively, we only have evidence that the element of the universe described by the attributes v does not have property P. Thus, the truth value of $p(v)$ is defined to be FALSE.
- $v \in \overline{P} \cap \overline{\neg P}$: in this case, we have contradictory evidences, i.e. an evidence that an element of the universe described by the attributes v has property P and an evidence that it does not have the property P. This is an uncertain information and we use the additional truth value BOUNDARY to denote it.
- $v \notin \overline{P} \cup \overline{\neg P}$: in this case, we have no evidence whether the element of the universe described by the attributes v has property P. We then use another truth value called UNKNOWN.

3.2 Is Belnap Logic Suitable for Rough Reasoning?

The truth values emerging from our discussion have been studied in the literature outside of the rough set context for defining four-valued logic. A standard reference is the well-known Belnap's logic [2]. We now recall its basic principles and we discuss whether it is suitable for rough set reasoning.

The Belnap logic is defined by considering a distributive *bilattice* of truth values and introducing logical connectives corresponding to the operations in the bilattice.

Bilattices have been introduced in [21, 22]. They generalize the notion of Kripke structures (see, e.g., [23]). A *bilattice* is a structure $B = \langle U, \leq_t, \leq_k \rangle$ such that U is a non-empty set, $\leq_t$ and $\leq_k$ are partial orderings each making set U a lattice. Moreover, there is usually a useful connection between both orderings.

We follow the usual convention that $\wedge_t$ and $\vee_t$ stand respectively for the meet and join, with respect to $\leq_t$. The symbols $\wedge_k$ and $\vee_k$ stand respectively for the meet and join, with respect $\leq_k$. Operations $\wedge_t$ and $\vee_t$ are also called the *conjunction* and *disjunction*, and $\wedge_k$ and $\vee_k$ are often designated as the *consensus* and *accept all* operators, respectively.

The bilattice used in Belnap's logic is shown in Fig 1. In the *knowledge ordering*, $\leq_k$, UNKNOWN is the least value, reflecting total lack of knowledge. Each of the values TRUE and FALSE provide more information than UNKNOWN. Finally, the INCONSISTENT value corresponds to the situation when there is evidence for both TRUE and FALSE.[1] The *truth ordering* $\leq_t$ (see Fig 1) has TRUE as its largest element, and FALSE as its smallest element.

Example 3. Assume that a family owns two cars: a and e. We want to check if the family has a safe car. This corresponds to the logical value of the expression

$$\mathrm{safe}(a) \vee_t \mathrm{safe}(e) \, . \tag{1}$$

---

[1] Observe that INCONSISTENT is replaced in our approach by BOUNDARY, which is closer to intuitions from rough set theory.

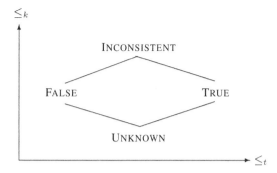

Fig. 1. The bilattice corresponding to Belnap's logic

The truth values of $\texttt{safe}(a)$ and $\texttt{safe}(e)$ are determined by the results of the tests, as specified in Example 2. Thus $\texttt{safe}(a)$ has the value BOUNDARY and $\texttt{safe}(e)$ has the value UNKNOWN. If the join operation $\vee_t$ is defined by Belnap's logic, as shown in Fig 1, then

$$\text{INCONSISTENT} \vee_t \text{UNKNOWN} = \text{TRUE} .$$

This contradicts our intuitions. We know that the safety of car a is unclear, since the results of both safety tests are contradictory, and we know nothing about safety of car e.

Asking instead if all cars of the family are safe,

$$\texttt{safe}(a) \wedge_t \texttt{safe}(e) , \tag{2}$$

would in Belnap's logic result in the answer FALSE. However, we really do not know whether both cars are safe because we do not have any information about the safety of car e. In contrast to the answer obtained in the Belnap's logic, UNKNOWN seems to be a more intuitive answer in this case.   ◁

The example above shows that the truth ordering of Fig 1, and consequently Belnap's logic are not suitable for rough set-based reasoning. On the other hand, the knowledge ordering of Fig. 1 is adequate for our purposes. Indeed, the values TRUE and FALSE show that only one kind of evidence, either positive or negative, is known while the value BOUNDARY indicates existence of contradictory evidence, both positive and negative.

3.3 A Four-Valued Logic for Rough Set Reasoning

We now define a four-valued logic suitable for rough set-based reasoning by modifying the bilattice of Fig.1. As discussed in Section 3.2, only the truth ordering is to be changed. We will use the new truth ordering to define conjunction ($\wedge_t$) as the greatest lower bound in this ordering. The ordering should preserve the usual meaning of conjunction for the truth values TRUE and FALSE. Intuitively, the value UNKNOWN represents the lack of information. Thus, the result of its conjunction with any other truth value is accepted here to be UNKNOWN. A new information may arrive, replacing UNKNOWN by either TRUE, or FALSE, or BOUNDARY, providing in each case a different result. On the other hand, BOUNDARY represents existence of contradictory information. Its conjunction with TRUE would not remove this contradiction. Thus, we define the result of such a conjunction to be BOUNDARY. It also seems natural, that the conjunction of FALSE with TRUE or BOUNDARY gives FALSE. Consequently the truth ordering, $\leq_t$, is redefined in our framework as

$$\text{UNKNOWN} \leq_t \text{FALSE} \leq_t \text{BOUNDARY} \leq_t \text{TRUE} . \tag{3}$$

The new structure $\mathcal{R} = \langle U, \leq_t, \leq_k \rangle$, where U is the universe of objects of interest, $\leq_t$ is the truth ordering defined in (3), and $\leq_k$ is the knowledge ordering as in the Belnap's logic, gives the meaning of the logical connectives and is used in our approach.

Example 4. Referring to Example 3, we then compute the logical values associated with the queries (1) and (2) by considering the new truth ordering above.
    The first query, (1) of Example 3,

$$\text{BOUNDARY} \vee_t \text{UNKNOWN} ,$$

returns the logical BOUNDARY which better corresponds to the intuitions.
    For the second query, (2) of Example 3, we have that

$$\text{BOUNDARY} \wedge_t \text{UNKNOWN} = \text{UNKNOWN} .$$

In contrast to Belnap's logic, it is not excluded that some cars of the family of Example 3 are safe, but to be sure we need to obtain some information about the safety of car e. So, the answer UNKNOWN adequately reflects our intuitions.   ◁

The proposition below shows that there is a connection between the knowledge ordering and the truth ordering. In this sense, the structure $\mathcal{R}$ can then be seen as a bilattice.

Proposition 1. *Consider the bilattice $\mathcal{R} = \langle U, \leq_t, \leq_k \rangle$ and that $x, y \in U$. The operation $\wedge_t$ is monotonic with respect to $\leq_k$ on both arguments, i.e. if $x \leq_k y$ then, for every $z \in U$, we have $(z \wedge_t x) \leq_k (z \wedge_t y)$ and $(x \wedge_t z) \leq_k (y \wedge_t z)$.*

Proof. Table 2 shows the result. Operation $\wedge_t$ is obviously commutative.   ◁

We now define formally the logic underlying our work, called *Rough Logic*.

Definition 2. *Consider the following negation operation $\neg$.*

$$\neg\text{TRUE} \overset{\text{def}}{=} \text{FALSE}, \quad \neg\text{FALSE} \overset{\text{def}}{=} \text{TRUE},$$
$$\neg\text{BOUNDARY} \overset{\text{def}}{=} \text{BOUNDARY}, \quad \neg\text{UNKNOWN} \overset{\text{def}}{=} \text{UNKNOWN}.$$

The propositional four-valued logic defined by the bilattice $\mathcal{R}$ together with negation $\neg$ is called the Rough Logic.   ◁

Table 2. The table considered in the proof of Proposition 1

z	x	y	$z \wedge_t x$	$z \wedge_t y$
BOUNDARY	UNKNOWN	TRUE	UNKNOWN	BOUNDARY
BOUNDARY	UNKNOWN	FALSE	UNKNOWN	FALSE
BOUNDARY	FALSE	BOUNDARY	FALSE	BOUNDARY
BOUNDARY	TRUE	BOUNDARY	BOUNDARY	BOUNDARY
BOUNDARY	UNKNOWN	BOUNDARY	UNKNOWN	BOUNDARY

z	x	y	$z \wedge_t x$	$z \wedge_t y$
FALSE	UNKNOWN	TRUE	UNKNOWN	FALSE
FALSE	UNKNOWN	FALSE	UNKNOWN	FALSE
FALSE	FALSE	BOUNDARY	FALSE	FALSE
FALSE	TRUE	BOUNDARY	FALSE	FALSE
FALSE	UNKNOWN	BOUNDARY	UNKNOWN	FALSE

z	x	y	$z \wedge_t x$	$z \wedge_t y$
TRUE	UNKNOWN	TRUE	UNKNOWN	TRUE
TRUE	UNKNOWN	FALSE	UNKNOWN	FALSE
TRUE	FALSE	BOUNDARY	FALSE	BOUNDARY
TRUE	TRUE	BOUNDARY	TRUE	BOUNDARY
TRUE	UNKNOWN	BOUNDARY	UNKNOWN	BOUNDARY

z	x	y	$z \wedge_t x$	$z \wedge_t y$
UNKNOWN	UNKNOWN	TRUE	UNKNOWN	UNKNOWN
UNKNOWN	UNKNOWN	FALSE	UNKNOWN	UNKNOWN
UNKNOWN	FALSE	BOUNDARY	UNKNOWN	UNKNOWN
UNKNOWN	TRUE	BOUNDARY	UNKNOWN	UNKNOWN
UNKNOWN	UNKNOWN	BOUNDARY	UNKNOWN	UNKNOWN

4   Rough Datalog Language

We now define a rule language, called *Rough Datalog*, such that its semantics is based on the Rough Logic. Intuitively, Rough Datalog corresponds to the usual logic programming language Datalog. While predicates in the latter denote crisp relations, in Rough Datalog a predicate p denotes a rough relation P. Thus, Rough Datalog caters for uncertainty in the knowledge.

A *rough literal* is any expression of the form $p(t_1, \ldots, t_n)$ or $\neg p(t_1, \ldots, t_n)$. In Rough Datalog, knowledge is represented in the form of rough clauses,

$$H : -\ B_1, \ldots, B_n.$$

where H and every B_i $(0 \leq i \leq n)$ is a rough literal. A rough clause with the empty body (i.e. $n = 0$) is called a *rough fact*. A *rough program* $\mathcal{P}$ is a finite set of rough clauses.

Rough clauses are used to specify rough relations as explained next. Intuitively, a rough clause is to be understood as the knowledge inequality $\leq_k$ stating that the truth value of the body is less than or equal to the truth value of the head. The comma symbol "," is interpreted as the meet in the truth ordering $\leq_t$. Notice that the arguments of $\leq_k$ are the truth values UNKNOWN, BOUNDARY, TRUE, or FALSE but the logical value associated with a rough clause is either TRUE or FALSE. Information obtained from different rough clauses with heads referring to the same rough relation P (i.e. p or $\neg p$ occurs in the head) is combined using the knowledge join operation $\vee_k$.

Example 5. The following rough clauses belong to an exemplary rough program $\mathcal{P}$.

(1) ¬useful(a) :- red(a), squared(a).
   "Object a is not useful if it is red and squared."
(2) squared(a) :- useful(a). —*"Object a is squared if it is useful."*
(3) ¬squared(a). —*"Object a is not squared."*   ◁

4.1 Semantics of Rough Datalog Programs

We now define notions of four-valued interpretation and model, extend the knowledge ordering to interpretations and show that each rough program has the least model in this ordering.

Let $\mathcal{P}$ be a rough program and L be the set of all constant symbols occurring in $\mathcal{P}$. Then, the *Herbrand base* $\mathcal{H}_{\mathcal{P}}$ is the set of all literals whose predicate symbols occur in $\mathcal{P}$ and whose arguments belong to L.

A four-valued *interpretation* $\mathcal{I}$ of a rough program $\mathcal{P}$ associates with each atom a occurring in $\mathcal{P}$ a logical value

$$\mathcal{I}(a) \in \{\text{UNKNOWN}, \text{TRUE}, \text{FALSE}, \text{BOUNDARY}\}$$

and $\neg\mathcal{I}(a) = \mathcal{I}(\neg a)$.

The notion of interpretation extends naturally to conjunction (disjunction) of literals. Let $l_1, \ldots, l_n$, with $n > 0$, be rough literals.

$$\mathcal{I}(l_1 \wedge_t \cdots \wedge_t l_n) = \mathcal{I}(l_1) \wedge_t \cdots \wedge_t \mathcal{I}(l_n) .$$

Definition 3. *An interpretation $\mathcal{I}$ of a rough program $\mathcal{P}$ is any subset of the* Herbrand base $\mathcal{H}_{\mathcal{P}}$. *Moreover, the rough relation $\mathcal{I}(p)$ is defined as*

$$\mathcal{I}(p) = \langle \overline{\mathcal{I}(p)}, \overline{\neg\mathcal{I}(p)} \rangle = \langle \{t \mid p(t) \in \mathcal{I}\}, \ \{t \mid \neg p(t) \in \mathcal{I}\} \rangle .$$ ◁

Intuitively, an interpretation associates each predicate p occurring in a program $\mathcal{P}$ with a rough set. Notice that $\neg\mathcal{I}(p) = \mathcal{I}(\neg p)$. Moreover, we have that

– $\mathcal{I}(p(t)) = \text{UNKNOWN}$, if $t \notin \overline{\mathcal{I}(p)} \cup \overline{\neg\mathcal{I}(p)}$.
– $\mathcal{I}(p(t)) = \text{FALSE}$, if $t \in \overline{\neg\mathcal{I}(p)}$.
– $\mathcal{I}(p(t)) = \text{TRUE}$, if $t \in \overline{\mathcal{I}(p)}$.
– $\mathcal{I}(p(t)) = \text{BOUNDARY}$, if $t \in \underline{\mathcal{I}(p)}$.

Notice that we only consider variable-free rough programs. However, the results presented below can be also extended to rough programs with variables.

An interpretation $\mathcal{I}$ of a rough program $\mathcal{P}$ *satisfies* a rough clause $H :\!- B. \in \mathcal{P}$ if $\mathcal{I}(B) \leq_k \mathcal{I}(H)$. A *model* $\mathcal{M}$ of $\mathcal{P}$ is any interpretation that satisfies every rough clause belonging to $\mathcal{P}$.

Notice also that the Herbrand base $\mathcal{H}_\mathcal{P}$ is a model of any rough program $\mathcal{P}$. In this model the truth value of every literal is BOUNDARY. However, usually a program has more models. For comparing them we introduce a partial order on interpretations based on the knowledge ordering relation, $\leq_k$.

Definition 4. *Let $\mathcal{I}_1 \subseteq \mathcal{H}_\mathcal{P}$ and $\mathcal{I}_2 \subseteq \mathcal{H}_\mathcal{P}$ be two interpretations. Then, $\mathcal{I}_1 \leq_k \mathcal{I}_2$, if and only if $\mathcal{I}_1(l) \leq_k \mathcal{I}_2(l)$, for every literal $l \in \mathcal{H}_\mathcal{P}$.* ◁

It can be checked that the knowledge ordering on interpretations corresponds to set inclusion.

Proposition 2. *$\mathcal{I}_1 \leq_k \mathcal{I}_2$ if and only if $\mathcal{I}_1 \subseteq \mathcal{I}_2$.* ◁

We show now that there is the least model for every rough program.

Proposition 3. *Let $\mathcal{P}$ be a rough program. Then, $\mathcal{P}$ has the least model with respect to $\leq_k$.*

Proof. To prove that $\mathcal{P}$ has a least model with respect to $\leq_k$, we show that the intersection of all models of $\mathcal{P}$ is also a model of $\mathcal{P}$.

Let $\mathcal{M} = \bigcap_i^n M_i$, where $\{M_1, \ldots, M_n\}$ $(n \geq 1)$ is the set of all models of $\mathcal{P}$. Notice that, by Proposition 2, $\mathcal{M} \leq_k M_i$, with $M_i \in \{M_1, \ldots, M_n\}$. We prove that $\mathcal{M}$ is a model of $\mathcal{P}$. For this we have to show that, for any clause $H :\!- B. \in \mathcal{P}$, we have $\mathcal{M}(H) \geq_k \mathcal{M}(B)$. We prove this by cases, considering possible truth values of the body of a clause.

(a) If $\mathcal{M}(B) = $ UNKNOWN then $\mathcal{M}$ satisfies the rough clause, since UNKNOWN is the least element in the knowledge ordering.

(b) If $\mathcal{M}(B) = $ TRUE then $W(B) \geq_t$ BOUNDARY, for every model W of $\mathcal{P}$. Hence, $W(H) \geq_k$ TRUE, for every model W of $\mathcal{P}$. Consequently, $\mathcal{M}(H) \geq_k$ TRUE because the literal occurring in the head belongs to every model W. We conclude then that $\mathcal{M}$ satisfies the rough clause.

(c) If $\mathcal{M}(B) = $ FALSE then B includes a literal l that is FALSE in some model of $\mathcal{P}$ and l is either FALSE or BOUNDARY in the other models. Obviously, no literal occurring in B can be UNKNOWN in any model. Consequently, $\mathcal{M}(H) \geq_k$ FALSE because $\neg H$ belongs to every model W. We conclude then that $\mathcal{M}$ satisfies the rough clause.

(d) If $\mathcal{M}(B) = $ BOUNDARY then $W(B) = $ BOUNDARY, for every model W of $\mathcal{P}$. Notice that if $\mathcal{I}(B) = $ BOUNDARY, for some interpretation $\mathcal{I}$ of $\mathcal{P}$, then we have that either $\mathcal{I}(l) = $ TRUE or $\mathcal{I}(l) = $ BOUNDARY, for every literal l in the body B. Hence, $W(H) = $ BOUNDARY, for every model W of $\mathcal{P}$. Consequently, $\mathcal{M}(H) = $ BOUNDARY because $\{H, \neg H\} \subseteq W$, for every model W. We conclude then that $\mathcal{M}$ satisfies the rough clause. ◁

The semantics of a rough program $\mathcal{P}$ is captured by its least model, with respect to $\leq_k$.

Example 6. Consider again the rough program of Example 5. Its least model is $M =$ $\{\neg\texttt{squared}(a)\}$. Hence, $\texttt{useful}(a)$ and $\texttt{red}(a)$ are UNKNOWN, while $\texttt{squared}(a)$ is FALSE.   ◁

4.2   A Fixpoint Characterization of the Least Model

We now give a fixpoint characterization of the least model which makes it possible to compute the semantics of a program. We define an operator on interpretations, considered as sets of literals. We show that the operator is monotonic with respect to set inclusion. Thus, it has the least fixpoint (with respect to set inclusion) which can be obtained by iterations of the operator starting with the empty interpretation. We also show that the least fixpoint is a model. Taking into account Proposition 2, we can then conclude that the least fixpoint is also the least model of the program with respect to knowledge ordering. In the following definition if l is a negative literal of the form $\neg a$, then $\neg l$ denotes a.

Definition 5. *Let $\mathcal{P}$ be a rough program. A total function $T_{\mathcal{P}}$ mapping interpretations into interpretations is defined as follows:*

$$
\begin{aligned}
T_{\mathcal{P}}(\mathcal{I}) = \ &\{l \mid l\,\text{:-}\, B.\ \in \mathcal{P} \text{ and } \mathcal{I}(B) = \text{TRUE}\} &\cup \\
&\{\neg l \mid l\,\text{:-}\, B.\ \in \mathcal{P} \text{ and } \mathcal{I}(B) = \text{FALSE}\} &\cup \\
&\{l, \neg l \mid l\,\text{:-}\, B.\ \in \mathcal{P} \text{ and } \mathcal{I}(B) = \text{BOUNDARY}\}\,.
\end{aligned}
$$
                                                                                      ◁

Thus, the set $T_{\mathcal{P}}(\mathcal{I})$ consists of the heads of the rough clauses whose bodies are TRUE or BOUNDARY in $\mathcal{I}$ and, the negated heads of the rules whose bodies are FALSE or BOUNDARY in $\mathcal{I}$. Such a way to gather heads of rules corresponds to defining the result by the disjunction of heads w.r.t. knowledge ordering $\leq_k$.

Proposition 4. *Given a rough program $\mathcal{P}$, the operator $T_{\mathcal{P}}$ is monotonic with respect to set inclusion.*

Proof. The bodies of the program clauses are conjunctions of atoms. By Proposition 1 the conjunction is monotonic with respect to knowledge ordering. Hence by Proposition 2, it is also monotonic with respect to set inclusion of the interpretations. Thus, $\mathcal{I} \subseteq T_{\mathcal{P}}(\mathcal{I})$, for every interpretation $\mathcal{I}$.   ◁

The proposition above guarantees that $T_{\mathcal{P}}$ has a least fixpoint (with respect to set inclusion), denoted as $\text{LFP}(T_{\mathcal{P}})$.

Proposition 5. *Given a rough program $\mathcal{P}$, the $\text{LFP}(T_{\mathcal{P}})$ coincides with the least model of $\mathcal{P}$.*

Proof. It is easy to see that the interpretation $\mathcal{I} = \text{LFP}(T_{\mathcal{P}})$ is a model of $\mathcal{P}$. Assume the contrary. Then, there exists a clause $H\,\text{:-}\, B.$ such that $\mathcal{I}(H) <_k \mathcal{I}(B)$. The possible cases are as follows.

- $\mathcal{I}(B) = $ TRUE and $\mathcal{I}(H) \leq_k$ FALSE.
- $\mathcal{I}(B) = $ FALSE and $\mathcal{I}(H) \leq_k$ TRUE.
- $\mathcal{I}(B) = $ BOUNDARY and $\mathcal{I}(H) <_k$ BOUNDARY.

In the first two cases, we immediately obtain the contradiction with the assumption $\mathcal{I} = \text{LFP}(T_{\mathcal{P}})$, since $T_{\mathcal{P}}(\mathcal{I})$ would then include, respectively, the literal H ($\neg H$). A similar contradiction is obtained for the third case, since $\mathcal{I}(B) = \text{BOUNDARY}$ means that $T_{\mathcal{P}}(\mathcal{I})$ would then include both literals H and $\neg H$. In any case, we conclude that $\mathcal{I}$ is not a fixpoint of $T_{\mathcal{P}}$.

It remains to prove that the model $\text{LFP}(T_{\mathcal{P}})$ is the least model in the knowledge ordering. This follows directly from Proposition 2. $\triangleleft$

Proposition 5 shows that the least model of a program $\mathcal{P}$ can be computed by applying iteratively operator $T_{\mathcal{P}}$, starting from the empty interpretation until the fixpoint is reached. Notice that in the empty interpretation, all literals of the Herbrand base have the truth value UNKNOWN.

We show below a simple example of a rough program, based on a classical example from logic programming, and it illustrates the use of $T_{\mathcal{P}}$ for computation of its semantics.

Example 7. Consider the rough program consisting of the following rough clauses.

  (1) `fly(tweety) :- bird(tweety).`
  (2) `bird(tweety) :- penguin(tweety).`
  (3) `¬fly(tweety) :- penguin(tweety).`
  (4) `¬dangerous(tweety) :- red(tweety), fly(tweety).`
  (5) `penguin(tweety).`
  (6) `red(tweety).`

Application of $T_{\mathcal{P}}$ to the empty interpretation gives

$$\mathcal{I}_1 = \texttt{penguin(tweety), red(tweety)}.$$

Further iterations of T_p give

$$\mathcal{I}_2 = \mathcal{I}_1 \cup \{\texttt{bird(tweety)}, \texttt{¬fly(tweety)}\},$$
$$\mathcal{I}_3 = \mathcal{I}_2 \cup \{\texttt{fly(tweety)}\},$$
$$\mathcal{I}_4 = \mathcal{I}_1 \cup \{\texttt{dangerous(tweety)}, \texttt{¬dangerous(tweety)}\},$$
$$\mathcal{I}_5 = \mathcal{I}_4.$$

Thus, we conclude that `tweety` belongs to the lower approximations of the rough relations Bird, Penguin and Red and it belongs to the boundary region of rough relations Fly and Dangerous. $\triangleleft$

5   A Query Language and Its Implementation

In this section we describe a query language for rough programs. We start by defining the notions of rough query and answer. Then, we briefly describe how the query language can be implemented in a logic programming as queries to a definite logic program. Existing systems like Prolog [24], XSB [25], or SModels [26, 27] can then be used to compute the answers. We assume that the reader is familiar with the basics of logic programming [28].

Definition 6. *A rough query is a pair* $\langle :\text{-} l_1, \ldots, l_n\,,\ \mathcal{P}\rangle$, *with* $n \geq 1$, *where* $\mathcal{P}$ *is a rough program and each* l_i *is a (variable-free) rough literal.*   ◁

We need now to define the notion of answer to a rough query.

Definition 7. *Let* $\langle :\text{-} l_1, \ldots, l_n\,,\ \mathcal{P}\rangle$ *be a rough query. The* answer *to the rough query is defined as the logical value of* $LFP(T_\mathcal{P})(l_1 \wedge_t \cdots \wedge_t l_n)$.   ◁

Example 8. Consider the rough program of Example 7. The answer to the rough query

$$\langle :\text{-} \texttt{bird(tweety)}, \mathcal{P}\rangle$$

is TRUE, while the answer to the rough query

$$\langle :\text{-} \texttt{fly(tweety)}, \texttt{penguin(tweety)}\,,\ \mathcal{P}\rangle$$

is BOUNDARY.   ◁

Rough programs can be compiled to definite logic programs as described below. A definite logic program is a non-empty set of clauses $H\text{:-} A_1, \cdots, A_n.$, where each A_i is an atom, $(0 \leq i \leq n)$. Clauses can informally be understood as implications: if every atom A_i is true then H must also be true. Therefore, the comma symbol "," is interpreted as conjunction. Notice that predicates in a logic program denote crisp relations and each atom is either TRUE or FALSE.

Any fact (a clause of the form $H\text{:-} .$) remains unchanged.

Let $C \equiv H\text{:-} l_1, \cdots, l_n.$, where $n \geq 1$, be a rough clause and φ be a function transforming C into a non-empty set of clauses such that $\varphi(C) = \{C\} \cup \phi(C)$, where

$$\phi(C) = \{\neg H\text{:-} l'_1, \cdots, l'_n. \mid (\forall 1 \leq i \leq n : l'_i \in \{l_i, \neg l_i\}) \text{ and} \atop H\text{:-} l'_1, \cdots, l'_n. \not\equiv C\}\,. \tag{4}$$

Hence, a rough program is compiled to a definite logic program by applying function φ to each rough clause, i.e. $\varphi(\mathcal{P}) = \bigcup_{C \in \mathcal{P}} \varphi(C)$. We assume that, in the compiled programs, $\neg p$ is treated as a new predicate symbol and $\neg\neg p$ is replaced with p, for any symbol p.

Informally, the main idea underlying the compilation of rough programs is that the body of a rough clause is associated with TRUE, if all literals occurring in it are TRUE. The body of a rough clause is associated with FALSE, if $\neg l$ is TRUE, for at least one literal l occurring in the body, and all other literals in the body are provable. The body of a rough clause is associated with BOUNDARY, if we can prove that it is TRUE and FALSE. If for some literal l in the body, it is neither possible to prove l nor $\neg l$ then the body of a rough clause is associated with UNKNOWN. It can be easily seen that the least model of $\mathcal{P}$ coincides with the least model of the definite logic program $\varphi(\mathcal{P})$.

Remark 2. The transformation expressed by formula (4) results in the exponential blow up of the number of clauses. Namely, if a body of a rule consists of n literals then we have 2^n resulting clauses. However, in practice n is rather small. Moreover, the transformation we have provided is the simplest one and we only intend to show that the required compilation can be done.   ◁

Definite logic programs can also be queried. A query for a definite logic program has the same syntax as a rough query. However in logic programming, queries are answered YES or NO depending whether the query is provable or not. A rough query $\langle Q, \mathcal{P} \rangle$ can be compiled to several queries to $\varphi(\mathcal{P})$. Thus,

- If the query $\langle Q, \varphi(\mathcal{P}) \rangle$ is answered YES and all queries $\langle Q', \varphi(\mathcal{P}) \rangle$ are answered NO, with $Q' \in \phi(Q)$, then the answer to the rough query $\langle Q, \mathcal{P} \rangle$ is TRUE.
- If the query $\langle Q, \varphi(\mathcal{P}) \rangle$ is answered NO and some query $\langle Q', \varphi(\mathcal{P}) \rangle$ is answered YES, with $Q' \in \phi(Q)$, then the answer to the rough query $\langle Q, \mathcal{P} \rangle$ is FALSE.
- If the query $\langle Q, \varphi(\mathcal{P}) \rangle$ is answered YES and some query $\langle Q', \varphi(\mathcal{P}) \rangle$ is also answered YES, with $Q' \in \phi(Q)$, then the answer to the rough query $\langle Q, \mathcal{P} \rangle$ is BOUNDARY.
- Otherwise, the answer to the rough query $\langle Q, \mathcal{P} \rangle$ is UNKNOWN.

6 Conclusions

In the paper we have presented a four-valued logic which we found adequate for approximate reasoning based on Pawlak's ideas of approximations. The four-valued approach reflects intuitions of fusing information from various, possibly independent data sources.

We have proposed a database language involving approximate concepts and provided its formal semantics. Lifting the level of description from approximations to sets/relations themselves facilitates the use of the language as well as the import of rules from other databases, including those based on two-valued and three-valued logics. A corresponding query language and its implementation have also been discussed.

As noticed in Remark 2, the transformation defined by formula (4) is rather inefficient. We plan to address its improvement in our future work.

References

1. Pawlak, Z.: Information systems – theoretical foundations. Information Systems **6** (1981) 205–218
2. Belnap, N.: A useful four-valued logic. In Eptein, G., Dunn, J., eds.: Modern Uses of Many Valued Logic, Reidel (1977) 8–37
3. Belnap, N.: How a computer should think. In Ryle, G., ed.: Contemporary Aspects of Philosophy, Stocksfield, Oriel Press (1977) 30–55
4. Pawlak, Z.: Rough sets. International Journal of Computer and Information Sciences **11** (1982) 341–356
5. Pawlak, Z.: Rough logic. Bull. Polish Acad. Sci. Tech **35** (1987) 253–258
6. Pawlak, Z.: Rough Sets. Theoretical Aspects of Reasoning about Data. Kluwer Academic Publishers, Dordrecht (1991)
7. Skowron, A., Stepaniuk, J.: Tolerance approximation spaces. Fundamenta Informaticae **27** (1996) 245–253
8. Słowiński, R., Vanderpooten, D.: Similarity relation as a basis for rough approximations. In Wang, P., ed.: Advances in Machine Intelligence & Soft Computing, Raleigh NC, Bookwrights (1997) 17–33

9. Słowiński, R., Vanderpooten, D.: A generalized definition of rough approximations based on similarity. IEEE Trans. on Data and Knowledge Engineering **12(2)** (2000) 331–336
10. Doherty, P., Łukaszewicz, W., Szałas, A.: Tolerance spaces and approximative representational structures. In Günter, A., Kruse, R., Neumann, B., eds.: Proc. 26th German Conf. on AI, KI'2003. Volume 2821 of LNAI., Springer-Verlag (2003) 475–489
11. Doherty, P., Łukaszewicz, W., Szałas, A.: Approximate databases and query techniques for agents with heterogenous perceptual capabilities. In: Proceedings of the 7th Int. Conf. on Information Fusion, FUSION'2004. (2004) 175–182
12. Doherty, P., Szałas, A.: On the correspondence between approximations and similarity. In Tsumoto, S., Slowinski, R., Komorowski, J., Grzymala-Busse, J., eds.: Proceedings of 4th International Conference on Rough Sets and Current Trends in Computing, RSCTC'2004. Volume 3066 of LNAI., Springer-Verlag (2004) 143–152
13. Doherty, P., Łukaszewicz, W., Skowron, A., Szałas, A.: Knowledge Representation Techniques. A Rough Set Approach. Volume 202 of Studies in Fuziness and Soft Computing. Springer-Verlag (2006)
14. Andersson, R., Vitória, A., Małuszyński, J., Komorowski, J.: Rosy: A rough knowledge base system. In: RSFDGrC (2). Volume 3642 of Lecture Notes in Computer Science., Springer (2005) 48–58
15. Vitória, A.: A framework for reasoning with rough sets. Transactions on Rough Sets IV **3700** (2005) 178–276
16. Abiteboul, S., Hull, R., Vianu, V.: Foundations of Databases. Addison-Wesley Pub. Co. (1996)
17. Doherty, P., Łukaszewicz, W., Szałas, A.: CAKE: A computer-aided knowledge engineering technique. In van Harmelen, F., ed.: Proc. 15th European Conference on Artificial Intelligence, ECAI'2002, Amsterdam, IOS Press (2002) 220–224
18. Doherty, P., Magnusson, M., Szałas, A.: Approximate databases: A support tool for approximate reasoning. Journal of Applied Non-Classical Logics **16** (2006) 87–118 Special issue on Implementation of logics.
19. Pawlak, Z.: A treatise on rough sets. Transactions on Rough Sets IV **3700** (2005) 1–17
20. Vitória, A., Damásio, C., Małuszyński, J.: Query answering for rough knowledge bases. In Wang, G., Liu, Q., Yao, Y., Skowron, A., eds.: Proceedings of 9th Internatinal Conference on Rough Sets, Fuzzy Sets, Data Mining and Granular Computing. Volume 2639 of LNCS., Springer-Verlag (2003) 197–204
21. Ginsberg, M.: Multi-valued logics. In: Proceedings of AAAI-86, Fifth National Conference on Artificial Intelligence. (1986) 243–247
22. Ginsberg, M.: Multivalued logics: a uniform approach to reasoning in ai. Computational Intelligence **4** (1988) 256–316
23. Fitting, M.: Bilattices are nice things. In: Proc. PhiLog Conference on Self-Reference, Copenhagen, The Danish Network for Philosophical Logic and Its Applications (2002)
24. Deransart, P., Ed-Bali, A., Cervoni, L.: Prolog: The Standard Reference Manual. Springer-Verlag (1996)
25. : XSB system. (Available at http://xsb.sourceforge.net/)
26. Niemelä, I., Simons, P.: Smodels - an implementation of stable model and the well-founded semantics for normal logic programs. In Dix, J., Furbach, U., Nerode, A., eds.: Proc. of the 4th Int. Conf. on Logic Programming and Nonmonotonic Reasoning (LPNMR'97). Volume 1265 of LNAI., Springer-Verlag (1997) 420–429
27. Simons, P.: Smodels system. (Available at http://www.tcs.hut.fi/Software/smodels/)
28. Nilsson, U., Małuszynski, J.: Logic, Programming and Prolog, 2nd edition. John Wiley & Sons, http://www.ida.liu.se/~ulfni/lpp/copyright.html (1995)

On Representation and Analysis of Crisp and Fuzzy Information Systems

Alicja Mieszkowicz-Rolka and Leszek Rolka

Department of Avionics and Control,
Rzeszów University of Technology,
ul. W. Pola 2, 35-959 Rzeszów, Poland
{alicjamr, leszekr}@prz.edu.pl

Abstract. This paper proposes an approach to representation and analysis of information systems with fuzzy attributes, which combines the variable precision fuzzy rough set (VPFRS) model with the fuzzy flow graph method. An idea of parameterized approximation of crisp and fuzzy sets is presented. A single ε-approximation, which is based on the notion of fuzzy rough inclusion function, can be used to express the crisp approximations in the rough set and variable precision rough set (VPRS) model. A unified form of the ε-approximation is particularly important for defining a consistent VPFRS model. The introduced fuzzy flow graph method enables alternative description of decision tables with fuzzy attributes. The generalized VPFRS model and fuzzy flow graphs, taken together, can be applied to determining a system of fuzzy decision rules from process data.

1 Introduction

Two important paradigms, developed in the recent decades, can be successfully used for modelling and analyzing decision processes performed by a human operator: the rough set theory introduced by Pawlak [19] and the theory of fuzzy sets proposed by Zadeh [34].

The idea of combining fuzzy sets with rough sets was realized by two independent approaches. The method given by Nakamura [18] consists in application of the classical rough set theory to a crisp representation of fuzzy sets. In contrast to that, Dubois and Prade [6] introduced a novel concept of fuzzy rough sets, suitable for expressing vagueness represented in fuzzy sets, and coarseness characteristic of rough sets. The concept of Dubois and Prade has been widely used and developed, see, e.g., [8,12,25].

A significant parameterized extension of the crisp rough set theory is the variable precision rough set (VPRS) model proposed by Ziarko [35]. It bases on the idea of relaxation of strong inclusion requirements. The VPRS model helps to overcome problems caused by errors and noise, which are present in data obtained from real decision processes. More recently, a probabilistic interpretation of the VPRS model was developed, see e.g., [11,29,36]. The original VPRS model and many other extensions of crisp rough sets can be expressed in

J.F. Peters et al. (Eds.): Transactions on Rough Sets VI, LNCS 4374, pp. 191–210, 2007.
© Springer-Verlag Berlin Heidelberg 2007

the framework of a generalized theory. The rough mereology of Polkowski and Skowron [24] presents an alternative generalized approach to rough sets, which is based on the mereology of Leśniewski. The idea of relaxation of strong inclusion requirements was also applied to fuzzy rough sets [7,33].

Another useful method, introduced and studied by Pawlak [20,21,22], is a hybrid approach to decision algorithms, which combines the idea of flow graphs with the crisp rough set model. It was shown [9] that every decision algorithm can be associated with a flow graph.

We emphasize the problem of obtaining a set of relevant fuzzy decision rules from recorded process data and decision examples. This is a crucial step in applications of fuzzy inference systems [14,32]. The used data can be represented in the form of a decision table with fuzzy attributes. To analyze efficiently this kind of decision table, we adapt and combine all three approaches mentioned above: fuzzy rough sets, variable precision rough set model and flow graphs.

First of all, we present a generalized version of our variable precision fuzzy rough set (VPFRS) model [16,17], which was introduced with the aim to enable analysis of fuzzy decision tables obtained from dynamic processes. There are many ways of performing basic operations on fuzzy sets. In order to get a consistent VPFRS model, we propose a unified parameterized approach to approximation of crisp and fuzzy sets. Basing on the notion of rough and fuzzy rough inclusion function, a definition of a single ε-approximation is given.

Secondly, we propose a fuzzy flow graph approach, which is suitable for representing and analyzing fuzzy decision systems. The connection of the flow graph approach with fuzzy inference systems is discussed, The problem of a correct choice of fuzzy connectives, with the aim to retain the flow conservation equations, is considered. Furthermore, we give new definitions of the path's certainty and strength, by respecting only the relevant part of the flow and disregarding the flow components which come from other paths.

Finally, we show that the VPFRS model can be effectively used for a simpler representation and easier selection of fuzzy decision rules with the help of fuzzy flow graphs.

We start with a formal description of fuzzy information systems.

2   Fuzzy Information Systems

In the classical concept of (crisp) sets with sharp boundaries, any element x of an universe U belongs or does not belong to a given subset of U. In contrast to that, the notion of fuzzy sets admits of partial membership. Any fuzzy set F can be defined by assigning to every element $x \in U$ a membership degree $\mu_F(x) \in [0, 1]$ in the set F. Thus, we get a membership function μ_F which describes the fuzzy set F.

In a crisp information system, a set of attributes Q is used to characterize the elements of an universe U. Each element x of the universe U is described by a combination of attributes values. Only one attribute value of each attribute $q \in Q$ can be assigned to a given element $x \in U$.

In order to generalize the notion of information system, we use a set of fuzzy attributes with linguistic values expressed by membership functions. Several linguistic values of every attribute $q \in Q$ can be assigned to an element $x \in U$. In other words, an element x can belong, to a non-zero membership degree, to many fuzzy sets representing linguistic values of an attribute q. We introduce a formal definition of a fuzzy information system.

Definition 1. *A fuzzy information system is the 4-tuple* $S = \langle X, Q, L, f \rangle$, *where*

U – *is a nonempty set, called the universe,*
Q – *is a finite set of fuzzy attributes,*
L – *is a set of fuzzy (linguistic) values of attributes,* $L = \bigcup_{q \in Q} L_q$,
$\quad L_q$ *is the set of linguistic values of an attribute* $q \in Q$,
f – *is an information function,* $f \colon U \times L \to [0, 1]$,
$\quad f(x, l) \in [0, 1]$ *for every* $l \in L$ *and every* $x \in U$.

In practice, we use fuzzy decision tables, which constitute a special form of fuzzy information systems with two disjoint groups of condition and decision attributes, respectively.

To give a formal description of decision tables, we assume a finite universe U with N elements: $U = \{x_1, x_2, \ldots, x_N\}$. Attributes are divided into a subset of n condition attributes: $C = \{c_1, c_2, \ldots, c_n\}$, and a subset of m decision attributes: $D = \{d_1, d_2, \ldots, d_m\}$.

Every fuzzy attribute is associated with a set of linguistic values. We denote by $V_i = \{V_{i1}, V_{i2}, \ldots, V_{in_i}\}$ the family of linguistic values of a condition attribute c_i, and by $W_j = \{W_{j1}, W_{j2}, \ldots, W_{jm_j}\}$ the family of linguistic values of a decision attribute d_j, where n_i and m_j, is the number of the linguistic values of the i-th condition and the j-th decision attribute, respectively, $i = 1, 2, \ldots, n$, and $j = 1, 2, \ldots, m$.

For any element $x \in U$, its membership degrees in all linguistic values of the condition attribute c_i (or decision attribute d_j) should be determined. This is performed in the process called fuzzification, using the recorded crisp value of a particular attribute of the element x. The fuzzy value of an attribute, for a given element x, is a fuzzy set on the domain of all linguistic values of that attribute.

We denote by $V_i(x)$ the fuzzy value of the condition attribute c_i for any $x \in U$, as a fuzzy set on the domain of the linguistic values of c_i

$$V_i(x) = \{\mu_{V_{i1}}(x)/V_{i1},\ \mu_{V_{i2}}(x)/V_{i2},\ \ldots,\ \mu_{V_{in_i}}(x)/V_{in_i}\}\,. \tag{1}$$

$W_j(x)$ denotes the fuzzy value of the decision attribute d_j for any $x \in U$, as a fuzzy set on the domain of the linguistic values of d_j

$$W_j(x) = \{\mu_{W_{j1}}(x)/W_{j1},\ \mu_{W_{j2}}(x)/W_{j2},\ \ldots,\ \mu_{W_{jm_j}}(x)/W_{jm_j}\}\,. \tag{2}$$

When the linguistic values of all attributes have the form of singletons or disjoint intervals on the original domain of attributes, we get a classical crisp decision table. In such a case, only one linguistic value can be assigned to each condition and decision attribute of an element $x \in U$.

Furthermore, we assume, for any element $x \in U$, that all linguistic values $V_i(x)$ and $W_j(x)$ ($i = 1, 2, \ldots, n$, $j = 1, 2, \ldots m$) satisfy the requirements

$$\text{power}(V_i(x)) = \sum_{k=1}^{n_i} \mu_{V_{ik}}(x) = 1, \qquad \text{power}(W_j(x)) = \sum_{k=1}^{m_j} \mu_{W_{jk}}(x) = 1. \quad (3)$$

The requirements (3) will be used in section 4 for introducing a generalized flow graph approach, which can be applied to analysis of fuzzy information systems.

3   Variable Precision Fuzzy Rough Set Model

3.1   Parameterized Crisp Rough Sets

The rough set theory, proposed by Pawlak [19], is based on the observation that any crisp subset of an universe U can be characterized with respect to an indiscernibility (equivalence) relation $R \subseteq U \times U$. Those classes of indiscernible elements $x \in U$, which are "completely in accordance" with a given set $A \subseteq U$, form the lower approximation of A. Indiscernibility classes, which are "partially in accordance" with A, form the upper approximation of A. A set is called exact, if its lower and upper approximations are equal to each other, otherwise the set is called rough.

The lower approximation $\underline{R}(A)$ and upper approximation $\overline{R}(A)$ of a crisp set A are defined formally as follows

$$\underline{R}(A) = \{x \in U : [x]_R \subseteq A\}, \quad (4)$$

$$\overline{R}(A) = \{x \in U : [x]_R \cap A \neq \emptyset\}, \quad (5)$$

where $[x]_R$ denotes an indiscernibility class which contains the element $x \in U$.

Observe that the above definitions are constructed using two operations on sets: inclusion and intersection. Let us define the lower and upper approximations, utilizing only the notion of set inclusion.

Definition 2. *Given an indiscernibility relation R, the lower approximation $\underline{R}(A)$ and upper approximation $\overline{R}(A)$ of a crisp set A are defined as follows*

$$\underline{R}(A) = \{x \in U : \forall S \subseteq [x]_R \wedge S \neq \emptyset, \; S \subseteq A\}, \quad (6)$$

$$\overline{R}(A) = \{x \in U : \exists S \subseteq [x]_R \wedge S \neq \emptyset, \; S \subseteq A\}. \quad (7)$$

The definitions (6) and (7) emphasize two extreme (ideal) cases of approximation obtained by applying the indiscernibility relation R.

The need for defining the lower and upper approximations in a unified way becomes clearer, when we consider the approximation of fuzzy sets. This is because there is no single method of performing basic operations on fuzzy sets. Using only one fuzzy connective is especially important for creating a consistent variable precision fuzzy rough set (VPFRS) model.

Now, let us recall the concept of crisp variable precision rough set (VPRS) model, introduced by Ziarko [35]. In order to cope with inconsistency of information systems, caused by noise and errors in data, it is necessary to admit of some level of misclassification, especially in the case of large information systems. This can be done by relaxing strong inclusion requirements, basing on a modified relation of set inclusion. We explain the VPRS concept using the notion of inclusion degree, $incl(A, B)$, of a nonempty (crisp) set A in a (crisp) set B, defined as follows

$$incl(A, B) = \frac{card(A \cap B)}{card(A)} . \tag{8}$$

The inclusion degree should be constrained by applying a lower limit l and an upper limit u, introduced in the extended version of VPRS [13], which satisfy the requirement

$$0 \leq l < u \leq 1 . \tag{9}$$

We assume a non-probabilistic interpretation of the VPRS model. The probabilistic rough set approach [27,36], introduced recently, is a generalization of the VPRS model, which bases on conditional probability of inclusion.

Using the limits l and u, which satisfy the constraint (9), one can introduce the notions of u-lower and l-upper approximation of any subset A of the universe U by an indiscernibility relation R.

The u-lower approximation of A by R is a set defined as follows

$$\underline{R}_u(A) = \{x \in U : incl([x]_R, A) \geq u\} , \tag{10}$$

where $[x]_R$ denotes an indiscernibility class of R containing the element x.

The l-upper approximation of A by R is a set defined as follows

$$\overline{R}_l(A) = \{x \in U : incl([x]_R, A) > l\} . \tag{11}$$

Observe that the definitions (10) and (11) use the same notion of inclusion degree and can be interpreted as a weakened form of (6) and (7).

To extend the crisp VPRS model to a parameterized rough set and fuzzy rough set model, we only apply the degree of set inclusion as the basic notion. For a general treatment of the problem, we adapt the idea of rough inclusion function, given by Skowron and Stepaniuk [26], which is defined on the Cartesian product of the powersets $\mathbb{P}(U)$ of the universe U

$$\nu : \mathbb{P}(U) \times \mathbb{P}(U) \to [0, 1] . \tag{12}$$

We assume that the first parameter represents a nonempty set, and the rough inclusion function should be monotonic with respect to the second parameter

$$\nu(X, Y) \leq \nu(X, Z) \quad \text{for any } Y \subseteq Z, \quad \text{where } X, Y, Z \subseteq U .$$

The inclusion degree (8) is an example of rough inclusion function (12).

Using the rough inclusion function ν, the lower and upper approximations of a crisp set A can be defined by

$$\underline{R}(A) = \{x \in U : \nu([x]_R, A) = 1\} , \tag{13}$$

$$\overline{R}(A) = \{x \in U : \nu([x]_R, A) > 0\} \,. \tag{14}$$

We want to go beyond the standard rough set perspective and introduce a parameterized single form of approximation of crisp sets.

Definition 3. *Given an indiscernibility relation R, the ε-approximation $R_\varepsilon(A)$ of a crisp set A is defined as follows*

$$R_\varepsilon(A) = \{x \in U : \nu([x]_R, A) \geq \varepsilon\} \,, \tag{15}$$

where $\varepsilon \in (0, 1]$.

The ε-approximation R_ε can be used for expressing any kind of approximation, due to the following properties:

$$
\begin{array}{llll}
\text{(P1)} & R_\varepsilon(A) = \underline{R}(A) & \text{for } \varepsilon = 1 \,, \\
\text{(P2)} & R_\varepsilon(A) = \overline{R}(A) & \text{for } \varepsilon = 0+ \,, \\
\text{(P3)} & R_\varepsilon(A) = \underline{R}_u(A) & \text{for } \varepsilon = u \,, \\
\text{(P4)} & R_\varepsilon(A) = \overline{R}_l(A) & \text{for } \varepsilon = l+ \,.
\end{array}
$$

Although, we have a single notion of ε-approximation, we are still able to determine a pair of approximations, by using a pair of appropriate values of the ε parameter. However, we are mainly interested in determining the consistent part of an analyzed information system. Hence, the lower approximation is the most important notion used for reasoning about data.

The concept of VPRS has turned out to be efficient in applications of the rough set theory to real decision processes [16], e.g. when analyzing the control of dynamic systems, characterized by large decision tables. In such a case the determination of the u-lower approximation (10) should be repeated for different (decreasing) values of the parameter u.

When considering a series of n ε-approximations of a set A, the following property is satisfied due to monotonicity of the inclusion function

$$\text{(P5)} \quad R_{\varepsilon_1}(A) \subseteq R_{\varepsilon_2}(A) \subseteq \ldots \subseteq R_{\varepsilon_n}(A) \quad \text{for} \quad \varepsilon_1 \geq \varepsilon_2 \geq \ldots \geq \varepsilon_n \,.$$

3.2   Parameterized Fuzzy Rough Sets

Our goal is to propose a unified approach to parameterized approximation of fuzzy sets. To this end, we generalize the notion of crisp ε-approximation and adapt the widely used concept of fuzzy rough sets of Dubois and Prade. In consequence, a consistent form of variable precision fuzzy rough set model will be obtained, suitable for analysis of large fuzzy information systems.

Let us recall the definition of fuzzy rough set, introduced by Dubois and Prade [6]. For a given fuzzy set A and a fuzzy partition $\Phi = \{F_1, F_2, \ldots, F_n\}$ on the universe U, the membership functions of the lower and upper approximations of A by Φ are defined as follows

$$\mu_{\underline{\Phi}(A)}(F_i) = \inf_{x \in U} \mathrm{I}(\mu_{F_i}(x), \mu_A(x)) \,, \tag{16}$$

$$\mu_{\overline{\Phi}(A)}(F_i) = \sup_{x \in U} \mathrm{T}(\mu_{F_i}(x), \mu_A(x)), \tag{17}$$

where T and I denote a T-norm operator and an implicator, respectively.

The pair of sets $(\underline{\Phi}F, \overline{\Phi}F)$ is called a fuzzy rough set.

In order to extend the approach given in previous subsection, we need to consider the problem of determining the degree of inclusion of one fuzzy set into another. This problem has been often discussed (see, e.g., [1,2,5,7,15]). Many different measures of fuzzy set inclusion were considered. Among many proposals, implication operators were applied to determination of set inclusion. Sinha-Dougherty [4] proposed an axiomatic approach, which can be formulated using the generalized Łukasiewicz implicators.

We propose a different idea of set inclusion in comparison with various solutions given in the literature. It consists in determination of inclusion with respect to particular elements of sets. This leads to a detailed description of inclusion. In consequence, we get a fuzzy set rather than a number. This method is particulary useful for elaborating an effective variable precision fuzzy rough set model.

A fuzzy set, which describes the inclusion of a fuzzy set A in a fuzzy set B, determined with respect to particular elements of the set A, constitutes the basic notion of our VPFRS model. The obtained fuzzy set will be called the inclusion set of A in B, and denoted by $\mathrm{INCL}(A, B)$.

There are many possibilities of defining the inclusion set. We apply to this end an implication operator denoted by I.

Definition 4. *The implication-based inclusion set* $\mathrm{INCL}(A, B)$ *of a nonempty fuzzy set A in a fuzzy set B is defined as follows*

$$\mu_{\mathrm{INCL}(A,B)}(x) = \begin{cases} \mathrm{I}(\mu_A(x), \mu_B(x)) & \text{if } \mu_A(x) > 0, \\ 0 & \text{otherwise}. \end{cases} \tag{18}$$

By assuming that $\mu_{\mathrm{Incl}(A,B)}(x) = 0$, for $\mu_A(x) = 0$, we take into account only the support of the set A. For the sake of simplicity of the computational algorithm, it is not necessary to consider inclusion for all elements of the universe.

Furthermore, we can require that the degree of inclusion with respect to x should be equal to 1, if the inequality $\mu_A(x) \leq \mu_B(x)$ for that x is satisfied

$$\mathrm{I}(\mu_A(x), \mu_B(x)) = 1, \quad \text{if } \mu_A(x) \leq \mu_B(x). \tag{19}$$

The requirement (19) is always satisfied by residual implicators.

In order to define a suitable fuzzy counterpart of the rough inclusion function (12), we apply the notions of α-cut, power (cardinality) and support of a fuzzy set. Given a fuzzy subset A of the universe U, the α-cut of A, denoted by A_α, is a crisp set defined as follows

$$A_\alpha = \{x \in U : \ \mu_A(x) \geq \alpha\} \quad \text{for} \quad \alpha \in [0, 1]. \tag{20}$$

For a finite fuzzy set A with n elements, power(A) and support(A) are given by

$$\mathrm{power}(A) = \sum_{i=1}^{n} \mu_A(x_i), \quad \mathrm{support}(A) = \{x : \mu_A(x_i) > 0\}. \tag{21}$$

Using the above notions, we define the fuzzy rough inclusion function on the Cartesian product of the families $\mathbb{F}(U)$ of all fuzzy subsets of the universe U

$$\nu_\alpha : \mathbb{F}(U) \times \mathbb{F}(U) \to [0,1] \,. \tag{22}$$

Definition 5. *The fuzzy rough α-inclusion function $\nu_\alpha(A,B)$ of any nonempty fuzzy set A in a fuzzy set B is defined as follows, for any $\alpha \in (0,1]$*

$$\nu_\alpha(A,B) = \frac{\mathrm{power}(A \cap \mathrm{INCL}(A,B)_\alpha)}{\mathrm{power}(A)} \,, \tag{23}$$

The value $\nu_\alpha(A,B)$ expresses how many elements of the nonempty fuzzy set A belong, at least to the degree α, to the fuzzy set B.

First, we prove monotonicity of the proposed fuzzy rough inclusion function.

Theorem 1. *Implication-based fuzzy rough inclusion function ν_α is monotonic with respect to the second parameter, for any $\alpha \in (0,1]$*

$$\nu_\alpha(X,Y) \le \nu_\alpha(X,Z) \quad \text{for any } Y \subseteq Z, \quad \text{where } X, Y, Z \subseteq \mathbb{F}(U) \,.$$

Proof. According to the definition of a fuzzy subset [14], for $Y \subseteq Z$, we have $\mu_Y(x) \le \mu_Z(x), \forall\, x \in U$. Since every R-implicator, S-implicator and QL-implicator is right monotonic [25], it holds that: $\mu_{\mathrm{I}(X,Y)}(x) \le \mu_{\mathrm{I}(X,Z)}(x), \forall\, x \in U$. Thus, using the definition (18), we get

$$\mu_{\mathrm{INCL}(X,Y)}(x) \le \mu_{\mathrm{INCL}(X,Z)}(x), \quad \forall\, x \in U \,.$$

Finally, for any $\alpha \in (0,1]$, we can easy show that

$$\frac{\mathrm{power}(X \cap \mathrm{INCL}(X,Y)_\alpha)}{\mathrm{power}(X)} \le \frac{\mathrm{power}(X \cap \mathrm{INCL}(X,Z)_\alpha)}{\mathrm{power}(X)} \,.$$

Hence $\nu_\alpha(X,Y) \le \nu_\alpha(X,Z)$. $\qquad\qquad\qquad\qquad\qquad\qquad\qquad\square$

Furthermore, we can show that the rough inclusion function used in formulae (10) and (11) is a special case of the fuzzy rough inclusion function (23), when we use the implication-based inclusion set.

Theorem 2. *For any nonempty crisp set A, any crisp set B, and for $\alpha \in (0,1]$, the implication-based inclusion function $\nu_\alpha(A,B)$ is equal to the inclusion degree incl(A,B).*

Proof. We show that for any crisp sets A and B, the inclusion set $\mathrm{Incl}(A,B)$ is equal to the crisp intersection $A \cap B$. The membership function of any crisp set X is given by

$$\mu_X(x) = \begin{cases} 1 & \text{for } x \in X \\ 0 & \text{for } x \notin X \,. \end{cases} \tag{24}$$

Every implicator I satisfies the conditions: $\mathrm{I}(1,1) = \mathrm{I}(0,1) = \mathrm{I}(0,0) = 1$, and $\mathrm{I}(1,0) = 0$.

Thus, applying the definition (18), we get

$$\mu_{\mathrm{Incl}(A,B)}(x) = \mu_{A\cap B}(x) = \begin{cases} 1 & \text{if } x \in A \text{ and } x \in B \\ 0 & \text{otherwise}. \end{cases} \qquad (25)$$

For any finite crisp set X, and any $\alpha \in (0,1]$, by formulae and (20), (21) and (24) we get: $\mathrm{power}(X) = \mathrm{card}(X)$, and $X_\alpha = X$.

Consequently, applying (25), we finally have

$$\frac{\mathrm{power}(A \cap \mathrm{Incl}(A,B)_\alpha)}{\mathrm{power}(A)} = \frac{\mathrm{card}(A \cap B)}{\mathrm{card}(A)}.$$

Hence, we proved that $\nu_\alpha(A,B) = \mathrm{incl}(A,B)$, for any $\alpha \in (0,1]$.   □

We want to formulate the fuzzy rough approximation in a general way. Therefore, we introduce a function called **res**, defined on the Cartesian product $\mathbb{P}(U)\times\mathbb{F}(U)$, where $\mathbb{P}(U)$ denotes the powerset of the universe U, and $\mathbb{F}(U)$ the family of all fuzzy subsets of the universe U, respectively

$$\mathrm{res} : \mathbb{P}(U) \times \mathbb{F}(U) \to [0,1]. \qquad (26)$$

We require that

$\mathrm{res}(\emptyset, Y) = 0$,
$\mathrm{res}(X,Y) \in \{0,1\}$,   if Y is a crisp set,
$\mathrm{res}(X,Y) \le \mathrm{res}(X,Z)$ for any $Y \subseteq Z$,   where $X \in \mathbb{P}(U)$, and $Y,Z \in \mathbb{F}(U)$.

The form of the function **res** can be chosen depending on requirements of a considered application. For a given crisp set X and fuzzy set Y, the value of function $\mathrm{res}(X,Y)$ should express the resulting membership degree in the set Y, taking into account not all elements of the universe, but only the elements of the set X. When we apply the limit-based approach, according to Dubois and Prade, we obtain the following form of the function **res**

$$\mathrm{res}(X,Y) = \inf_{x\in X} \mu_Y(x). \qquad (27)$$

In the definition (27) of the function **res**, only one (limit) value of membership degree of elements in the set Y is taken into account. However, this means that we disregard the character (shape) of the membership function. Basing on a single value of membership degree is not always acceptable, especially in the case of large information systems. Hence, we can use the opportunity of giving another definitions of **res**, in which many values of membership degree are considered.

Now, we introduce the notion of generalized fuzzy rough ε-approximation.

Definition 6. *For $\varepsilon \in (0,1]$, the ε-approximation $\Phi_\varepsilon(A)$ of a fuzzy set A, by a fuzzy partition $\Phi = \{F_1, F_2, \ldots, F_n\}$, is a fuzzy set on the domain Φ with membership function expressed as*

$$\mu_{\Phi_\varepsilon(A)}(F_i) = \mathrm{res}(S_\varepsilon(F_i, A), \mathrm{INCL}(F_i, A)), \qquad (28)$$

where

$$S_\varepsilon(F_i, A) = \text{support}(F_i \cap \text{INCL}(F_i, A)_{\alpha_\varepsilon}),$$

$$\alpha_\varepsilon = \sup\{\alpha \in [0, 1]: \ \nu_\alpha(F_i, A) \geq \varepsilon\}.$$

The set $S_\varepsilon(F_i, A)$ is equal to support of the intersection of the class F_i with the part of $\text{INCL}(F_i, A)$, which contains those elements of the approximating class F_i which are included in A at least to the degree α_ε. The resulting membership $\mu_{\Phi_\varepsilon(A)}(F_i)$ is determined using only the elements from $S_\varepsilon(F_i, A)$ instead of the whole class F_i. This is accomplished by applying the function res.

It can be easy shown that applying the definition (27) of the function res leads to a simple form of the ε-approximation (28)

$$\mu_{\Phi_\varepsilon(A)}(F_i) = \sup\{\alpha \in [0, 1]: \ \nu_\alpha(F_i, A) \geq \varepsilon\}. \tag{29}$$

In contrast to the approximations (16) and (17), which use two different fuzzy connectives, we have a single unified definition of fuzzy rough approximation. In this way we obtain a consistent variable precision fuzzy rough set model. Thus, we are able to compare approximations determined for various values of the parameter ε.

3.3   Analysis of Fuzzy Decision Tables

In the analysis of fuzzy decision tables, two fuzzy partitions are generated with the help of a suitable similarity relation. The partition obtained with respect to condition attributes is used for approximation of fuzzy similarity classes obtained with respect to decision attributes. It is necessary to address the problem of comparing objects described by fuzzy sets. This issue has been widely studied in the literature, see, for example, [3,7,8]. In our considerations, we apply a symmetric T-transitive fuzzy similarity relation [3], which is defined by means of the distance between the compared elements. In the following, we only give formulae for condition attributes. We apply the notation given in section 2.

If we need to compare any two elements x and y of the universe U with respect to the condition attribute c_i, $i = 1, 2, \ldots, n$, then the similarity between x and y could be expressed using a T-similarity relation based on the Łukasiewicz T-norm [7].

$$S_{c_i}(x, y) = 1 - \max_{k=1,n_i} |\mu_{V_{ik}}(x) - \mu_{V_{ik}}(y)|. \tag{30}$$

In order to evaluate the similarity $S_C(x, y)$, with respect to all condition attributes C, we must aggregate the results obtained for particular attributes c_i, $i = 1, 2, \ldots, n$. This can be done by using the T-norm operator min as follows

$$S_C(x, y) = \min_{i=1,n} S_{c_i}(x, y) = \min_{i=1,n} \left(1 - \max_{k=1,n_i} |\mu_{V_{ik}}(x) - \mu_{V_{ik}}(y)|\right). \tag{31}$$

By calculating the similarity for all pairs of elements of the universe U, we obtain a symmetric similarity matrix. If the value of similarity between the elements x and y is equal to 1, they belong to the same similarity class. In that case two rows of the similarity matrix should be merged into one fuzzy set with membership degrees equal to 1 for x and y. In consequence, we get a family of fuzzy similarity classes $\widetilde{C} = \{C_1, C_2, \ldots, C_{\tilde{n}}\}$, for condition attributes C and a family of fuzzy similarity classes $\widetilde{D} = \{D_1, D_2, \ldots, D_{\tilde{m}}\}$, for decision attributes D.

In the next step, we determine fuzzy rough approximations of elements of the family $\widetilde{D}$ by the family $\widetilde{C}$, using the parameterized fuzzy rough set model.

To determine the consistency of fuzzy decision tables and significance of attributes, we apply a generalized measure of ε-approximation quality [17]. For the family $\widetilde{D} = \{D_1, D_2, \ldots, D_{\tilde{m}}\}$ and the family $\widetilde{C} = \{C_1, C_2, \ldots, C_{\tilde{n}}\}$ the ε-approximation quality of $\widetilde{D}$ by $\widetilde{C}$ is defined as follows

$$\gamma_{\widetilde{C}_\varepsilon}(\widetilde{D}) = \frac{\text{power}(\text{Pos}_{\widetilde{C}_\varepsilon}(\widetilde{D}))}{\text{card}(U)}, \tag{32}$$

where

$$\text{Pos}_{\widetilde{C}_\varepsilon}(\widetilde{D}) = \bigcup_{D_j \in \widetilde{D}} \omega(\widetilde{C}_\varepsilon(D_j)) \cap D_j .$$

The fuzzy extension ω denotes a mapping from the domain $\widetilde{C}$ into the domain of the universe U, which is expressed for any fuzzy set X by

$$\mu_{\omega(X)}(x) = \mu_X(C_i), \quad \text{if } \mu_{C_i}(x) = 1 . \tag{33}$$

The definition (32) is based on the generalized notion of positive region. For any fuzzy set X and a similarity relation R, the positive region of X is defined as follows

$$\text{Pos}_{R_\varepsilon}(X) = X \cap \omega(R_\varepsilon(X)) . \tag{34}$$

In the definition of the positive region (34), we take into account only those elements of the ε-approximation, for which there is no contradiction between the set X and the approximating similarity classes.

4   Fuzzy Flow Graphs

In addition to the VPFRS model, we want to introduce fuzzy flow graphs as a second tool for analysis of fuzzy information systems. The idea of applying flow graphs in the framework of crisp rough sets, for discovering the statistical properties of decision algorithms, was proposed by Pawlak [20,21,22]. We should start with recalling the basic notions of the crisp flow graph approach.

A flow graph is given in the form of directed acyclic final graph $G = (\mathcal{N}, \mathcal{B}, \varphi)$, where $\mathcal{N}$ is a set of nodes, $\mathcal{B} \subseteq \mathcal{N} \times \mathcal{N}$ is a set of directed branches, $\varphi \colon \mathcal{B} \to \mathrm{R}^+$ is a flow function with values in the set of non-negative reals R^+.

For any $(X, Y) \in \mathcal{B}$, X is an input of Y and Y is an output of X. The quantity $\varphi(X, Y)$ is called the throughflow from X to Y.

$I(X)$ and $O(X)$ denote an input and an output of X, respectively. The input $I(G)$ and output $O(G)$ of a graph G are defined by

$$I(G) = \{X \in \mathcal{N} \colon I(X) = \emptyset\}, \qquad O(G) = \{X \in \mathcal{N} \colon O(X) = \emptyset\} . \tag{35}$$

Every node $X \in \mathcal{N}$ of a flow graph G is characterized by its inflow

$$\varphi_+(X) = \sum_{Y \in I(X)} \varphi(Y, X), \tag{36}$$

and by its outflow

$$\varphi_-(X) = \sum_{Y \in O(X)} \varphi(X, Y). \tag{37}$$

For any internal node X, the equality $\varphi_+(X) = \varphi_-(X) = \varphi(X)$ is satisfied. The quantity $\varphi(X)$ is called the flow of the node X.

The flow for the whole graph G is defined by

$$\varphi(G) = \sum_{x \in I(G)} \varphi_-(X) = \sum_{x \in O(G)} \varphi_+(X). \tag{38}$$

By using the flow $\varphi(G)$, the normalized throughflow $\sigma(X, Y)$ and the normalized flow $\sigma(X)$ are determined as follows

$$\sigma(X, Y) = \frac{\varphi(X, Y)}{\varphi(G)}, \qquad \sigma(X) = \frac{\varphi(X)}{\varphi(G)}. \tag{39}$$

For every branch of a flow graph G the certainty factor is defined by

$$\mathrm{cer}(X, Y) = \frac{\sigma(X, Y)}{\sigma(X)}. \tag{40}$$

The coverage factor for every branch of a flow graph G is defined by

$$\mathrm{cov}(X, Y) = \frac{\sigma(X, Y)}{\sigma(Y)}. \tag{41}$$

The certainty and coverage factors satisfy the following properties

$$\sum_{Y \in O(X)} \mathrm{cer}(X, Y) = 1, \qquad \sum_{X \in I(Y)} \mathrm{cov}(X, Y) = 1. \tag{42}$$

The measures of certainty (40) and coverage (41) are useful for analysis of decision algorithms [10].

Now, we consider the issue of applying flow graphs to representation and analysis of fuzzy decision algorithms. We use decision tables with fuzzy values of attributes, presented in section 2. All possible decision rules, generated by the Cartesian product of sets of linguistic values of the attributes, will be examined. According to notation used in section 2, we obtain $r = \prod_{i=1}^{n} n_i \prod_{j=1}^{m} m_j$ possible rules. The k-th decision rule, denoted by R_k, is expressed as follows

$$R_k: \ \begin{array}{l} \text{IF } c_1 \text{ is } V_1^k \text{ AND } c_2 \text{ is } V_2^k \ \dots \ \text{AND } c_n \text{ is } V_n^k \\ \text{THEN } d_1 \text{ is } W_1^k \text{ AND } d_2 \text{ is } W_2^k \ \dots \ \text{AND } d_m \text{ is } W_m^k \end{array} \tag{43}$$

where $k = 1, 2, \dots, r$, $V_i^k \in V_i$, $i = 1, 2, \dots n$, $W_j^k \in W_j$, $j = 1, 2, \dots, m$.

When we use the fuzzy Cartesian products $C^k = V_1^k \times V_2^k \times \ldots \times V_n^k$ and $D^k = W_1^k \times W_2^k \times \ldots \times W_m^k$, the k-th decision rule can be expressed in the form of a fuzzy implication, denoted here by $C^k \to D^k$.

It is necessary to select a subset of decision rules which are relevant to the considered decision process. This can be done by determining to what degree any element $x \in U$, corresponding to a single row of the decision table, confirms particular decision rules. We calculate the truth value of the decision rule's antecedent and the truth value of the decision rule's consequent, by determining the conjunction of the respective membership degrees of x in the linguistic values of attributes.

If we take a decision table with crisp attributes, a decision rule can be confirmed for some x, if the result of conjunction is equal to 1, both for the rule's premise and the rule's conclusion. Otherwise, the element x does not confirm the considered decision rule. The set of elements $x \in U$, which confirm a decision rule, is called the support of the decision rule.

To determine the confirmation degree of fuzzy decision rules, a T-norm operator need to be applied. By $\mathrm{cd}(x,k)$, we denote the confirmation degree of the k-th decision rule by the element $x \in U$

$$\mathrm{cd}(x,k) = \mathrm{T}(\mathrm{cda}(x,k), \mathrm{cdc}(x,k)), \tag{44}$$

where $\mathrm{cda}(x,k)$ denotes the confirmation degree of the decision rule's antecedent

$$\mathrm{cda}(x,k) = \mathrm{T}(\mu_{V_1^k}(x), \mu_{V_2^k}(x), \ldots, \mu_{V_n^k}(x)), \tag{45}$$

and $\mathrm{cdc}(x,k)$ the confirmation degree of the decision rule's consequent

$$\mathrm{cdc}(x,k) = \mathrm{T}(\mu_{W_1^k}(x), \mu_{W_2^k}(x), \ldots, \mu_{W_m^k}(x)). \tag{46}$$

Through determining the confirmation degrees (45), (46) and (44), we generate the following fuzzy sets on the domain U:

the support of the decision rule's antecedent

$$\mathrm{support}(\mathrm{cda}(x,k)) = \{\mathrm{cda}(x_1,k)/x_1,\ \mathrm{cda}(x_2,k)/x_2,\ \ldots,\ \mathrm{cda}(x_N,k)/x_N\}, \tag{47}$$

the support of the decision rule's consequent

$$\mathrm{support}(\mathrm{cdc}(x,k)) = \{\mathrm{cdc}(x_1,k)/x_1,\ \mathrm{cdc}(x_2,k)/x_2,\ \ldots,\ \mathrm{cda}(x_N,k)/x_N\}, \tag{48}$$

and the support of the decision rule R_k, respectively

$$\mathrm{support}(R_k) = \{\mathrm{cd}(x_1,k)/x_1,\ \mathrm{cd}(x_2,k)/x_2,\ \ldots,\ \mathrm{cd}(x_N,k)/x_N\}. \tag{49}$$

The introduced notions (47), (48) and (49) will be used for defining strength, certainty, and coverage factors of a decision rule.

Now, let us explain the way of constructing fuzzy flow graphs on the basis of a decision table with fuzzy attributes.

Every fuzzy attribute is represented by a layer of nodes. The nodes of a layer correspond to linguistic values of a given attribute.

We denote by $\tilde{X}$ a fuzzy set on the universe U, which describes membership degree of particular elements $x \in U$ in the linguistic value represented by X. The membership degrees of all x in the set $\tilde{X}$ can be found in a respective column of the considered decision table.

Let us pick out such two attributes, which are represented by two consecutive layers of the flow graph. We denote by X a linguistic value of the first attribute, and by Y a linguistic value of the second attribute. In the case of crisp flow graphs, the flow between nodes X and Y is equal to the number of elements of the universe U, which are characterized by the combination of attribute values X and Y. In consequence, a particular element $x \in U$ can only be assigned to a unique path in the flow graph. In a fuzzy information system, however, every element of the universe can belong to several linguistic values, and it can be assigned to several paths in the flow graph.

It is possible to determine the flow distribution in the crisp flow graph by using the operations of set intersection and set cardinality. To obtain the flow $\varphi(X,Y)$ for the branch (X,Y) of a fuzzy flow graph, we have to calculate power of the intersection of fuzzy sets $\tilde{X}$ and $\tilde{Y}$. Many definitions of fuzzy intersection (T-norm operator) are known. In order to satisfy the flow conservation equations, it is necessary to use the T-norm operator prod for determining the intersection of sets. Furthermore, we should assume that the linguistic values of attributes satisfy the requirement (3). We conclude the above discussion with the following theorem.

Theorem 3. *Let* S *be a fuzzy information systems with the linguistic values of attributes satisfying the requirement (3), and let* $\cap$ *denote a fuzzy intersection operator based on the* T-*norm* prod. *The following properties are satisfied for the flow graph, which represents the information system* S :

(G1) *the inflow for any output or internal layer node* X *is given by*

$$\varphi_+(X) = \text{power}(\tilde{X}) = \sum_{Y \in I(X)} \varphi(Y,X) = \sum_{Y \in I(X)} \text{power}(\tilde{X} \cap \tilde{Y}), \qquad (50)$$

(G2) *the outflow for any input or internal layer node* X *is given by*

$$\varphi_-(X) = \text{power}(\tilde{X}) = \sum_{Y \in O(X)} \varphi(X,Y) = \sum_{Y \in O(X)} \text{power}(\tilde{X} \cap \tilde{Y}), \qquad (51)$$

(G3) *for any internal layer node* X, *it holds that*

$$\varphi_+(X) = \varphi_-(X). \qquad (52)$$

The properties (G1), (G2) and (G3) do not hold in general, if we use another T-norm operator, e.g. min. In the special case of crisp decision tables, the formulae (50) and (51) become equivalent to (36) and (37).

The layers corresponding to condition attributes can be merged into a single layer, which contains nodes representing all possible combinations of linguistic values of the condition attributes. We can also merge all the layers corresponding to decision attributes. Let us denote by X^*, a node of the resulting layer obtained for condition attributes and by Y^*, a node of the resulting layer obtained for decision attributes. The node X^* corresponds to antecedent of some decision rule R_k. Support of the antecedent of the decision rule R_k is determined with the help of formula (47).

The decision rule R_k is represented by the branch (X^*, Y^*). Power of the support of the rule R_k is equal to the flow between the nodes X^* and Y^*, which is obtained using formula (49)

$$\varphi(X^*, Y^*) = \text{power}(\text{support}(R_k)) \,. \tag{53}$$

By applying the formulae (47), (48) and (49), we can determine, for every decision rule R_k, the certainty factor $\text{cer}(X^*, Y^*)$, the coverage factor $\text{cov}(X^*, Y^*)$, and the strength of the rule $\sigma(X^*, Y^*)$

$$\text{cer}(X^*, Y^*) = \text{cer}(R_k) = \frac{\text{power}(\text{support}(R_k))}{\text{power}(\text{support}(\text{cda}(x, k)))} \,, \tag{54}$$

$$\text{cov}(X^*, Y^*) = \text{cov}(R_k) = \frac{\text{power}(\text{support}(R_k))}{\text{power}(\text{support}(\text{cdc}(x, k)))} \,, \tag{55}$$

$$\sigma(X^*, Y^*) = \text{strength}(R_k) = \frac{\text{power}(\text{support}(R_k))}{\text{card}(U)} \,. \tag{56}$$

It is possible to represent any decision rule by a sequence of nodes $[X_1 \ldots X_n]$, namely by a path from the 1-th to the n-th layer of the flow graph G. For a given path $[X_1 \ldots X_n]$, the resulting certainty and strength can be defined. In contrast to the definitions presented in [20,21,22], in which the statistical properties of flow are taken into account, we propose a different form of the path's certainty and strength

$$\text{cer}[X_1 \ldots X_n] = \prod_{i=1}^{n-1} \text{cer}(X_1 \ldots X_i, X_{i+1}) \,, \tag{57}$$

$$\sigma[X_1 \ldots X_n] = \sigma(X_1) \, \text{cer}[X_1 \ldots X_n] \,, \tag{58}$$

where

$$\text{cer}(X_1 \ldots X_i, X_{i+1}) = \frac{\text{power}(\widetilde{X}_1 \cap \widetilde{X}_2 \cap \ldots \cap \widetilde{X}_{i+1})}{\text{power}(\widetilde{X}_1 \cap \widetilde{X}_2 \cap \ldots \cap \widetilde{X}_i)} \,. \tag{59}$$

The resulting certainty (57) of the path $[X_1 \ldots X_n]$, expresses what part of the flow of the starting node X_1 reaches the final node X_n, passing through all nodes of the path.

5   Examples

Let us analyze a fuzzy decision table (Table 1) with condition attributes c_1 and c_2 and one decision attribute d. All attributes have three linguistic values.

Table 1. Decision table with fuzzy attributes

	c_1			c_2			d		
	V_{11}	V_{12}	V_{13}	V_{21}	V_{22}	V_{23}	W_{11}	W_{12}	W_{13}
x_1	0.1	0.9	0.0	0.0	0.9	0.1	0.0	1.0	0.0
x_2	0.8	0.2	0.0	1.0	0.0	0.0	0.0	0.1	0.9
x_3	0.0	0.2	0.8	0.0	0.2	0.8	0.9	0.1	0.0
x_4	0.1	0.9	0.0	0.0	0.9	0.1	0.0	1.0	0.0
x_5	0.0	0.8	0.2	0.8	0.2	0.0	0.0	0.1	0.9
x_6	0.8	0.2	0.0	0.0	0.2	0.8	1.0	0.0	0.0
x_7	0.1	0.9	0.0	0.0	0.9	0.1	0.1	0.9	0.0
x_8	0.0	0.1	0.9	0.8	0.2	0.0	0.0	0.0	1.0
x_9	0.0	0.2	0.8	0.0	0.2	0.8	0.9	0.1	0.0
x_{10}	0.1	0.9	0.0	0.1	0.9	0.0	0.0	0.9	0.1

First, we apply the variable precision fuzzy rough set approach. Using similarity relation in the form (31), we determine similarity matrices with respect to condition and decision attributes. By merging identical rows of the similarity matrix, we get 9 condition similarity classes and and 6 decision similarity classes. We calculate ε-approximation quality using the Łukasiewicz implication operator. The results are presented in table 2.

Table 2. ε-approximation quality for different values of parameter ε

Method	Removed attribute	$\gamma_{\widetilde{C}_\varepsilon}(\widetilde{D})$			
		$\varepsilon = 1$	$\varepsilon = 0.9$	$\varepsilon = 0.85$	$\varepsilon = 0.8$
	none	0.830	0.900	0.900	0.910
Ł-inf	c_1	0.820	0.880	0.880	0.910
	c_2	0.250	0.250	0.410	0.450

We can state that the considered information system has a high consistency. The condition attribute c_1 can be omitted from the decision table without a significant decrease of the ε-approximation quality.

In the next step, the flow graph method will be applied. We use the same labels for both the linguistic values of the attributes and the corresponding nodes of the flow graph. As stated in previous section, the T-norm operator prod should be used in our calculations. The obtained fuzzy flow graph has a very simple form, because there is only one condition attribute c_2 and one decision attribute d. Values of the normalized flow between nodes of the condition layer and nodes of the decision layer are given in Table 3.

Table 3. Normalized flow between nodes of condition and decision layers

	$\sigma(V_{2i}, W_{1j})$			
	W_{11}	W_{12}	W_{13}	Σ
V_{21}	0.000	0.027	0.243	0.270
V_{22}	0.065	0.348	0.047	0.460
V_{23}	0.225	0.045	0.000	0.270
Σ	0.290	0.420	0.290	1.000

We see that the flow conservation equations (50) and (51), are satisfied, for example,

$$\sigma_-(V_{21}) = \frac{\text{power}(\widetilde{V}_{21})}{\text{card}(U)} = \sum_{i=1}^{3} \sigma(V_{21}, W_{1i}) = 0.270,$$

$$\sigma_+(W_{11}) = \frac{\text{power}(\widetilde{W}_{11})}{\text{card}(U)} = \sum_{i=1}^{3} \sigma(V_{2i}, W_{11}) = 0.290.$$

Let us determine the certainty and coverage factors for branches between the layers according to formulae (54), (55). The results are given in Tables 4 and 5.

Table 4. Certainty factor for branches between condition and decision layers

	$\text{cer}(V_{2i}, W_{1j})$			
	W_{11}	W_{12}	W_{13}	Σ
V_{21}	0.0000	0.1000	0.9000	1.0000
V_{22}	0.1413	0.7565	0.1022	1.0000
V_{23}	0.8333	0.1667	0.0000	1.0000

Table 5. Coverage factor for branches between condition and decision layers

	$\text{cov}(V_{2i}, W_{1j})$		
	W_{11}	W_{12}	W_{13}
V_{21}	0.0000	0.0643	0.8379
V_{22}	0.2241	0.8286	0.1621
V_{23}	0.7759	0.1071	0.0000
Σ	1.0000	1.0000	1.0000

Fuzzy decision rules with the largest values of certainty factor (Table 6) can be included in the final fuzzy inference system. The respective values of coverage factor are useful for explaining the selected decision rules. Only 3 decision rules

Table 6. Decision rules with the largest value of certainty factor

decision rule	certainty	coverage	strength [%]
$V_{21} \rightarrow W_{13}$	0.9000	0.8379	24.30
$V_{22} \rightarrow W_{12}$	0.7565	0.8286	34.80
$V_{23} \rightarrow W_{11}$	0.8333	0.7759	22.50

could be generated from our decision table. Owing to the application of the VPFRS approach, we got a simple fuzzy flow graph.

Let us construct a flow graph without a prior reduction of attributes. We merge the layers corresponding to condition attributes c_1 and c_2 to a resulting layer, which represents all possible linguistic values in the antecedences of decision rules.

We determine the degrees of satisfaction of the rules' antecedences for particular elements $x \in U$. For the antecedence represented by $V_{12}V_{22}$, we get:

$$\widetilde{V_{12}V_{22}} = \tilde{V}_{12} \cap \tilde{V}_{22} = \{\ 0.81/x_1,\ \ 0.00/x_2,\ \ 0.04/x_3,\ \ 0.81/x_4,\ \ 0.16/x_5,\ \ 0.04/x_6,$$
$$0.81/x_7,\ \ 0.02/x_8,\ \ 0.04/x_9,\ \ 0.81/x_{10}\},$$

$$\varphi(V_{12}, V_{22}) = \text{power}(\widetilde{V_{12}V_{22}}) = 3.54,\ \ \sigma(V_{12}, V_{22}) = \frac{\varphi(V_{12}, V_{22})}{\text{card}U} = 0.354.$$

Table 7. Decision rules with the largest certainty factor (full information system)

decision rule	certainty	coverage	strength [%]
$V_{11}V_{21} \rightarrow W_{13}$	0.8901	0.2486	7.21
$V_{11}V_{23} \rightarrow W_{11}$	0.9567	0.2210	6.41
$V_{12}V_{21} \rightarrow W_{13}$	0.8366	0.2914	8.45
$V_{12}V_{22} \rightarrow W_{12}$	0.8763	0.7386	31.02
$V_{13}V_{21} \rightarrow W_{13}$	0.9818	0.2979	8.64
$V_{13}V_{23} \rightarrow W_{11}$	0.9000	0.3972	11.52

Finally, we determine the normalized throughflow, certainty and coverage factors for branches between of the resulting condition and decision layers. Decision rules with the largest value of certainty factor are given in Table 7. We can observe that the attribute c_1 is superfluous in the obtained decision rules.

6   Conclusions

Information systems with crisp and fuzzy attributes can be effectively analyzed by a hybrid approach which combines the variable precision fuzzy rough set (VPFRS) model with fuzzy flow graphs. The VPFRS model can be defined in a unified way with the help of a single notion of ε-approximation. This allows to avoid the inconsistency of the VPFRS model caused by different forms of fuzzy connectives. The proposed fuzzy flow graph method is suitable for representing and analyzing decision tables with fuzzy attributes. Every fuzzy attribute can be represented by a layer of a flow graph. All nodes of a layer correspond to linguistic values of an attribute. A fuzzy decision table can be reduced by applying the VPFRS approach prior to using the fuzzy flow graph method for determining a system of fuzzy decision rules.

References

1. Bandler, W., Kohout, L.: Fuzzy Power Sets and Fuzzy Implication Operators. Fuzzy Sets and Systems **4** (1980) 13–30
2. Burillo, P., Frago, N., Fuentes, R.: Inclusion Grade and Fuzzy Implication Operators. Fuzzy Sets and Systems **114** (2000) 417–429
3. Chen, S.M., Yeh, M.S., Hsiao, P.Y.: A Comparison of Similarity Measures of Fuzzy Values. Fuzzy Sets and Systems **72** (1995) 79–89
4. Cornelis, C., Van der Donck, C., Kerre, E.: Sinha-Dougherty Approach to the Fuzzification of Set Inclusion Revisited. Fuzzy Sets and Systems **134** (2003) 283–295
5. De Baets, B., De Meyer, H., Naessens, H.: On Rational Cardinality-based Inclusion Measures. Fuzzy Sets and Systems **128** (2002) 169–183
6. Dubois, D., Prade, H.: Putting Rough Sets and Fuzzy Sets Together. [30] 203–232
7. Fernández Salido, J.M., Murakami, S.: Rough Set Analysis of a General Type of Fuzzy Data Using Transitive Aggregations of Fuzzy Similarity Relations. Fuzzy Sets and Systems **139** (2003) 635–660
8. Greco, S., Matarazzo, B., Słowiński, R.: Rough Set Processing of Vague Information Using Fuzzy Similarity Relations. In: Calude, C.S., Paun, G., (eds.): Finite Versus Infinite — Contributions to an Eternal Dilemma. Springer-Verlag, Berlin Heidelberg New York (2000) 149–173
9. Greco, S., Pawlak, Z., Słowiński, R.: Generalized Decision Algorithms, Rough Inference Rules, and Flow Graphs. In: Alpigini, J., Peters, J.F., Skowron, A., Zhong, N., (eds.): Rough Sets and Current Trends in Computing. Lecture Notes in Artificial Intelligence, Vol. 2475. Springer-Verlag, Berlin Heidelberg New York (2002) 93–104
10. Greco, S., Pawlak, Z., Słowiński, R.: Bayesian Confirmation Measures within Rough Set Approach. [31] 264–273
11. Greco, S., Matarazzo, B., Słowiński, R.: Rough Membership and Bayesian Confirmation Measures for Parameterized Rough Sets. [28] 314–324
12. Inuiguchi, M.: Generalizations of Rough Sets: From Crisp to Fuzzy Cases. [31] 26–37
13. Katzberg, J.D., Ziarko, W.: Variable Precision Extension of Rough Sets. Fundamenta Informaticae **27** (1996) 155–168
14. Klir, G.J., Folger, T.A.: Fuzzy Sets, Uncertainty, and Information. Prentice Hall, Englewood, New Jersey (1988)
15. Lin, T.Y.: Coping with Imprecision Information — Fuzzy Logic. Downsizing Expo, Santa Clara Convention Center (1993)
16. Mieszkowicz-Rolka, A., Rolka, L.: Variable Precision Rough Sets: Evaluation of Human Operator's Decision Model. In: Sołdek, J., Drobiazgiewicz, L., (eds.): Artificial Intelligence and Security in Computing Systems. Kluwer Academic Publishers, Boston Dordrecht London (2003) 33–40

17. Mieszkowicz-Rolka, A., Rolka, L.: Variable Precision Fuzzy Rough Sets Model in the Analysis of Process Data. [28] 354–363
18. Nakamura, A.: Application of Fuzzy-Rough Classifications to Logics. [30] 233–250
19. Pawlak, Z.: Rough Sets: Theoretical Aspects of Reasoning about Data. Kluwer Academic Publishers, Boston Dordrecht London (1991)
20. Pawlak, Z.: Decision Algorithms, Bayes' Theorem and Flow Graphs. In: Rutkowski, L., Kacprzyk, J., (eds.): Advances in Soft Computing. Physica-Verlag, Heidelberg (2003) 18–24
21. Pawlak, Z.: Flow Graphs and Data Mining. [23] 1–36
22. Pawlak, Z.: Rough Sets and Flow Graphs. [28] 1–11
23. Peters, J.F., et al., (eds.): Transactions on Rough Sets III. Lecture Notes in Computer Science (Journal Subline), Vol. 3400. Springer-Verlag, Berlin Heidelberg New York (2005)
24. Polkowski, L.: Toward Rough Set Foundations. Mereological Approach. [31] 8–25
25. Radzikowska, A.M., Kerre, E.E.: A Comparative Study of Fuzzy Rough Sets. Fuzzy Sets and Systems 126 (2002) 137–155
26. Skowron, A., Stepaniuk, J.: Tolerance Approximation Spaces. Fundamenta Informaticae 27 (1996) 245–253
27. Ślęzak, D., Ziarko, W.: Variable Precision Bayesian Rough Set Model. In: Wang, G., Liu, Q., Yao, Y., Skowron, A., (eds.): Rough Sets, Fuzzy Sets, Data Mining, and Granular Computing. Lecture Notes in Artificial Intelligence, Vol. 2639. Springer-Verlag, Berlin Heidelberg New York (2003) 312–315
28. Ślęzak, D., et al., (eds.): Rough Sets and Current Trends in Computing. Lecture Notes in Artificial Intelligence, Vol. 3641. Springer-Verlag, Berlin Heidelberg New York (2005)
29. Ślęzak, D.: Rough Sets and Bayes Factor. [23] 202–229
30. Słowiński, R., (ed.): Intelligent Decision Support: Handbook of Applications and Advances of the Rough Sets Theory. Kluwer Academic Publishers, Boston Dordrecht London (1992)
31. Tsumoto, S., et al., (eds.): Rough Sets and Current Trends in Computing. Lecture Notes in Artificial Intelligence, Vol. 3066. Springer-Verlag, Berlin Heidelberg New York (2004)
32. Yager, R.R., Filev, D.P.: Essentials of Fuzzy Modelling and Control. John Wiley & Sons, Inc., New York (1994)
33. Liu, W.N., Yao, J., Yao, Y.: Rough Approximations under Level Fuzzy Sets. [31] 78–83
34. Zadeh, L.: Fuzzy Sets. Information and Control 8 (1965) 338–353
35. Ziarko, W.: Variable Precision Rough Sets Model. Journal of Computer and System Sciences 46 (1993) 39–59
36. Ziarko, W.: Probabilistic Rough Sets. [28] 283–293

On Partial Covers, Reducts and Decision Rules with Weights

Mikhail Ju. Moshkov[1], Marcin Piliszczuk[2], and Beata Zielosko[3]

[1] Institute of Computer Science, University of Silesia
39, Będzińska St., Sosnowiec, 41-200, Poland
`moshkov@us.edu.pl`
[2] ING Bank Śląski S.A., 34, Sokolska St., Katowice, 40-086, Poland
`marcin.piliszczuk@ingbank.pl`
[3] Institute of Computer Science, University of Silesia
39, Będzińska St., Sosnowiec, 41-200, Poland
`zielosko@us.edu.pl`

Abstract. In the paper the accuracy of greedy algorithms with weights for construction of partial covers, reducts and decision rules is considered. Bounds on minimal weight of partial covers, reducts and decision rules based on an information on greedy algorithm work are studied. Results of experiments with software implementation of greedy algorithms are described.

Keywords: partial covers, partial reducts, partial decision rules, weights, greedy algorithms.

1 Introduction

The paper is devoted to consideration of partial decision-relative reducts (we will omit often words "decision-relative") and partial decision rules for decision tables on the basis of partial cover investigation.

Rough set theory [11,17] often deals with decision tables containing noisy data. In this case exact reducts and rules can be "overlearned" i.e. depend essentially on noise. If we see constructed reducts and rules as a way of knowledge representation [16] then instead of large exact reducts and rules it is more appropriate to work with relatively small partial ones. In [12] Zdzisław Pawlak wrote that "the idea of an approximate reduct can be useful in cases when a smaller number of condition attributes is preferred over accuracy of classification".

Last years in rough set theory partial reducts, partial decision rules and partial covers are studied intensively [6,7,8,9,10,13,19,20,21,22,23,24,27]. Approximate reducts are investigated also in extensions of rough set model such as VPRS (variable precision rough sets) [26] and α-RST (alpha rough set theory) [14].

We study the case where each subset, used for covering, has its own weight, and we must minimize the total weight of subsets in partial cover. The same situation is with partial reducts and decision rules: each conditional attribute has its own weight, and we must minimize the total weight of attributes in partial

J.F. Peters et al. (Eds.): Transactions on Rough Sets VI, LNCS 4374, pp. 211–246, 2007.
© Springer-Verlag Berlin Heidelberg 2007

reduct or decision rule. If weight of each attribute characterizes time complexity of attribute value computation then we try to minimize total time complexity of computation of attributes from partial reduct or partial decision rule. If weight characterizes a risk of attribute value computation (as in medical or technical diagnosis) then we try to minimize total risk, etc.

In rough set theory various problems can be represented as set cover problems with weights:

- problem of construction of a reduct [16] or partial reduct with minimal total weight of attributes for an information system;
- problem of construction of a decision-relative reduct [16] or partial decision-relative reduct with minimal total weight of attributes for a decision table;
- problem of construction of a decision rule or partial decision rule with minimal total weight of attributes for a row of a decision table (note that this problem is closely connected with the problem of construction of a local reduct [16] or partial local reduct with minimal total weight of attributes);
- problem of construction of a subsystem of a given system of decision rules which "covers" the same set of rows and has minimal total weight of rules (in the capacity of a rule weight we can consider its length).

So the study of covers and partial covers is of some interest for rough set theory. In this paper we list some known results on set cover problems which can be useful in applications and obtain certain new results.

From results obtained in [20,22] it follows that the problem of construction of partial cover with minimal weight is NP-hard. Therefore we must consider polynomial approximate algorithms for minimization of weight of partial covers.

In [18] a greedy algorithm with weights for partial cover construction was investigated. This algorithm is a generalization of well known greedy algorithm with weights for exact cover construction [2]. The algorithm from [18] is a greedy algorithm with one threshold which gives the exactness of constructed partial cover.

Using results from [9] (based on results from [3,15] and technique created in [20,22]) on precision of polynomial approximate algorithms for construction of partial cover with minimal cardinality and results from [18] on precision of greedy algorithm with one threshold we show that under some natural assumptions on the class NP the greedy algorithm with one threshold is close to best polynomial approximate algorithms for construction of partial cover with minimal weight. However we can try to improve results of the work of greedy algorithm with one threshold for some part of set cover problems with weight.

We generalize greedy algorithm with one threshold [18], and consider greedy algorithm with two thresholds. First threshold gives the exactness of constructed partial cover, and the second one is an interior parameter of the considered algorithm. We prove that for the most part of set cover problems there exist a weight function and values of thresholds such that the weight of partial cover constructed by greedy algorithm with two thresholds is less than the weight of partial cover constructed by greedy algorithm with one threshold.

We describe two polynomial algorithms which always construct partial covers that are not worse than the one constructed by greedy algorithm with one threshold, and for the most part of set cover problems there exists a weight function and a value of first threshold such that the weight of partial covers constructed by the considered two algorithms is less than the weight of partial cover constructed by greedy algorithm with one threshold.

Information on greedy algorithm work can be used for obtaining lower bounds on minimal cardinality of partial covers [9]. We fix some kind of information on greedy algorithm work, and find unimprovable lower bound on minimal weight of partial cover depending on this information. Obtained results show that this bound is not trivial and can be useful for investigation of set cover problems.

There exist bounds on precision of greedy algorithm without weights for partial cover construction which do not depend on the cardinality of covered set [1,6,7,8]. We obtain similar bound for the case of weight.

The most part of the results obtained for partial covers is generalized on the case of partial decision-relative reducts and partial decision rules for decision tables which, in general case, are inconsistent (a decision table is inconsistent if it has equal rows with different decisions). In particular, we show that

- Under some natural assumptions on the class NP greedy algorithms with weights are close to best polynomial approximate algorithms for minimization of total weight of attributes in partial reducts and partial decision rules.
- Based on information receiving during greedy algorithm work it is possible to obtain nontrivial lower bounds on minimal total weight of attributes in partial reducts and partial decision rules.
- There exist polynomial modifications of greedy algorithms which for a part of decision tables give better results than usual greedy algorithms.

Obtained results will further to more wide use of greedy algorithms with weighs and their modifications in rough set theory and applications.

This paper is, in some sense, an extension of [9] on the case of weights which are not equal to 1. However, problems considered in this paper (and proofs of results) are more complicated than the ones considered in [9]. Bounds obtained in this paper are sometimes more weak than the corresponding bounds from [9]. We must note also that even if all weights are equal to 1 then results of the work of greedy algorithms considered in this paper can be different from the results of the work of greedy algorithms considered in [9]. For example, for case of reducts the number of chosen attributes is the same, but last attributes can differ.

The paper consists of five sections. In Sect. 2 partial covers are studied. In Sect. 3 partial tests (partial superreducts) and partial reducts are investigated. In Sect. 4 partial decision rules are considered. Sect. 5 contains short conclusions.

2 Partial Covers

2.1 Main Notions

Let $A = \{a_1, \ldots, a_n\}$ be a nonempty finite set. Elements of A are enumerated by numbers $1, \ldots, n$ (in fact we fix a linear order on A). Let $S = \{B_i\}_{i \in \{1,\ldots,m\}} =$

$\{B_1, \ldots, B_m\}$ be a family of subsets of A such that $B_1 \cup \ldots \cup B_m = A$. We will assume that S can contain equal subsets of A. The pair (A, S) will be called a *set cover problem*. Let w be a *weight function* which corresponds to each $B_i \in S$ a natural number $w(B_i)$. The triple (A, S, w) will be called a *set cover problem with weights*. Note that in fact weight function w is given on the set of indexes $\{1, \ldots, m\}$. But, for simplicity, we are writing $w(B_i)$ instead of $w(i)$.

Let I be a subset of $\{1, \ldots, m\}$. The family $P = \{B_i\}_{i \in I}$ will be called a *subfamily* of S. The number $|P| = |I|$ will be called the *cardinality* of P. Let $P = \{B_i\}_{i \in I}$ and $Q = \{B_i\}_{i \in J}$ be subfamilies of S. The notation $P \subseteq Q$ will mean that $I \subseteq J$. Let us denote $P \cup Q = \{B_i\}_{i \in I \cup J}$, $P \cap Q = \{B_i\}_{i \in I \cap J}$, and $P \setminus Q = \{B_i\}_{i \in I \setminus J}$.

A subfamily $Q = \{B_{i_1}, \ldots, B_{i_t}\}$ of the family S will be called a *partial cover* for (A, S). Let α be a real number such that $0 \le \alpha < 1$. The subfamily Q will be called an α-*cover* for (A, S) if $|B_{i_1} \cup \ldots \cup B_{i_t}| \ge (1 - \alpha)|A|$. For example, 0.01-cover means that we must cover at least 99% of elements from A. Note that a 0-cover is usual (exact) cover. The number $w(Q) = \sum_{j=1}^{t} w(B_{i_j})$ will be called the *weight* of the partial cover Q. Let us denote by $C_{\min}(\alpha) = C_{\min}(\alpha, A, S, w)$ the minimal weight of α-cover for (A, S).

Let α and γ be real numbers such that $0 \le \gamma \le \alpha < 1$. Let us describe a *greedy algorithm with two thresholds* α and γ.

Let us denote $N = \lceil |A|(1 - \gamma) \rceil$ and $M = \lceil |A|(1 - \alpha) \rceil$. Let we make $i \ge 0$ steps and choose subsets $B_{j_1}, \ldots, B_{j_i}$. Let us describe the step number $i + 1$.

Let us denote $D = B_{j_1} \cup \ldots \cup B_{j_i}$ (if $i = 0$ then $D = \emptyset$). If $|D| \ge M$ then we finish the work of the algorithm. The family $\{B_{j_1}, \ldots, B_{j_i}\}$ is the constructed α-cover. Let $|D| < M$. Then we choose a subset $B_{j_{i+1}}$ from S with minimal number j_{i+1} for which $B_{j_{i+1}} \setminus D \ne \emptyset$ and the value

$$\frac{w(B_{j_{i+1}})}{\min\{|B_{j_{i+1}} \setminus D|, N - |D|\}}$$

is minimal. Pass to the step number $i + 2$.

Let us denote by $C_{\text{greedy}}^{\gamma}(\alpha) = C_{\text{greedy}}^{\gamma}(\alpha, A, S, w)$ the weight of α-cover constructed by the considered algorithm for the set cover problem with weights (A, S, w).

Note that greedy algorithm with two thresholds α and α coincides with the greedy algorithm with one threshold α considered in [18].

2.2   Some Known Results

In this subsection we assume that the weight function has values from the set of positive real numbers.

For natural m denote $H(m) = 1 + \ldots + 1/m$. It is known that

$$\ln m \le H(m) \le \ln m + 1 .$$

Consider some results for the case of exact covers where $\alpha = 0$. In this case $\gamma = 0$. First results belong to Chvátal.

Theorem 1. (Chvátal [2]) *For any set cover problem with weights (A, S, w) the inequality $C^0_{\text{greedy}}(0) \leq C_{\min}(0)H(|A|)$ holds.*

Theorem 2. (Chvátal [2]) *For any set cover problem with weights (A, S, w) the inequality $C^0_{\text{greedy}}(0) \leq C_{\min}(0)H(\max_{B_i \in S}|B_i|)$ holds.*

Chvátal proved in [2] that the bounds from Theorems 1 and 2 are almost unimprovable.

Consider now some results for the case where $\alpha \geq 0$ and $\gamma = \alpha$. First upper bound on $C^\alpha_{\text{greedy}}(\alpha)$ was obtained by Kearns.

Theorem 3. (Kearns [5]) *For any set cover problem with weights (A, S, w) and any α, $0 \leq \alpha < 1$, the inequality $C^\alpha_{\text{greedy}}(\alpha) \leq C_{\min}(\alpha)(2H(|A|) + 3)$ holds.*

This bound was improved by Slavík.

Theorem 4. (Slavík [18]) *For any set cover problem with weights (A, S, w) and any α, $0 \leq \alpha < 1$, the inequality $C^\alpha_{\text{greedy}}(\alpha) \leq C_{\min}(\alpha)H(\lceil(1-\alpha)|A|\rceil)$ holds.*

Theorem 5. (Slavík [18])) *For any set cover problem with weights (A, S, w) and any α, $0 \leq \alpha < 1$, the inequality $C^\alpha_{\text{greedy}}(\alpha) \leq C_{\min}(\alpha)H(\max_{B_i \in S}|B_i|)$ holds.*

Slavík proved in [18] that the bounds from Theorems 4 and 5 are unimprovable.

2.3 On Polynomial Approximate Algorithms for Minimization of Partial Cover Weight

In this subsection we consider three theorems which follow immediately from Theorems 13–15 [9].

Let $0 \leq \alpha < 1$. Consider the following problem: for given set cover problem with weights (A, S, w) it is required to find an α-cover for (A, S) with minimal weight.

Theorem 6. *Let $0 \leq \alpha < 1$. Then the problem of construction of α-cover with minimal weight is NP-hard.*

From this theorem it follows that we must consider polynomial approximate algorithms for minimization of α-cover weight.

Theorem 7. *Let $\alpha \in \mathbb{R}$ and $0 \leq \alpha < 1$. If $NP \not\subseteq DTIME(n^{O(\log \log n)})$ then for any ε, $0 < \varepsilon < 1$, there is no polynomial algorithm that for a given set cover problem with weights (A, S, w) constructs an α-cover for (A, S) which weight is at most $(1 - \varepsilon)C_{\min}(\alpha, A, S, w)\ln|A|$.*

Theorem 8. *Let α be a real number such that $0 \leq \alpha < 1$. If $P \neq NP$ then there exists $\delta > 0$ such that there is no polynomial algorithm that for a given set cover problem with weights (A, S, w) constructs an α-cover for (A, S) which weight is at most $\delta C_{\min}(\alpha, A, S, w)\ln|A|$.*

From Theorem 4 it follows that $C^{\alpha}_{\text{greedy}}(\alpha) \leq C_{\min}(\alpha)(1 + \ln|A|)$. From this inequality and from Theorem 7 it follows that under the assumption $NP \not\subseteq DTIME(n^{O(\log\log n)})$ greedy algorithm with two thresholds α and α (in fact greedy algorithm with one threshold α from [18]) is close to best polynomial approximate algorithms for minimization of partial cover weight. From the considered inequality and from Theorem 8 it follows that under the assumption $P \neq NP$ greedy algorithm with two thresholds α and α is not far from best polynomial approximate algorithms for minimization of partial cover weight.

However we can try to improve the results of the work of greedy algorithm with two thresholds α and α for some part of set cover problems with weights.

2.4   Comparison of Greedy Algorithms with One and Two Thresholds

The following example shows that if for greedy algorithm with two thresholds α and γ we will use γ such that $\gamma < \alpha$ we can obtain sometimes better results than in the case $\gamma = \alpha$.

Example 1. Consider a set cover problem (A, S, w) such that $A = \{1, 2, 3, 4, 5, 6\}$, $S = \{B_1, B_2\}$, $B_1 = \{1\}$, $B_2 = \{2, 3, 4, 5, 6\}$, $w(B_1) = 1$ and $w(B_2) = 4$. Let $\alpha = 0.5$. It means that we must cover at least $M = \lceil(1 - \alpha)|A|\rceil = 3$ elements from A. If $\gamma = \alpha = 0.5$ then the result of the work of greedy algorithm with thresholds α and γ is the 0.5-cover $\{B_1, B_2\}$ which weight is equal to 5. If $\gamma = 0 < \alpha$ then the result of the work of greedy algorithm with thresholds α and γ is the 0.5-cover $\{B_2\}$ which weight is equal to 4.

In this subsection we show that under some assumptions on $|A|$ and $|S|$ for the most part of set cover problems (A, S) there exist a weight function w and real numbers α, γ such that $0 \leq \gamma < \alpha < 1$ and $C^{\gamma}_{\text{greedy}}(\alpha, A, S, w) < C^{\alpha}_{\text{greedy}}(\alpha, A, S, w)$. First, we consider criterion of existence of such w, α and γ (see Theorem 9). First part of the proof of this criterion is based on a construction similar to considered in Example 1.

Let A be a finite nonempty set and $S = \{B_1, \ldots, B_m\}$ be a family of subsets of A. We will say that the family S is 1-*uniform* if there exists a natural number k such that $|B_i| = k$ or $|B_i| = k+1$ for any nonempty subset B_i from S. We will say that S is *strongly* 1-*uniform* if S is 1-uniform and for any subsets $B_{l_1}, \ldots, B_{l_t}$ from S the family $\{B_1 \setminus U, \ldots, B_m \setminus U\}$ is 1-uniform where $U = B_{l_1} \cup \ldots \cup B_{l_t}$.

Theorem 9. *Let (A, S) be a set cover problem. Then the following two statements are equivalent:*

1. *The family S is not strongly 1-uniform.*
2. *There exist a weight function w and real numbers α and γ such that $0 \leq \gamma < \alpha < 1$ and $C^{\gamma}_{\text{greedy}}(\alpha, A, S, w) < C^{\alpha}_{\text{greedy}}(\alpha, A, S, w)$.*

Proof. Let $S = \{B_1, \ldots, B_m\}$. Let the family S be not strongly 1-uniform. Let us choose minimal number of subsets $B_{l_1}, \ldots, B_{l_t}$ from the family S (it is possible that $t = 0$) such that the family $\{B_1 \setminus U, \ldots, B_m \setminus U\}$ is not 1-uniform where $U = B_{l_1} \cup \ldots \cup B_{l_t}$ (if $t = 0$ then $U = \emptyset$). Since $\{B_1 \setminus U, \ldots, B_m \setminus U\}$ is not 1-uniform, there exist two subsets B_i and B_j from S such that $|B_i \setminus U| > 0$ and $|B_j \setminus U| \geq |B_i \setminus U| + 2$. Let us choose real α and γ such that $M = \lceil |A|(1 - \alpha) \rceil = |U| + |B_i \setminus U| + 1$ and $N = \lceil |A|(1 - \gamma) \rceil = |U| + |B_i \setminus U| + 2$. It is clear that $0 \leq \gamma < \alpha < 1$. Let us define a weight function w as follows: $w(B_{l_1}) = \ldots = w(B_{l_t}) = 1$, $w(B_i) = |A| \cdot 2|B_i \setminus U|$, $w(B_j) = |A|(2|B_i \setminus U| + 3)$ and $w(B_r) = |A|(3|B_i \setminus U| + 6)$ for any B_r from S such that $r \notin \{i, j, l_1, \ldots, l_t\}$.

Let us consider the work of greedy algorithm with two thresholds α and α. One can show that during first t steps the greedy algorithm will choose subsets $B_{l_1}, \ldots, B_{l_t}$ (may be in an another order). It is clear that $|U| < M$. Therefore the greedy algorithm must make the step number $t + 1$. During this step the greedy algorithm will choose a subset B_k from S with minimal number k for which $B_k \setminus U \neq \emptyset$ and the value $p(k) = \frac{w(B_k)}{\min\{|B_k \setminus U|, M - |U|\}} = \frac{w(B_k)}{\min\{|B_k \setminus U|, |B_i \setminus U| + 1\}}$ is minimal.

It is clear that $p(i) = 2|A|$, $p(j) = (2 + \frac{1}{|B_i \setminus U| + 1})|A|$ and $p(k) > 3|A|$ for any subset B_k from S such that $B_k \setminus U \neq \emptyset$ and $k \notin \{i, j, l_1, \ldots, l_t\}$. Therefore during the step number $t + 1$ the greedy algorithm will choose the subset B_i. Since $|U| + |B_i \setminus U| = M - 1$, the greedy algorithm will make the step number $t + 2$ and will choose a subset from S which is different from $B_{l_1}, \ldots, B_{l_t}, B_i$. As the result we obtain $C^\alpha_{\text{greedy}}(\alpha, A, S, w) \geq t + |A| \cdot 2|B_i \setminus U| + |A|(2|B_i \setminus U| + 3)$.

Let us consider the work of greedy algorithm with two thresholds α and γ. One can show that during first t steps the greedy algorithm will choose subsets $B_{l_1}, \ldots, B_{l_t}$ (may be in an another order). It is clear that $|U| < M$. Therefore the greedy algorithm must make the step number $t + 1$. During this step the greedy algorithm will choose a subset B_k from S with minimal number k for which $B_k \setminus U \neq \emptyset$ and the value $q(k) = \frac{w(B_k)}{\min\{|B_k \setminus U|, N - |U|\}} = \frac{w(B_k)}{\min\{|B_k \setminus U|, |B_i \setminus U| + 2\}}$ is minimal.

It is clear that $q(i) = 2|A|$, $q(j) = (2 - \frac{1}{|B_i \setminus U| + 2})|A|$ and $q(k) \geq 3|A|$ for any subset B_k from S such that $B_k \setminus U \neq \emptyset$ and $k \notin \{i, j, l_1, \ldots, l_t\}$. Therefore during the step number $t + 1$ the greedy algorithm will choose the subset B_j. Since $|U| + |B_j \setminus U| > M$, the α-cover constructed by greedy algorithm will be equal to $\{B_{l_1}, \ldots, B_{l_t}, B_j\}$. As the result we obtain $C^\gamma_{\text{greedy}}(\alpha, A, S, w) = t + |A|(2|B_i \setminus U| + 3)$. Since $C^\alpha_{\text{greedy}}(\alpha, A, S, w) \geq t + |A| \cdot 2|B_i \setminus U| + |A|(2|B_i \setminus U| + 3)$ and $|B_i \setminus U| > 0$, we conclude that $C^\alpha_{\text{greedy}}(\alpha, A, S, w) > C^\gamma_{\text{greedy}}(\alpha, A, S, w)$.

Let the family S be strongly 1-uniform. Consider arbitrary weight function w for S and real numbers α and γ such that $0 \leq \gamma < \alpha < 1$. Let us show that $C^\gamma_{\text{greedy}}(\alpha, A, S, w) \geq C^\alpha_{\text{greedy}}(\alpha, A, S, w)$. Let us denote $M = \lceil |A|(1 - \alpha) \rceil$ and $N = \lceil |A|(1 - \gamma) \rceil$. If $M = N$ then $C^\gamma_{\text{greedy}}(\alpha, A, S, w) = C^\alpha_{\text{greedy}}(\alpha, A, S, w)$. Let $N > M$.

Let us apply the greedy algorithm with thresholds α and α to the set cover problem with weights (A, S, w). Let during the construction of α-cover this algorithm choose sequentially subsets $B_{g_1}, \ldots, B_{g_t}$. Let us apply now the greedy algorithm with thresholds α and γ to the set cover problem with weights (A, S, w). If during the construction of α-cover this algorithm chooses sequentially subsets

$B_{g_1}, \ldots, B_{g_t}$ then $C^\gamma_{\text{greedy}}(\alpha, A, S, w) = C^\alpha_{\text{greedy}}(\alpha, A, S, w)$. Let there exist a non-negative integer r, $0 \leq r \leq t-1$, such that during first r steps the considered algorithm chooses subsets $B_{g_1}, \ldots, B_{g_r}$, but at the step number $r+1$ the algorithm chooses a subset B_k such that $k \neq g_{r+1}$. Let us denote $B_{g_0} = \emptyset$, $D = B_{g_0} \cup \ldots \cup B_{g_r}$ and $J = \{i : i \in \{1, \ldots, m\}, B_i \setminus D \neq \emptyset\}$. It is clear that $g_{r+1}, k \in J$. For any $i \in J$ denote $p(i) = \frac{w(B_i)}{\min\{|B_i \setminus D|, M - |D|\}}$, $q(i) = \frac{w(B_i)}{\min\{|B_i \setminus D|, N - |D|\}}$.

Since $k \neq g_{r+1}$, we conclude that there exists $i \in J$ such that $p(i) \neq q(i)$. Therefore $|B_i \setminus D| > M - |D|$. Since S is strongly 1-uniform family, we have $|B_j \setminus D| \geq M - |D|$ for any $j \in J$. From here it follows, in particular, that $r+1 = t$, and $\{B_{g_1}, \ldots, B_{g_{t-1}}, B_k\}$ is an α-cover for (A, S).

It is clear that $p(g_t) \leq p(k)$. Since $|B_k \setminus D| \geq M - |D|$ and $|B_{g_t} \setminus D| \geq M - |D|$, we have $p(k) = \frac{w(B_k)}{M - |D|}$, $p(g_t) = \frac{w(B_{g_t})}{M - |D|}$. Therefore $w(B_{g_t}) \leq w(B_k)$.

Taking into account that $C^\gamma_{\text{greedy}}(\alpha, A, S, w) = w(B_{g_1}) + \ldots + w(B_{g_{t-1}}) + w(B_k)$ and $C^\alpha_{\text{greedy}}(\alpha, A, S, w) = w(B_{g_1}) + \ldots + w(B_{g_{t-1}}) + w(B_{g_t})$ we obtain $C^\gamma_{\text{greedy}}(\alpha, A, S, w) \geq C^\alpha_{\text{greedy}}(\alpha, A, S, w)$. □

Let us show that under some assumptions on $|A|$ and $|S|$ the most part of set cover problems (A, S) is not 1-uniform, and therefore is not strongly 1-uniform.

There is one-to-one correspondence between set cover problems and tables filled by numbers from $\{0, 1\}$ and having no rows filled by 0 only. Let $A = \{a_1, \ldots, a_n\}$ and $S = \{B_1, \ldots, B_m\}$. Then the problem (A, S) corresponds to the table with n rows and m columns which for $i = 1, \ldots, n$ and $j = 1, \ldots, m$ has 1 at the intersection of i-th row and j-th column if and only if $a_i \in B_j$.

A table filled by numbers from $\{0, 1\}$ will be called *SC-table* if this table has no rows filled by 0 only. For completeness of the presentation we consider here a statement from [9] with proof.

Lemma 1. *The number of SC-tables with n rows and m columns is at least*

$$2^{mn} - 2^{mn - m + \log_2 n} .$$

Proof. Let $i \in \{1, \ldots, n\}$. The number of tables in which the i-th row is filled by 0 only is equal to $2^{mn - m}$. Therefore the number of tables which are not SC-tables is at most $n 2^{mn - m} = 2^{mn - m + \log_2 n}$. Thus, the number of SC-tables is at least $2^{mn} - 2^{mn - m + \log_2 n}$. □

Lemma 2. *Let $n \in \mathbb{N}$, $n \geq 4$ and $k \in \{0, \ldots, n\}$. Then $C_n^k \leq C_n^{\lfloor n/2 \rfloor} < \frac{2^n}{\sqrt{n}}$.*

Proof. It is well known (see, for example, [25], p. 178) that $C_n^k \leq C_n^{\lfloor n/2 \rfloor}$. Let n be even and $n \geq 4$. It is known (see [4], p. 278) that $C_n^{\lfloor n/2 \rfloor} \leq \frac{2^n}{\sqrt{\frac{3n}{2} + 1}} < \frac{2^n}{\sqrt{n}}$.

Let n be odd and $n \geq 5$. Using well known equality $C_n^{\lfloor n/2 \rfloor} = C_{n-1}^{\lfloor n/2 \rfloor} + C_{n-1}^{\lfloor n/2 \rfloor - 1}$ and the fact, that $C_{n-1}^{\lfloor (n-1)/2 \rfloor} \geq C_{n-1}^k$ for any $k \in \{0, \ldots, n-1\}$, we obtain $C_n^{\lfloor n/2 \rfloor} \leq 2 C_{n-1}^{\lfloor (n-1)/2 \rfloor}$. Thus, $C_n^{\lfloor n/2 \rfloor} \leq \frac{2^n}{\sqrt{\frac{3(n-1)}{2} + 1}} < \frac{2^n}{\sqrt{\frac{3(n-1)}{3} + 1}} = \frac{2^n}{\sqrt{n}}$. Therefore for any $n \geq 4$ the inequality $C_n^{\lfloor n/2 \rfloor} < \frac{2^n}{\sqrt{n}}$ holds. □

Theorem 10. *Consider set cover problems* (A, S) *such that* $A = \{a_1, \ldots, a_n\}$ *and* $S = \{B_1, \ldots, B_m\}$. *Let* $n \geq 4$ *and* $m \geq \log_2 n + 1$. *Then the fraction of set cover problems which are not 1-uniform is at least* $1 - \frac{9^{\frac{m}{2}+1}}{n^{\frac{m}{2}-1}}$.

Proof. The considered fraction is at least $\frac{q-p}{q}$ where q is the number of SC-tables with n rows and m columns, and p is the number of tables with n rows and m columns filled by 0 and 1 for each of which there exists $k \in \{1, \ldots, n-1\}$ such that the number of units in each column belongs to the set $\{0, k, k+1\}$.

From Lemma 1 it follows that $q \geq 2^{mn} - 2^{mn-m+\log_2 n}$. It is clear that $p \leq \sum_{k=1}^{n-1}(C_n^k + C_n^{k+1} + 1)^m$. From Lemma 2 it follows that $C_n^{\lfloor n/2 \rfloor} \geq C_n^k$ for any $k \in \{1, \ldots, n\}$. Therefore $p \leq (n-1)\left(3C_n^{\lfloor n/2 \rfloor}\right)^m$. Using Lemma 2 we conclude that $3C_n^{\lfloor n/2 \rfloor} < \frac{2^n}{\sqrt{\frac{n}{9}}}$ for any $n \geq 4$. Therefore $p < \frac{(n-1)2^{mn}}{\left(\frac{n}{9}\right)^{m/2}}$. Thus, $\frac{q-p}{q} = 1 - \frac{p}{q} >$
$1 - \frac{(n-1)2^{mn}}{\left(\frac{n}{9}\right)^{m/2}\left(2^{mn} - 2^{mn-m+\log_2 n}\right)}$. Taking into account that $m \geq \log_2 n + 1$ we obtain
$\frac{q-p}{q} > 1 - \frac{2(n-1)}{\left(\frac{n}{9}\right)^{m/2}} > 1 - \frac{9^{\frac{m}{2}+1}}{n^{\frac{m}{2}-1}}$.   □

So if n is large enough and $m \geq \log_2 n + 1$ then the most part of set cover problems (A, S) with $|A| = n$ and $|S| = m$ is not 1-uniform.

For example, the fraction of set cover problems (A, S) with $|A| = 81$ and $|S| = 20$ which are not 1-uniform is at least $1 - \frac{1}{9^7} = 1 - \frac{1}{4782969}$.

2.5   Two Modifications of Greedy Algorithm

Results obtained in the previous subsection show that the greedy algorithm with two thresholds is of some interest. In this subsection we consider two polynomial modifications of greedy algorithm which allow to use advantages of greedy algorithm with two thresholds.

Let (A, S, w) be a set cover problem with weights and α be a real number such that $0 \leq \alpha < 1$.

1. Of course, it is impossible to consider effectively all γ such that $0 \leq \gamma \leq \alpha$. Instead of this we can consider all natural N such that $M \leq N \leq |A|$ where $M = \lceil |A|(1-\alpha) \rceil$ (see the description of greedy algorithm with two thresholds). For each $N \in \{M, \ldots, |A|\}$ we apply greedy algorithm with parameters M and N to set cover problem with weights (A, S, w) and after that choose an α-cover with minimal weight among constructed α-covers.
2. There exists also an another way to construct an α-cover which is not worse than the one obtained under consideration of all N such that $M \leq N \leq |A|$. Let us apply greedy algorithm with thresholds α and α to set cover problem with weights (A, S, w). Let the algorithm choose sequentially subsets $B_{g_1}, \ldots, B_{g_t}$. For each $i \in \{0, \ldots, t-1\}$ we find (if it is possible) a subset B_{l_i} from S with minimal weight $w(B_{l_i})$ such that $|B_{g_1} \cup \ldots \cup B_{g_i} \cup B_{l_i}| \geq M$, and form an α-cover $\{B_{g_1}, \ldots, B_{g_i}, B_{l_i}\}$ (if $i = 0$ then it will be the family $\{B_{l_0}\}$). After that among constructed α-covers $\{B_{g_1}, \ldots, B_{g_t}\}$, ...,

$\{B_{g_1}, \ldots, B_{g_i}, B_{l_i}\}$, ... we choose an α-cover with minimal weight. From Proposition 1 it follows that the constructed α-cover is not worse than the one constructed under consideration of all γ, $0 \leq \gamma \leq \alpha$, or (which is the same) all N, $M \leq N \leq |A|$.

Proposition 1. *Let (A, S, w) be a set cover problem with weights and α, γ be real numbers such that $0 \leq \gamma < \alpha < 1$. Let the greedy algorithm with two thresholds α and α, which is applied to (A, S, w), choose sequentially subsets $B_{g_1}, \ldots, B_{g_t}$. Let the greedy algorithm with two thresholds α and γ, which is applied to (A, S, w), choose sequentially subsets $B_{l_1}, \ldots, B_{l_k}$. Then either $k = t$ and $(l_1, \ldots, l_k) = (g_1, \ldots, g_t)$ or $k \leq t$, $(l_1, \ldots, l_{k-1}) = (g_1, \ldots, g_{k-1})$ and $l_k \neq g_k$.*

Proof. Let $S = \{B_1, \ldots, B_m\}$. Let us denote $M = \lceil |A|(1-\alpha) \rceil$ and $N = \lceil |A|(1-\gamma) \rceil$.

Let $(l_1, \ldots, l_k) \neq (g_1, \ldots, g_t)$. Since $\{B_{g_1}, \ldots, B_{g_{t-1}}\}$ is not an α-cover for (A, S), it is impossible that $k < t$ and $(l_1, \ldots, l_k) = (g_1, \ldots, g_k)$. Since $\{B_{g_1}, \ldots, B_{g_t}\}$ is an α-cover for (A, S), it is impossible that $k > t$ and $(l_1, \ldots, l_t) = (g_1, \ldots, g_t)$. Therefore there exists $i \in \{0, \ldots, t-1\}$ such that during first i steps algorithm with thresholds α and α and algorithm with thresholds α and γ choose the same subsets from S, but during the step number $i + 1$ the algorithm with threshold α and γ chooses a subset $B_{l_{i+1}}$ such that $l_{i+1} \neq g_{i+1}$.

Let us denote $B_{g_0} = \emptyset$, $D = B_{g_0} \cup \ldots \cup B_{g_i}$ and $J = \{j : j \in \{1, \ldots, m\}, B_j \setminus D \neq \emptyset\}$. It is clear that $g_{i+1}, l_{i+1} \in J$. For any $j \in J$ let $p(j) = \frac{w(B_j)}{\min\{|B_j \setminus D|, M - |D|\}}$ and $q(j) = \frac{w(B_j)}{\min\{|B_j \setminus D|, N - |D|\}}$. Since $N \geq M$, we have $p(j) \geq q(j)$ for any $j \in J$. Consider two cases.

Let $g_{i+1} < l_{i+1}$. In this case we have $p(g_{i+1}) \leq p(l_{i+1})$ and $q(g_{i+1}) > q(l_{i+1})$. Using inequality $p(g_{i+1}) \geq q(g_{i+1})$ we obtain $p(g_{i+1}) > q(l_{i+1})$ and $p(l_{i+1}) > q(l_{i+1})$. From last inequality it follows that $|B_{l_{i+1}} \setminus D| > M - |D|$.

Let $g_{i+1} > l_{i+1}$. In this case we have $p(g_{i+1}) < p(l_{i+1})$ and $q(g_{i+1}) \geq q(l_{i+1})$. Using inequality $p(g_{i+1}) \geq q(g_{i+1})$ we obtain $p(g_{i+1}) \geq q(l_{i+1})$ and $p(l_{i+1}) > q(l_{i+1})$. From last inequality it follows that $|B_{l_{i+1}} \setminus D| > M - |D|$.

So in any case we have $|B_{l_{i+1}} \setminus D| > M - |D|$. From this inequality it follows that after the step number $i+1$ the algorithm with thresholds α and γ must finish the work. Thus, $k = i + 1$, $k \leq t$, $(l_1, \ldots, l_{k-1}) = (g_1, \ldots, g_{k-1})$ and $l_k \neq g_k$. $\quad\square$

2.6   Lower Bound on $C_{\min}(\alpha)$

In this subsection we fix some information about the work of greedy algorithm with two thresholds and find the best lower bound on the value $C_{\min}(\alpha)$ depending on this information.

Let (A, S, w) be a set cover problem with weights and α, γ be real numbers such that $0 \leq \gamma \leq \alpha < 1$. Let us apply the greedy algorithm with thresholds α and γ to the set cover problem with weights (A, S, w). Let during the construction of α-cover the greedy algorithm choose sequentially subsets $B_{g_1}, \ldots, B_{g_t}$.

Let us denote $B_{g_0} = \emptyset$ and $\delta_0 = 0$. For $i = 1, \ldots, t$ denote $\delta_i = |B_{g_i} \setminus (B_{g_0} \cup \ldots \cup B_{g_{i-1}})|$ and $w_i = w(B_{g_i})$.

As information on the greedy algorithm work we will use numbers $M_C = M_C(\alpha, \gamma, A, S, w) = \lceil |A|(1 - \alpha) \rceil$ and $N_C = N_C(\alpha, \gamma, A, S, w) = \lceil |A|(1 - \gamma) \rceil$, and tuples $\Delta_C = \Delta_C(\alpha, \gamma, A, S, w) = (\delta_1, \ldots, \delta_t)$ and $W_C = W_C(\alpha, \gamma, A, S, w) = (w_1, \ldots, w_t)$.

For $i = 0, \ldots, t - 1$ denote

$$\rho_i = \left\lceil \frac{w_{i+1}(M_C - (\delta_0 + \ldots + \delta_i))}{\min\{\delta_{i+1}, N_C - (\delta_0 + \ldots + \delta_i)\}} \right\rceil .$$

Let us define parameter $\rho_C(\alpha, \gamma) = \rho_C(\alpha, \gamma, A, S, w)$ as follows:

$$\rho_C(\alpha, \gamma) = \max\{\rho_i : i = 0, \ldots, t - 1\} .$$

We will prove that $\rho_C(\alpha, \gamma)$ is the best lower bound on $C_{\min}(\alpha)$ depending on M_C, N_C, Δ_C and W_C. This lower bound is based on a generalization of the following simple reasoning: if we must cover M elements and the maximal cardinality of a subset from S is δ then we must use at least $\lceil \frac{M}{\delta} \rceil$ subsets.

Theorem 11. *For any set cover problem with weights (A, S, w) and any real numbers α, γ, $0 \leq \gamma \leq \alpha < 1$, the inequality $C_{\min}(\alpha, A, S, w) \geq \rho_C(\alpha, \gamma, A, S, w)$ holds, and there exists a set cover problem with weights (A', S', w') such that*

$$M_C(\alpha, \gamma, A', S', w') = M_C(\alpha, \gamma, A, S, w), N_C(\alpha, \gamma, A', S', w') = N_C(\alpha, \gamma, A, S, w)$$
$$\Delta_C(\alpha, \gamma, A', S', w') = \Delta_C(\alpha, \gamma, A, S, w), W_C(\alpha, \gamma, A', S', w') = W_C(\alpha, \gamma, A, S, w)$$
$$\rho_C(\alpha, \gamma, A', S', w') = \rho_C(\alpha, \gamma, A, S, w), C_{\min}(\alpha, A', S', w') = \rho_C(\alpha, \gamma, A', S', w') .$$

Proof. Let (A, S, w) be a set cover problem with weights, $S = \{B_1, \ldots, B_m\}$, and α, γ be real numbers such that $0 \leq \gamma \leq \alpha < 1$. Let us denote $M = M_C(\alpha, \gamma, A, S, w) = \lceil |A|(1 - \alpha) \rceil$ and $N = N_C(\alpha, \gamma, A, S, w) = \lceil |A|(1 - \gamma) \rceil$. Let $\{B_{l_1}, \ldots, B_{l_k}\}$ be an optimal α-cover for (A, S, w), i.e. $w(B_{l_1}) + \ldots + w(B_{l_k}) = C_{\min}(\alpha, A, S, w) = C_{\min}(\alpha)$ and $|B_{l_1} \cup \ldots \cup B_{l_k}| \geq M$.

Let us apply the greedy algorithm with thresholds α and γ to (A, S, w). Let during the construction of α-cover the greedy algorithm choose sequentially subsets $B_{g_1}, \ldots, B_{g_t}$. Let us denote $B_{g_0} = \emptyset$.

Let $i \in \{0, \ldots, t - 1\}$. Let us denote $D = B_{g_0} \cup \ldots \cup B_{g_i}$. It is clear that after i steps of greedy algorithm work in the set $B_{l_1} \cup \ldots \cup B_{l_k}$ at least $|B_{l_1} \cup \ldots \cup B_{l_k}| - |B_{g_0} \cup \ldots \cup B_{g_i}| \geq M - |D| > 0$ elements remained uncovered. After i-th step $p_1 = |B_{l_1} \setminus D|$ elements remained uncovered in the set B_{l_1}, ..., and $p_k = |B_{l_k} \setminus D|$ elements remained uncovered in the set B_{l_k}. We know that $p_1 + \ldots + p_k \geq M - |D| > 0$. Let, for the definiteness, $p_1 > 0, \ldots, p_r > 0, p_{r+1} = \ldots = p_k = 0$. For $j = 1, \ldots, r$ denote $q_j = \min\{p_j, N - |D|\}$. It is clear that $N - |D| \geq M - |D|$. Therefore $q_1 + \ldots + q_r \geq M - |D|$. Let us consider numbers $\frac{w(B_{l_1})}{q_1}, \ldots, \frac{w(B_{l_r})}{q_r}$. Let us show that at least one of these numbers is at most $\beta = \frac{w(B_{l_1}) + \ldots + w(B_{l_r})}{q_1 + \ldots + q_r}$. Assume the contrary. Then $w(B_{l_1}) + \ldots + w(B_{l_r}) = \frac{w(B_{l_1})q_1}{q_1} + \ldots + \frac{w(B_{l_r})q_r}{q_r} > (q_1 + \ldots + q_r)\beta = w(B_{l_1}) + \ldots + w(B_{l_r})$ which is impossible.

We know that $q_1 + \ldots + q_r \geq M - |D|$ and $w(B_{l_1}) + \ldots + w(B_{l_r}) \leq C_{\min}(\alpha)$. Therefore $\beta \leq \frac{C_{\min}(\alpha)}{M-|D|}$, and there exists $j \in \{1, \ldots, k\}$ such that $B_{l_j} \setminus D \neq \emptyset$ and $\frac{w(B_{l_j})}{\min\{|B_{l_j}\setminus D|, N-|D|\}} \leq \beta$. Hence $\frac{w(B_{g_{i+1}})}{\min\{|B_{g_{i+1}}\setminus D|, N-|D|\}} \leq \beta \leq \frac{C_{\min}(\alpha)}{M-|D|}$ and $C_{\min}(\alpha) \geq \frac{w(B_{g_{i+1}})(M-|D|)}{\min\{|B_{g_{i+1}}\setminus D|, N-|D|\}}$.

Taking into account that $C_{\min}(\alpha)$ is a natural number we obtain $C_{\min}(\alpha) \geq \left\lceil \frac{w(B_{g_{i+1}})(M-|D|)}{\min\{|B_{g_{i+1}}\setminus D|, N-|D|\}} \right\rceil = \rho_i$. Since last inequality holds for any $i \in \{0, \ldots, t-1\}$ and $\rho_C(\alpha, \gamma) = \rho_C(\alpha, \gamma, A, S, w) = \max\{\rho_i : i = 0, \ldots, t-1\}$, we conclude that $C_{\min}(\alpha) \geq \rho_C(\alpha, \gamma)$.

Let us show that this bound is unimprovable depending on M_C, N_C, Δ_C and W_C. Let us consider a set cover problem with weights (A', S', w') where $A' = A$, $S' = \{B_1, \ldots, B_m, B_{m+1}\}$, $|B_{m+1}| = M$, $B_{g_1} \cup \ldots \cup B_{g_{t-1}} \subseteq B_{m+1} \subseteq B_{g_1} \cup \ldots \cup B_{g_t}$, $w'(B_1) = w(B_1), \ldots, w'(B_m) = w(B_m)$ and $w'(B_{m+1}) = \rho_C(\alpha, \gamma)$. It is clear that $M_C(\alpha, \gamma, A', S', w') = M_C(\alpha, \gamma, A, S, w) = M$ and $N_C(\alpha, \gamma, A', S', w') = N_C(\alpha, \gamma, A, S, w) = N$. We show $\Delta_C(\alpha, \gamma, A', S', w') = \Delta_C(\alpha, \gamma, A, S, w)$ and $W_C(\alpha, \gamma, A', S', w') = W_C(\alpha, \gamma, A, S, w)$.

Let us show by induction on $i \in \{1, \ldots, t\}$ that for the set cover problem with weights (A', S', w') at the step number i the greedy algorithm with two thresholds α and γ will choose the subset B_{g_i}. Let us consider the first step. Let us denote $D = \emptyset$. It is clear that $\frac{w'(B_{m+1})}{\min\{|B_{m+1}\setminus D|, N-|D|\}} = \frac{\rho_C(\alpha, \gamma)}{M-|D|}$. From the definition of $\rho_C(\alpha, \gamma)$ it follows that $\frac{w'(B_{g_1})}{\min\{|B_{g_1}\setminus D|, N-|D|\}} = \frac{w(B_{g_1})}{\min\{|B_{g_1}\setminus D|, N-|D|\}} \leq \frac{\rho_C(\alpha, \gamma)}{M-|D|}$. Using this fact and the inequality $g_1 < m+1$ it is not difficult to prove that at the first step greedy algorithm will choose the subset B_{g_1}.

Let $i \in \{1, \ldots, t-1\}$. Let us assume that the greedy algorithm made i steps for (A', S', w') and chose subsets $B_{g_1}, \ldots, B_{g_i}$. Let us show that at the step $i+1$ the subset $B_{g_{i+1}}$ will be chosen. Let us denote $D = B_{g_1} \cup \ldots \cup B_{g_i}$. Since $B_{g_1} \cup \ldots \cup B_{g_i} \subseteq B_{m+1}$ and $|B_{m+1}| = M$, we have $|B_{m+1} \setminus D| = M - |D|$. Therefore $\frac{w'(B_{m+1})}{\min\{|B_{m+1}\setminus D|, N-|D|\}} = \frac{\rho_C(\alpha, \gamma)}{M-|D|}$. From the definition of the parameter $\rho_C(\alpha, \gamma)$ it follows that $\frac{w'(B_{g_{i+1}})}{\min\{|B_{g_{i+1}}\setminus D|, N-|D|\}} = \frac{w(B_{g_{i+1}})}{\min\{|B_{g_{i+1}}\setminus D|, N-|D|\}} \leq \frac{\rho_C(\alpha, \gamma)}{M-|D|}$. Using this fact and the inequality $g_{i+1} < m+1$ it is not difficult to prove that at the step number $i+1$ greedy algorithm will choose the subset $B_{g_{i+1}}$.

Thus, $\Delta_C(\alpha, \gamma, A', S', w') = \Delta_C(\alpha, \gamma, A, S, w)$ and $W_C(\alpha, \gamma, A', S', w') = W_C(\alpha, \gamma, A, S, w)$. Therefore $\rho_C(\alpha, \gamma, A', S', w') = \rho_C(\alpha, \gamma, A, S, w) = \rho_C(\alpha, \gamma)$. From been proven it follows that $C_{\min}(\alpha, A', S', w') \geq \rho_C(\alpha, \gamma, A', S', w')$. It is clear that $\{B_{m+1}\}$ is an α-cover for (A', S') and the weight of $\{B_{m+1}\}$ is equal to $\rho_C(\alpha, \gamma, A', S', w')$. Hence $C_{\min}(\alpha, A', S', w') = \rho_C(\alpha, \gamma, A', S', w')$. $\square$

Let us consider a property of the parameter $\rho_C(\alpha, \gamma)$ which is important for practical use of the bound from Theorem 11.

Proposition 2. *Let (A, S, w) be a set cover problem with weights and α, γ be real numbers such that $0 \leq \gamma \leq \alpha < 1$. Then $\rho_C(\alpha, \alpha, A, S, w) \geq \rho_C(\alpha, \gamma, A, S, w)$.*

Proof. Let $S = \{B_1, \ldots, B_m\}$, $M = \lceil |A|(1 - \alpha) \rceil$, $N = \lceil |A|(1 - \gamma) \rceil$, $\rho_C(\alpha, \alpha) = \rho_C(\alpha, \alpha, A, S, w)$ and $\rho_C(\alpha, \gamma) = \rho_C(\alpha, \gamma, A, S, w)$.

Let us apply the greedy algorithm with thresholds α and α to (A, S, w). Let during the construction of α-cover this algorithm choose sequentially subsets $B_{g_1}, \ldots, B_{g_t}$. Let us denote $B_{g_0} = \emptyset$. For $j = 0, \ldots, t - 1$ denote $D_j = B_{g_0} \cup \ldots \cup B_{g_j}$ and $\rho_C(\alpha, \alpha, j) = \left\lceil \frac{w(B_{g_{j+1}})(M - |D_j|)}{\min\{|B_{g_{j+1}} \setminus D_j|, M - |D_j|\}} \right\rceil$. Then $\rho_C(\alpha, \alpha) = \max\{\rho_C(\alpha, \alpha, j) : j = 0, \ldots, t - 1\}$.

Apply the greedy algorithm with thresholds α and γ to (A, S, w). Let during the construction of α-cover this algorithm choose sequentially subsets $B_{l_1}, \ldots, B_{l_k}$. From Proposition 1 it follows that either $k = t$ and $(l_1, \ldots, l_k) = (g_1, \ldots, g_t)$ or $k \leq t$, $(l_1, \ldots, l_{k-1}) = (g_1, \ldots, g_{k-1})$ and $l_k \neq g_k$. Let us consider these two cases separately. Let $k = t$ and $(l_1, \ldots, l_k) = (g_1, \ldots, g_t)$. For $j = 0, \ldots, t - 1$ denote $\rho_C(\alpha, \gamma, j) = \left\lceil \frac{w(B_{g_{j+1}})(M - |D_j|)}{\min\{|B_{g_{j+1}} \setminus D_j|, N - |D_j|\}} \right\rceil$. Then $\rho_C(\alpha, \gamma) = \max\{\rho_C(\alpha, \gamma, j) : j = 0, \ldots, t - 1\}$. Since $N \geq M$, we have $\rho_C(\alpha, \gamma, j) \leq \rho_C(\alpha, \alpha, j)$ for $j = 0, \ldots, t - 1$. Hence $\rho_C(\alpha, \gamma) \leq \rho_C(\alpha, \alpha)$. Let $k \leq t$, $(l_1, \ldots, l_{k-1}) = (g_1, \ldots, g_{k-1})$ and $l_k \neq g_k$. Let us denote $\rho_C(\alpha, \gamma, k - 1) = \left\lceil \frac{w(B_{l_k})(M - |D_{k-1}|)}{\min\{|B_{l_k} \setminus D_{k-1}|, N - |D_{k-1}|\}} \right\rceil$ and $\rho_C(\alpha, \gamma, j) = \left\lceil \frac{w(B_{g_{j+1}})(M - |D_j|)}{\min\{|B_{g_{j+1}} \setminus D_j|, N - |D_j|\}} \right\rceil$ for $j = 0, \ldots, k - 2$. Then $\rho_C(\alpha, \gamma) = \max\{\rho_C(\alpha, \gamma, j) : j = 0, \ldots, k - 1\}$. Since $N \geq M$, we have $\rho_C(\alpha, \gamma, j) \leq \rho_C(\alpha, \alpha, j)$ for $j = 0, \ldots, k - 2$. It is clear that $\frac{w(B_{l_k})}{\min\{|B_{l_k} \setminus D_{k-1}|, N - |D_{k-1}|\}} \leq \frac{w(B_{g_k})}{\min\{|B_{g_k} \setminus D_{k-1}|, N - |D_{k-1}|\}} \leq \frac{w(B_{g_k})}{\min\{|B_{g_k} \setminus D_{k-1}|, M - |D_{k-1}|\}}$. Thus, $\rho_C(\alpha, \gamma, k - 1) \leq \rho_C(\alpha, \alpha, k - 1)$ and $\rho_C(\alpha, \gamma) \leq \rho_C(\alpha, \alpha)$. $\square$

2.7 Upper Bounds on $C_{\text{greedy}}^{\gamma}(\alpha)$

In this subsection we study some properties of parameter $\rho_C(\alpha, \gamma)$ and obtain two upper bounds on the value $C_{\text{greedy}}^{\gamma}(\alpha)$ which do not depend directly on cardinality of the set A and cardinalities of subsets B_i from S.

Theorem 12. *Let (A, S, w) be a set cover problem with weights and α, γ be real numbers such that $0 \leq \gamma < \alpha < 1$. Then*

$$C_{\text{greedy}}^{\gamma}(\alpha, A, S, w) < \rho_C(\gamma, \gamma, A, S, w) \left(\ln \left(\frac{1 - \gamma}{\alpha - \gamma} \right) + 1 \right).$$

Proof. Let $S = \{B_1, \ldots, B_m\}$. Let us denote $M = \lceil |A|(1 - \alpha) \rceil$ and $N = \lceil |A|(1 - \gamma) \rceil$.

Let us apply the greedy algorithm with thresholds γ and γ to (A, S, w). Let during the construction of γ-cover the greedy algorithm choose sequentially subsets $B_{g_1}, \ldots, B_{g_t}$. Let us denote $B_{g_0} = \emptyset$, for $i = 0, \ldots, t - 1$ denote $D_i = B_{g_0} \cup \ldots \cup B_{g_i}$, and denote $\rho = \rho_C(\gamma, \gamma, A, S, w)$. Immediately from the definition of the parameter ρ it follows that for $i = 0, \ldots, t - 1$

$$\frac{w(B_{g_{i+1}})}{\min\{|B_{g_{i+1}} \setminus D_i|, N - |D_i|\}} \leq \frac{\rho}{N - |D_i|}. \tag{1}$$

Note that $\min\{|B_{g_{i+1}} \setminus D_i|, N - |D_i|\} = |B_{g_{i+1}} \setminus D_i|$ for $i = 0, \ldots, t - 2$ since $\{B_{g_0}, \ldots, B_{g_{i+1}}\}$ is not a γ-cover for (A, S). Therefore for $i = 0, \ldots, t - 2$ we have $\frac{w(B_{g_{i+1}})}{|B_{g_{i+1}} \setminus D_i|} \leq \frac{\rho}{N - |D_i|}$ and $\frac{N - |D_i|}{\rho} \leq \frac{|B_{g_{i+1}} \setminus D_i|}{w(B_{g_{i+1}})}$. Thus, for $i = 1, \ldots, t - 1$ during the step number i the greedy algorithm covers at least $\frac{N - |D_{i-1}|}{\rho}$ elements on each unit of weight. From (1) it follows that that for $i = 0, \ldots, t - 1$

$$w(B_{g_{i+1}}) \leq \frac{\rho \min\{|B_{g_{i+1}} \setminus D_i|, N - |D_i|\}}{N - |D_i|} \leq \rho . \qquad (2)$$

Assume that $\rho = 1$. Using (2) we obtain $w(B_{g_1}) = 1$. From this equality and (1) it follows that $|B_{g_1}| \geq N$. Therefore $\{B_{g_1}\}$ is an α-cover for (A, S), and $C^{\gamma}_{\text{greedy}}(\alpha) = 1$. It is clear that $\ln\left(\frac{1-\gamma}{\alpha-\gamma}\right) + 1 > 1$. Therefore the statement of the theorem holds if $\rho = 1$.

Assume now that $\rho \geq 2$. Let $|B_{g_1}| \geq M$. Then $\{B_{g_1}\}$ is an α-cover for (A, S). Using (2) we obtain $C^{\gamma}_{\text{greedy}}(\alpha) \leq \rho$. Since $\ln\left(\frac{1-\gamma}{\alpha-\gamma}\right) + 1 > 1$, we conclude that the statement of the theorem holds if $|B_{g_1}| \geq M$. Let $|B_{g_1}| < M$. Then there exists $q \in \{1, \ldots, t - 1\}$ such that $|B_{g_1} \cup \ldots \cup B_{g_q}| < M$ and $|B_{g_1} \cup \ldots \cup B_{g_{q+1}}| \geq M$.

Taking into account that for $i = 1, \ldots, q$ during the step number i the greedy algorithm covers at least $\frac{N - |D_{i-1}|}{\rho}$ elements on each unit of weight we obtain $N - |B_{g_1} \cup \ldots \cup B_{g_q}| \leq N\left(1 - \frac{1}{\rho}\right)^{w(B_{g_1}) + \ldots + w(B_{g_q})}$. Let us denote $k = w(B_{g_1}) + \ldots + w(B_{g_q})$. Then $N - N\left(1 - \frac{1}{\rho}\right)^k \leq |B_{g_1} \cup \ldots \cup B_{g_q}| \leq M - 1$. Therefore $|A|(1 - \gamma) - |A|(1 - \gamma)\left(1 - \frac{1}{\rho}\right)^k < |A|(1 - \alpha)$, $1 - \gamma - 1 + \alpha < (1 - \gamma)\left(\frac{\rho-1}{\rho}\right)^k$, $\left(\frac{\rho}{\rho-1}\right)^k < \frac{1-\gamma}{\alpha-\gamma}$, $\left(1 + \frac{1}{\rho-1}\right)^k < \frac{1-\gamma}{\alpha-\gamma}$, and $\frac{k}{\rho} < \ln\left(\frac{1-\gamma}{\alpha-\gamma}\right)$. To obtain last inequality we use known inequality $\ln\left(1 + \frac{1}{r}\right) > \frac{1}{r+1}$ which holds for any natural r. It is clear that $C^{\gamma}_{\text{greedy}}(\alpha) = k + w(B_{q+1})$. Using (2) we conclude that $w(B_{q+1}) \leq \rho$. Therefore $C^{\gamma}_{\text{greedy}}(\alpha) < \rho \ln\left(\frac{1-\gamma}{\alpha-\gamma}\right) + \rho$. $\square$

Corollary 1. *Let ε be a real number, and $0 < \varepsilon < 1$. Then for any α such that $\varepsilon \leq \alpha < 1$ the following inequalities hold:*

$$\rho_C(\alpha, \alpha) \leq C_{\min}(\alpha) \leq C^{\alpha-\varepsilon}_{\text{greedy}}(\alpha) < \rho_C(\alpha - \varepsilon, \alpha - \varepsilon)\left(\ln\frac{1}{\varepsilon} + 1\right) .$$

For example, if $\varepsilon = 0.01$ and $0.01 \leq \alpha < 1$ then $\rho_C(\alpha, \alpha) \leq C_{\min}(\alpha) \leq C^{\alpha-0.01}_{\text{greedy}}(\alpha) < 5.61\rho_C(\alpha - 0.01, \alpha - 0.01)$, and if $\varepsilon = 0.1$ and $0.1 \leq \alpha < 1$ then $\rho_C(\alpha, \alpha) \leq C_{\min}(\alpha) \leq C^{\alpha-0.1}_{\text{greedy}}(\alpha) < 3.31\rho_C(\alpha - 0.1, \alpha - 0.1)$.

The obtained results show that the lower bound $C_{\min}(\alpha) \geq \rho_C(\alpha, \alpha)$ is non-trivial.

Theorem 13. *Let (A, S, w) be a set cover problem with weights and α, γ be real numbers such that $0 \leq \gamma < \alpha < 1$. Then*

$$C^{\gamma}_{\text{greedy}}(\alpha, A, S, w) < C_{\min}(\gamma, A, S, w)\left(\ln\left(\frac{1-\gamma}{\alpha-\gamma}\right) + 1\right) .$$

Proof. From Theorem 12 it follows that $C^\gamma_{\text{greedy}}(\alpha, A, S, w) < \rho_C(\gamma, \gamma, A, S, w) \cdot$ $\left(\ln \left(\frac{1-\gamma}{\alpha-\gamma} \right) + 1 \right)$. The inequality $\rho_C(\gamma, \gamma, A, S, w) \le C_{\min}(\gamma, A, S, w)$ follows from Theorem 11. □

Corollary 2. $C^0_{\text{greedy}}(0.001) < 7.91 C_{\min}(0)$, $C^{0.001}_{\text{greedy}}(0.01) < 5.71 C_{\min}(0.001)$, $C^{0.1}_{\text{greedy}}(0.2) < 3.20 C_{\min}(0.1)$, $C^{0.3}_{\text{greedy}}(0.5) < 2.26 C_{\min}(0.3)$.

Corollary 3. *Let* $0 < \alpha < 1$. *Then* $C^0_{\text{greedy}}(\alpha) < C_{\min}(0) \left(\ln \frac{1}{\alpha} + 1 \right)$.

Corollary 4. *Let* ε *be a real number, and* $0 < \varepsilon < 1$. *Then for any* α *such that* $\varepsilon \le \alpha < 1$ *the inequalities* $C_{\min}(\alpha) \le C^{\alpha-\varepsilon}_{\text{greedy}}(\alpha) < C_{\min}(\alpha - \varepsilon) \left(\ln \frac{1}{\varepsilon} + 1 \right)$ *hold.*

3   Partial Tests and Reducts

3.1   Main Notions

Let T be a table with n rows labeled by nonnegative integers (decisions) and m columns labeled by attributes (names of attributes) $f_1, \ldots, f_m$. This table is filled by nonnegative integers (values of attributes). The table T is called a *decision table*. Let w be a *weight function* for T which corresponds to each attribute f_i a natural number $w(f_i)$.

Let us denote by $P(T)$ the set of unordered pairs of different rows of T with different decisions. We will say that an attribute f_i *separates* a pair of rows $(r_1, r_2) \in P(T)$ if rows r_1 and r_2 have different numbers at the intersection with the column f_i. For $i = 1, \ldots, m$ denote by $P(T, f_j)$ the set of pairs from $P(T)$ which the attribute f_i separates.

Let α be a real number such that $0 \le \alpha < 1$. A set of attributes $Q \subseteq \{f_1, \ldots, f_m\}$ will be called an α-*test* for T if attributes from Q separate at least $(1 - \alpha)|P(T)|$ pairs from the set $P(T)$. An α-test is called an α-*reduct* if each proper subset of the considered α-test is not α-test. If $P(T) = \emptyset$ then each subset of $\{f_1, \ldots, f_m\}$ is an α-test, and only empty set is an α-reduct.

For example, 0.01-test means that we must separate at least 99% of pairs from $P(T)$.

Note that 0-reduct is usual (exact) reduct. It must be noted also that each α-test contains at least one α-reduct as a subset.

The number $w(Q) = \sum_{f_i \in Q} w(f_i)$ will be called the *weight* of the set Q. If $Q = \emptyset$ then $w(Q) = 0$.

Let us denote by $R_{\min}(\alpha) = R_{\min}(\alpha, T, w)$ the minimal weight of α-reduct for T. It is clear that $R_{\min}(\alpha, T, w)$ coincides with the minimal weight of α-test for T.

Let α, γ be real numbers such that $0 \le \gamma \le \alpha < 1$. Let us describe a greedy algorithm with thresholds α and γ which constructs an α-test for given decision table T and weight function w.

If $P(T) = \emptyset$ then the constructed α-test is empty set. Let $P(T) \ne \emptyset$. Let us denote $M = \lceil |P(T)|(1 - \alpha) \rceil$ and $N = \lceil |P(T)|(1 - \gamma) \rceil$. Let we make $i \ge 0$

steps and construct a set Q containing i attributes (if $i = 0$ then $Q = \emptyset$). Let us describe the step number $i + 1$.

Let us denote by D the set of pairs from $P(T)$ separated by attributes from Q (if $i = 0$ then $D = \emptyset$). If $|D| \geq M$ then we finish the work of the algorithm. The set of attributes Q is the constructed α-test. Let $|D| < M$. Then we choose an attribute f_j with minimal number j for which $P(T, f_j) \setminus D \neq \emptyset$ and the value

$$\frac{w(f_j)}{\min\{|P(T, f_j) \setminus D|, N - |D|\}}$$

is minimal. Add the attribute f_j to the set Q. Pass to the step number $i + 2$.

Let us denote by $R^{\gamma}_{\text{greedy}}(\alpha) = R^{\gamma}_{\text{greedy}}(\alpha, T, w)$ the weight of α-test constructed by greedy algorithm with thresholds α and γ for given decision table T and weight function w.

3.2   Relationships Between Partial Covers and Partial Tests

Let (A, S, w) be a set cover problem with weights and α, γ be real numbers such that $0 \leq \gamma \leq \alpha < 1$. Let us apply the greedy algorithm with thresholds α and γ to (A, S, w). Let during the construction of α-cover the greedy algorithm choose sequentially subsets $B_{j_1}, \ldots, B_{j_t}$ from the family S. Let us denote $O_C(\alpha, \gamma, A, S, w) = (j_1, \ldots, j_t)$.

Let T be a decision table with m columns labeled by attributes $f_1, \ldots, f_m$, and with a nonempty set $P(T)$. Let w be a weight function for T. We correspond a set cover problem with weights $(A(T), S(T), u_w)$ to the considered decision table T and weight function w in the following way: $A(T) = P(T)$, $S(T) = \{B_1(T), \ldots, B_m(T)\}$ where $B_1(T) = P(T, f_1), \ldots, B_m(T) = P(T, f_m)$, $u_w(B_1(T)) = w(f_1), \ldots, u_w(B_m(T)) = w(f_m)$.

Let α, γ be real numbers such that $0 \leq \gamma \leq \alpha < 1$. Let us apply the greedy algorithm with thresholds α and γ to decision table T and weight function w. Let during the construction of α-test the greedy algorithm choose sequentially attributes $f_{j_1}, \ldots, f_{j_t}$. Let us denote $O_R(\alpha, \gamma, T, w) = (j_1, \ldots, j_t)$.

Let us denote $P(T, f_{j_0}) = \emptyset$. For $i = 1, \ldots, t$ denote $w_i = w(f_{j_i})$ and

$$\delta_i = |P(T, f_{j_i}) \setminus (P(T, f_{j_0}) \cup \ldots \cup P(T, f_{j_{i-1}}))| \ .$$

Let us denote $M_R(\alpha, \gamma, T, w) = \lceil |P(T)|(1 - \alpha) \rceil$, $N_R(\alpha, \gamma, T, w) = \lceil |P(T)|(1 - \gamma) \rceil$, $\Delta_R(\alpha, \gamma, T, w) = (\delta_1, \ldots, \delta_t)$ and $W_R(\alpha, \gamma, T, w) = (w_1, \ldots, w_t)$.

It is not difficult to prove the following statement.

Proposition 3. *Let T be a decision table with m columns labeled by attributes $f_1, \ldots, f_m$, $P(T) \neq \emptyset$, w be a weight function for T, and α, γ be real numbers such that $0 \leq \gamma \leq \alpha < 1$. Then*

$$|P(T)| = |A(T)| \ ,$$
$$|P(T, f_i)| = |B_i(T)|, \ \ i = 1, \ldots, m \ ,$$
$$O_R(\alpha, \gamma, T, w) = O_C(\alpha, \gamma, A(T), S(T), u_w) \ ,$$

$$M_R(\alpha, \gamma, T, w) = M_C(\alpha, \gamma, A(T), S(T), u_w) ,$$
$$N_R(\alpha, \gamma, T, w) = N_C(\alpha, \gamma, A(T), S(T), u_w) ,$$
$$\Delta_R(\alpha, \gamma, T, w) = \Delta_C(\alpha, \gamma, A(T), S(T), u_w) ,$$
$$W_R(\alpha, \gamma, T, w) = W_C(\alpha, \gamma, A(T), S(T), u_w) ,$$
$$R_{\min}(\alpha, T, w) = C_{\min}(\alpha, A(T), S(T), u_w) ,$$
$$R_{\text{greedy}}^{\gamma}(\alpha, T, w) = C_{\text{greedy}}^{\gamma}(\alpha, A(T), S(T), u_w) .$$

Let (A, S, w) be a set cover problem with weights where $A = \{a_1, \ldots, a_n\}$ and $S = \{B_1, \ldots, B_m\}$. We correspond a decision table $T(A, S)$ and a weight function v_w for $T(A, S)$ to the set cover problem with weights (A, S, w) in the following way. The table $T(A, S)$ contains m columns labeled by attributes $f_1, \ldots, f_m$ and $n+1$ rows filled by numbers from $\{0, 1\}$. For $i = 1, \ldots, n$ and $j = 1, \ldots, m$ at the intersection of i-th row and j-th column the number 1 stays if and only if $a_i \in B_j$. The row number $n + 1$ is filled by 0. First n rows are labeled by the decision 0. Last row is labeled by the decision 1. Let $v_w(f_1) = w(B_1), \ldots, v_w(f_m) = w(B_m)$.

For $i = \{1, \ldots, n + 1\}$ denote by r_i the i-th row. It is not difficult to see that $P(T(A, S)) = \{(r_1, r_{n+1}), \ldots, (r_n, r_{n+1})\}$. Let $i \in \{1, \ldots, n\}$ and $j \in \{1, \ldots, m\}$. One can show that the attribute f_j separates the pair (r_i, r_{n+1}) if and only if $a_i \in B_j$.

It is not difficult to prove the following statement.

Proposition 4. *Let (A, S, w) be a set cover problem with weights and α, γ be real numbers such that $0 \leq \gamma \leq \alpha < 1$. Then*

$$|P(T(A, S))| = |A| ,$$
$$O_R(\alpha, \gamma, T(A, S), v_w) = O_C(\alpha, \gamma, A, S, w) ,$$
$$M_R(\alpha, \gamma, T(A, S), v_w) = M_C(\alpha, \gamma, A, S, w) ,$$
$$N_R(\alpha, \gamma, T(A, S), v_w) = N_C(\alpha, \gamma, A, S, w) ,$$
$$\Delta_R(\alpha, \gamma, T(A, S), v_w) = \Delta_C(\alpha, \gamma, A, S, w) ,$$
$$W_R(\alpha, \gamma, T(A, S), v_w) = W_C(\alpha, \gamma, A, S, w) ,$$
$$R_{\min}(\alpha, T(A, S), v_w) = C_{\min}(\alpha, A, S, w) ,$$
$$R_{\text{greedy}}^{\gamma}(\alpha, T(A, S), v_w) = C_{\text{greedy}}^{\gamma}(\alpha, A, S, w) .$$

3.3 On Precision of Greedy Algorithm with Thresholds α and α

The following two statements are simple corollaries of results of Slavík (see Theorems 4 and 5) and Proposition 3.

Theorem 14. *Let T be a decision table, $P(T) \neq \emptyset$, w be a weight function for T, $\alpha \in \mathbb{R}$ and $0 \leq \alpha < 1$. Then $R_{\text{greedy}}^{\alpha}(\alpha) \leq R_{\min}(\alpha) H(\lceil (1 - \alpha)|P(T)|\rceil)$.*

Theorem 15. *Let T be a decision table with m columns labeled by attributes $f_1, \ldots, f_m$, $P(T) \neq \emptyset$, w be a weight function for T, and α be a real number such that $0 \leq \alpha < 1$. Then $R_{\text{greedy}}^{\alpha}(\alpha) \leq R_{\min}(\alpha) H\left(\max_{i \in \{1, \ldots, m\}} |P(T, f_i)|\right)$.*

3.4   On Polynomial Approximate Algorithms

In this subsection we consider three theorems which follows immediately from Theorems 26–28 [9].

Let $0 \leq \alpha < 1$. Let us consider the following problem: for given decision table T and weight function w for T it is required to find an α-test (α-reduct) for T with minimal weight.

Theorem 16. *Let $0 \leq \alpha < 1$. Then the problem of construction of α-test (α-reduct) with minimal weight is NP-hard.*

So we must consider polynomial approximate algorithms for minimization of α-test (α-reduct) weight.

Theorem 17. *Let $\alpha \in \mathbb{R}$ and $0 \leq \alpha < 1$. If $NP \not\subseteq DTIME(n^{O(\log \log n)})$ then for any ε, $0 < \varepsilon < 1$, there is no polynomial algorithm that for given decision table T with $P(T) \neq \emptyset$ and weight function w for T constructs an α-test for T which weight is at most $(1 - \varepsilon)R_{\min}(\alpha, T, w) \ln |P(T)|$.*

Theorem 18. *Let α be a real number such that $0 \leq \alpha < 1$. If $P \neq NP$ then there exists $\delta > 0$ such that there is no polynomial algorithm that for given decision table T with $P(T) \neq \emptyset$ and weight function w for T constructs an α-test for T which weight is at most $\delta R_{\min}(\alpha, T, w) \ln |P(T)|$.*

From Theorem 14 it follows that $R_{\text{greedy}}^{\alpha}(\alpha) \leq R_{\min}(\alpha)(1 + \ln |P(T)|)$. From this inequality and from Theorem 17 it follows that under the assumption $NP \not\subseteq DTIME(n^{O(\log \log n)})$ greedy algorithm with two thresholds α and α is close to best polynomial approximate algorithms for minimization of partial test weight. From the considered inequality and from Theorem 18 it follows that under the assumption $P \neq NP$ greedy algorithm with two thresholds α and α is not far from best polynomial approximate algorithms for minimization of partial test weight.

However we can try to improve the results of the work of greedy algorithm with two thresholds α and α for some part of decision tables.

3.5   Two Modifications of Greedy Algorithm

First, we consider binary diagnostic decision tables and prove that under some assumptions on the number of attributes and rows for the most part of tables there exist weight function w and numbers α, γ such that the weight of α-test constructed by greedy algorithm with thresholds α and γ is less than the weight of α-test constructed by greedy algorithm with thresholds α and α.

Binary means that the table is filled by numbers from the set $\{0, 1\}$ (all attributes have values from $\{0, 1\}$). *Diagnostic* means that rows of the table are labeled by pairwise different numbers (decisions). Let T be a binary diagnostic decision table with m columns labeled by attributes $f_1, \ldots, f_m$ and with n rows. We will assume that rows of T with numbers $1, \ldots, n$ are labeled by decisions $1, \ldots, n$ respectively. Therefore the number of considered tables is equal to 2^{mn}. Decision table will be called *simple* if it has no equal rows.

Theorem 19. *Let us consider binary diagnostic decision tables with m columns labeled by attributes $f_1, \ldots, f_m$ and $n \geq 4$ rows labeled by decisions $1, \ldots, n$. The fraction of decision tables T for each of which there exist a weight function w and numbers α, γ such that $0 \leq \gamma < \alpha < 1$ and $R^\gamma_{\text{greedy}}(\alpha, T, w) < R^\alpha_{\text{greedy}}(\alpha, T, w)$ is at least $1 - \frac{3^m}{n^{\frac{m}{2}-1}} - \frac{n^2}{2^m}$.*

Proof. We will say that a decision table T is not 1-uniform if there exist two attributes f_i and f_j of T such that $|P(T, f_i)| > 0$ and $|P(T, f_j)| \geq |P(T, f_i)| + 2$. Otherwise, we will say that T is 1-uniform. Using Theorem 9 and Proposition 3 we conclude that if T is not 1-uniform then there exist a weight function w and numbers α, γ such that $0 \leq \gamma < \alpha < 1$ and $R^\gamma_{\text{greedy}}(\alpha, T, w) < R^\alpha_{\text{greedy}}(\alpha, T, w)$.

We evaluate the number of simple decision tables which are 1-uniform.

Let us consider a simple decision table T which is 1-uniform. Let f_i be an attribute of T. It is clear that $|P(T, f_i)| = 0$ if and only if the number of units in the column f_i is equal to 0 or n. Let k, l be natural numbers such that $k, k + l \in \{1, \ldots, n - 1\}$, and $i, j \in \{1, \ldots, m\}$, $i \neq j$. Let the decision table T have k units in the column f_i and $k + l$ units in the column f_j. Then $|P(T, f_i)| = k(n - k) = kn - k^2$ and $|P(T, f_j)| = (k + l)(n - k - l) = kn - k^2 + l(n - 2k - l)$. Since T is 1-uniform, we have $l(n - 2k - l) \in \{0, 1, -1\}$.

Let $l(n - 2k - l) = 0$. Then $n - 2k - l = 0$ and $l = n - 2k$. Since l is a natural number, we have $k < n/2$.

Let $l(n - 2k - l) = 1$. Since l, n and k are natural numbers, we have $l = 1$ and $n - 2k - 1 = 1$. Therefore $k = \frac{n}{2} - 1$. Since k is a natural number, we have n is even.

Let $l(n - 2k - l) = -1$. Since l, n and k are natural numbers, we have $l = 1$ and $n - 2k - 1 = -1$. Therefore $k = \frac{n}{2}$. Since k is a natural number, we have n is even.

Let n be odd. Then there exists natural k such that $1 \leq k < \frac{n}{2}$ and the number of units in each column of T belongs to the set $\{0, n, k, n - k\}$. Therefore the number of considered tables is at most $\sum_{k=1}^{\lfloor n/2 \rfloor}(C_n^k + C_n^{n-k} + 2)^m$. Since $n \geq 4$, we have $2 \leq C_n^{\lfloor n/2 \rfloor}$. Using Lemma 2 we conclude that the number of 1-uniform simple tables is at most $\sum_{k=1}^{\lfloor n/2 \rfloor}\left(3C_n^{\lfloor n/2 \rfloor}\right)^m < n\left(\frac{3 \cdot 2^n}{\sqrt{n}}\right)^m$.

Let n be even. Then there exists natural k such that $1 \leq k < \frac{n}{2} - 1$ and the number of units in each column of T belongs to the set $\{0, n, k, n - k\}$, or the number of units in each column belongs to the set $\{0, n, \frac{n}{2} - 1, \frac{n}{2}, \frac{n}{2} + 1\}$. Therefore the number of considered tables is at most $\sum_{k=1}^{\lfloor n/2 \rfloor - 2}(C_n^k + C_n^{n-k} + 2)^m + (C_n^{n/2-1} + C_n^{n/2} + C_n^{n/2+1} + 2)^m$. It is well known (see, for example, [25], page 178) that $C_n^r < C_n^{n/2}$ for any $r \in \{1, \ldots, n\} \setminus \{n/2\}$. Therefore the number of 1-uniform tables is at most $n\left(3C_n^{n/2}\right)^m$. Using Lemma 2 we conclude that (as in the case of odd n) the number of 1-uniform simple tables is less than $n\left(\frac{3 \cdot 2^n}{\sqrt{n}}\right)^m = \frac{2^{mn}3^m}{n^{\frac{m}{2}-1}}$. The number of tables which are not simple is at most $n^2 2^{mn-m}$. Therefore the number of tables which are not 1-uniform is at least

$2^{mn} - \frac{2^{mn}3^m}{n^{\frac{m}{2}-1}} - n^2 2^{mn-m}$. Thus, the fraction, considered in the statement of the theorem, is at least $1 - \frac{3^m}{n^{\frac{m}{2}-1}} - \frac{n^2}{2^m}$.   □

So if $m \geq 4$ and n, $\frac{2^m}{n^2}$ are large enough then for the most part of binary diagnostic decision tables there exist weight function w and numbers α, γ such that the weight of α-test constructed by greedy algorithm with thresholds α and γ is less than the weight of α-test constructed by greedy algorithm with thresholds α and α.

The obtained results show that the greedy algorithm with two thresholds α and γ is of some interest. Now we consider two polynomial modifications of greedy algorithm which allow to use advantages of greedy algorithm with two thresholds α and γ.

Let T be a decision table, $P(T) \neq \emptyset$, w be a weight function for T and α be a real number such that $0 \leq \alpha < 1$.

1. It is impossible to consider effectively all γ such that $0 \leq \gamma \leq \alpha$. Instead of this we can consider all natural N such that $M \leq N \leq |P(T)|$ where $M = \lceil |P(T)|(1 - \alpha) \rceil$ (see the description of greedy algorithm with two thresholds). For each $N \in \{M, \dots, |P(T)|\}$ we apply greedy algorithm with parameters M and N to T and w and after that choose an α-test with minimal weight among constructed α-tests.

2. There exists also an another way to construct an α-test which is not worse than the one obtained under consideration of all N such that $M \leq N \leq |P(T)|$. Let us apply greedy algorithm with thresholds α and α to T and w. Let the algorithm choose sequentially attributes $f_{j_1}, \dots, f_{j_t}$. For each $i \in \{0, \dots, t-1\}$ we find (if it is possible) an attribute f_{l_i} of T with minimal weight $w(f_{l_i})$ such that the set $\{f_{j_1}, \dots, f_{j_i}, f_{l_i}\}$ is an α-test for T (if $i = 0$ then it will be the set $\{f_{l_0}\}$). After that among constructed α-tests $\{f_{j_1}, \dots, f_{j_t}\}$, ..., $\{f_{j_1}, \dots, f_{j_i}, f_{l_i}\}$, ... we choose an α-test with minimal weight. From Proposition 5 it follows that the constructed α-test is not worse than the one constructed under consideration of all γ, $0 \leq \gamma \leq \alpha$, or (which is the same) all N, $M \leq N \leq |P(T)|$.

Next statement follows immediately from Propositions 1 and 3.

Proposition 5. *Let T be a decision table, $P(T) \neq \emptyset$, w be a weight function for T and α, γ be real numbers such that $0 \leq \gamma < \alpha < 1$. Let the greedy algorithm with two thresholds α and α, which is applied to T and w, choose sequentially attributes $f_{g_1}, \dots, f_{g_t}$. Let the greedy algorithm with two thresholds α and γ, which is applied to T and w, choose sequentially attributes $f_{l_1}, \dots, f_{l_k}$. Then either $k = t$ and $(l_1, \dots, l_k) = (g_1, \dots, g_t)$ or $k \leq t$, $(l_1, \dots, l_{k-1}) = (g_1, \dots, g_{k-1})$ and $l_k \neq g_k$.*

3.6 Bounds on $R_{\min}(\alpha)$ and $R_{\text{greedy}}^{\gamma}(\alpha)$

First, we fix some information about the work of greedy algorithm with two thresholds and find the best lower bound on the value $R_{\min}(\alpha)$ depending on this information.

Let T be a decision table such that $P(T) \neq \emptyset$, w be a weight function for T, and α, γ be real numbers such that $0 \leq \gamma \leq \alpha < 1$. Let us apply the greedy algorithm with thresholds α and γ to the decision table T and the weight function w. Let during the construction of α-test the greedy algorithm choose sequentially attributes $f_{g_1}, \ldots, f_{g_t}$.

Let us denote $P(T, f_{g_0}) = \emptyset$ and $\delta_0 = 0$. For $i = 1, \ldots, t$ denote $\delta_i = |P(T, f_{g_i}) \setminus (P(T, f_{g_0}) \cup \ldots \cup P(T, f_{g_{i-1}}))|$ and $w_i = w(f_{g_i})$.

As information on the greedy algorithm work we will use numbers $M_R = M_R(\alpha, \gamma, T, w) = \lceil |P(T)|(1 - \alpha) \rceil$ and $N_R = N_R(\alpha, \gamma, T, w) = \lceil |P(T)|(1 - \gamma) \rceil$, and tuples $\Delta_R = \Delta_R(\alpha, \gamma, T, w) = (\delta_1, \ldots, \delta_t)$ and $W_R = W_R(\alpha, \gamma, T, w) = (w_1, \ldots, w_t)$.

For $i = 0, \ldots, t - 1$ denote

$$\rho_i = \left\lceil \frac{w_{i+1}(M_R - (\delta_0 + \ldots + \delta_i))}{\min\{\delta_{i+1}, N_R - (\delta_0 + \ldots + \delta_i)\}} \right\rceil .$$

Let us define parameter $\rho_R(\alpha, \gamma) = \rho_R(\alpha, \gamma, T, w)$ as follows:

$$\rho_R(\alpha, \gamma) = \max\{\rho_i : i = 0, \ldots, t - 1\} .$$

We will show that $\rho_R(\alpha, \gamma)$ is the best lower bound on $R_{\min}(\alpha)$ depending on M_R, N_R, Δ_R and W_R. Next statement follows from Theorem 11 and Propositions 3 and 4.

Theorem 20. *For any decision table T with $P(T) \neq \emptyset$, any weight function w for T, and any real numbers α, γ, $0 \leq \gamma \leq \alpha < 1$, the inequality $R_{\min}(\alpha, T, w) \geq \rho_R(\alpha, \gamma, T, w)$ holds, and there exist a decision table T' and a weight function w' for T' such that*

$$M_R(\alpha, \gamma, T', w') = M_R(\alpha, \gamma, T, w), \ N_R(\alpha, \gamma, T', w') = N_R(\alpha, \gamma, T, w) ,$$

$$\Delta_R(\alpha, \gamma, T', w') = \Delta_R(\alpha, \gamma, T, w), \ W_R(\alpha, \gamma, T', w') = W_R(\alpha, \gamma, T, w) ,$$

$$\rho_R(\alpha, \gamma, T', w') = \rho_R(\alpha, \gamma, T, w), \ R_{\min}(\alpha, T', w') = \rho_R(\alpha, \gamma, T', w') .$$

Let us consider a property of the parameter $\rho_R(\alpha, \gamma)$ which is important for practical use of the bound from Theorem 20. Next statement follows from Propositions 2 and 3.

Proposition 6. *Let T be a decision table with $P(T) \neq \emptyset$, w be a weight function for T, $\alpha, \gamma \in \mathbb{R}$ and $0 \leq \gamma \leq \alpha < 1$. Then $\rho_R(\alpha, \alpha, T, w) \geq \rho_R(\alpha, \gamma, T, w)$.*

Now we study some properties of parameter $\rho_R(\alpha, \gamma)$ and obtain two upper bounds on the value $R^\gamma_{\text{greedy}}(\alpha)$ which do not depend directly on cardinality of the set $P(T)$ and cardinalities of subsets $P(T, f_i)$.

Next statement follows from Theorem 12 and Proposition 3.

Theorem 21. *Let T be a decision table with $P(T) \neq \emptyset$, w be a weight function for T and α, γ be real numbers such that $0 \leq \gamma < \alpha < 1$. Then*

$$R^\gamma_{\text{greedy}}(\alpha, T, w) < \rho_R(\gamma, \gamma, T, w)\left(\ln\left(\frac{1 - \gamma}{\alpha - \gamma}\right) + 1\right) .$$

Corollary 5. *Let $\varepsilon \in \mathbb{R}$ and $0 < \varepsilon < 1$. Then for any α, $\varepsilon \leq \alpha < 1$, the inequalities $\rho_C(\alpha, \alpha) \leq R_{\min}(\alpha) \leq R_{\text{greedy}}^{\alpha-\varepsilon}(\alpha) < \rho_R(\alpha - \varepsilon, \alpha - \varepsilon) \left(\ln \frac{1}{\varepsilon} + 1 \right)$ hold.*

For example, $\left(\ln \frac{1}{0.01} + 1 \right) < 5.61$ and $\left(\ln \frac{1}{0.1} + 1 \right) < 3.31$. The obtained results show that the lower bound $R_{\min}(\alpha) \geq \rho_R(\alpha, \alpha)$ is nontrivial.

Next statement follows from Theorem 13 and Proposition 3.

Theorem 22. *Let T be a decision table with $P(T) \neq \emptyset$, w be a weight function for T and α, γ be real numbers such that $0 \leq \gamma < \alpha < 1$. Then*

$$R_{\text{greedy}}^{\gamma}(\alpha, T, w) < R_{\min}(\gamma, T, w) \left(\ln \left(\frac{1 - \gamma}{\alpha - \gamma} \right) + 1 \right) \ .$$

Corollary 6. $R_{\text{greedy}}^{0}(0.001) < 7.91 R_{\min}(0)$, $R_{\text{greedy}}^{0.001}(0.01) < 5.71 R_{\min}(0.001)$, $R_{\text{greedy}}^{0.1}(0.2) < 3.20 C_{\min}(0.1)$, $R_{\text{greedy}}^{0.3}(0.5) < 2.26 R_{\min}(0.3)$.

Corollary 7. *Let $0 < \alpha < 1$. Then $R_{\text{greedy}}^{0}(\alpha) < R_{\min}(0) \left(\ln \frac{1}{\alpha} + 1 \right)$.*

Corollary 8. *Let ε be a real number, and $0 < \varepsilon < 1$. Then for any α such that $\varepsilon \leq \alpha < 1$ the inequalities $R_{\min}(\alpha) \leq R_{\text{greedy}}^{\alpha-\varepsilon}(\alpha) < R_{\min}(\alpha - \varepsilon) \left(\ln \frac{1}{\varepsilon} + 1 \right)$ hold.*

3.7   Results of Experiments for α-Tests and α-Reducts

In this subsection we will consider only binary decision tables with binary decision attributes.

First Group of Experiments. First group of experiments is connected with study of quality of greedy algorithm with one threshold (where $\gamma = \alpha$ or, which is the same, $N = M$), and comparison of quality of greedy algorithm with one threshold and first modification of greedy algorithm (where for each $N \in \{M, \ldots, |P(T)|\}$ we apply greedy algorithm with parameters M and N to decision table and weight function and after that choose an α-test with minimal weight among constructed α-tests).

We generate randomly 1000 decision tables T and weight functions w such that T contains 10 rows and 10 conditional attributes $f_1, \ldots, f_{10}$, and $1 \leq w(f_i) \leq 1000$ for $i = 1, \ldots, 10$.

For each $\alpha \in \{0.0, 0.1, \ldots, 0.9\}$ we find the number of pairs (T, w) for which greedy algorithm with one threshold constructs an α-test with minimal weight (an optimal α-test), i.e. $R_{\text{greedy}}^{\alpha}(\alpha, T, w) = R_{\min}(\alpha, T, w)$. This number is contained in the row of Table 1 labeled by "Opt".

We find the number of pairs (T, w) for which first modification of greedy algorithm constructs an α-test which weight is less than the weight of α-test constructed by greedy algorithm with one threshold, i.e. there exists γ such that $0 \leq \gamma < \alpha$ and $R_{\text{greedy}}^{\gamma}(\alpha, T, w) < R_{\text{greedy}}^{\alpha}(\alpha, T, w)$. This number is contained in the row of Table 1 labeled by "Impr".

Also we find the number of pairs (T, w) for which first modification of greedy algorithm constructs an optimal α-test which weight is less than the weight of

α-test constructed by greedy algorithm with one threshold, i.e. there exists γ such that $0 \leq \gamma < \alpha$ and $R^{\gamma}_{\text{greedy}}(\alpha, T, w) = R_{\min}(\alpha, T, w) < R^{\alpha}_{\text{greedy}}(\alpha, T, w)$. This number is contained in the row of Table 1 labeled by "Opt+".

Table 1. Results of first group of experiments with α-tests

α	0.0	0.1	0.2	0.3	0.4	0.5	0.6	0.7	0.8	0.9
Opt	409	575	625	826	808	818	950	981	992	1000
Impr	0	42	47	33	24	8	6	5	2	0
Opt+	0	22	28	24	22	5	6	5	2	0

The obtained results show that the percentage of pairs for which greedy algorithm with one threshold finds an optimal α-test grows almost monotonically (with local minimum near to 0.4–0.5) from 40.9% up to 100%. The percentage of problems for which first modification of greedy algorithm can improve the result of the work of greedy algorithm with one threshold is less than 5%. However, sometimes (for example, if $\alpha = 0.3$ or $\alpha = 0.7$) the considered improvement is noticeable.

Second Group of Experiments. Second group of experiments is connected with comparison of quality of greedy algorithm with one threshold and first modification of greedy algorithm.

We make 25 experiments (row "Nr" in Table 2 contains the number of experiment). Each experiment includes the work with three randomly generated families of pairs (T, w) (1000 pairs in each family) such that T contains n rows and m conditional attributes, and w has values from the set $\{1, \ldots, v\}$.

If the column "n" contains one number, for example "40", it means that $n = 40$. If this row contains two numbers, for example "30–120", it means that for each of 1000 pairs we choose the number n randomly from the set $\{30, \ldots, 120\}$. The same situation is for the column "m".

If the column "α" contains one number, for example "0.1", it means that $\alpha = 0.1$. If this column contains two numbers, for example "0.2–0.4", it means that we choose randomly the value of α such that $0.2 \leq \alpha \leq 0.4$.

For each of the considered pairs (T, w) and number α we apply greedy algorithm with one threshold and first modification of greedy algorithm. Column "#i", $i = 1, 2, 3$, contains the number of pairs (T, w) from the family number i for each of which the weight of α-test, constructed by first modification of greedy algorithm, is less than the weight of α-test constructed by greedy algorithm with one threshold. In other words, in column "#i" we have the number of pairs (T, w) from the family number i such that there exists γ for which $0 \leq \gamma < \alpha$ and $R^{\gamma}_{\text{greedy}}(\alpha, T, w) < R^{\alpha}_{\text{greedy}}(\alpha, T, w)$. The column "avg" contains the number $\frac{\#1 + \#2 + \#3}{3}$.

In experiments 1–3 we consider the case where the parameter v increases. In experiments 4–8 the parameter α increases. In experiments 9–12 the parameter m increases. In experiments 13–16 the parameter n increases. In experiments

Table 2. Results of second group of experiments with α-tests

Nr	n	m	v	α	#1	#2	#3	avg
1	1–50	1–50	1–10	0–1	1	2	3	2.00
2	1–50	1–50	1–100	0–1	5	6	13	8.00
3	1–50	1–50	1–1000	0–1	10	8	11	9.67
4	1–50	1–50	1–1000	0–0.2	16	20	32	22.67
5	1–50	1–50	1–1000	0.2–0.4	23	8	12	14.33
6	1–50	1–50	1–1000	0.4–0.6	7	6	5	6.00
7	1–50	1–50	1–1000	0.6–0.8	3	5	3	3.67
8	1–50	1–50	1–1000	0.8–1	1	0	0	0.33
9	50	1–20	1–1000	0–0.2	19	11	22	17.33
10	50	20–40	1–1000	0–0.2	26	24	24	24.67
11	50	40–60	1–1000	0–0.2	21	18	23	20.67
12	50	60–80	1–1000	0–0.2	13	18	22	17.67
13	1–20	30	1–1000	0–0.2	27	26	39	30.67
14	20–40	30	1–1000	0–0.2	34	37	35	35.33
15	40–60	30	1–1000	0–0.2	22	26	23	23.67
16	60–80	30	1–1000	0–0.2	19	14	14	15.67
17	10	10	1–1000	0.1	36	42	50	42.67
18	10	10	1–1000	0.2	33	53	46	44.00
19	10	10	1–1000	0.3	43	25	45	37.67
20	10	10	1–1000	0.4	30	18	19	22.33
21	10	10	1–1000	0.5	10	10	13	11.00
22	10	10	1–1000	0.6	12	13	7	10.67
23	10	10	1–1000	0.7	3	13	6	7.33
24	10	10	1–1000	0.8	5	2	6	4.33
25	10	10	1–1000	0.9	0	0	0	0

17–25 the parameter α increases. The results of experiments show that the value of #i can change from 0 to 53. It means that the percentage of pairs for which first modification of greedy algorithm is better than the greedy algorithm with one threshold can change from 0% to 5.3%.

Third Group of Experiments. Third group of experiments is connected with investigation of quality of lower bound $R_{\min}(\alpha) \geq \rho_R(\alpha, \alpha)$.

We choose natural n, m, v and real α, $0 \leq \alpha < 1$. For each chosen tuple (n, m, v, α) we generate randomly 30 pairs (T, w) such that T contains n rows and m conditional attributes, and w has values from the set $\{1, ..., v\}$. After that we find values of $R^{\alpha}_{\text{greedy}}(\alpha, T, w)$ and $\rho_R(\alpha, \alpha, T, w)$ for each of generated 30 pairs. Note that $\rho_R(\alpha, \alpha, T, w) \leq R_{\min}(\alpha, T, w) \leq R^{\alpha}_{\text{greedy}}(\alpha, T, w)$. Finally, we find mean values of $R^{\alpha}_{\text{greedy}}(\alpha, T, w)$ and $\rho_R(\alpha, \alpha, T, w)$ for generated 30 pairs.

Results of experiments can be found in Figs. 1 and 2. In these figures mean values of $\rho_R(\alpha, \alpha, T, w)$ are called "average lower bound" and mean values of $R^{\alpha}_{\text{greedy}}(\alpha, T, w)$ are called "average upper bound".

In Fig. 1 (left-hand side) one can see the case when $n \in \{1000, 2000, \ldots, 5000\}$, $m = 30$, $v = 1000$ and $\alpha = 0.01$.

In Fig. 1 (right-hand side) one can see the case when $n = 1000$, $m \in \{10, 20, \ldots, 100\}$, $v = 1000$ and $\alpha = 0.01$.

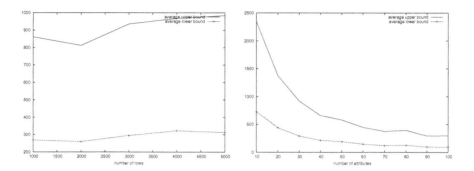

Fig. 1. Results of third group of experiments with α-tests (n and m are changing)

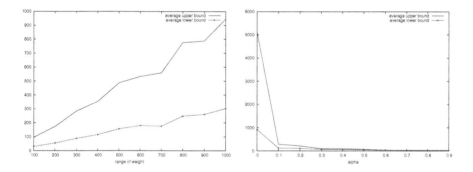

Fig. 2. Results of third group of experiments with α-tests (v and α are changing)

In Fig. 2 (left-hand side) one can see the case when $n = 1000$, $m = 30$, $v \in \{100, 200, \ldots, 1000\}$ and $\alpha = 0.01$.

In Fig. 2 (right-hand side) one can see the case when $n = 1000$, $m = 30$, $v = 1000$ and $\alpha \in \{0.0, 0.1, \ldots, 0.9\}$.

Results of experiments show that the considered lower bound is nontrivial and can be useful in investigations.

4   Partial Decision Rules

In this section we omit reasoning on relationships between partial covers and partial decision rules including reductions of one problem to another (description of such reductions can be found in [9]) and two propositions similar to Propositions 3 and 4.

4.1   Main Notions

Let T be a table with n rows labeled by nonnegative integers (decisions) and m columns labeled by attributes (names of attributes) $f_1, \ldots, f_m$. This table is filled by nonnegative integers (values of attributes). The table T is called a *decision table*. Let w be a *weight function* for T which corresponds to each attribute f_i a natural number $w(f_i)$. Let $r = (b_1, \ldots, b_m)$ be a row of T labeled by a decision d.

Let us denote by $U(T, r)$ the set of rows from T which are different from r and are labeled by decisions different from d. We will say that an attribute f_i *separates* rows r and $r' \in U(T, r)$ if rows r and r' have different numbers at the intersection with the column f_i. For $i = 1, \ldots, m$ denote by $U(T, r, f_i)$ the set of rows from $U(T, r)$ which attribute f_i separates from the row r.

Let α be a real number such that $0 \le \alpha < 1$. A decision rule

$$f_{i_1} = b_{i_1} \wedge \ldots \wedge f_{i_t} = b_{i_t} \to d \tag{3}$$

is called an α-*decision rule* for T and r if attributes $f_{i_1}, \ldots, f_{i_t}$ separate from r at least $(1 - \alpha)|U(T, r)|$ rows from $U(T, r)$. The number $\sum_{j=1}^{t} w(f_{i_j})$ is called the *weight* of the considered decision rule.

If $U(T, r) = \emptyset$ then for any $f_{i_1}, \ldots, f_{i_t} \in \{f_1, \ldots, f_m\}$ the rule (3) is an α-decision rule for T and r. Also, the rule (3) with empty left-hand side (when $t = 0$) is an α-decision rule for T and r. The weight of this rule is equal to 0.

For example, 0.01-decision rule means that we must separate from r at least 99% of rows from $U(T, r)$. Note that 0-rule is usual (exact) rule. Let us denote by $L_{\min}(\alpha) = L_{\min}(\alpha, T, r, w)$ the minimal weight of α-decision rule for T and r.

Let α, γ be real numbers such that $0 \le \gamma \le \alpha < 0$. Let us describe a *greedy algorithm with thresholds* α and γ which constructs an α-decision rule for given T, r and weight function w. Let $r = (b_1, \ldots, b_m)$, and r be labeled by the decision d.

The right-hand side of constructed α-decision rule is equal to d. If $U(T, r) = \emptyset$ then the left-hand side of constructed α-decision rule is empty. Let $U(T, r) \ne \emptyset$. Let us denote $M = \lceil |U(T, r)|(1 - \alpha) \rceil$ and $N = \lceil |U(T, r)|(1 - \gamma) \rceil$. Let we make $i \ge 0$ steps and construct a decision rule R with i conditions (if $i = 0$ then the left-hand side of R is empty). Let us describe the step number $i + 1$.

Let us denote by D the set of rows from $U(T, r)$ separated from r by attributes belonging to R (if $i = 0$ then $D = \emptyset$). If $|D| \ge M$ then we finish the work of the algorithm, and R is the constructed α-decision rule. Let $|D| < M$. Then we choose an attribute f_j with minimal number j for which $U(T, r, f_j) \setminus D \ne \emptyset$ and the value

$$\frac{w(f_j)}{\min\{|U(T, r, f_j) \setminus D|, N - |D|\}}$$

is minimal. Add the condition $f_j = b_j$ to R. Pass to the step number $i + 2$.

Let us denote by $L_{\text{greedy}}^{\gamma}(\alpha) = L_{\text{greedy}}^{\gamma}(\alpha, T, r, w)$ the weight of α-decision rule constructed by the considered algorithm for given table T, row r and weight function w.

4.2   On Precision of Greedy Algorithm with Thresholds α and α

The following two statements are simple corollaries of results of Slavík (see Theorems 4 and 5).

Theorem 23. *Let T be a decision table, r be a row of T, $U(T,r) \neq \emptyset$, w be a weight function for T, and α be a real number such that $0 \leq \alpha < 1$. Then $L^{\alpha}_{\text{greedy}}(\alpha) \leq L_{\min}(\alpha)H\left(\lceil(1-\alpha)|U(T,r)|\rceil\right)$.*

Theorem 24. *Let T be a decision table with m columns labeled by attributes $f_1, \ldots, f_m$, r be a row of T, $U(T,r) \neq \emptyset$, w be a weight function for T, $\alpha \in \mathbb{R}$ and $0 \leq \alpha < 1$. Then $L^{\alpha}_{\text{greedy}}(\alpha) \leq L_{\min}(\alpha)H\left(\max_{i \in \{1,\ldots,m\}} |U(T,r,f_i)|\right)$.*

4.3   On Polynomial Approximate Algorithms

In this subsection we consider three theorems which follow immediately from Theorems 39–41 [9].

Let $0 \leq \alpha < 1$. Let us consider the following problem: for given decision table T, row r of T and weight function w for T it is required to find an α-decision rule for T and r with minimal weight.

Theorem 25. *Let $0 \leq \alpha < 1$. Then the problem of construction of α-decision rule with minimal weight is NP-hard.*

So we must consider polynomial approximate algorithms for minimization of α-decision rule weight.

Theorem 26. *Let $\alpha \in \mathbb{R}$ and $0 \leq \alpha < 1$. If $NP \not\subseteq DTIME(n^{O(\log \log n)})$ then for any ε, $0 < \varepsilon < 1$, there is no polynomial algorithm that for given decision table T, row r of T with $U(T,r) \neq \emptyset$ and weight function w for T constructs α-decision rule for T and r which weight is at most $(1-\varepsilon)L_{\min}(\alpha,T,r,w) \ln |U(T,r)|$.*

Theorem 27. *Let α be a real number such that $0 \leq \alpha < 1$. If $P \neq NP$ then there exists $\delta > 0$ such that there is no polynomial algorithm that for given decision table T, row r of T with $U(T,r) \neq \emptyset$ and weight function w for T constructs α-decision rule for T and r which weight is at most $\delta L_{\min}(\alpha,T,r,w) \ln |U(T,r)|$.*

From Theorem 23 it follows that $L^{\alpha}_{\text{greedy}}(\alpha) \leq L_{\min}(\alpha)(1 + \ln |U(T,r)|)$. From this inequality and from Theorem 26 it follows that under the assumption $NP \not\subseteq DTIME(n^{O(\log \log n)})$ greedy algorithm with two thresholds α and α is close to best polynomial approximate algorithms for minimization of partial decision rule weight. From the considered inequality and from Theorem 27 it follows that under the assumption $P \neq NP$ greedy algorithm with two thresholds α and α is not far from best polynomial approximate algorithms for minimization of partial decision rule weight.

However we can try to improve the results of the work of greedy algorithm with two thresholds α and α for some part of decision tables.

4.4 Two Modifications of Greedy Algorithm

First, we consider binary diagnostic decision tables and prove that under some assumptions on the number of attributes and rows for the most part of tables for each row there exist weight function w and numbers α, γ such that the weight of α-decision rule constructed by greedy algorithm with thresholds α and γ is less than the weight of α-decision rule constructed by greedy algorithm with thresholds α and α.

Binary means that the table is filled by numbers from the set $\{0, 1\}$ (all attributes have values from $\{0, 1\}$). *Diagnostic* means that rows of the table are labeled by pairwise different numbers (decisions). Let T be a binary diagnostic decision table with m columns labeled by attributes $f_1, \ldots, f_m$ and with n rows. We will assume that rows of T with numbers $1, \ldots, n$ are labeled by decisions $1, \ldots, n$ respectively. Therefore the number of considered tables is equal to 2^{mn}. Decision table will be called *simple* if it has no equal rows.

Theorem 28. *Let us consider binary diagnostic decision tables with m columns labeled by attributes $f_1, \ldots, f_m$ and $n \geq 5$ rows labeled by decisions $1, \ldots, n$. The fraction of decision tables T for each of which for each row r of T there exist a weight function w and numbers α, γ such that $0 \leq \gamma < \alpha < 1$ and $L_{\text{greedy}}^{\gamma}(\alpha, T, r, w) < L_{\text{greedy}}^{\alpha}(\alpha, T, r, w)$ is at least*

$$1 - \frac{n3^m}{(n-1)^{\frac{m}{2}-1}} - \frac{n^2}{2^m} .$$

Proof. Let T be a decision table and r be a row of T with number $s \in \{1, \ldots, n\}$.

We will say that a decision table T is 1-uniform relatively r if there exists natural p such that for any attribute f_i of T if $|U(T, r, f_i)| > 0$ then $|U(T, r, f_i)| \in \{p, p+1\}$. Using reasoning similar to the proof of Theorem 9 one can show that if T is not 1-uniform relatively r then there exist a weight function w and numbers α, γ such that $0 \leq \gamma < \alpha < 1$ and $L_{\text{greedy}}^{\gamma}(\alpha, T, r, w) < L_{\text{greedy}}^{\alpha}(\alpha, T, r, w)$.

We evaluate the number of decision tables which are not 1-uniform relatively each row. Let $(\delta_1, \ldots, \delta_m) \in \{0, 1\}^m$. First, we evaluate the number of simple decision tables for which $r = (\delta_1, \ldots, \delta_m)$ and which are 1-uniform relatively r. Let us consider such a decision table T. It is clear that there exists $p \in \{1, \ldots, n-2\}$ such that for $i = 1, \ldots, m$ the column f_i contains exactly 0 or p or $p+1$ numbers $\neg\delta_i$. Therefore the number of considered decision tables is at most $\sum_{p=1}^{n-2} \left(C_{n-1}^{p} + C_{n-1}^{p+1} + 1 \right)^m$. Using Lemma 2 we conclude that this number is at most $(n-2) \left(3 C_{n-1}^{\lfloor (n-1)/2 \rfloor} \right)^m < (n-1) \left(\frac{3 \cdot 2^{n-1}}{\sqrt{n-1}} \right)^m = \frac{2^{mn-m} 3^m}{(n-1)^{\frac{m}{2}-1}}$. There are 2^m variants for the choice of the tuple $(\delta_1, \ldots, \delta_m)$ and n variants for the choice of the number s of row r. Therefore the number of simple decision tables which are 1-uniform relatively at least one row is at most $n 2^m \frac{2^{mn-m} 3^m}{(n-1)^{\frac{m}{2}-1}} =$

$\frac{n2^{mn}3^m}{(n-1)^{\frac{m}{2}-1}}$. The number of tables which are not simple is at most $n^2 2^{mn-m}$. Hence the number of tables which are not 1-uniform for each row is at least $2^{mn} - \frac{n2^{mn}3^m}{(n-1)^{\frac{m}{2}-1}} - n^2 2^{mn-m}$. Thus, the fraction, considered in the statement of the theorem, is at least $1 - \frac{n3^m}{(n-1)^{\frac{m}{2}-1}} - \frac{n^2}{2^m}$. $\square$

So if $m \geq 6$ and n, $\frac{2^m}{n^2}$ are large enough then for the most part of binary diagnostic decision tables for each row there exist weight function w and numbers α, γ such that the weight of α-decision rule constructed by greedy algorithm with thresholds α and γ is less than the weight of α-decision rule constructed by greedy algorithm with thresholds α and α.

The obtained results show that the greedy algorithm with two thresholds α and γ is of some interest. Now we consider two polynomial modifications of greedy algorithm which allow to use advantages of greedy algorithm with two thresholds α and γ.

Let T be a decision table with m columns labeled by attributes $f_1, \ldots, f_m$, $r = (b_1, \ldots, b_m)$ be a row of T labeled by decision d, $U(T, r) \neq \emptyset$, w be a weight function for T and α be a real number such that $0 \leq \alpha < 1$.

1. It is impossible to consider effectively all γ such that $0 \leq \gamma \leq \alpha$. Instead of this we can consider all natural N such that $M \leq N \leq |U(T, r)|$ where $M = \lceil |U(T, r)|(1 - \alpha) \rceil$ (see the description of greedy algorithm with two thresholds). For each $N \in \{M, \ldots, |U(T, r)|\}$ we apply greedy algorithm with parameters M and N to T, r and w and after that choose an α-decision rule with minimal weight among constructed α-decision rules.
2. There exists also an another way to construct an α-decision rule which is not worse than the one obtained under consideration of all N such that $M \leq N \leq |U(T, r)|$. Let us apply greedy algorithm with thresholds α and α to T, r and w. Let the algorithm choose sequentially attributes $f_{j_1}, \ldots, f_{j_t}$. For each $i \in \{0, \ldots, t-1\}$ we find (if it is possible) an attribute f_{l_i} of T with minimal weight $w(f_{l_i})$ such that the rule $f_{j_1} = b_{j_1} \wedge \ldots \wedge f_{j_i} = b_{j_i} \wedge f_{l_i} = b_{l_i} \rightarrow d$ is an α-decision rule for T and r (if $i = 0$ then it will be the rule $f_{l_0} = b_{l_0} \rightarrow d$). After that among constructed α-decision rules $f_{j_1} = b_{j_1} \wedge \ldots \wedge f_{j_t} = b_{j_t} \rightarrow d$, $\ldots$, $f_{j_1} = b_{j_1} \wedge \ldots \wedge f_{j_i} = b_{j_i} \wedge f_{l_i} = b_{l_i} \rightarrow d$, $\ldots$ we choose an α-decision rule with minimal weight. From Proposition 7 it follows that the constructed α-decision rule is not worse than the one constructed under consideration of all γ, $0 \leq \gamma \leq \alpha$, or (which is the same) all N, $M \leq N \leq |U(T, r)|$.

Using Propositions 1 one can prove the following statement.

Proposition 7. *Let T be a decision table, r be a row of T, $U(T, r) \neq \emptyset$, w be a weight function for T and α, γ be real numbers such that $0 \leq \gamma < \alpha < 1$. Let the greedy algorithm with two thresholds α and α, which is applied to T, r and w, choose sequentially attributes $f_{g_1}, \ldots, f_{g_t}$. Let the greedy algorithm with two thresholds α and γ, which is applied to T, r and w, choose sequentially attributes $f_{l_1}, \ldots, f_{l_k}$. Then either $k = t$ and $(l_1, \ldots, l_k) = (g_1, \ldots, g_t)$ or $k \leq t$, $(l_1, \ldots, l_{k-1}) = (g_1, \ldots, g_{k-1})$ and $l_k \neq g_k$.*

4.5   Bounds on $L_{\min}(\alpha)$ and $L^{\gamma}_{\text{greedy}}(\alpha)$

First, we fix some information about the work of greedy algorithm with two thresholds and find the best lower bound on the value $L_{\min}(\alpha)$ depending on this information.

Let T be a decision table, r be a row of T such that $U(T,r) \neq \emptyset$, w be a weight function for T, and α, γ be real numbers such that $0 \leq \gamma \leq \alpha < 1$. Let us apply the greedy algorithm with thresholds α and γ to the decision table T, row r and the weight function w. Let during the construction of α-decision rule the greedy algorithm choose sequentially attributes $f_{g_1}, \ldots, f_{g_t}$.

Let us denote $U(T, r, f_{g_0}) = \emptyset$ and $\delta_0 = 0$. For $i = 1, \ldots, t$ denote $\delta_i = |U(T, r, f_{g_i}) \setminus (U(T, r, f_{g_0}) \cup \ldots \cup U(T, r, f_{g_{i-1}}))|$ and $w_i = w(f_{g_i})$. As information on the greedy algorithm work we will use numbers $M_L = M_L(\alpha, \gamma, T, r, w) = \lceil |U(T, r)|(1 - \alpha) \rceil$, $N_L = N_L(\alpha, \gamma, T, r, w) = \lceil |U(T, r)|(1 - \gamma) \rceil$ and tuples $\Delta_L = \Delta_L(\alpha, \gamma, T, r, w) = (\delta_1, \ldots, \delta_t)$, $W_L = W_L(\alpha, \gamma, T, r, w) = (w_1, \ldots, w_t)$.

For $i = 0, \ldots, t - 1$ denote

$$\rho_i = \left\lceil \frac{w_{i+1}(M_L - (\delta_0 + \ldots + \delta_i))}{\min\{\delta_{i+1}, N_L - (\delta_0 + \ldots + \delta_i)\}} \right\rceil .$$

Let us define parameter $\rho_L(\alpha, \gamma) = \rho_L(\alpha, \gamma, T, r, w)$ as follows:

$$\rho_L(\alpha, \gamma) = \max\{\rho_i : i = 0, \ldots, t - 1\} .$$

We will show that $\rho_L(\alpha, \gamma)$ is the best lower bound on $L_{\min}(\alpha)$ depending on M_L, N_L, Δ_L and W_L. Using Theorem 11 one can prove the following statement.

Theorem 29. *For any decision table T, any row r of T with $U(T, r) \neq \emptyset$, any weight function w for T, and any real numbers α, γ, $0 \leq \gamma \leq \alpha < 1$, the inequality $L_{\min}(\alpha, T, r, w) \geq \rho_L(\alpha, \gamma, T, r, w)$ holds, and there exist a decision table T', a row r' of T' and a weight function w' for T' such that*

$$M_L(\alpha, \gamma, T', r', w') = M_L(\alpha, \gamma, T, r, w), \quad N_L(\alpha, \gamma, T', r', w') = N_L(\alpha, \gamma, T, r, w) ,$$
$$\Delta_L(\alpha, \gamma, T', r', w') = \Delta_L(\alpha, \gamma, T, r, w), \quad W_L(\alpha, \gamma, T', r', w') = W_L(\alpha, \gamma, T, r, w) ,$$
$$\rho_L(\alpha, \gamma, T', r', w') = \rho_L(\alpha, \gamma, T, r, w), \quad L_{\min}(\alpha, T', r', w') = \rho_L(\alpha, T', r', w') .$$

Let us consider a property of the parameter $\rho_L(\alpha, \gamma)$ which is important for practical use of the bound from Theorem 29. Using Proposition 2 one can prove the following statement.

Proposition 8. *Let T be a decision table, r be a row of T with $U(T, r) \neq \emptyset$, w be a weight function for T, and α, γ be real numbers such that $0 \leq \gamma \leq \alpha < 1$. Then $\rho_L(\alpha, \alpha, T, r, w) \geq \rho_L(\alpha, \gamma, T, r, w)$.*

Now we study some properties of parameter $\rho_L(\alpha, \gamma)$ and obtain two upper bounds on the value $L^{\gamma}_{\text{greedy}}(\alpha)$ which do not depend directly on cardinality of the set $U(T, r)$ and cardinalities of subsets $U(T, r, f_i)$.

Using Theorem 12 one can prove the following statement.

Theorem 30. *Let T be a decision table, r be a row of T with $U(T,r) \neq \emptyset$, w be a weight function for T, $\alpha, \gamma \in \mathbb{R}$ and $0 \leq \gamma < \alpha < 1$. Then $L^{\gamma}_{\text{greedy}}(\alpha, T, r, w) < \rho_L(\gamma, \gamma, T, r, w) \left(\ln \left(\frac{1-\gamma}{\alpha-\gamma} \right) + 1 \right)$.*

Corollary 9. *Let $\varepsilon \in \mathbb{R}$ and $0 < \varepsilon < 1$. Then for any α, $\varepsilon \leq \alpha < 1$, the inequalities $\rho_L(\alpha, \alpha) \leq L_{\min}(\alpha) \leq L^{\alpha-\varepsilon}_{\text{greedy}}(\alpha) < \rho_L(\alpha - \varepsilon, \alpha - \varepsilon) \left(\ln \frac{1}{\varepsilon} + 1 \right)$ hold.*

For example, $\left(\ln \frac{1}{0.01} + 1 \right) < 5.61$ and $\left(\ln \frac{1}{0.1} + 1 \right) < 3.31$. The obtained results show that the lower bound $L_{\min}(\alpha) \geq \rho_L(\alpha, \alpha)$ is nontrivial.

Using Theorem 13 one can prove the following statement.

Theorem 31. *Let T be a decision table, r be a row of T with $U(T,r) \neq \emptyset$, w be a weight function for T, $\alpha, \gamma \in \mathbb{R}$ and $0 \leq \gamma < \alpha < 1$. Then $L^{\gamma}_{\text{greedy}}(\alpha, T, r, w) < L_{\min}(\gamma, T, r, w) \left(\ln \left(\frac{1-\gamma}{\alpha-\gamma} \right) + 1 \right)$.*

Corollary 10. *$L^0_{\text{greedy}}(0.001) < 7.91 L_{\min}(0)$, $L^{0.001}_{\text{greedy}}(0.01) < 5.71 L_{\min}(0.001)$, $L^{0.1}_{\text{greedy}}(0.2) < 3.20 L_{\min}(0.1)$, $L^{0.3}_{\text{greedy}}(0.5) < 2.26 L_{\min}(0.3)$.*

Corollary 11. *Let $0 < \alpha < 1$. Then $L^0_{\text{greedy}}(\alpha) < L_{\min}(0) \left(\ln \frac{1}{\alpha} + 1 \right)$.*

Corollary 12. *Let ε be a real number, and $0 < \varepsilon < 1$. Then for any α such that $\varepsilon \leq \alpha < 1$ the inequalities $L_{\min}(\alpha) \leq L^{\alpha-\varepsilon}_{\text{greedy}}(\alpha) < L_{\min}(\alpha - \varepsilon) \left(\ln \frac{1}{\varepsilon} + 1 \right)$ hold.*

4.6   Results of Experiments for α-Decision Rules

In this subsection we will consider only binary decision tables T with binary decision attributes.

First Group of Experiments. First group of experiments is connected with study of quality of greedy algorithm with one threshold (where $\gamma = \alpha$ or, which is the same, $N = M$), and comparison of quality of greedy algorithm with one threshold and first modification of greedy algorithm (where for each $N \in \{M, \ldots, |U(T,r)|\}$ we apply greedy algorithm with parameters M and N to decision table, row and weight function and after that choose an α-decision rule with minimal weight among constructed α-decision rules).

We generate randomly 1000 decision tables T, rows r and weight functions w such that T contains 40 rows and 10 conditional attributes $f_1, \ldots, f_{10}$, r is the first row of T, and $1 \leq w(f_i) \leq 1000$ for $i = 1, \ldots, 10$.

For each $\alpha \in \{0.1, \ldots, 0.9\}$ we find the number of triples (T, r, w) for which greedy algorithm with one threshold constructs an α-decision rule with minimal weight (an optimal α-decision rule), i.e. $L^{\alpha}_{\text{greedy}}(\alpha, T, r, w) = L_{\min}(\alpha, T, r, w)$. This number is contained in the row of Table 3 labeled by "Opt".

We find the number of triples (T, r, w) for which first modification of greedy algorithm constructs an α-decision rule which weight is less than the weight of α-decision rule constructed by greedy algorithm with one threshold, i.e. there exists γ such that $0 \leq \gamma < \alpha$ and $L_{\text{greedy}}^{\gamma}(\alpha, T, r, w) < L_{\text{greedy}}^{\alpha}(\alpha, T, r, w)$. This number is contained in the row of Table 3 labeled by "Impr".

Also we find the number of triples (T, r, w) for which first modification of greedy algorithm constructs an optimal α-decision rule which weight is less than the weight of α-decision rule constructed by greedy algorithm with one threshold, i.e. there exists γ such that $0 \leq \gamma < \alpha$ and $L_{\text{greedy}}^{\gamma}(\alpha, T, r, w) = L_{\min}(\alpha, T, r, w) < L_{\text{greedy}}^{\alpha}(\alpha, T, r, w)$. This number is contained in the row of Table 3 labeled by "Opt+".

Table 3. Results of first group of experiments with α-decision rules

α	0.0	0.1	0.2	0.3	0.4	0.5	0.6	0.7	0.8	0.9
Opt	434	559	672	800	751	733	866	966	998	1000
Impr	0	31	51	36	22	27	30	17	1	0
Opt+	0	16	35	28	17	26	25	13	1	0

The obtained results show that the percentage of triples for which greedy algorithm with one threshold finds an optimal α-decision rule grows almost monotonically (with local minimum near to 0.4–0.5) from 43.4% up to 100%. The percentage of problems for which first modification of greedy algorithm can improve the result of the work of greedy algorithm with one threshold is less than 6%. However, sometimes (for example, if $\alpha = 0.3$, $\alpha = 0.6$ or $\alpha = 0.7$) the considered improvement is noticeable.

Second Group of Experiments. Second group of experiments is connected with comparison of quality of greedy algorithm with one threshold and first modification of greedy algorithm.

We make 25 experiments (row "Nr" in Table 4 contains the number of experiment). Each experiment includes the work with three randomly generated families of triples (T, r, w) (1000 triples in each family) such that T contains n rows and m conditional attributes, r is the first row of T, and w has values from the set $\{1, \ldots, v\}$.

If the column "n" contains one number, for example "40", it means that $n = 40$. If this row contains two numbers, for example "30–120", it means that for each of 1000 triples we choose the number n randomly from the set $\{30, \ldots, 120\}$. The same situation is for the column "m".

If the column "α" contains one number, for example "0.1", it means that $\alpha = 0.1$. If this column contains two numbers, for example "0.2–0.4", it means that we choose randomly the value of α such that $0.2 \leq \alpha \leq 0.4$.

For each of the considered triples (T, r, w) and number α we apply greedy algorithm with one threshold and first modification of greedy algorithm. Column "#i", $i = 1, 2, 3$, contains the number of triples (T, r, w) from the family number i for each of which the weight of α-decision rule, constructed by first modification of greedy algorithm, is less than the weight of α-decision rule constructed by

Table 4. Results of second group of experiments with α-decision rules

Nr	n	m	v	α	#1	#2	#3	avg
1	1–100	1–100	1–10	0–1	4	2	4	3.33
2	1–100	1–100	1–100	0–1	7	14	13	11.33
3	1–100	1–100	1–1000	0–1	19	13	15	15.67
4	1–100	1–100	1–1000	0–0.2	20	39	22	27.00
5	1–100	1–100	1–1000	0.2–0.4	28	29	28	28.33
6	1–100	1–100	1–1000	0.4–0.6	22	23	34	26.33
7	1–100	1–100	1–1000	0.6–0.8	7	6	4	5.67
8	1–100	1–100	1–1000	0.8–1	0	1	0	0.33
9	100	1–30	1–1000	0–0.2	35	38	28	33.67
10	100	30–60	1–1000	0–0.2	47	43	31	40.33
11	100	60–90	1–1000	0–0.2	45	51	36	44.00
12	100	90–120	1–1000	0–0.2	37	40	55	44.00
13	1–30	30	1–1000	0–0.2	11	8	9	9.33
14	30–60	30	1–1000	0–0.2	20	22	35	25.67
15	60–90	30	1–1000	0–0.2	30	33	34	32.33
16	90–120	30	1–1000	0–0.2	40	48	38	42.00
17	40	10	1–1000	0.1	31	39	34	34.67
18	40	10	1–1000	0.2	37	39	47	41.00
19	40	10	1–1000	0.3	35	30	37	34.00
20	40	10	1–1000	0.4	27	20	27	24.67
21	40	10	1–1000	0.5	32	32	36	33.33
22	40	10	1–1000	0.6	28	26	24	26.00
23	40	10	1–1000	0.7	10	12	10	10.67
24	40	10	1–1000	0.8	0	2	0	0.67
25	40	10	1–1000	0.9	0	0	0	0

greedy algorithm with one threshold. In other words, in column "#i" we have the number of triples (T, r, w) from the family number i such that there exists γ for which $0 \leq \gamma < \alpha$ and $L^{\gamma}_{\text{greedy}}(\alpha, T, r, w) < L^{\alpha}_{\text{greedy}}(\alpha, T, r, w)$. The column "avg" contains the number $\frac{\#1 + \#2 + \#3}{3}$.

In experiments 1–3 we consider the case where the parameter v increases. In experiments 4–8 the parameter α increases. In experiments 9–12 the parameter m increases. In experiments 13–16 the parameter n increases. In experiments 17–25 the parameter α increases. The results of experiments show that the value of #i can change from 0 to 55. It means that the percentage of triples for which the first modification of greedy algorithm is better than the greedy algorithm with one threshold can change from 0% to 5.5%.

Third Group of Experiments. Third group of experiments is connected with investigation of quality of lower bound $L_{\min}(\alpha) \geq \rho_L(\alpha, \alpha)$.

We choose natural n, m, v and real α, $0 \leq \alpha < 1$. For each chosen tuple (n, m, v, α) we generate randomly 30 triples (T, r, w) such that T contains

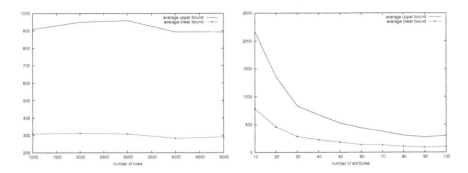

Fig. 3. Results of third group of experiments with rules (n and m are changing)

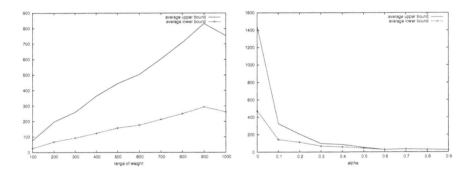

Fig. 4. Results of third group of experiments with rules (v and α are changing)

n rows and m conditional attributes, r is the first row of T, and w has values from the set $\{1, ..., v\}$. After that we find values of $L^\alpha_{\text{greedy}}(\alpha, T, r, w)$ and $\rho_L(\alpha, \alpha, T, r, w)$ for each of generated 30 triples. Note that $\rho_L(\alpha, \alpha, T, r, w) \le L_{\min}(\alpha, T, r, w) \le L^\alpha_{\text{greedy}}(\alpha, T, r, w)$. Finally, for generated 30 triples we find mean values of $L^\alpha_{\text{greedy}}(\alpha, T, r, w)$ and $\rho_L(\alpha, \alpha, T, r, w)$.

Results of experiments can be found in Figs. 3 and 4. In these figures mean values of $\rho_L(\alpha, \alpha, T, r, w)$ are called "average lower bound" and mean values of $L^\alpha_{\text{greedy}}(\alpha, T, r, w)$ are called "average upper bound".

In Fig. 3 (left-hand side) one can see the case when $n \in \{1000, 2000, \ldots, 5000\}$, $m = 30$, $v = 1000$ and $\alpha = 0.01$.

In Fig. 3 (right-hand side) one can see the case when $n = 1000$, $m \in \{10, 20, \ldots, 100\}$, $v = 1000$ and $\alpha = 0.01$.

In Fig. 4 (left-hand side) one can see the case when $n = 1000$, $m = 30$, $v \in \{100, 200, \ldots, 1000\}$ and $\alpha = 0.01$.

In Fig. 4 (right-hand side) one can see the case when $n = 1000$, $m = 30$, $v = 1000$ and $\alpha \in \{0.0, 0.1, \ldots, 0.9\}$.

Results of experiments show that the considered lower bound is nontrivial and can be useful in investigations.

5 Conclusions

The paper is devoted (mainly) to theoretical and experimental analysis of greedy algorithms with weights and their modifications for partial cover, reduct and decision rule construction. Obtained results will further to more wide use of such algorithms in rough set theory and its applications.

In the further investigations we are planning to generalize the obtained results to the case of decision tables which can contain missing values, continuous attributes, and discrete attributes with large number of values.

Acknowledgments

The authors are greatly indebted to Andrzej Skowron for stimulating discussions and to anonymous reviewers for helpful remarks and suggestions.

References

1. Cheriyan, J., Ravi, R.: Lecture Notes on Approximation Algorithms for Network Problems (1998) http://www.math.uwaterloo.ca/~jcheriya/lecnotes.html
2. Chvátal, V.: A greedy heuristic for the set-covering problem. *Mathematics of Operations Research* 4 (1979) 233–235.
3. Feige, U.: A threshold of ln n for approximating set cover (Preliminary version). Proceedings of 28th Annual ACM Symposium on the Theory of Computing (1996) 314–318.
4. Gavrilov G.P., Sapozhenko, A.A.: Problems and Exercises in Discrete Mathematics (third edition). Fizmatlit, Moscow, 2004 (in Russian).
5. Kearns, M.J.: The Computational Complexity of Machine Learning. MIT Press, Cambridge, Massachussetts, 1990.
6. Moshkov, M.Ju.: Greedy algorithm for set cover in context of knowledge discovery problems. In: Proceedings of the International Workshop on Rough Sets in Knowledge Discovery and Soft Computing (ETAPS 2003 Satellite Event). Warsaw, Poland. *Electronic Notes in Theoretical Computer Science* 82 (2003).
7. Moshkov, M.Ju.: On greedy algorithm for partial cover construction. In: Proceedings of the Fourteenth International Workshop Design and Complexity of Control Systems. Nizhny Novgorod, Russia (2003) 57 (in Russian).
8. Moshkov, M.Ju., Piliszczuk, M., Zielosko, B.: Greedy algorithm for construction of partial covers. In: Proceedings of the Fourteenth International Conference Problems of Theoretical Cybernetics. Penza, Russia (2005) 103 (in Russian).
9. Moshkov, M.Ju., Piliszczuk, M., Zielosko, B.: On partial covers, reducts and decision rules. LNCS Transactions on Rough Sets, Springer-Verlag (submitted).
10. Nguyen, H.S., Ślęzak, D.: Approximate reducts and association rules - correspondence and complexity results. In: Proceedings of the Seventh International Workshop on Rough Sets, Fuzzy Sets, Data Mining, and Granular-Soft Computing. Yamaguchi, Japan. Lecture Notes in Artificial Intelligence 1711, Springer-Verlag (1999) 137–145.
11. Pawlak, Z.: Rough Sets – Theoretical Aspects of Reasoning about Data. Kluwer Academic Publishers, Dordrecht, Boston, London, 1991.

12. Pawlak, Z.: Rough set elements. In: Polkowski, L., Skowron, A. (Eds.), Rough Sets in Knowledge Discovery 1. Methodology and Applications (Studies in Fuzziness and Soft Computing 18). Phisica-Verlag. A Springer-Verlag Company (1998) 10–30.
13. Piliszczuk, M.: On greedy algorithm for partial reduct construction. In: Proceedings of Concurrency, Specification and Programming Workshop 2. Ruciane-Nida, Poland (2005) 400–411.
14. Quafafou, M.: α-RST: a generalization of rough set theory. *Information Sciences* 124 (2000) 301–316.
15. Raz, R., Safra, S.: A sub-constant error-probability low-degree test, and sub-constant error-probability PCP characterization of NP. In: Proceedings of 29th Annual ACM Symposium on the Theory of Computing (1997) 475–484.
16. Skowron, A.: Rough sets in KDD. In: Proceedings of the 16-th World Computer Congress (IFIP'2000). Beijing, China (2000) 1–14.
17. Skowron, A., Rauszer, C.: The discernibility matrices and functions in information systems. In: Slowinski, R. (Ed.), Intelligent Decision Support. Handbook of Applications and Advances of the Rough Set Theory. Kluwer Academic Publishers, Dordrecht, Boston, London (1992) 331–362.
18. Slavík, P.: Approximation algorithms for set cover and related problems. Ph.D. thesis. University of New York at Buffalo (1998).
19. Ślęzak, D.: Approximate reducts in decision tables. In: Proceedings of the Congress Information Processing and Management of Uncertainty in Knowledge-based Systems 3. Granada, Spain (1996) 1159–1164.
20. Ślęzak, D.: Normalized decision functions and measures for inconsistent decision tables analysis. *Fundamenta Informaticae* 44 (2000) 291–319.
21. Ślęzak, D.: Approximate decision reducts. Ph.D. thesis. Warsaw University (2001) (in Polish).
22. Ślęzak, D.: Approximate entropy reducts. *Fundamenta Informaticae* 53 (2002) 365–390.
23. Ślęzak, D., Wróblewski, J.: Order-based genetic algorithms for the search of approximate entropy reducts. In: Proceedings of the International Conference Rough Sets, Fuzzy Sets, Data Mining, and Granular Computing. Chongqing, China. Lecture Notes in Artificial Intelligence 2639, Springer-Verlag (2003) 308–311.
24. Wróblewski, J.: Ensembles of classifiers based on approximate reducts. *Fundamenta Informaticae* 47 (2001) 351–360.
25. Yablonskii, S.V.: Introduction into Discrete Mathematics (fourth edition). Vishaya Shkola, Moscow, 2003 (in Russian).
26. Ziarko, W.: Analysis of uncertain information in the framework of variable precision rough sets. *Foundations of Computing and Decision Sciences* 18 (1993) 381–396.
27. Zielosko, B.: On partial decision rules. In: Proceedings of Concurrency, Specification and Programming Workshop 2. Ruciane-Nida, Poland (2005) 598–609.

A Personal View on AI, Rough Set Theory and Professor Pawlak

Toshinori Munakata

Computer and Information Science Department, Cleveland State University
Cleveland, Ohio 44115, U.S.A.
t.munakata@csuohio.edu

It is an honor to contribute my short article to this special issue commemorating the life and work of Professor Zdzisław Pawlak. In this article I would like to discuss my encounters with the field of artificial intelligence (AI) in general, and how I see rough set theory and Professor Zdzisław Pawlak in this context. I have been fortunate to know some of the greatest scholars in the AI field. There are many of them, but if I had to choose the three I admire most, they are: Professors Zdzisław Pawlak, Lotfi Zadeh and Herbert A. Simon. There are common characteristics among all of them. Although they are the most prominent of scholars, all are frank and easy and pleasant to talk with. All are professionally active at ages where ordinary people would have long since retired.

I became interested in the field of AI in the mid 70s. I have observed many ups and downs of the field in terms of the popularity since then - a common phenomena in any field. The AAAI (American Association for Artificial Intelligence) was inaugurated and the first issue of the AI Magazine was published in the spring 1980. The timing of the birth of rough set theory was soon after this event. At this time many people in the world were becoming interested in the field of AI, while there were only a handful researchers when the field started in the 1950s. In the spring of 1986, the first issue of the *IEEE Expert* (now *IEEE Intelligent Systems*) was inaugurated. I served as an Associate Editor of this magazine for two terms from 1987 to 1991. In terms of public popularity AI was flourishing in this eras.

During many years of the 70s and 80s, I observed that despite media hype and claims for break-thorough technologies, most AI techniques were not practical. Here "practical" means "having real-world commercial and industrial applications on an everyday basis." For example, I could not find cases where machine learning techniques discussed in textbooks such as "learning from examples" and "learning from analogy" were actually employed at industrial plants or commercial banks. The same were true for other AI techniques such as blackboard and neural networks. After Minsky's pessimistic view on the field, the U.S. government funding ceased, and only a handful researchers remained active in the field. The field revived in the mid to late 80s, and became quite popular. However, I could not find a single case where neural networks were actually used every day for commercial and industrial applications. For all of these observations I could be wrong because there could have been exceptions I was not aware of, but I was certain that these exceptions were few, if any.

J.F. Peters et al. (Eds.): Transactions on Rough Sets VI, LNCS 4374, pp. 247–252, 2007.
© Springer-Verlag Berlin Heidelberg 2007

This situation of impracticality of most AI techniques appeared to start to change around 1990. That is, many AI techniques were becoming truly practical and their application domains much broader. Of course, there were practical AI techniques before 1990. Robotics was one. The first major industrial fuzzy control was implemented in Denmark in 1982, followed by the famous Hitachi's Sendai subway control in 1986. There had been practical expert systems in the 80s. However, the repertories of AI techniques and their application domains were becoming much more extensive around 1990.

With this background, in March 1993 I made a proposal to the *Communications of the ACM* (*CACM*) for a Special Issue entitled "Commercial and Industrial AI." The *CACM* was a primary computer science magazine subscribed by some 85,000 professionals worldwide at that time. Its readers went far beyond the computer science discipline including fields such as engineering, social science and education. The proposal was accepted and I selected the most practical or promising AI areas with the help of many experts. The Special Issue was published in March 1994 [3] and was divided into four major sections with 11 articles. They are: I. "Knowledge Engineering Systems" with two articles – an article on general expert systems and an article on case-based reasoning. II. "Perception, Understanding, and Action" with three articles on vision, speech and robotics. III. "Fuzzy Systems" with two articles – an overview and a soft computing article by Professor Zadeh. IV. "Models of the Brain and Evolution" with four articles - two articles by Rumelhart, Widrow, et al., an article on neural networks in Japan, and an article on genetic algorithms. There were many behind-the-scene stories and one of them was that my original plan was to have only one article by Rumelhart, et al. After much delay, they had an article twice as long as originally planned, and I suggested splitting the manuscript into two parts.

In the Guest Editors Introduction, I wrote:

The practical application of artificial intelligence (AI) has been the center of controversy for many years. Certainly, if we mean AI to be a realization of real human intelligence in the machine, its current state may be considered primitive. In this sense, the name artificial "intelligence" can be misleading. However, when AI is looked at as "advanced computing," it can be seen as much more. In the past few years, the repertory of AI techniques has evolved and expanded, and applications have been made in everyday commercial and industrial domains. AI applications today span the realms of manufacturing, consumer products, finance, management, and medicine. Implementation of the correct AI technique in an application is often a must to stay competitive. Truly profitable AI techniques are even kept secret.

Many of the statements I wrote here are still basically true today. This Special Issue turned out to be a big hit. The ACM printed 1,000 extra copies of this issue for back orders, but they sold out less than a month. A person from a Japanese company wanted to purchase a box of fifty copies of this issue, but it was too late. The issue became one of the most cited *CACM*, for not only within computer science but also some unexpected places such as *Scientific American*, the Berkeley Law School, the Stanford Philosophy Department, etc. The US Air Force has issued interesting predictions in the past. Around 1947, the Air Force predicted likely technologies for the next 50 years. They included jet rather than propeller powered, and supersonic airplanes. They became the reality by 1997. In 1997, they issued predictions for the next 50 years, i.e., by 2047. Between these major predictions, they published future perspectives for shorter time ranges. My Special Issue was cited in a document within a report entitled "Air Force 2025" that describes 30-year predictions by the US Air Force [1].

When I was preparing my first *CACM* Special Issue, I knew there were other AI areas that were not covered. As soon as the first issue was nearly complete, I started working on a follow-up Special Issue entitled "New Horizons in Commercial and Industrial AI." In the "Editorial Pointers" in the first issue, Executive Editor Diane Crawford wrote: "He has assembled some of the foremost minds in AI to author and/or review the 11 articles presented here. If that weren't enough, he's already digging into another proposed issue for *Communications* to appear early next year, where he hopes to address new horizons and applications in other AI-related fields."

For the first Special Issue I received many responses. One of them was a letter from Professor Herbert A. Simon of Carnegie-Mellon University, a prominent scholar in AI with a Turing Award and a Nobel Prize in economics. Basically, he stated: "The first issue was well done, although if I were the Guest Editor I would have had less emphasis on neural networks and included an article on machine learning." He suggested placing an article on machine learning in the second issue. I replied to him saying I had already planned that and asked him to write one, and subsequently he co-authored an article. I was lucky to be able to have close contact with Professor Simon. When IBM's Deep Blue defeated the human chess champion Garry Kasparov in 1997, he and I co-authored a commentary article on the significance of this event on AI [14]. He was a pleasant person to talk with. He was a fulltime professor and active until two weeks before his death in 2001 at age 84. For the second Special Issue I had planned to include certain topics from a very early stage. They included symbolic machine learning, natural language processing (e.g., machine translation) and logic programming. Also, I wanted to include articles addressing the commonsense problem, although I did not expect that this area would have many commercial or industrial applications in the near term.

At a later stage of preparation of the second issue, I searched for additional areas appropriate for the issue, and found rough set theory. I was not familiar with this area, but from what I found I thought it was a promising technique, appropriate for the second issue. Perhaps it could complement other AI techniques. I contacted Professor Pawlak and asked him whether he was interested in contributing an article to such a *CACM* Special Issue. These were my first encounters with rough set theory and Professor Pawlak. He kindly accepted my invitation and contributed an article co-authored with Professors Jerzy GrzymalaBusse, Roman Slowinski and Wojciech Ziarko [8]. This was my first acquaintance with the rough set community. As said earlier, *CACM* has a large number of audience worldwide and its impact is high. I don't know how much the appearance of this article has influenced the promotion of this theory, but I think at least it helped to introduce the term "rough sets" worldwide.

Incidentally, when I studied practically successful symbolic machine learning techniques for the first time, such as ID3, I was a bit disappointed. From the term "learning," I expected some elements of human-like learning. For example, given a specific experience the machine would abstract it, generalize it and be able to use it for similar circumstances in the future. I did not see such human-like learning in ID3. Rather, it simply classifies data based on entropy in information theory. The characteristics of the target data seemed to be too simple. Perhaps the term "learning" was misleading, and probably I expected too much on what we could do from this kind of data. Both ID3 and rough sets can learn from data, but probably ID3 had attracted more attention than rough sets in the scientific community, at least during the 80s and 90s. Why? One reason might be that ID3 appears to have been in the main stream in the machine learning community, and had received more support from its early introduction. Professor Simon was one of it's supporters, and he was a great scientist as well as a good salesman to promote his beliefs. For example, he called a software system he developed a "general problem solver," which implied, with a bit of exaggeration, the system would solve every problem on earth. He was also an optimist. In the late 1950s he predicted that a computer would defeat a human chess champion within 10 years. We waited 10 years, another 10 years, and so forth for the next 40 years. In contrast, Professor Pawlak was a humble and modest scientist and perhaps not particularly a good salesman. In my opinion, rough set theory was not as widely recognized in the AI and CS fields as it should have been.

After my first encounter with the rough set community through my *CACM* second special issue, I have been fortunate to be able to work in this field together with these people. I attended several rough set related conferences after my first encounter [4, 5, 6]. To promote rough sets, I could think of two among many possibilities. One was to have promotional articles in journals of large audience like the *CACM*. Another area was to have a rough set application with a high social impact. For the latter, rough control might be a good candidate, I thought. Fuzzy set theory became a hot topic after Hitachi successfully applied fuzzy logic to Sendai subway control. I tried to push rough control, and I was Chair of the rough control interest group. The basic idea of rough control is to employ rough sets to automatically generate input-to-output control rules [7, 9]. The idea was not particularly new, but breakthrough applications would place rough set theory in the spotlight. A long time rough set activist Professor T.Y. Lin financially supported me for this endeavor. Although we have not observed a major breakthrough yet, I think possibilities are still there. In the communication with Professor Pawlak, he suggested presenting a co-authored conference paper [13]. When I published an AI book from Springer, I included a chapter for rough sets [10]. When I served as Guest Editor for third time for *CACM* Special Section on knowledge discovery [11], I asked Professor Ziarko to contribute an article.

When a young colleague approached me to work on a data mining article, I suggested employing rough sets [2]. I am currently working on another article on a rough set application with a young assistant professor.

Although we all saddened by the recent death of Professor Pawlak, I think he was fortunate to observe that his theory has been widely recognized in the scientific community worldwide. This was not necessarily the case for many great scholars in the past. During my sabbatical in the fall of 2002, I traveled to Poland, visiting Professors Slowinski, Skowron and Pawlak, and received a warm welcome. This was the last time I saw Professor Pawlak.

What are the future prospects of rough sets? No one knows, but the following is my speculation. Despite it's founder's death, the community will grow – there will be more researchers worldwide and more theoretical and application developments. But, growth in the field may level out eventually, unless we achieve major breakthroughs. As in the case of other machine learning techniques and AI in general, we don't know what, when or if such breakthroughs may come. Targeting to extremely large volumes of data (e.g., terabytes) and/or massively parallel computing alone do not look very promising, as we have observed similar attempts such as the Cyc and the Connection Machine. For knowledge discovery techniques such as rough sets, there may be a limit when we deal only with decision tables. Perhaps we should also look at other formats of data as well as other types of data, for example, non-text, comprehensive types of information, such as symbolic, visual, audio, etc. Also, the use of huge background knowledge, in a manner similar to human thought, would be necessary and effective. Human-computer interactions would also enhance the discovery processes. Other totally different domains are non-silicon based new computing paradigms. I am currently working on my fourth Special Section for the *Communications of the ACM* as a guest editor on this subject [12]. These approaches may lead to a new dimension of information processing in a wide range of application domains including rough sets. As with other scientific developments in history, such as alchemy and the first airplane, a breakthrough may come in a totally unexpected form.

References

1. Clarence E. Carter, et al. The Man In The Chair: Cornerstone Of Global Battlespace Dominance, *Air Force 2025*, 1996.
2. Brenda Mak and T. Munakata. "Rule extraction from expert heuristics: A comparative study of rough sets with neural networks and ID3," *European Journal of Operational Research*, 136(1), pp. 212-229, 2002.
3. Toshinori Munakata. "Guest Editor's Introduction," in the Special Issue "Commercial and Industrial AI Applications," *Communications of the ACM*, 37(3), pp. 2325, 1994.
4. Toshinori Munakata. "Commercial and Industrial AI: Where it is now, and where it will be," an invited talk at *the Third International Workshop on Rough Sets and Soft Computing (RSSC'94)*, San Jose, CA, Nov., 1012, pp. 5155, 1994.

5. Toshinori Munakata. "Rough Control: A Perspective," *Workshop on Rough Sets and Data Mining, the 1995 ACM Computer Science Conference*, Nashville, TN, March 2, 1995.

6. Toshinori Munakata. "Commercial and Industrial AI and a Future Perspective on Rough Sets," in *Soft Computing: the Third International Workshop on Rough Sets and Soft Computing (RSSC94)*, T.Y. Lin and A.M. Wildberger (Eds.), the Society of Computer Simulation, pp. 219222, 1995.

7. Toshinori Munakata. "Rough Control: Basic Ideas and Applications," *Workshop on Rough Set Theory, the Second Annual Joint Conference on Information Sciences*, Wrightsville Beach, NC, Sept. 28 Oct. 1, 1995.

8. Toshinori Munakata. "Guest Editor's Introduction," in the Special Issue "New Horizons for Commercial and Industrial AI," *Communications of the ACM*, 38(11), pp. 28-31, 1995.

9. Toshinori Munakata. "Rough Control: A Perspective," in T.Y. Lin and N. Cercone (Eds.), *Rough Sets and Data Mining: Analysis of Imprecise Data*, Kluwer Academic, pp. 7788, 1997.

10. Toshinori Munakata. *Fundamentals of the New Artificial Intelligence: Beyond Traditional Paradigms*, Springer-Verlag, 1998, 2007.

11. Toshinori Munakata. "Guest Editor's Introduction," in the Special Section "Knowledge Discovery," *Communications of the ACM*, 42(11), pp. 26-29, 1999.

12. Toshinori Munakata. "Guest Editor's Introduction," in the Special Section "Beyond Silicon: New Computing Paradigms," *Communications of the ACM*, to appear in 2007.

13. Zdzisław Pawlak and Toshinori Munakata. "Rough Control: Application of Rough Set Theory to Control," *EUFIT '96 the Fourth European Congress on Intelligent Techniques and Soft Computing*, Aachen, Germany, Sept. 25, 1996, pp. 209-217.

14. Herbert A. Simon and Toshinori Munakata. "Kasparov vs. Deep Blue: The Aftermath - AI Lessons," *Communications of the ACM*, 40(8), pp. 23-25, 1997.

Formal Topology and Information Systems

Piero Pagliani[1] and Mihir K. Chakraborty[2]

[1] Research Group on Knowledge and Communication Models
Via Imperia, 6. 00161 Roma, Italy
`p.pagliani@agora.stm.it`
[2] Department of Pure Mathematics, University of Calcutta
35, Ballygunge Circular Road, Calcutta-700019, India
`mihirc99@vsnl.com`

Abstract. Rough Set Theory may be considered as a formal interpretation of observation of phenomena. On one side we have objects and on the other side we have properties. This is what we call a *Property System*. Observing is then the act of perceiving and then interpreting the binary relation (of satisfaction) between the two sides. Of course, the set of properties can be given a particular structure. However, from a pure "phenomenological" point of view, a structure is given by the satisfaction relation we observe. So it is a result and not a precondition. Phenomena, in general, do not give rise to topological systems but to pre-topological systems. In particular, "interior" and "closure" operators are not continuous with respect to joins, so that they can "miss" information. To obtain continuous operators we have to lift the abstraction level of Property Systems by synthesizing relations between objects and properties into systems of relations between objects and objects. Such relations are based on the notion of a *minimal amount of information* that is carried by an item. This way we can also account for *Attribute Systems*, that is, systems in which we have attributes instead of properties and items are evaluated by means of attribute values. But in order to apply our mathematical machinery to Attribute Systems we have to transform them into Property Systems in an appropriate manner.

Keywords: approximation spaces, formal topology, Galois adjunctions, rough sets, information quanta, information systems, pointless topology, pretopology.

1 Introduction

Rough Sets arise from information systems in which items are evaluated against a set of attributes or properties.

In Computer Science properties are often interpreted as "open subsets" of some topological space. M. Smyth pioneered this interpretation in 1983 when he observed that semi-decidable properties are analogous to open sets in a topological space (cf. [28]). This intuition was developed by distinguished scholars such as D. Scott who introduced *Domain Theory* and the so-called *Scott Topology* to study continuous approximating maps between structures of information called *domains*.

J.F. Peters et al. (Eds.): Transactions on Rough Sets VI, LNCS 4374, pp. 253–297, 2007.
© Springer-Verlag Berlin Heidelberg 2007

This research is paralleled, in a sense, by logical studies such as Cohen's forcing and Kripke models, where the notion of approximable sets of properties (or approximable information) is the core of the construction.

W. Lawvere showed that these constructions can be synthesized into the notion of a *topos* as the abstract form of continuously variable sets.

S. Vickers combined the logical and the Computer Science approach. In [33], a prominent role is played by *topological systems* where just the formal properties of open sets are considered, without mentioning points (pointless topology).

Indeed, this approach originates in Stone's and Birkhoff's representation theorems where the notion of an *abstract point* is *de facto*, introduced. And an abstract point is nothing else but a bunch of properties (once we interpret the elements of a lattice as properties).

Influenced by P. Martin-Löf's *Intuitionistic Type Theory*, G. Sambin undertook his own way to deal with pointless topology, and specifically pointless pretopology, as related to Logic (namely Linear Logic), which led to the notion of a *Formal Topology* (see [25]) which later on has been presented as a result of a construction arising from binary relations between the *concrete* side (points) and the *abstract* side (properties) of an observational system called a *Basic Pair* (cf. [26]). As a matter of fact, the interrelations between concrete points and abstract properties is considered by Vickers, too. However in Formal Topology one does not impose any pre-established structure on the set of properties, not even that suggested by "Observation Logic" in Vicker's approach, which makes a system of observations into a *frame*[1].

In [18] it was noted that the properties of the operators of Formal Topology may be deduced from the fact that they are based on constructors which enjoy adjointness relations. The pretopological approach was applied to account for approximation operators arising from families of Property Systems in [19] and, finally, to generalize approximation operators arising from single Property Systems ([21] and [22]). Moreover, this machinery was applied to Attribute Systems too, by transforming them into Property Systems (cf. [20]).

The latter researches were, in turn, influenced by A. Skowron, J. Stepaniuk and T. Y. Lin's pioneering investigations which have shown that neighborhood systems may account for natural organizations of information systems (for this topic and its applications the reader is referred to [11] and [29]).

Moreover, it must be noticed that neighborhood systems give rise to pretopological operators which turn into topological operators just under particular conditions. Therefore, we claim that pre-topology is a most natural setting for approximation operators as induced by information systems. Also, this claim fits with recent suggestions on the role of "true borders" (hence non topological) to account for a more dynamic approach to data analysis and granular computing (see, for instance [19] and for "rough topologies" see [24]).

Specifically, the present investigation is induced by the observation that, from a very general point of view, Rough Set Theory arises from a sort of "phenomenological" approach to data analysis with two peculiar characteristics:

---

[1] A *frame* is a lattice with finite meets distributing over arbitrary joins.

- *data is analyzed statically at a given point in time of a possibly evolving observation activity;*
- *as a consequence, the analyzed data provides us only with an approximated picture of the domain of interest.*

We shall see that the status of an observation system at a certain point in time is essentially a triple $\mathbf{P} = \langle G, M, \Vdash \rangle$, that we call a *Property system* were G is a set of objects (also called "points"), M a set of properties (also called "formal neighborhoods") and $\Vdash \subseteq G \times M$ is intended as a fulfillment relation[2]. From the concept of an "observation" we shall define a family of basic "perception constructors" mapping sets of objects into sets of properties, called *intensional constructors*, and sets of properties into sets of objects, called *extensional constructors*. We show that some pairs of constructors from opposite sides, fulfill *adjunction properties*. That is, one behaves in a particular way with respect to properties if and only if the other behaves in a mirror way with respect to objects. Hence, adjunction properties state a sort of "dialectic" relationship, or mutual relationship, between perception constructors, which is exactly what we require in view of a "phenomenological" approach.

Adjunction properties make some combinations of these basic constructors into generalized closure and generalized interior operators. Particularly, some combinations happen to be pre-topological operators in the sense of Sambin's Formal Topology. Actually, we shall see that they are generalizations of the approximation operators provided by Rough Set Theory.

However, for they are pretopological and not topological, these approximation operators are not continuous, that is, they exhibit "jumps" in the presence of set-theoretical operations. Therefore we synthesize the structuring properties of a *Property system*, $\mathbf{P}$, into a second level informational structure $\langle G, G, R_{\mathbf{P}} \rangle$, called an *Information Quantum Relational System - IQRS*, where $R_{\mathbf{P}}$ is a relation between objects - hence no longer between objects and properties - embedding the relevant informational patterns of $\mathbf{P}$. In IQRSs adjointness makes second level approximation operators fulfill nice properties. Also, this way we shall be able to account for *Attribute systems* after appropriately transforming them into Property Systems[3].

This study aims at presenting the state-of-the-art of a conception of Rough Set Theory as a convergence of different approaches and different techniques, such as Formal Topology, duality and representation theory, Quantum Logic, adjoint functors and so on, as sketched in the following figure:

---

[2] Property Systems may be also regarded as "Chu Spaces" ([37]). However Chu Spaces have additional features, namely a pair of covariant and contravariant functors, which links spaces together. For Chu Spaces, see the WWW site edited by V. Pratt, http://boole.stanford.edu/chuguide.html.

[3] The term "Quantum" refers to the fact that a basic information grains is given by the minimal amount of information which is organised by $R_{\mathbf{P}}$ around an item what, technically, is linked to the notion of a *quantum of information at a location* once one substitute "item" for "location" - cf. [4].

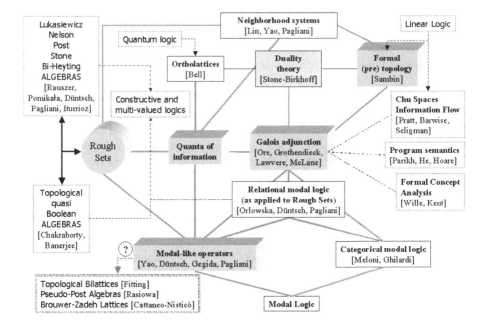

Fig. 1. A Rough Set connection

More precisely, in the present paper we shall deal with the boxed topics along the flow displayed by bold lines. Dotted arrows display the relationships between logical systems and some topics connected to Rough Sets Theory. Dotted lines show some interesting links between some techniques used in the present paper and topics connected with data and information analysis. Bold arrows display some well-established links between logico-algebraic systems and Rough Set systems, while the dotted arrow marked with "?" suggests links between the modal-style approach applied in the paper and logico-algebraic interpretations to be explored[4].

2   Formal Relationships Between "Objects" and "Observables"

Observation is a dynamic process aimed at getting more and more information about a domain. The larger the information, the finer the picture that we have about the elements of the domain. Using topological terms, an observation process makes it possible to move from a trivial topology on the domain, in

---

[4] We have proved that Rough Set Systems are semi-simple Nelson algebras (or equivalently, three-valued Łukasiewicz algebras) (cf. [14] and [15]). What algebraic systems arise from this generalisation has to be studied yet. *Brouwer-Zadeh Lattices, Bilattices* and *Semi-Post algebras* may provide some hints.

which everything is indistinguishable to a topology in which any single element is sharply separable from all the other elements (say a discrete topology or a Hausdorff space). In this case we can "name" each single element of the domain by means of its characteristic properties.

However, in this respect, observation is an asymptotic process. What usually happens is that at a certain point in time we stop our observation process, at least temporarily, and analyse the stock of pieces of information we have collected so far. In a sense we consider a "flat slice" of the observation process.

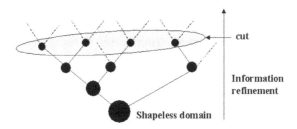

Fig. 2. A process of differentiation via observations

Therefore, our slice is basically composed by:

(a) a set G of 'objects'; (b) a set M of 'observable properties'; (c) a relation between G and M, denoted with the symbol $\Vdash$. Given $g \in G$ and $m \in M$ we shall say that if $g \Vdash m$, then g enjoys property m, or that g induces the *observable property* m.

We have also to assume that $\Vdash$ *is defined for all the elements of* G and M because we consider immaterial any property which cannot help making any distinction among objects and, symmetrically, if an object g does not manifest any property, then it is a "non-object" from a phenomenological point of view[5]. We encode everything in the following definition:

Definition 1. *A triple* $\langle G, M, \Vdash \rangle$ *where* G *and* M *are finite sets,* $\Vdash \subseteq G \times M$ *is a relation such that for all* $g \in G$ *there is* $m \in M$ *such that* $g \Vdash m$*, and for all* $m \in M$ *there is* $g \in G$ *such that* $g \Vdash m$*, is called a* property system *or a* P-system.

Among *P-systems* we distinguish:

a) *Functional systems*, or *FP-systems*, where $\Vdash$ is functional in the sense that for any element $g \in G$, $g \Vdash m$ and $g \Vdash m'$ implies $m = m'$.

---

[5] The symbols "G" and "M" are after the German terms "*Gegenstände*" ("objects") and, respectively, "*Merkmale* ("properties"). A "Gegenstand" is what stays in front of a subject, while the German term "Object" means an interpreted "Gegenstand". These are the terms used in Formal Concept Analysis and we adopt them for their philosophical meaning.

b) *Dichotomic systems* or *DP-systems*, if for all $p \in M$ there is $\bar{p} \in M$ such that for all $g \in G$, $g \Vdash p$ if and only if $g \not\Vdash \bar{p}$.

Functional and dichotomic systems enjoy particular classification properties[6]. Moreover we shall also consider *Deterministic attribute systems*:

Definition 2. *A structure of the form* $\langle G, At, \{V_a\}_{a \in At}, \rangle$*, where* G*,* At *and* V_a *are sets (of objects, attributes and, respectively, attribute-values) and for each* $a \in At$*,* $a : G \longmapsto At_a$*, is called a* deterministic Attribute System *or an A-system[7].*

From now on we assume that $\mathbf{P}$ always denotes a *Property System* $\langle G, M, \Vdash \rangle$ and that $\mathbf{A}$ denotes an *Attribute System* $\langle G, At, \{V_a\}_{a \in At}, \rangle$. Moreover, we shall use the following notation:

If $f : A \longmapsto B$ and $g : B \longmapsto C$ are functions, then:

(a) with $(f \circ g)(x)$ or, equivalently, $g(f(x))$ we denote the *composition of g after f*;
(b) $f^{\rightarrow} : \wp(A) \longmapsto \wp(B)$; $f^{\rightarrow}(X) = \{f(a) : a \in X\}$ - denotes the *image* of X via f;
(c) $f^{\leftarrow} : \wp(B) \longmapsto \wp(A)$; $f^{\leftarrow}(Y) = \{a : f(a) \in Y\}$ - denotes the *pre-image of Y via f*;
(d) the set $f^{\rightarrow}(A)$ is denoted with Im_f; 1_A, denotes the *identity function on A*;
(e) the map $f^o : A \longmapsto Im_f$; $f^o(a) = f(a)$ denotes the *corestriction* of f to Im_f and the map $f_o : Im_f \longmapsto B$; $f_o(b) = b$ denotes the *inclusion* of Im_f into B;
(f) the equivalence relation $k_f = \{\langle a, a' \rangle : f(a) = f(a')\}$ is called the *kernel* of f or the *fibred product* $A \times_B A$ obtained by pulling back f along itself.

2.1   Ideal Observation Situations

If $\langle G, M, \Vdash \rangle$ is an *FP-system*, we are in a privileged position for classifying objects, for the reasons we are going to explain.

The "best" case is when $\Vdash$ is an injective function. Indeed in this case the converse relation $\Vdash^{\smile}$ (or, also, $\Vdash^{-1}$) is a function, too, and we are able to distinguish, sharply, each object. In mathematical words we can compute the *retraction of* $\Vdash$.

Definition 3. *Let* $f : A \longmapsto B$ *be a function. Then a morphism* $r : B \longmapsto A$ *is called a* retraction of f, *if* $f \circ r = 1_A$.

But, first of all, this is an unusual situation, from a practical point of view. Further, "observation" and "interpretation" is a modeling activity. Thus, from an epistemological point of view we may wonder if the best model of a horse is really a horse.

Actually, "modeling" means "abstracting" and "abstracting" means "losing something", some quality or characteristic. Thus the situation depicted in Figures 3 and 4 cannot be but the result of some reduction process.

---

[6] Indeed *FP-systems* and *DP-systems* are closely linked together, as we shall see.
[7] The traditional term in Rough Set Theory is "Information System".

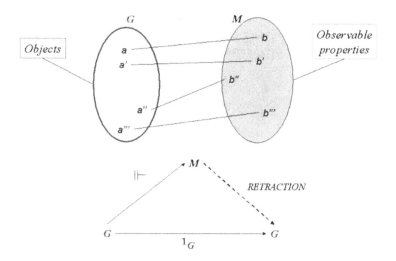

Fig. 3. An ideal situation: the existence of retractions

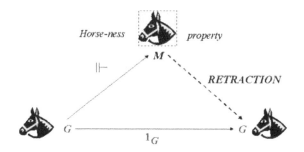

Fig. 4. Is a horse the best model of a horse?

A second optimal situation is when $\Vdash$ is a surjective function (what always happens of *FP-systems*). Indeed, by reading back $\Vdash$ we obtain an equivalence relation $E_{\Vdash}$, so that any element of G will belong to one and just one equivalence class modulo $E_{\Vdash}$, without ambiguity and borderline situations, what is a perfect case of a classification. Indeed, $E_{\Vdash}$ is the kernel of $\Vdash$ and it induces a classification of the elements of G through *properties* in M. This is tantamount to the construction of *stalk spaces*, or *espace etalé*, through *fibers* (or *stalks, sorts*). This means that $\Vdash$ has a *section* or a *co-retraction*.

Definition 4. *Let $f : A \longmapsto B$ be a function. Then a morphism $s : B \longmapsto A$ is called a* section *or* co-retraction *of f, if $s \circ f = 1_B$.*

We can interpret f as a way to list or parametrise (some of) the elements of B, through the elements of A.

In turn the notions of a section and a retraction are special cases of a more fundamental concept: a *divisor*.

Definition 5. *Let $f : A \longmapsto B$, $g : B \longmapsto C$ and $h : A \longmapsto C$ be three functions. Then, g is called a* right divisor *of h by f and f is called a* left divisor *of h by g if $h = f \circ g$, that is, if the following diagram commutes:*

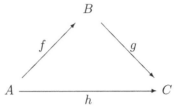

From the above definitions, we immediately deduce that if s is a section of f, then f is a retraction of s; and vice-versa. Moreover, it is not difficult to verify that f does not have any section if it is not *surjective* on B (otherwise, how would it be possible to obtain 1_B?). Intuitively if there is a $b \in B$ that is not f-image of any $a \in A$, then b would be associated with a void sort. Vice-versa, a function f does not have any retraction if f is not *injective* in B. In fact, if $f(a) = f(a') = b$, for $a \neq a'$, then any morphism from B to A either maps b onto a and forgets a', or it maps b onto a' and forgets a, because of unicity of the image, and we could not obtain 1_A (thus, for any function $f : A \longmapsto A$, f_o is a section with retraction f^o).

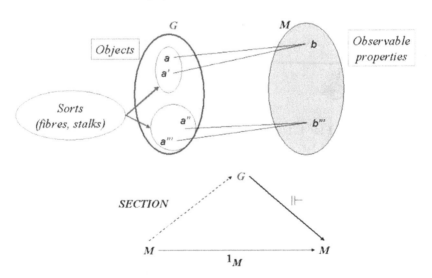

Fig. 5. An almost ideal situation: the existence of co-retractions

If $r : A \longmapsto B$ is a retraction of a function $h : B \longmapsto A$, then $r \circ h$ is an idempotent endomorphism of A: $(r \circ h) \circ (r \circ h) = r \circ (h \circ r) \circ h = r \circ 1_B \circ h = r \circ h$.

It follows that if $s : B \longmapsto A$ is a section of $f : A \longmapsto B$, then $f \circ s$ is an idempotent endomorphism in A, provided s is onto, because f is a retraction of s (see above). Clearly, if $a = s(b)$, then $s(f(a)) = s(f(s(b))) = s(1_B(b)) = s(b) = a$. Hence, any image of the section s is a fixed point of the endomorphism $f \circ s$.

Thus sections, retractions and kernels of a function f make it possible to organise the domain of f into sharp classification categories (groups) and to single out representatives of such categories.

3   Categorizing Through Relational *P-systems*

On the contrary, if we deal with generic relational *P-systems*, not necessarily functional, it hardly happens to obtain sharp classifications, at least without a proper manipulation of the given *P-system* that, in turn, may be or may not be an appropriate maneuver. It follows that the identity relation in the definition of left and right divisors must be weakened to an inequality relation "$\geq$" or "$\leq$". Therefore, to deal with generic cases we need a more subtle mathematical machinery.

Such a machinery is based on the notion of an "approximation". However, this notion depends on another one. Indeed, we cannot speak of "approximation" without comparing a result with a goal and this comparison depends on the *granularity* of the target and of the instruments to get it. For instance, in a document search system, in general we face a situation in which queries refer to a set of possible answers and not to single objects. Otherwise we would not have "queries" but "selections" (the realm of sections and retractions). In other words, objects constitute, in principle, a finer domain than those obtained by any modeling or interpretation activity.

So we can distinguish an *extensional granulation*, related to objects, and an *intensional granulation*, related to properties, and assume that the extensional granulation is finer than the intensional one. Thus, when we have to determine a point on the extensional scale by means of the intensional ruler, we hardly will be able to get a precise determination. We can approximate it. But in order to be able to have "best approximations" the intensional granulation and its relationships with the extensional granulation must fulfill non trivial properties. First of all we need an order. Suppose X is a set of candidate results of a query. Then we generally do not have a selection criterion to single out elements of X. But if the elements of X are ordered in some way, we can use this order to choose, for instance, the least or the largest element in X, if any.

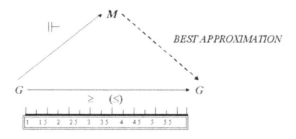

Fig. 6. A usual *P-system* needs a scale

But this is not enough. In fact dealing with functions (either surjective or injective) is, in a sense, a lucky event in this business which happens only if we are able to completely reduce a structure to a simpler one. This means that generally we cannot have sections and retractions, so that we cannot directly manipulate pre-images of single elements of the codomain or images of single elements of the domain of some "connecting" function. On the contrary we have to manipulate some kinds of subsets of the domain and co-domain which, we hope, embed enough ordering features to compute approximations.

Having this picture in mind, we underline what follows. From an observational point of view the only relationships between objects are induced by the fulfillment relation $\Vdash$ and they are grouping relationships so that we can compare subsets of objects (or properties) but not, directly, objects (or properties). In other words in this paper we assume that there is no relation (hence any order) either between objects or between properties. Hence the result of an approximation activity is, generally, a "type" not a "token"[8]. It follows that we shall move from the level of pure P-systems $\langle G, M, \Vdash \rangle$ to that of $Perception$ $systems$ $\langle \wp(G), \wp(M), \{\phi_i\}_{i \in I}\rangle$ where ϕ_i is a map from $\wp(G)$ to $\wp(M)$ or from $\wp(M)$ to $\wp(G)$.

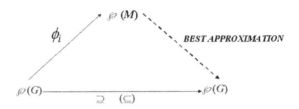

Fig. 7. Approximation deals with types, not with tokens

4   Concrete and Formal Observation Spaces

Given a P-system, the first, and obliged, step we have to do is "observing", in the elementary sense of "perceiving" the manifested properties. Thus if **P** is a P-system let us define an '$observation$ $function$' $obs : G \mapsto \wp(M)$, by setting

$$m \in obs(g) \Leftrightarrow g \Vdash m. \tag{1}$$

Technically, obs is what is called a $constructor$ for it builds-up a set from a point. Indeed, for each point g, $obs(g) = \{m \in M : g \Vdash m\}$. We shall call $obs(g)$ the '$intension$ of g'. In fact, any element g appears through the series of its observable properties, so that $obs(g)$ is actually the $intensional$ $description$ of g. The intension of a point g is, therefore, its description through the observable properties listed in M. We shall also say that if $g \Vdash m$ (i. e. if $m \in obs(g)$), then m is an observable property $connected$ with g and that g belongs to the $field$ of m.

---

[8] By the way, note that in [3], classification is achieved at "type" level.

Symmetrically we can introduce a "*sub*stance function" $sub : M \mapsto \wp(G)$ defined by setting

$$g \in sub(m) \Leftrightarrow g \Vdash m. \tag{2}$$

This symmetry reflects the intuition that a point can be *intensionally* conceived as the set of properties it is connected with, just as a property may be *extensionally* conceived as the set of points belonging to its field. Dually to *obs*, given a property $m \in M$, $sub(m) = \{g \in G : g \Vdash m\}$, so that $sub(m)$ is the 'extension', or the field, of m.

The link between these two functions is the relation $\Vdash$:

$$g \in sub(m) \Leftrightarrow m \in obs(g) \Leftrightarrow g \Vdash m, \forall g \in G, \forall m \in M \tag{3}$$

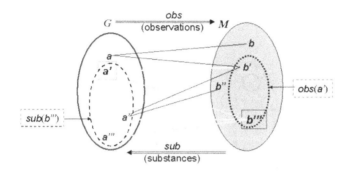

Fig. 8. A first level perception process

We now notice that since the set M is given and fixed, any $P - system$ will provide only *partial* observations of the members of G so that a single point x possibly fails to be uniquely described by its intension $obs(x)$.

We shall also say that $obs(x)$ is an intensional *approximation* of a 'partially describable' member x of G and claim that if $obs(x) = obs(y)$, then x and y cannot be *discerned* by means of the observable approximating properties (or "partial descriptions") at hand, so that x and y will be said to be *indiscernible* in the given P-*system*. If x and y are indiscernible they will collapse into the same intentional description.

Indeed, if *obs* fails to be injective then we know that it cannot have a retraction and this means that the identity $1_{\wp(G)}$ cannot be determined by means of the properties at our disposal (that is, the subsets of M mapped by *obs*), so that a "loss of identity" literally happens.

However, we can define, by means of *obs* and *sub* some approximation operators.

4.1   The Basic Perception Constructors

The second step after observing, is an initial interpretation of what we have observed.

Thus we shall introduce the "perception constructors" that are induced by a P-*system*.

These constructors will make it possible to define different kinds of structures over $\wp(G)$ and $\wp(M)$. Since such structures are defined as extensions of the two functions *obs* and *sub* and since, in turn, these two functions are linked by the relation (3), it is clear that any structurization on points will have a dual structurization on observables, and vice-versa.

Definition 6 (Basic contructors). *Let* $\mathbf{P} = \langle G, M, \Vdash \rangle$ *be a P-system. Then:*

- $\langle e \rangle : \wp(M) \longmapsto \wp(G); \langle e \rangle(Y) = \{g \in G : \exists m (m \in Y \ \& \ g \in sub(m))\};$
- $[e] : \wp(M) \longmapsto \wp(G); [e](Y) = \{g \in G : \forall m (g \in sub(m) \Longrightarrow m \in Y)\};$
- $[[e]] : \wp(M) \longmapsto \wp(G); [[e]](Y) = \{g \in G : \forall m (m \in Y \Longrightarrow g \in sub(m))\};$
- $\langle i \rangle : \wp(G) \longmapsto \wp(M); \langle i \rangle(X) = \{m \in M : \exists g (g \in X \ \& \ m \in obs(g))\}$
- $[i] : \wp(G) \longmapsto \wp(M); [i](X) = \{m \in M : \forall g (m \in obs(g) \Longrightarrow g \in X)\};$
- $[[i]] : \wp(G) \longmapsto \wp(M); [[i]](X) = \{m \in M : \forall g (g \in X \Longrightarrow m \in obs(g))\}.$

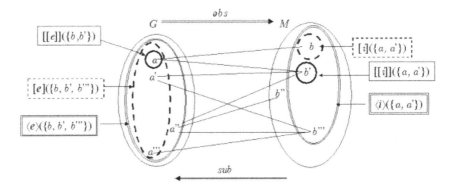

Fig. 9. Basic constructors derived from a basic pair

An intuitive interpretation of the above functions is in order. As for the constructors decorated with 'e' (because the result of the function is an *extent*), we notice that if we want to extend function *sub* from elements to subsets of M, we have essentially two choices: a "disjunctive" or "existential" extension and a "conjunctive" or "universal" extension. The former is $\langle e \rangle$ while the latter is $[[e]]$. Obviously, $\langle e \rangle = sub^{\rightarrow} = obs^{\leftarrow}$. It is not difficult to see that $[e]$ is the dual of $\langle e \rangle$, hence it is the "co-existential extension" of *sub* (the dual of $[[e]]$ is not discussed in this paper).

Given $Y \subseteq B$, the set $[[e]](Y)$ collects the points that fulfill *at least all* the properties from Y (and, possibly, others), while $\langle e \rangle(Y)$ gives the set of points which fulfill *at least one* property from Y. Finally, $[e](Y)$ collects the points which fulfill *at most all* the properties from Y (but possibly not all the properties in Y). The same considerations apply symmetrically to the operators decorated with 'i' (because the result of these functions is an *intent*). Indeed $\langle i \rangle$ and $[[i]]$ are the disjunctive and, respectively, conjunctive extensions to subsets of G of the function *obs* and $\langle i \rangle = obs^{\rightarrow} = sub^{\leftarrow}$. More precisely, $\langle i \rangle(X)$ collects the set of properties that are fulfilled *at least by one* point of X, while $[[i]](X)$ collects the set of properties that are fulfilled *at least by all* the points of X, that is, the properties *com-

mon at least to all the points of X. Finally, $[i](X)$, the "co-existential extension" of *obs*, gives the set of properties that are fulfilled at most by all of the points in X. In particular, $[i](\{x\})$ is the set of properties which uniquely characterize x.

We summarize these remarks in the following table:

	... property/ies in Y	... point/s in X
at least one ...	$\langle e \rangle(Y)$	$\langle i \rangle(X)$
at least all ...	$[[e]](Y)$	$[[i]](X)$
at most all ...	$[e](Y)$	$[i](X)$

If one of the above or following operators on sets, say Op, is applied to a singleton $\{x\}$ we shall also write $Op(x)$ instead of the correct $Op(\{x\})$, if there is no risk of confusion.

4.2 A Modal Reading of the Basic Constructors

At this point a modal reading of the basic constructors is in order. This will be formalized in Section 6. Indeed, we can read these constructors by means of operators taken from extended forms of modal logic, namely, *possibility*, *necessity* and *sufficiency*.

OPERATOR	$x \in X$	$b \in B$	EXAMPLE READING
$\langle i \rangle(X) = B$	if $x \in X$ then it is **possible** that x enjoys elements in B	b is enjoyed by some element collected in X	there are examples of elements in X that enjoy b
$[i](X) = B$	to enjoy elements in B it is **necessary** to be in X	b is enjoyed by at most all the element of X	there are not examples of elements enjoying b that are not in X
$[[i]](X) = B$	to enjoy elements in B it is **sufficient** to be in X	b is enjoyed by at least all the elements of X	there are not examples of elements of X that do not enjoy b

REMARKS: Sufficiency was introduced in modal logic by [10]. Recently it was discussed in [8] and in [16] from an informational point of view. Sufficiency happens to be the fundamental operator to define Formal Concepts, which are pairs of the form $\langle [[e]][[i]](X), [[i]](X) \rangle$, for $X \subseteq G$ (see [34]). From the point of view of pointless topology, the operators $[i]$, $\langle i \rangle$, $[e]$ and $\langle e \rangle$ have been studied by [26], (where the notation "$\Box$", "$\Diamond$", "*rest*" and, respectively, "*ext*" is used) and in [19]. From an informational point of view they have been investigated in [8], [7], [36], [21] and [22]. It is worth noticing that variations of concept lattices have been introduced by means of these operators, namely "object oriented concepts" of the form $(\langle e \rangle[i](X), [i](X))$ (by Y. Y. Yao) and "property oriented concepts" of the form $\langle [e]\langle i \rangle(X), \langle i \rangle(X) \rangle$ (by Düntsch and Gegida). From *Proposition* 2 and *Corollary* 3 below one can easily deduce some of the properties discussed in the quoted papers.

A pictorial description of the above modal reading follows:

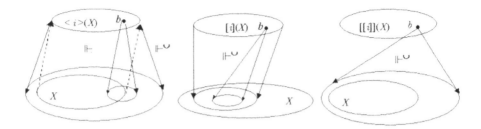

Fig. 10. Possibility, Necessity, Sufficiency

The above functions are linked by some structural relationships.

First, recall that our operators are defined on Boolean algebras of type $\langle \wp(A), \cap, \cup, A, -, \emptyset \rangle$, where A is either G or M, so that negation coincides with set-theoretical complementation.

We say that if an operator Op^o is obtained by negating all of the defining subformulas in an operator Op, and by further applying the contraposition law according to negations (or, equivalently, by first putting the definition in disjunctive normal form), then Op^o and Op are called *opposite* or *orthogonal* (to each other), or "o" in symbols[9]. If $Op^d(X) = \sim Op(\sim X)$ then Op^d is called the *dual* of Op and we denote the relation of duality with "d". Furthermore we can easily observe that functions decorated with e and functions decorated with i are symmetric with respect to the relation $\Vdash$, and we denote this fact with "s". The following table summarizes these relationships between basic operators (some of these connections are well known in literature: cf. [26], [7] and [8] - but see also the literature about Galois connections):

	$\langle e \rangle$	$\langle i \rangle$	$[e]$	$[i]$	$[[e]]$	$[[i]]$
$\langle e \rangle$	=	s	d	sd	od	ods
$\langle i \rangle$	s	=	sd	d	ods	od
$[e]$	d	sd	=	s	o	os
$[i]$	sd	d	s	=	os	o
$[[e]]$	od	ods	o	os	=	s
$[[i]]$	ods	od	os	o	s	=

Obviously, symmetric functions fulfill the same formal properties, opposite functions fulfill opposite properties, while dual and symmetric-dual operators fulfill dual properties.

5   Fundamental Properties of the Basic Constructors

Let us investigate the fundamental properties of basic constructors. We carry on this job in a more general dimension concerning binary relations at large.

---

[9] So, for instance, if $\alpha \implies \beta$ appears in a defining formula, of Op, then in Op^o we have $\sim \beta \implies \sim \alpha$.

Definition 7. *Let $R \subseteq A \times B$ and $Q \subseteq C \times D$ be binary relations, $X \subseteq A$, $Y \subseteq B$, $x \in A$, $y \in B$. Then we define[10]:*

1. $R^{\smile} = \{\langle y, x \rangle : \langle x, y \rangle \in R\}$ - *the* inverse relation *of R.*
2. $\langle R^{\smile} \rangle(X) = \{y \in B : \exists x \in X(\langle x, y \rangle \in R)\}$ - *the* left Peirce product *of R and X. We shall also call $\langle R^{\smile} \rangle(X)$ the R- neighborhood of X. In particular, if X is a singleton $\{x\}$, then we shall usually write $\langle R^{\smile} \rangle(x)$ instead of $\langle R^{\smile} \rangle(\{x\})$.*
3. $\langle R \rangle(Y) = \{x \in A : \exists y \in Y(\langle x, y \rangle \in R)\}$ - *the* left Pierce product *of $R^{\smile}$ and Y. Clearly, $\langle R \rangle(Y)$ is the $R^{\smile}$-neighborhood of Y.*
4. $[R^{\smile}](X) = \{y \in B : \forall x(\langle x, y \rangle \in R \implies x \in X)\}$ - *the* right residual *of R and X.*
5. $[R](Y) = \{x \in A : \forall y(\langle x, y \rangle \in R \implies y \in Y)\}$ - *the* right residual *of $R^{\smile}$ and X.*
6. $[[R^{\smile}]](X) = \{y \in B : \forall x(x \in X \implies \langle x, y \rangle \in R)\}$ - *the* left residual *of X and $R^{\smile}$.*
7. $[[R]](Y) = \{x \in A : \forall y(y \in Y \implies \langle x, y \rangle \in R)\}$ - *the* left residual *of X and R.*
8. $R \otimes Q = \{\langle a, d \rangle : \exists z \in B \cap C(\langle a, z \rangle \in R \ \& \ \langle z, d \rangle \in Q)\}$ - *the* right composition *of R with Q or the* left composition *of Q with R. If defined, $R \otimes Q \subseteq A \times D$.*

Lemma 1. *Let $R \subseteq A \times B$. Then for any $X \subseteq A, Y \subseteq B, a \in A, b \in B$:*

1. (a) $b \in [R^{\smile}](X)$ iff $\langle R \rangle(b) \subseteq X$; (b) $a \in [R](Y)$ iff $\langle R^{\smile} \rangle(a) \subseteq Y$;
2. (a) $a \in [[R]](Y)$ iff $Y \subseteq \langle R^{\smile} \rangle(a)$; (b) $b \in [[R^{\smile}]](X)$ iff $X \subseteq \langle R \rangle(b)$;
3. (a) $[[R]](\emptyset) = A$, (b) $[[R^{\smile}]](\emptyset) = B$, (c) $\langle R^{\smile} \rangle(\emptyset) = \langle R \rangle(\emptyset) = \emptyset$, (d) $[R^{\smile}](A) = B$. (e) If R is onto then $\langle R^{\smile} \rangle(A) = B$ and $[R^{\smile}](\emptyset) = \emptyset$; (f) $[R](B) = A$. (g) If $R^{\smile}$ is onto then $\langle R \rangle(B) = A$ and $[R](\emptyset) = \emptyset$.
4. If X and Y are singletons, then (a) $\langle R^{\smile} \rangle(X) = [[R^{\smile}]](X)$; (b) $\langle R \rangle(Y) = [[R]](Y)$;
5. (a) If R is onto, $[R^{\smile}](X) \subseteq \langle R^{\smile} \rangle(X)$; (b) If $R^{\smile}$ is onto $[R](Y) \subseteq \langle R \rangle(Y)$;
6. If R is a functional relation then $[R](Y) = \langle R \rangle(Y)$.
7. If $R^{\smile}$ is a functional relation then $[R^{\smile}](X) = \langle R^{\smile} \rangle(X)$.

Proof. (**1**) (a) By definition $b \in [R^{\smile}](X)$ iff $\forall a(\langle a, b \rangle \in R \implies a \in X)$ iff $\langle R \rangle(b) \subseteq X$.(b) By symmetry. (**2**) from (1) by swapping the position of the

---

[10] As the reader will probably note, the operations we denote with $\langle R^{\smile} \rangle(X)$ and $\langle R \rangle(Y)$ are often denoted with $R(X)$ and, respectively, $R^{\smile}(Y)$. Moreover, the left composition of R with Q is usually denoted with $R; Q$ in mathematical literature. However, there are several reasons which suggest to adopt the following symbols, mostly depending on both logic and relational algebra. In particular logical reasons are related to Kripke models for Modal Logics and, as to the left composition, to Linear Logic. Apart from symbols, $\langle R \rangle(Y)$ coincides with what in [26] is called the "*extension*" of Y along $R^{\smile}$ and $[R](Y)$ the "*restriction*" of Y along $R^{\smile}$. In *Formal Concept Analysis* $[[R]](Y)$ is the "(derived) extent" of Y, while $[[R^{\smile}]](X)$ is called the "(derived) intent" of X. The terminology used here is that of Relation Algebra and connected topics, (strictly speaking, residuals are defined provided X and Y are right ideal elements - see for instance [16]).

relations $\in$ and R. (**3**) (a), (b) and (c) are obvious. (d) For any $b \in B$, either $\langle a, b \rangle \in R$ for some $a \in A$ or the premise of the implication defining the operator $[R^\smile]$ is false. (e) If R is surjective then for all $a \in A$ there is a $b \in B$ such that $\langle a, b \rangle \in R$. Moreover, in $[R^\smile](\emptyset)$ the consequence is always false. Similar proofs for (f) and (g). (**4**) Applied on singletons the definitions of $[[\alpha]]$ and $\langle \alpha \rangle$ operators trivially coincide, for $\alpha = R$ or $\alpha = R^\smile$. (**5**) For all $b \in B, b \in [R^\smile](X)$ iff $\langle R \rangle(b) \subseteq X$ iff (for isotonicity of $\langle R^\smile \rangle$) $\langle R^\smile \rangle(\langle R \rangle(b)) \subseteq \langle R^\smile \rangle(X)$. But $b \in \langle R^\smile \rangle(\langle R \rangle(b))$. Hence $b \in \langle R \rangle(X)$. Symmetrically for $[R]$ and $\langle R \rangle$. (**6**) If R is a functional relation, by definition $R^\smile$ is onto, thus from point (5) $[R](Y) \subseteq \langle R \rangle(Y)$ for any $Y \subseteq B$. Suppose $x \in \langle R \rangle(Y)$ and $x \notin [R](Y)$. Then there is $y \in Y$ such that $\langle x, y \rangle \in R$ and there is a $y' \notin Y$ such that $\langle x, y \rangle \in R$ and $\langle x, y' \rangle \in R$. Hence R is not functional. (**7**) It is an instance of (6).

5.1   Solving the Divisor Inequalities

Now we come back for a while to the divisor diagram of *Definition* 5. Our instance of this diagram reads as in *Figure* 3. Thus we have to understand under what conditions we can have "best approximating" maps.

Therefore, suppose in general ϕ is a function which maps subsets of a set A into subsets of a set B (possibly in dependence on how the elements of A are related *via* a binary relation $R \subseteq A \times B$ with the members of B). If $\phi(X) \supseteq Y$ we can say that X approximates Y from above *via* ϕ. The smallest of these X can therefore be thought of as a "best approximation from above" *via* ϕ, for its image is the closest to Y.

In order to get such a best approximation, if any, we should take $\bigcap \phi^\leftarrow(\uparrow Y)$, where $\uparrow Y = \{Y' \subseteq B : Y' \supseteq Y\}$. In fact $\bigcap \phi^\leftarrow(\uparrow Y) = \bigcap\{X : \phi(X) \supseteq Y\}$.

Dually, if we take $\bigcup \phi^\leftarrow(\downarrow Y)$, where $\downarrow Y = \{Y' \subseteq B : Y' \subseteq Y\}$, we should obtain a "best approximation from below" of Y, if any, *via* ϕ, because $\bigcup \phi^{\leftarrow 1}(\downarrow Y) = \bigcup\{X : \phi(X) \subseteq Y\}$.

To be sure, this approach is successful if $\phi(\bigcap \phi^\leftarrow(\uparrow Y)) \supseteq Y$ and, dually, $\phi(\bigcup \phi^\leftarrow(\downarrow Y)) \subseteq Y$. So we now shall examine, in an abstract setting, the conditions under which the above operations are admissible and behave as expected.

Indeed, we have a mathematical result which states rigorously these informal intuitions[11].

---

[11] We remind that in a preordered set **O**:

(a) $\uparrow X = \{y : \exists x(x \in X \ \& \ x \preccurlyeq y)\} = \langle \preccurlyeq^\smile \rangle(X)$ is called the *order filter* generated by X. In particular $\forall p \in A, \uparrow p = \uparrow \{p\}$ is called the *principal order filter generated by p*. If **O** is partially ordered $p = min(\uparrow p)$, where, given a set X, $min(X)$ is the *minimum* element of X.

(b) $\downarrow X = \{y : \exists x(x \in X \ \& \ y \preccurlyeq x)\} = \langle \preccurlyeq \rangle(X)$ is called the *order ideal* generated by X. In particular, $\forall p \in A, \downarrow p = \downarrow \{p\}$ is called the *principal order ideal generated by p*. If **O** is partially ordered $p = max(\downarrow p)$, where, given a set X, $max(X)$ is the *maximum* element of X.

From now on $\mathbf{O} = \langle A, \leq \rangle$ and $\mathbf{O}' = \langle A', \leq' \rangle$ will denote preordered or partial ordered sets. Furthermore, with $\mathbf{L}$ and $\mathbf{L}'$ we shall denote two arbitrary complete and bounded lattices $\mathbf{L} = \langle L, \vee, \wedge, 0, 1 \rangle$ and, respectively, $\mathbf{L}' = \langle L', \vee', \wedge', 0', 1' \rangle$.

Proposition 1. *Let* **A** *and* **B** *be partially ordered sets,* ϕ *a functor (i. e. an isotone map) between* **A** *and* **B**. *Then, the following conditions are equivalent:*

1. *(a) there exists a functor* $\psi : \mathbf{B} \longmapsto \mathbf{A}$ *such that* $\phi \circ \psi \geq_{\mathbf{A}} 1_{\mathbf{A}}$ *and* $\psi \circ \phi \leq_{\mathbf{B}} 1_{\mathbf{B}}$;
   (a') for all $b \in B, \phi^{\leftarrow}(\downarrow b)$ *is a principal order ideal of* **A**.
2. *(b) there exists a functor* $\vartheta : \mathbf{A} \longmapsto \mathbf{B}$ *such that* $\vartheta \circ \phi \geq_{\mathbf{B}} 1_{\mathbf{B}}$ *and* $\phi \circ \vartheta \leq_{\mathbf{A}} 1_{\mathbf{A}}$;
   (b') for all $a \in A, \vartheta^{\leftarrow}(\uparrow a)$ *is a principal filter of* B.

The proof can be found in [5].

5.2 Galois Adjunctions and Galois Connections

The conditions stated in *Proposition* 1 define a basic mathematical notion which is at the core of our construction (NOTE: the following materials are known and we have included them to render completeness to this paper).

Definition 8. *Let* $\sigma : \mathbf{O} \longmapsto \mathbf{O}'$ *and* $\iota : \mathbf{O}' \longmapsto \mathbf{O}$ *be two maps between partial ordered sets. Then we say that* ι *and* σ *fulfill an adjointness relation if the following holds:*

$$\forall p \in O, \forall p' \in O', \iota(p') \leq p \ \textit{if and only if} \ p' \leq' \sigma(p) \tag{4}$$

If the above conditions hold, then σ *is called the* upper adjoint *of* ι *and* ι *is called the* lower adjoint *of* σ. *This fact is denoted by*

$$\mathbf{O}' \dashv^{\iota,\sigma} \mathbf{O} \tag{5}$$

and we shall say that the two maps form an adjunction *between* **O** *and* **O**'. *If the two preorders are understood we shall denote it with* $\iota \dashv \sigma$, *too*[12].

When an adjointness relationship holds between two preordered structures we say that the pair $\langle \sigma, \iota \rangle$ forms a *Galois adjunction* or an *axiality*. This name is after the notion of a *Galois connection* which is defined by means of a similar but covariant condition where, indeed, ι and σ are antitone:

$$\forall p \in O, \forall p' \in O', \iota(p) \geq' p' \ \textit{if and only if} \ p \leq \sigma(p') \tag{6}$$

We read this fact by saying that the pair $\langle \sigma, \iota \rangle$ forms a *Galois connection* or a *polarity*. Clearly, a Galois connection is a Galois adjoint with the right category **O** turned into its opposite $\mathbf{O}^{op}$. In other words, $\langle \sigma, \iota \rangle$ is a polarity if and only if $\mathbf{O}' \dashv^{\iota,\sigma} \mathbf{O}^{op}$.

---

[12] Sometimes in mathematical literature, the lower adjoint is called "left adjoint" and the upper adjoint is called "right adjoint". However, the reader must take care of the fact that we have two levels of duality. The first swaps the partial order ($\leq$ into $\geq$ and vice-versa). The second swaps the order of application of the functors ($\iota \circ \sigma$ into $\sigma \circ \iota$, and the other way around) and the position of the two structures (by the way, we notice that in usual literature given a map ϕ, the upper residual is denoted with ϕ_* and the lower residual is denoted with ϕ^*).

We now state without proof a number of properties fulfilled by adjoint maps.

Proposition 2. *Let* $\sigma : \mathbf{O} \longmapsto \mathbf{O}'$ *and* $\iota : \mathbf{O}' \longmapsto \mathbf{O}$ *be mappings,* $p \in O$ *and* $p' \in O'$. *Then,*
(a) *the following statements are equivalent:*
    (a.1) $\mathbf{O}' \dashv^{\iota,\sigma} \mathbf{O}$;
    (a.2) $\sigma\iota(p') \geq' p'$ *and* $\iota\sigma(p) \leq p$, *and both* ι *and* σ *are isotone;*
    (a.3) σ *is isotone and* $\iota(p') = min(\sigma^{\leftarrow}(\uparrow p'))$;
    (a.4) ι *is isotone and* $\sigma(p) = max(\iota^{\leftarrow}(\downarrow p))$;
If $\mathbf{O}' \dashv^{\iota,\sigma} \mathbf{O}$, *then:*
(b) σ *preserves all the existing* infs *and* ι *preserves all the existing* sups;
(c) $\iota = \iota\sigma\iota$, $\sigma = \sigma\iota\sigma$;
(d) $\sigma\iota$ *and* $\iota\sigma$ *are idempotent.*
(e) σ *is surjective iff* $\iota(p') = min(\sigma^{\leftarrow}(\{p'\}))$ *iff* $\sigma\iota(p') = p'$ *iff* ι *is injective.*
(f) σ *is injective iff* $\sigma(p) = max(\iota^{\leftarrow}(\{p\}))$ *iff* $\iota\sigma(p) = p$ *iff* ι *is surjective.*

Notice that (e) and (f) are the reader's digest of the story about retraction and coretraction we told in Section 2.1.

Now we have a good stock of results in order to "implement" a sufficiently large body of useful operators, actually those operators which will constitute the backbone of all the present story.

Definition 9. *Let* $\phi : \mathbf{O} \mapsto \mathbf{O}$ *be an operator on a partially ordered set and* $\vartheta : \mathbf{L} \mapsto \mathbf{L}'$ *be an operator between two lattices. Then,*
(1) ϕ *is a* projection *operator on* $\mathbf{O}$ *iff it is isotone and idempotent;*
(2) a projection operator is a closure *operator iff it is increasing;*
(3) a projection operator is an interior *operator iff it is decreasing;*
(4) ϑ *is a* modal *operator iff it is normal (i. e.* $\vartheta(0) = 0'$*) and additive;*
(5) a closure operator ϑ *on a lattice is* topological *iff it is modal;*
(6) ϑ *is a* co-modal *operator iff it is co-normal (i. e.* $\vartheta(1) = 1'$*) and multiplicative;*
(7) an interior operator ϑ *on a lattice is* topological *iff it is co-modal;*
(8) ϑ *is an* anti-modal *operator iff it is anti-normal (i. e.* $\vartheta(0) = 1'$*) and anti-additive (i. e.* $\vartheta(x \vee y) = \vartheta(x) \wedge' \vartheta(y)$*).*

Notice that in our definition of modal operators we do not require $\mathbf{L} = \mathbf{L}'$.

Then from *Proposition* 2 we immediately obtain:

Corollary 1. *Let* $\mathbf{O}' \dashv^{\iota,\sigma} \mathbf{O}$ *and* $\mathbf{O}' \dashv^{\varepsilon,\varsigma} \mathbf{O}^{op}$ *hold (hence the latter is a Galois connection between* $\mathbf{O}$ *and* $\mathbf{O}'$*). Then,*

1. (a) $\sigma\iota$ *is a closure operator on* $\mathbf{O}'$; *(b)* $\iota\sigma$ *is an interior operator on* $\mathbf{O}$.
2. (a) $\varsigma\varepsilon$ *is a closure operator on* $\mathbf{O}'$; *(b)* $\varepsilon\varsigma$ *is a closure operator on* $\mathbf{O}$.

It is worth underlining that none of these operators needs to be topological.

Moreover given the above adjointness situations we can underline what follows:

(m1) σ is half of a co-modal operator: it lacks co-normality;

(m2) ι is half of a modal operator: it lacks normality;

(m3) ϵ and ς are half of an anti-modal operator: they lack anti-normality.

The lack of properties concerning preservation of operations may be partially amended when we restrict domains to the families of fixed points of the operators $\sigma\iota, \iota\sigma, \epsilon\varsigma$ and $\varsigma\epsilon$.

To this end the following two results are fundamental:

Lemma 2. *Let* $\phi : \mathbf{O} \longmapsto \mathbf{O}$ *be a map. Then,*
(a) *if* ϕ *is closure then* $\mathbf{O} \dashv^{\phi^o, \phi_o} Im_\phi$; (b) *if* ϕ *is interior then* $Im_\phi \dashv^{\phi_o, \phi^o} \mathbf{O}$.

Corollary 2. *Let* $\phi : \mathbf{L} \longmapsto \mathbf{L}$ *be a map. Then,*

1. *if* ϕ *is closure then* ϕ^o *is additive,* $\phi_o(Im_\phi)$ *is closed under infs, and* $\mathbf{Sat}_\phi(L) = \langle Im_\phi, \wedge, \sqcup, 1 \rangle$, *where for all* $x, y \in Im_\phi$, $x \sqcup y = \phi(x \vee y)$, *is a lattice;*

2. *if* ϕ *is interior then* ϕ^o *is multiplicative,* $\phi_o(Im_\phi)$ *is closed under sups, and* $\mathbf{Sat}_\phi(L) = \langle Im_\phi, \sqcap, \vee, 0 \rangle$, *where for all* $x, y \in Im_\phi$, $x \sqcap y = \phi(x \wedge y)$, *is a lattice.*

We want to point out that if ϕ is closure then sups in $\mathbf{L}$ and sups in Im_ϕ may differ. Hence, although for all $x \in L, \phi(x) = \phi^o(x)$ and ϕ^o is additive, nonetheless ϕ in general is not sup-preserving so that $\phi_o(Im_\phi)$ is not closed under sups (dually if ϕ is interior).

These results give the following proposition (where we have just to notice that turning $\mathbf{L}^{op}$ upside-down interiors turns into closures, sups into infs and viceversa):

Proposition 3. *Let* $\mathbf{L}' \dashv^{\iota, \sigma} \mathbf{L}$ *and* $\mathbf{L}' \dashv^{\epsilon, \varsigma} \mathbf{L}^{op}$ *hold. Then:*

1. $\mathbf{Sat}_{\iota\sigma}(\mathbf{L}) = \langle Im_{\iota\sigma}, \sqcap, \vee, 0 \rangle$, *where for all* $x, y \in Im_{\iota\sigma}$, $x \sqcap y = \iota\sigma(x \wedge y)$, *is a lattice;*

2. $\mathbf{Sat}_{\sigma\iota}(\mathbf{L}') = \langle Im_{\sigma\iota}, \wedge', \sqcup, 1' \rangle$, *where for all* $x, y \in Im_{\sigma\iota}$, $x \sqcup y = \sigma\iota(x \vee' y)$, *is a lattice;*

3. $\mathbf{Sat}_{\varsigma\epsilon}(\mathbf{L}') = \langle Im_{\varsigma\epsilon}, \wedge', \sqcup, 1' \rangle$, *where for all* $x, y \in Im_{\varsigma\epsilon}$, $x \sqcup y = \varsigma\epsilon(x \vee' y)$, *is a lattice;*

4. $\mathbf{Sat}_{\epsilon\varsigma}(\mathbf{L}) = \langle Im_{\epsilon\varsigma}, \wedge, \sqcup, 1 \rangle$, *where for all* $x, y \in Im_{\epsilon\varsigma}$, $x \sqcup y = \epsilon\varsigma(x \vee y)$, *is a lattice.*

6   Formal Operators on Points and on Observables

Now let us come back to our basic constructors.

Proposition 4 (Fundamental relationships). *Let A and B be two sets, $X \subseteq A, Y \subseteq B, R \subseteq A \times B$ a relation and $\hat{f} \subseteq A \times B$ a functional relation. Then the following holds:*

1. (a) $\langle R \rangle(Y) \subseteq X$ *iff* $Y \subseteq [R^\smile](X)$; (b) $\langle R^\smile \rangle(X) \subseteq Y$ *iff* $X \subseteq [R](Y)$.
2. $Y \subseteq [[R^\smile]](X)$ *iff* $X \subseteq [[R]](Y)$;

3. $\langle \hat{f}^{\smile} \rangle(X) \subseteq Y$ iff $X \subseteq \langle \hat{f} \rangle(Y)$;
4. The operators $\langle R \rangle$, $\langle R^{\smile} \rangle$, $[R]$ and $[R^{\smile}]$ are isotone; $[[R]]$ and $[[R^{\smile}]]$ are antitone.

Proof. (1) (a) $\langle R \rangle(Y) \subseteq X$ iff $R^{\smile}(y) \subseteq X, \forall y \in Y$ iff (from *Lemma* 1.(1)) $Y \subseteq [R^{\smile}](X)$. (b) By symmetry. (2) $X \subseteq [[R]](Y)$ iff $\forall x \in X(Y \subseteq R(x))$ (from *Lemma* 1.(2)), iff $\forall x \in X, \forall y \in Y(y \in R(x))$ iff $\forall y \in Y(X \subseteq R^{\smile}(y))$ iff $\forall y \in Y(y \in [[R^{\smile}]](X))$ iff $Y \subseteq [[R^{\smile}]](X)$ (in view of *Lemma* 1.(1).(b)). (3) Directly from *Proposition* 1. (4) Easily from the position of the subformula "$y \in Y$" and "$x \in X$" in the definitions.

From the above discussion we trivially have:

$$\langle \Vdash^{\smile} \rangle = \langle i \rangle \Big| \langle \Vdash \rangle = \langle e \rangle \Big| [\Vdash^{\smile}] = [i] \Big| [\Vdash] = [e] \Big| [[\Vdash^{\smile}]] = [[i]] \Big| [[\Vdash]] = [[e]]$$

Therefore it is clear that if given a *P-system*, **P**, we set $\mathbf{M} = \langle \wp(M), \subseteq \rangle$ and $\mathbf{G} = \langle \wp(G), \subseteq \rangle$, in view of *Proposition* 4 the following adjointness properties hold:

(a) $\mathbf{M} \dashv^{\langle e \rangle, [i]} \mathbf{G}$; (b) $\mathbf{G} \dashv^{\langle i \rangle, [e]} \mathbf{M}$; (c) $\mathbf{M} \dashv^{[[e]], [[i]]} \mathbf{G}^{\mathbf{op}}$; (d) $\mathbf{G} \dashv^{[[i]], [[e]]} \mathbf{M}^{\mathbf{op}}$.

The lack of properties involving top and bottom elements, such as "normality" and "co-normality", for generic adjoint functions, is quite obvious since they depend on the adjoint structures. But in the case of the basic constructors *Lemma* 1.(3) and *Proposition* 2 immediately prove that $\langle e \rangle$ and $\langle i \rangle$ are modal operators, $[e]$ and $[i]$ are co-modal operators and, finally, $[[e]]$ and $[[i]]$ are anti-modal operators.

Moreover, in view of these adjunction properties, some sequences of constructors with alternate decorations provide a number of useful operators on $\wp(G)$ and $\wp(M)$. Indeed axiality says that if one operator lowers an element then its conjugate operator lifts it, and vice-versa, so that by combining them either we obtain the maximum of the lowering elements or the minimum of the lifting elements of a given argument.

Definition 10. *Let $\langle G, M, \Vdash \rangle$ be a P-system. Then:*

- $int : \wp(G) \longmapsto \wp(G); int(X) = \langle e \rangle([i](X))$.
- $cl : \wp(G) \longmapsto \wp(G); cl(X) = [e](\langle i \rangle(X))$.
- $est : \wp(G) \longmapsto \wp(G); est(X) = [[e]]([[i]](X))$.
- $\mathcal{A} : \wp(M) \longmapsto \wp(M); \mathcal{A}(Y) = [i](\langle e \rangle(Y))$.
- $\mathcal{C} : \wp(M) \longmapsto \wp(M); \mathcal{C}(Y) = \langle i \rangle([e](Y))$.
- $\mathcal{ITS} : \wp(M) \longmapsto \wp(M); \mathcal{ITS}(Y) = [[i]]([[e]](Y))$.

The above operators inherit 'd', 'o', 's', 'sd', 'os', 'od' and 'ods' reciprocal relationships from the outermost constructors which define them.

Proposition 5. *In any P-system $\langle G, M, \Vdash \rangle$, for any $X \subseteq G, Y \subseteq M, g \in G, m \in M$:*

1. (a) $m \in \mathcal{A}(Y)$ iff $\langle e \rangle(m) \subseteq \langle e \rangle(Y)$, (b) $g \in cl(X)$ iff $\langle i \rangle(g) \subseteq \langle i \rangle(X)$;
2. (a) $g \in int(X)$ iff $\langle i \rangle(g) \cap [i](X) \neq \emptyset$, (b) $m \in \mathcal{C}(Y)$ iff $\langle e \rangle(m) \cap [e](Y) \neq \emptyset$;
3. (a) $g \in est(X)$ iff $[[i]](X) \subseteq \langle i \rangle(g)$, (b) $m \in \mathcal{ITS}(Y)$ iff $[[e]](Y) \subseteq \langle e \rangle(m)$.

Proof. (**1**) (a) By definition $m \in \mathcal{A}(Y)$ iff $m \in [i]\,(\langle e\rangle(Y))$. Hence from *Lemma* 1.(1), $m \in \mathcal{A}(Y)$ iff $\langle e\rangle(m) \subseteq \langle e\rangle(Y)$. (b) By symmetry. (**2**) (a) $g \in int(X)$ iff $g \in \langle e\rangle([i]\,(X))$ iff $g \in \langle e\rangle(\{m : \langle e\rangle(m) \subseteq X\})$, iff $\langle i\rangle(g) \cap \{m : \langle e\rangle(m) \subseteq X\} \neq \emptyset$, iff $\langle i\rangle(g) \cap [i]\,(X) \neq \emptyset$. (b) By symmetry. (**3**) (a) Directly from *Lemma* 1.(2) and the definition of "*est*". (b) By symmetry[13].

Therefore, $g \in est(X)$ if and only if g fulfills *at least all* the properties that are shared *by all* the elements of X. In this sense $est(X)$ is the *extent* of the set of properties that characterises X as a whole. Symmetrically, $m \in \mathcal{ITS}(Y)$ if and only if m is fulfilled by *at least all* the objects that enjoy *all* the properties from Y. In this sense $\mathcal{ITS}(Y)$ is the *intent* of the set of objects that are characterised by Y as a whole.

In order to understand the meaning of the other operators, let us notice that the elements of M can be interpreted as "*formal neighborhoods*"[14]. In fact, in topological terms a neighborhood of a point x is a collection of points that are linked with x by means of some *nearness* relation. For a member m of M is associated, *via* $\langle e\rangle$ with a subset X of G, m may be intended as a 'proxy' of X itself. Thus if X is a *concrete* neighborhood of a point x, then m may be intended as a *formal* neighborhood of x, on the basis of the observation that the nearness relation represented by X states that two points are close to each other if they both fulfill property m[15]. It follows that $obs(g)$ is the family of formal neighborhoods of g (symmetrically for $sub(m)$ we have the concrete neighborhoods of m). This is the intuitive content of the following gift of the adjointness relationships between basic constructors:

Interior operators	$int, \mathcal{C}$
Closure operators	$cl, \mathcal{A}, est, \mathcal{ITS}$

In view of the observation after *Corollary* 1 one easily notices that none of the above operators needs to be topological.

6.1   Fundamental Properties of the Formal Perception Operators

Definition 11. *Let* **P** *be any P-system. Then we define the following families of fixpoints of the operators induced by* **P**:

1. $\Omega_{int}(\mathbf{P}) = \{X \subseteq G : int(X) = X\}$; $\Gamma_{cl}(\mathbf{P}) = \{X \subseteq G : cl(X) = X\}$;
2. $\Gamma_{est}(\mathbf{P}) = \{X \subseteq G : est(X) = X\}$; $\Omega_{\mathcal{A}}(\mathbf{P}) = \{Y \subseteq M : \mathcal{A}(Y) = Y\}$;
3. $\Gamma_{\mathcal{C}}(\mathbf{P}) = \{Y \subseteq M : \mathcal{C}(Y) = Y\}$; $\Gamma_{\mathcal{ITS}}(\mathbf{P}) = \{Y \subseteq M : \mathcal{ITS}(Y) = Y\}$.

---

[13] Moreover, $[[i]](X) = \{m : X \subseteq \langle e\rangle(m)\} = \{m : \forall x \in X(x \Vdash m)\}$. Henceforth $[[i]](X) \subseteq \langle i\rangle(g)$ iff $g \Vdash m$ for all m such that $x \Vdash m$, for any member x of X, that is, iff $\forall m \in M((\forall x \in X(x \Vdash m)) \Longrightarrow g \Vdash m)$.

[14] Indeed, this is the framework in which the operators cl, int, $\mathcal{A}$ and $\mathcal{C}$ are introduced, although not by means of adjointness properties, by the Padua School of Formal Topology (see [26]).

[15] This interpretation is close to the approach of [11].

It is understood that the partial order between saturated subsets is inherited from the category they are derived from. Thus, for instance, we shall have $\langle \Gamma_{\mathcal{C}}(M), \subseteq \rangle$.

Proposition 6. *Let* **P** *be a P-system. Then the following are complete lattices:*

1. $\mathbf{Sat}_{int}(\mathbf{P}) = \langle \Omega_{int}(\mathbf{P}), \cup, \wedge, \emptyset, G \rangle$, *where* $\bigwedge_{i \in I} X_i = int(\bigcap_{i \in I} X_i)$;
2. $\mathbf{Sat}_{\mathcal{A}}(\mathbf{P}) = \langle \Omega_{\mathcal{A}}(\mathbf{P}), \vee, \cap, \emptyset, M \rangle$, *where* $\bigvee_{i \in I} Y_i = \mathcal{A}(\bigcup_{i \in I} Y_i)$;
3. $\mathbf{Sat}_{cl}(\mathbf{P}) = \langle \Gamma_{cl}(\mathbf{P}), \vee, \cap, \emptyset, G \rangle$, *where* $\bigvee_{i \in I} X_i = cl(\bigcup_{i \in I} X_i)$;
4. $\mathbf{Sat}_{\mathcal{C}}(\mathbf{P}) = \langle \Gamma_{\mathcal{C}}(\mathbf{P}), \cup, \wedge, \emptyset, M \rangle$, *where* $\bigwedge_{i \in I} Y_i = \mathcal{C}(\bigcap_{i \in I} Y_i)$;
5. $\mathbf{Sat}_{est}(\mathbf{P}) = \langle \Gamma_{est}(\mathbf{P}), \cap, \vee, est(\emptyset), G \rangle$, *where* $\bigvee_{i \in I} X_i = est(\bigcup_{i \in I} X_i)$;
6. $\mathbf{Sat}_{\mathcal{ITS}}(\mathbf{P}) = \langle \Gamma_{\mathcal{ITS}}(\mathbf{P}), \cap, \vee, \mathcal{ITS}(\emptyset), M \rangle$, *where* $\bigvee_{i \in I} Y_i = \mathcal{ITS}(\bigcup_{i \in I} Y_i)$.

Proof. Much work has already been done in *Proposition 3*. We just need to justify the choice of top and bottom elements. To this end, remember that in any *P-system* both $\Vdash$ and $\Vdash^{\smile}$ are onto. Hence in view of *Lemma 1.(3)*. $int(G) = \langle e \rangle [i](G) = \langle \Vdash] [\Vdash^{\smile}](G) = \langle \Vdash \rangle (M) = G$, and analogously for the other operators. The only difference is for $\mathcal{ITS}$ and est because $[[\Vdash]](\emptyset) = G$ but $[[\Vdash^{\smile}]](G) = \{m : \langle \Vdash \rangle (m) = G\} \supseteq \emptyset$, dually for $[[\Vdash]][[\Vdash^{\smile}]](\emptyset)$.

Lemma 3. *Let* **P** *be a P-system. Then for all* $X \subseteq G, Y \subseteq M$,

$X \in \Omega_{int}(\mathbf{P})$ iff $X = \langle e \rangle(Y')$	$X \in \Gamma_{cl}(\mathbf{P})$ iff $X = [e](Y')$	$X \in \Gamma_{est}(\mathbf{P})$ iff $X = [[e]](Y')$
$Y \in \Omega_{\mathcal{A}}(\mathbf{P})$ iff $Y = [i](X')$	$Y \in \Gamma_{\mathcal{C}}(\mathbf{P})$ iff $Y = \langle i \rangle(X')$	$Y \in \Gamma_{\mathcal{ITS}}(\mathbf{P})$ iff $Y = [[i]](X')$

for some $Y' \subseteq M, X' \subseteq G$.

Proof. If $X = \langle e \rangle(Y')$ then $X = \langle e \rangle [i] \langle e \rangle (Y')$, from *Proposition 2.(c)*. Therefore, by definition of int, $X = int(\langle e \rangle(Y')) = int(X)$. Vice-versa, if $X = int(X)$, then $X = \langle e \rangle [i](X)$. Hence, $X = \langle e \rangle(Y')$ for $Y' = [i](X)$. The other cases are proved in the same way, by exploiting the appropriate equations of *Proposition 2.(c)*.

Corollary 3. *Let* **P** *be a P-system. Then the following are isomorphisms:*

1. (a) $\langle e \rangle : \mathbf{Sat}_{\mathcal{A}}(\mathbf{P}) \longmapsto \mathbf{Sat}_{int}(\mathbf{P})$; (b) $[i] : \mathbf{Sat}_{int}(\mathbf{P}) \longmapsto \mathbf{Sat}_{\mathcal{A}}(\mathbf{P})$;
2. (a) $[e] : \mathbf{Sat}_{\mathcal{C}}(\mathbf{P}) \longmapsto \mathbf{Sat}_{cl}(\mathbf{P})$; (b) $\langle i \rangle : \mathbf{Sat}_{cl}(\mathbf{P}) \longmapsto \mathbf{Sat}_{\mathcal{C}}(\mathbf{P})$;
3. *The following are anti-isomorphisms (where* $-$ *is the set-theoretical complementation):*
   (a) $[[i]] : \mathbf{Sat}_{est}(\mathbf{P}) \longmapsto \mathbf{Sat}_{\mathcal{ITS}}(\mathbf{P})$; $[[e]] : \mathbf{Sat}_{\mathcal{ITS}}(\mathbf{P}) \longmapsto \mathbf{Sat}_{est}(\mathbf{P})$;
   (b) $- : \mathbf{Sat}_{cl}(\mathbf{P}) \longmapsto \mathbf{Sat}_{int}(\mathbf{P})$; $- : \mathbf{Sat}_{\mathcal{C}}(\mathbf{P}) \longmapsto \mathbf{Sat}_{\mathcal{A}}(\mathbf{P})$.

Proof. Let us notice, at once, that the proof for an operator requires the proof for its adjoint operator. Then, let us prove **(1).(a)** and (b) together: First, let us prove bijection for $\langle e \rangle$ and $[i]$. From *Lemma 3* the codomain of $\langle e \rangle$ is $\Omega_{int}(\mathbf{P})$ and the codomain of $[i]$ is $\Omega_{\mathcal{A}}(\mathbf{P})$. Moreover, for all $X \in \Omega_{int}(\mathbf{P})$, $X = \langle e \rangle [i](X)$ and for all $Y \in \Omega_{\mathcal{A}}(\mathbf{P})$, $Y = [i] \langle e \rangle (Y)$. From the adjointness properties we have:

(i) $\langle e \rangle$ is surjective on $\Omega_{int}(\mathbf{P})$ and (ii) $[i]$ is injective from $\Omega_{int}(\mathbf{P})$.
(iii) $\langle e \rangle$ is injective from $\Omega_{\mathcal{A}}(\mathbf{P})$ and (iv) $[i]$ is surjective onto $\Omega_{\mathcal{A}}(\mathbf{P})$.

Moreover, if $[i]$ is restricted to $\Omega_{int}(\mathbf{P})$, then its codomain is the set $H = \{Y : Y = [i](X)\ \&\ X \in \Omega_{int}(\mathbf{P})\}$. Clearly, $H \subseteq \Omega_{\mathcal{A}}(\mathbf{P})$. In turn, if $\langle e \rangle$ is restricted to $\Omega_{\mathcal{A}}(\mathbf{P})$, then its codomain is the set $K = \{X : X = \langle e \rangle(Y)\ \&\ Y \in \Omega_{\mathcal{A}}(\mathbf{P})\}$. Clearly $K \subseteq \Omega_{int}(G)$. Therefore, (i) and (iii) give that $\langle e \rangle$ is bijective if restricted to $\Omega_{\mathcal{A}}(\mathbf{P})$, while (ii) and (iv) give that $[i]$ is a bijection whenever restricted to $\Omega_{int}(\mathbf{P})$[16].

Now we have to show that $\langle e \rangle$ and $[i]$ preserve joins and meets. For $\langle e \rangle$ we proceed as follows: (v) $\langle e \rangle(\bigvee_{i \in I}(\mathcal{A}(Y_i))) =_{def} \langle e \rangle(\mathcal{A}(\bigcup_{i \in I}(\mathcal{A}(Y_i)))$. But $\langle e \rangle \mathcal{A} = \langle e \rangle$, from *Proposition* 2.(c). Moreover, $\langle e \rangle$ distributes over unions. Hence the right side of (v) equals to $\bigcup_{i \in I}\langle e \rangle(\mathcal{A}(Y_i))$. But in view of *Proposition* 6, the union of extensional open subsets is concrete open and from *Lemma* 3 $\langle e \rangle(\mathcal{A}(Y_i))$ belongs indeed to $\in \Omega_{int}(\mathbf{P})$, so that the right side of (v) turns into $int(\bigcup_{i \in I}\langle e \rangle(\mathcal{A}(Y_i))) =_{def} \bigvee_{i \in I}\langle e \rangle(\mathcal{A}(Y_i))$.

(vi) $\langle e \rangle(\bigwedge_{i \in I}\mathcal{A}(Y_i)) = \langle e \rangle(\bigcap_{i \in I}[i]\langle e \rangle(Y_i))$. Since $[i]$ distributes over intersections, the right side of (vi) turns into $\langle e \rangle[i](\bigcap_{i \in I}\langle e \rangle(Y_i)) = int(\bigcap_{i \in I}\langle e \rangle(Y_i))$. But $\langle e \rangle = \langle e \rangle \mathcal{A}$, so that the last term is exactly $\bigwedge_{i \in I}\langle e \rangle(\mathcal{A}(Y_i))$. Since $[i]$ is the inverse of $\langle e \rangle$, *qua* isomorphism, we have that $[i]$ preserves meets and joins, too. As to (**2**) the results come by symmetry.

(**3**) (a) As in the above proof by noticing that in polarities the right structure is reversed upside-down (we can optimize a passage by noticing that $[[e]]$ and $[[i]]$ are both upper and lower adjoints). (b) By duality between the operators.

6.2   Pre-topological Approximation Spaces

Now we are in position to show how the above mathematical machinery may be used to generalise the upper and lower approximation operators provided by Rough Set Theory.

Given $X \subseteq G$ we know that $[e]\langle i \rangle(X) \supseteq X$ and $\langle e \rangle[i](X) \subseteq X$.
We can interpret these relationships by saying that

– cl is an *upper approximation* of the identity map on $\wp(G)$;
– int is a *lower approximation* of the identity map on $\wp(G)$.

More precisely, $\langle i \rangle(X) = min([e]^{\leftarrow}(\uparrow X)) = min\{X' \subseteq G : [e](X') \supseteq X\}$, it follows that $[e]\langle i \rangle(X)$ (i. e. cl) is the *best approximation from above* to X *via* function $[e]$.

Dually, $[i](X) = max(\langle e \rangle^{\leftarrow}(\downarrow X)) = max\{X' \subseteq G : \langle e \rangle(X') \subseteq X\}$. Hence, $\langle e \rangle[i](X)$ (i. e. int) is the *best approximation from below* to X, *via* function $\langle e \rangle$. Of course, if $\langle i \rangle$ is injective (or, equivalently, $[e]$ is surjective), then we can exactly

---

[16] As side results, we have: (i) $\Omega_{\mathcal{A}}(\mathbf{P}) = H$ and (ii) $\Omega_{int}(\mathbf{P}) = K$. This is not surprising, because if $Y \in \Omega_{\mathcal{A}}(\mathbf{P})$ then $Y = [i]\langle e \rangle(Z)$ for some $Z \subseteq M$ and $\langle e \rangle(Z) \in \Omega_{int}(\mathbf{P})$, any $Z \subseteq M$. Vice-versa, if $X \in \Omega_{int}(\mathbf{P})$, then $X = \langle e \rangle(Z)$. Hence $[i](X) = [i]\langle e \rangle(Z)$ belongs to $\Omega_{\mathcal{A}}(\mathbf{P})$. Symmetrically for (ii).

reach X from above by means of $[e]$. The element that must be mapped is, indeed, $\langle i \rangle(X)$. Dually, if $[i]$ is injective (or $\langle e \rangle$ is surjective), then we can exactly reach X from below by means of $\langle e \rangle$ applied to $[i](X)$.

7   Information, Concepts and Formal Operators

So far we have discussed a number of instruments that act on either abstraction sides we are dealing with, that is, points and properties. Indeed, the introduced operators are based on well-defined mathematical properties, such as adjointness, and feature proper informational and conceptual interpretations.

Also, the use of the terms "open" and "closed" is not an abuse because, on the contrary, these operators translate the usual topological definitions of an interior and, respectively, a closure of a set $X \subseteq G$, into the language of observation systems, provided the elements of M are interpreted as formal neighborhoods.

For instance, the usual definition tells us that for any subset X of G, a point a belongs to the interior of X if and only if there is a neighbourhood of a included in X. If the elements of the set M are intended as formal neighbourhoods, then the relation $a \Vdash b$ (hence, $a \in \langle e \rangle(b)$) says that b is a formal neighborhood of a and $\langle e \rangle(b) \subseteq X$ says that the extension of this neighbourhood b is included in X. But this is precisely a reading of $a \in \langle e \rangle[i](X)$, because in view of the adjunction properties, $\langle e \rangle(b) \subseteq X$ if and only if $b \in [i](X)$.

Thus we have made a further step beyond M. Smyth's seminal observation that semi-decidable properties are analogous to open sets in a topological space, with the aid of the interpretation of basic constructors elaborated by the Padua School on Formal Topology[17].

Moreover, we have seen that int and cl provide us with lower and upper approximations of any set $X \subseteq G$.

But are we really happy with this machinery? The answer is "yes and no". Yes, for we have found a mathematically sound way to deal with approximations which enjoy a reliable intuitive interpretation. No, for both int and cl are discontinuous (non topological) operators because int is not multiplicative and cl is not additive, so that we have to face "jumps" which can be too wide and make us miss information.

EXAMPLE 1
Here we give an example of a *P-system* and its induced operators:

$\Vdash$	b	b_1	b_2	b_3
a	1	1	0	0
a_1	0	1	0	1
a_2	0	1	1	1
a_3	0	0	0	1

---

[17] See for instance [25] and [27].

Let us try and compute some instances of basic formal operators on system **P**:

1) Extensional operators:
$int(\{a, a_1\}) = \langle e \rangle [i](\{a, a_1\}) = \langle e \rangle(\{b\}) = \{a\}; \; int(\{a_2\}) = \langle e \rangle(\{b_2\}) = \{a_2\}.$
$cl(\{a, a_1\}) = [e] \langle i \rangle(\{a, a_1\}) = [e](\{b, b_1, b_3\}) = \{a, a_1, a_3\}; \; cl(\{a, a_2\}) = [e](M) = G.$
$est(\{b_1, b_2\}) = [[i]][[e]](\{b_1, b_2\}) = [[i]](\{a_2\}) = \{b_1, b_2, b_3\}.$
$int(int(\{a, a_1\})) = int(\{a\}) = \langle e \rangle [i](\{a\}) = \langle e \rangle(\{b\}) = \{a\}.$
$cl(cl(\{a, a_1\})) = cl(\{a, a_1, a_3\}) = [e] \langle i \rangle(\{a, a_1, a_3\}) = [e](\{b, b_1, b_3\}) = \{a, a_1, a_3\}.$
$est(est(\{b_1, b_2\})) = est(\{b_1, b_2, b_3\}) = [[i]][[e]](\{b_1, b_2, b_3\}) = [[i]](\{a_2\}) = \{b_1, b_2, b_3\}.$
Thus, this is also an example of the fact that int is **decreasing** while cl and est are **increasing** and all of them are **idempotent**. Moreover, one can see that $int(\{a, a_1\}) \cup int(\{a_2\}) = \{a, a_2\} \subseteq \{a, a_1, a_2\} = int(\{a, a_1\} \cup \{a_2\})$ and $cl(\{a, a_1\}) \cap cl(\{a, a_2\}) = \{a, a_1, a_2\} \supseteq \{a\} = cl(\{a, a_1\} \cap \{a, a_2\}).$

2) Intensional operators:
$\mathcal{A}(\{b, b_1\}) = [i] \langle e \rangle(\{b, b_2\}) = [i](\{a, a_1, a_2\}) = \{b, b_1, b_2\}.$
$\mathcal{C}(\{b_2, b_3\}) = \langle i \rangle [e](\{b_2, b_3\}) = \langle i \rangle(\{a_3\}) = \{b_3\}.$
$\mathcal{ITS}(\{b_1, b_2\}) = [[i]][[e]](\{b_1, b_2\}) = [[i]](\{a_2\}) = \{b_1, b_2, b_3\}.$
$\mathcal{A}(\mathcal{A}(\{b, b_1\})) = \mathcal{A}(\{b, b_1, b_2\}) = [i] \langle e \rangle(\{b, b_1, b_2\}) = [i](\{a, a_1, a_2\}) = \{b, b_1, b_2\}.$
$\mathcal{C}(\mathcal{C}(\{b_2, b_3\})) = \mathcal{C}(\{b_3\}) = \langle i \rangle [e](\{b_3\}) = \langle i \rangle(\{a_3\}) = \{b_3\}.$
$\mathcal{ITS}(\mathcal{ITS}(\{b_1, b_2\})) = \mathcal{ITS}(\{b_3\}) = [[i]][[e]](\{b_1, b_2, b_3\}) = [[i]](\{a_2\}) = \{b_1, b_2, b_3\}.$
Thus, this is also an example of the fact that $\mathcal{C}$ is **decreasing** while $\mathcal{A}$ and $\mathcal{ITS}$ are **increasing** and all of them are **idempotent**.

Let us now visualise the lattices of saturated sets:

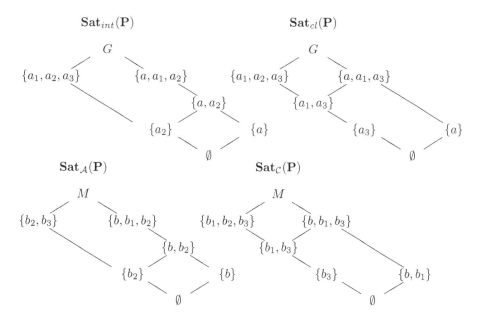

$\mathbf{Sat}_{\mathcal{ITS}}(\mathbf{P})$ $\qquad\qquad\qquad$ $\mathbf{Sat}_{est}(\mathbf{P})$

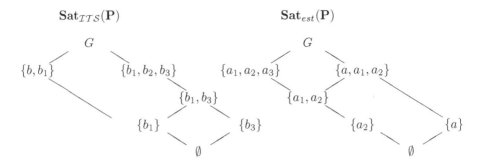

Pay attention that $\mathbf{Sat}_{\mathcal{ITS}}(\mathbf{P})$ and $\mathbf{Sat}_{\mathcal{A}}(\mathbf{P})$ have the same shape just by chance (idem for $\mathbf{Sat}_{est}(\mathbf{P})$ and $\mathbf{Sat}_{cl}(\mathbf{P})$).

The non topological nature of these operators is openly visible in the above pictures. For instance in $\mathbf{Sat}_{int}(\mathbf{P})$ we do not have the intersection of $\{a_1, a_2, a_3\}$ and $\{a, a_1, a_2\}$. Hence int does not distribute over intersections. in $\mathbf{Sat}_{cl}(\mathbf{P})$ we lack the union of $\{a_3\}$ and $\{a\}$, so that cl does not distribute over unions.

However we have a few results which will be useful.

Proposition 7. *Let θ and ϕ be two dual basic operators. Then,*
(a) θ is a closure operator if and only if ϕ is an interior operator;
(b) θ is topological if and only if ϕ is topological.

Proof. (a) Trivially, since complementation reverses the order. (b) Suppose θ is additive, then $\phi(X \cap Y) = -\theta - (X \cap Y) = -\theta(-X \cup -Y) = -(\theta(-X) \cup \theta(-Y)) = -\theta(-X) \cap -\theta(-Y) = \phi(X) \cap \phi(Y)$. Dually for the opposite implication.

In order to try and solve the above issue, we must notice that any answer and solution depends on the nature of the *P-system* at hand. Generally, the nature of points is not really important. More important is the nature of properties. And, even more important is the nature of the operator supposed to better represent the basic perception act.

7.1 Choosing the Initial Perception Act

We have assumed that our first act of knowledge is a grouping act, a sort of "data abstraction". However, we can basically perform this act in two opposite ways: either collect around an object g the elements which fulfills at least all the properties (or attribute-values) of g, or the elements fulfilling at most all the properties (or attribute-values) of g. Otherwise stated, in the first case we collect the objects which are characterised at least as g by the properties (attributes) at hand, while in the second case we collect the objects which are characterised at most as g. However if we consider attribute-values the two conditions collapse (see later on).

Moreover notice that the grouping rule just asserted does not imply any form of symmetry. Indeed, g' could manifest all the properties of g but also additional properties that are not manifested by g. To put it another way, we are claiming that our basic grouping act is not based on the notion "to manifest *exactly* the same properties", but on the notion "to manifest *at least* (or *at most*) the same properties". Indeed, the set of properties which are manifested by g, is the attracting phenomenon around which we form our perception. Thus from an analytical point of view we have just to focus on these properties and not to take into account additional properties. In the present paragraph we shall see that the former notion is subsumed by the latter. That is, if we define this way the basic "cells" of our categorisation process, we shall be able to cover a wider range of cases.

Let $\langle G, At, \{V_a\}_{a \in At} \rangle$ be an *Attribute System*. Then for all $g \in G$ define:
$Q_g = \{g' : \forall a \in At, \forall x \in V_a((a(g) = x) \Longrightarrow (a(g') = x))\}$

Let $\langle G, M, \Vdash \rangle$ be a *Property System*. Then for all $g \in G$ define:
$Q_g = \{g' : \forall p \in M(p \in \langle i \rangle(g) \Longrightarrow p \in \langle i \rangle(g'))\}$

Q_g will be called the *quantum of information at g*.

In view of the previous discussion we adopt quanta of information because they reflect the idea "g is perceived together g' whenever it manifests *at least* the same properties as g".

Therefore, given (the properties manifested by) an object g the perception cell organised around g should be Q_g which should be referred to as the "*minimum perceptibilium at location g*", for it is not possible to perceive g without perceiving (the manifestations of) the other members of that "perception parcel". Therefore, we call such perception parcel a *quantum of perception* or a *quantum of information at g*. This terminology drew its inspiration from [4] and this term expresses a sort of programme that may be epitomized by the slogan "**any information is a quantum of information**".

As to quanta of information from a *Property System*, we can elaborate a little further (we shall resume *Attribute Systems* later on).

Proposition 8. *Let $\langle G, M, \Vdash \rangle$ be a P-system and $g, g' \in G$. Then,*

1. $Q_g = est(g)$; $g' \in Q_g$ iff $\langle i \rangle(g) \subseteq \langle i \rangle(g')$
2. $g' \in Q_g$ iff for all $X \in \Gamma_{est}, g \in X \Longrightarrow g' \in X$;
3. $g' \in Q_g$ iff for all $p \in M, g \in \langle e \rangle(p) \Longrightarrow g' \in \langle e \rangle(p)$;
4. $g' \in Q_g$ iff $g \in cl(g')$.

Proof. **(1)** Indeed, $g' \in est(g)$ iff $\langle i \rangle(g') \supseteq [[i]](g)$. But $[[i]](g) = \langle i \rangle(g)$, whence $g' \in est(g)$ if and only if $p \in \langle i \rangle(g) \Longrightarrow p \in \langle i \rangle(g')$ if and only if $g' \in Q_g$ if and only if $\langle i \rangle(g) \subseteq \langle i \rangle(g')$.
(2) ($\Leftarrow$) Suppose $X \in \Gamma_{est}$ and $g \in X \Longrightarrow g' \in X$. Then $\langle i \rangle(g) \supseteq [[i]](X)$ implies $\langle i \rangle(g') \supseteq [[i]](X)$. Since this happens for all *est*-saturated X, it happens for $[[i]](g)$ too and we trivially obtain $\langle i \rangle(g') \supseteq \langle i \rangle(g)$, so that $g' \in Q_g$. ($\Rightarrow$) If $g' \in Q_g$ then $\langle i \rangle(g') \supseteq \langle i \rangle(g)$. If, moreover, $g \in X$, for $X = est(X)$, then $\langle i \rangle(g) \supseteq [[i]](X)$. By transitivity, $\langle i \rangle(g') \supseteq [[i]](X)$, whence $g' \in X$ too. **(3)**

Indeed, $g \in \langle e \rangle(x)$ if and only if $g \Vdash x$ if and only if $x \in \langle i \rangle(g)$. Hence for all $p \in M, g \in \langle e \rangle(p) \implies g' \in \langle e \rangle(p)$ if and only if for all $p \in M, p \in \langle i \rangle(g) \implies p \in \langle i \rangle(g')$. (4) So, $g' \in Q_g$ if and only if $\langle i \rangle(g) \subseteq \langle i \rangle(g')$ if and only if $g \in cl(g')$.

Indeed, these results, trivial consequences of *Lemma* 1 and *Proposition* 4, formally state that g' is perceived together with g if and only if it fulfills at least all the properties fulfilled by g. Moreover, a quantum of perception at location g is the *universal extension* of function *sub* to the set of properties $\Vdash (g)$ fulfilled by g. In turn, since we start from a singleton $\{g\}$, $\Vdash (g)$ is both a universal and an existential extension of function *obs*.

When we have to move from grouping maneuvers around a single object to grouping maneuvers around two or more objects we have essentially two kinds of choice: universal extensions from singletons to a generic set X and existential extensions. The existential extension is defined as

$$Q_X^\cup = \bigcup_{x \in X} Q_x \tag{7}$$

This is not the sole choice, but for the very reasons discussed so far, we shall adopt it. Moreover, it makes a uniform treatment of both *P-systems* and *A-systems* possible.

As for universal extensions we briefly discuss only the following alternative:

$$Q_X^\otimes = [[e]][[i]](X) = est(X). \tag{8}$$

The superscript $\otimes$ underlines the fact that in $est(X)$ we consider the properties which glue the elements of X together. Otherwise stated, we extract from $\langle i \rangle(X)$ those properties P' which are shared by all the elements of X. Then we make exactly the same thing with respect to P'. Thus, according to this universal extension, an object g belongs to $Q_X^\otimes$ if g fulfills all the properties fulfilled by all the elements of X.

Whenever we need to distinguish the system inducing a quantum of information, we shall use the name of the system as an exponent. The same for any operator (for instance we shall write $Q_g^{\mathbf{P}}$ and $int^{\mathbf{P}}$ if needed). Moreover, if $\mathbf{O} = \langle X, R \rangle$, we set $-\mathbf{O} = \langle X, -R \rangle$ and $\mathbf{O}^\smile = \langle X, R^\smile \rangle$.

7.2   Information Quantum Relational Systems

Now that we have chosen the basic mechanisms (basis and step) leading from atomic perception (or "elementary perception cells") to complex perception, let us analyse what kinds of a relation arise between elements of G from these grouping maneuvers.

Let us then set the following definition:

Definition 12 (Information Quantum Relational System). *Let* $\mathbf{S}$ *be an A-system or a P-system over a set of points* G. *Let* R *be a binary relation on* G. *We say that* R *is induced by* $\mathbf{S}$ *whenever the following holds, for all* $g, g' \in G$:

$$\langle g, g' \rangle \in R \text{ iff } g' \in Q_g.$$

We call R the information quantum relation *- or* i-quantum relation *or* quantum relation *in short - induced by* **S** *and it will be denoted as* $R_\mathbf{S}$. *Moreover,* $\mathbf{Q}(\mathbf{S})$ *will denote the relational system* $\langle G, R_\mathbf{S} \rangle$, *called the* Informational Quantum Relational System - IQRS, *or* Quantum Relational System *in short, induced by* **S**. *Finally we set* $\Omega_Q(\mathbf{S}) = \{R_\mathbf{S}(X) : X \subseteq G\}$[18].

Since $g \in Q_{g'}$ says that g fulfills at least all the properties fulfilled by g', then $\langle g', g \rangle \in R_\mathbf{S}$ has the same meaning.

Clearly the properties of i-quantum relations depend on the patterns of objects induced by the given systems. However, they uniformly fulfill some basic properties.

In view of the additivity property of generalised quanta, we can confine our attention to i-quanta at a location.

Lemma 4. *In any A-system* **S** *over a set G, for all $g, g', g'' \in G$:*

1. (a) $g \in Q_g$ *(q-reflexivity)*; (b) $g'' \in Q_{g'}$ & $g' \in Q_g \implies g'' \in Q_g$ *(q-transitivity)*.
2. *If* **S** *is an A-system, a functional or a dichotomic P-system then $g' \in Q_g$ implies $g \in Q_{g'}$ (AFD-q-symmetry).*

Proof. The first two statements are obvious consequences of transitivity and reflexivity of the relation $\subseteq$. Notice that antisymmetry does not hold because of the obvious fact that $g' \in Q_g$ and $g \in Q'_g$ does not imply $g' = g$. Now, let **S** be an *A-system*. Suppose $g' \in Q_g$ and $a(g) \neq x$, then $a(g) = x'$ for some $x' \neq x$ so that $a(g') = x'$, because $g' \in Q_g$, whence $a(g') \neq x$ too. Therefore, $g' \in Q_g$ implies $g \in Q_{g'}$, so that the induced relation is also symmetric. If **S** is a functional *P-system* then we trivially obtain the proof from definitions and the fact that $\langle i \rangle(g)$ is a singleton. Finally, if **S** is dichotomic and $g' \in Q_g$, then g' fulfills at least the same properties as g. Now, if $g' \Vdash p$ while $g \not\Vdash p$, then $g \Vdash \bar{p}$, where $\bar{p}$ is a complementary copy of p. But $g' \not\Vdash \bar{p}$, since it fulfills p. Hence we cannot have $\langle i \rangle(g) \subseteq \langle i \rangle(g')$, whence $g' \notin Q_g$. Contradiction.

So notice that in *A-systems* the universal quantification over attribute-values hides a bi-implication because the set of attribute-values of g' and that of g must coincide in order to have $g' \in Q_g$.

As an immediate consequence of the above result we have:

Proposition 9. *Let* **S** *be an A-system or a P-system. Then:*

1. *The i-quantum relation $R_\mathbf{S}$ induced by* **S** *is a preorder;*
2. *If* **S** *is an A-system or an FP or DP system, then $R_\mathbf{S}$ is an equivalence relation;*
3. *If* **S** *is an FP-system then $R_\mathbf{S} = \Vdash \otimes \Vdash^\smile$ and $g' \in Q_g$ iff $g' \in [g]_{k_\Vdash}$.*

---

[18] I-quanta and i-quantum relations from *A-systems* were introduced in [13], with different names. If the entire set M is considered as a multi-valued property, then i-quantum relations coincide with the so-called "forward inclusion relations" introduced in [12].

Proof. We just have to prove statement (3). In view of *Proposition* 8.(4) we just need to show that if $\Vdash$ is a map then for all $x \in G$, $cl(\{x\}) = [x]_{\kappa_{\Vdash}}$. From *Proposition* 7.1, $a \in cl(\{a'\})$ if and only if $\langle i\rangle(\{a\}) \subseteq \langle i\rangle(\{a'\})$. Therefore, if $\Vdash$ happens to be a map, we have that $a \in cl(\{a'\})$ if and only if $\langle i\rangle(\{a\}) = \langle i\rangle(\{a'\})$, since exactly one value is admitted. We can conclude that for all $x \in G$, $cl(\{x\}) = \{x' : \langle i\rangle(x') = \langle i\rangle(x)\} = \langle e\rangle\langle i\rangle(\{x\}) = [x]_{k_{\Vdash}}$ (that is, the kernel of $\Vdash$).

EXAMPLE 2

Consider the *P-system* **P** of EXAMPLE 1. Let us compute some quanta of information of **P**: $Q_a = \{a\}, Q_{a'} = \{a', a''\}, Q_{a''} = \{a''\}, Q_{a'''} = \{a', a'', a'''\}, Q_{\{a,a'\}} = \{a, a', a''\}$ and so on addictively. Thus $\langle a', a''\rangle \in R_{\mathbf{S}}$ because $a'' \in Q_{a'}$ but the opposite does not hold.
Here below the i-quantum relations $R_{\mathbf{P}}$ and $R_{\mathbf{Q(P)}}$ are displayed:

$R_{\mathbf{P}}$	a	a'	a''	a'''	$R_{\mathbf{Q(P)}}$	a	a'	a''	a'''
a	1	0	0	0	a	1	0	0	0
a'	0	1	1	0	a'	0	1	0	1
a''	0	0	1	0	a''	0	1	1	1
a'''	0	1	1	1	a'''	0	0	0	1

It is easy to verify that both of the above relations are reflexive and transitive, and $R_{\mathbf{P}}^{\smile} = R_{\mathbf{Q(P)}}$. Moreover one can see that, for instance, $Q_{a'}^{\mathbf{Q(P)}} = \{a', a'''\}$ or $Q_{a''}^{\mathbf{Q(P)}} = \{a', a'', a'''\}$. Indeed we have that $a' \in Q_{a'''}$ and $a''' \in Q_{a'}^{\mathbf{Q(P)}}$, or $a'' \in Q_{a'}$ whereas $a' \in Q_{a''}^{\mathbf{Q(P)}}$, and so on.

Now consider the *A-system* $\mathbf{A} = \langle G = \{a, a_1, a_2, a_3\}, At = \{A, B, C\}, \{V_A = \{0, 1, 3\}, V_B = \{b, c, f\}, V_C = \{\alpha, \delta\}\}\rangle$ such that $A(a) = A(a_2) = 1, A(a_1) = 0, A(a_3) = 3, B(a) = B(a_2) = b, B(a_1) = c, B(a_3) = f, C(a) = C(a_1) = C(a_2) = \alpha, C(a_3) = \delta$. We have: $Q_a^{\mathbf{A}} = Q_{a_2}^{\mathbf{A}} = \{a, a_2\}, Q_{a_1}^{\mathbf{A}} = \{a_1\}, Q_{a_3}^{\mathbf{A}} = \{a_3\}$.
The resulting i-quantum relation $R_{\mathbf{A}} = \{\langle a, a_2\rangle, \langle a_2, a\rangle, \langle a, a\rangle, \langle a_2, a_2\rangle, \langle a_1, a_1\rangle, \langle a_3, a_3\rangle\}$ is an equivalence relation.

From the above results we obtain immediately some interesting consequences about functional *P-systems*:

Corollary 4. *Let* **P** *be an FP-system. Then,*

(a) *cl is a topological closure operator;* (b) *int is a topological interior operator.*

Proof. From *Proposition* 8.(4) and *Proposition* 9.(3) we have that $cl(x) = [x]_{k_{\Vdash}}$. But $k_{\Vdash}$ is the kernel of $\Vdash$ and the kernel of a function is a congruence. It follows by induction that $cl(X) \cup cl(Y) = [X]_{k_{\Vdash}} \cup [Y]_{k_{\Vdash}} = [X \cup Y]_{k_{\Vdash}} = cl(X \cup Y)$. Hence cl is additive. Since int is dual of cl we immediately obtain that int is multiplicative.

We now list some results in terms of IQRSs. Since i-quantum relations are preorders, it is useful to prove some general facts about this kind of relations:

Proposition 10. *Let* $\mathbf{O} = \langle X, R \rangle$ *be any preordered set. Then for any* $x, y \in X$ *the following are equivalent:*

(1) $y \in \langle R^{\smile} \rangle(x)$; (2) $\langle R^{\smile} \rangle(y) \subseteq \langle R^{\smile} \rangle(x)$; (3) $x \in Q_y^{\mathbf{O}}$; (4) $x \in \langle R \rangle(y)$; (5)
$$y \in Q_x^{\mathbf{O}^{\smile}}.$$

Proof. (1 $\Longleftrightarrow$ 2) $y \in R(x)$ iff $\langle x, y \rangle \in R$. Suppose $\langle y, y' \rangle \in R$. Since R is transitive, for all $y' \in X$, $\langle y, y' \rangle \in R \Longrightarrow \langle x, y' \rangle \in R$ so that $\langle R^{\smile} \rangle(x) \supseteq \langle R^{\smile} \rangle(y)$. Conversely, since R is reflexive, $y \in \langle R \rangle(y)$ holds. Thus if $\langle R^{\smile} \rangle(x) \supseteq \langle R^{\smile} \rangle(y)$ then $y \in \langle R^{\smile} \rangle(x)$. All the other equivalences are obvious consequences or even just definitions.

Corollary 5. *Let* $\mathbf{S}$ *be an A-system or a P-system over a set* G. *Then,*

$$g' \in Q_g^{\mathbf{S}} \text{ iff } g' \in \langle R_{\mathbf{S}}^{\smile} \rangle(g) \text{ iff } g \in Q_{g'}^{\mathbf{Q(S)}} \text{ iff } g' \in Q_g^{\mathbf{Q(Q(S))}} \text{ iff } g \in \langle R_{\mathbf{S}} \rangle(g') \text{ iff }$$
$$g \in Q_{g'}^{-\mathbf{S}}.$$

Proof. The first equivalence is just a definition. Now, $g' \in \langle R_{\mathbf{S}}^{\smile} \rangle(g)$ iff $\langle R_{\mathbf{S}}^{\smile} \rangle(g') \subseteq \langle R_{\mathbf{S}}^{\smile} \rangle(g)$ iff $g \in Q_{g'}^{\mathbf{Q(S)}}$ iff $g \in \langle R_{\mathbf{Q(S)}}^{\smile} \rangle(g')$ iff $\langle R_{\mathbf{Q(S)}}^{\smile} \rangle(g) \subseteq \langle R_{\mathbf{Q(S)}}^{\smile} \rangle(g')$ iff $g' \in Q_g^{\mathbf{Q(Q(S))}}$. From this we have that $Q_g^{\mathbf{Q(S)}} = \langle R_{\mathbf{S}} \rangle(g)$ so that in view of trivial set-theoretic considerations, ($X \subseteq Y$ iff $-Y \subseteq -X$) we obtain the last two equivalences.

These equivalences show that IQRSs of level higher than 1 do not provide any further information.

Corollary 6. *If* $\mathbf{S}$ *is an A-system, an FP-system or a DP-system over a set* G, *then for all* $g, g' \in G, X \subseteq G$, (a) $g' \in Q_g^{\mathbf{S}}$ *iff* $g' \in Q_g^{\mathbf{Q(S)}}$; (b) $\langle R_{\mathbf{S}}^{\smile} \rangle(X) = \langle R_{\mathbf{Q(S)}}^{\smile} \rangle(X)$.

Moreover, since a *P-system* is a generic relational system we have that all facts valid for *P-systems* are valid for any relational system.

The notion of a quantum of information is asymmetric for *P-systems*, because if g' fulfills strictly more properties than g, we have $g' \in Q_g$ but $g \notin Q_{g'}$. On the contrary it is symmetric in the case of *A-systems* and dichotomic or functional *P-systems*.

8 Higher Level Operators

Let $\mathbf{S}$ be an *A-system* and let $\mathbf{Q(S)} = \langle G, G, R_{\mathbf{S}} \rangle$ be its induced IQRS. What kinds of patterns of data can we collect by applying our operators to these derivative systems?

First of all, since in IQRSs there is no longer the distinction between objects and properties and intension or extension, it is better we change once more our symbols and notation:

The operator	defined as	turns into
$\langle i \rangle$	$\langle i \rangle(X) = \{g : \exists g'(g' \in X \ \& \ \langle g', g \rangle \in R_{\mathbf{S}})\}$	$\langle R_{\mathbf{S}}^{\smile} \rangle$
$\langle e \rangle$	$\langle e \rangle(X) = \{g : \langle R_{\mathbf{S}}^{\smile} \rangle(g) \cap X \neq \emptyset\}$	$\langle R_{\mathbf{S}} \rangle$
$[i]$	$[i](X) = \{g : \langle R_{\mathbf{S}} \rangle(g) \subseteq X\}$	$[R_{\mathbf{S}}^{\smile}]$
$[e]$	$[e](X) = \{g : \langle R_{\mathbf{S}}^{\smile} \rangle(g) \subseteq X\}$	$[R_{\mathbf{S}}]$

Let us call the above operators decorated with $R_{\mathbf{S}}$ "quantum operators" (notice that in this context $[[R_{\mathbf{S}}]]$ and $[[R_{\mathbf{S}}^{\smile}]]$ are not quantum operators).

Quantum operators behave in a very particular way, because, we remind, they fulfill adjoint properties. Namely $\langle R_{\mathbf{S}} \rangle \dashv [R_{\mathbf{S}}^{\smile}]$ and $\langle R_{\mathbf{S}}^{\smile} \rangle \dashv [R_{\mathbf{S}}]$.

Actually, the following results apply to any preorder.

Proposition 11. *Let* $\mathbf{Q}(\mathbf{S}) = \langle G, G, R_{\mathbf{S}} \rangle$ *be a IQRS. Let* O_i *and* O_j *be any two adjoint quantum operators from the set* $\{\langle R_{\mathbf{S}}^{\smile} \rangle, \langle R_{\mathbf{S}} \rangle, [R_{\mathbf{S}}^{\smile}], [R_{\mathbf{S}}]\}$. *Then*

(a) $O_i O_j = O_j$; (b) $O_j O_j = O_j$; (c) the fixpoints of O_i and O_j coincide.

Proof. (a) (i) In view of *Proposition 5*, for all $g \in G$ and $X \subseteq G$, $g \in [R_{\mathbf{S}}]\langle R_{\mathbf{S}}^{\smile} \rangle$ (X) iff $\langle R_{\mathbf{S}}^{\smile} \rangle(g) \subseteq \langle R_{\mathbf{S}}^{\smile} \rangle(X)$, iff (from *Proposition 10*) $g \in \langle R_{\mathbf{S}}^{\smile} \rangle(X)$.
One can prove $g \in [R_{\mathbf{S}}^{\smile}]\langle R_{\mathbf{S}} \rangle(X)$ iff $g \in \langle R_{\mathbf{S}} \rangle(X)$ similarly.
(ii) In view again of *Proposition 5*, $a \in \langle R_{\mathbf{S}} \rangle[R_{\mathbf{S}}^{\smile}](X)$ iff $\langle R_{\mathbf{S}}^{\smile} \rangle(a) \cap [R_{\mathbf{S}}^{\smile}](X) \neq \emptyset$, iff there is a' such that $a \in \langle R_{\mathbf{S}} \rangle(a')$ and $a' \in [R_{\mathbf{S}}^{\smile}](X)$. But $a' \in [R_{\mathbf{S}}^{\smile}](X)$ iff $\langle R_{\mathbf{S}} \rangle(a') \subseteq X$ iff $\langle R_{\mathbf{S}} \rangle(a) \subseteq X$. Hence, $a \in \langle R_{\mathbf{S}} \rangle[R_{\mathbf{S}}^{\smile}](X)$ iff $a \in [R_{\mathbf{S}}^{\smile}](X)$.
(b) From point (a) and *Proposition 6.3*, $O_j O_j = O_i O_j O_i O_j = O_i O_j = O_j$.
(c) Let $X = O_j(X)$. Then from point (a) $O_i(X) = O_i O_j(X) = O_j(X) = X$.

Definition 13. *Let* $\mathbf{S}$ *be an A-system or a P-system. With* $\Omega_Q(\mathbf{S})$ *we shall denote the family* $\{Q_X^{\mathbf{S}} : X \subseteq G\}$. *With* $\mathbf{Q}_n(\mathbf{S})$ *we denote the n-nested application of the functor* $\mathbf{Q}$ *to* $\mathbf{S}$.

In view of these results we can prove a number of properties.

Lemma 5. *For any P-system* $\mathbf{P}$, (a) $\Omega_{int}(\mathbf{P}) \subseteq \Omega_Q(\mathbf{P})$; (b) $\Gamma_{cl}(\mathbf{P}) \subseteq \Omega_{int}$ $(\mathbf{Q}(\mathbf{P}))$.

Proof. (a) Assume $X \subsetneq Q_X$. Thus we must have some x such that $x \notin X$ but $x \in Q_X$. Thus there is $g \in X$ such that $\Vdash (x) \supseteq \Vdash (g)$, so that for all $m \in M$ such that $g \Vdash m$ surely $\Vdash^{\smile} (m) \nsubseteq X$. It follows that $g \notin int^{\mathbf{P}}(X)$ and, hence, $int^{\mathbf{P}}(X) \neq X$. (b) We remind that $x \in cl^{\mathbf{P}}(X)$ iff $\langle i \rangle(x) \subseteq \langle i \rangle(X)$ iff $\Vdash^{\smile} (x) \subseteq \Vdash^{\smile} (X)$. Moreover, if $x \in X, \langle i \rangle(x) \subseteq \langle i \rangle(X)$. Now suppose $X \neq int^{\mathbf{Q}(\mathbf{P})}(X)$. Then there is $x \in X$ such that $R_{\mathbf{P}}(x) \nsubseteq X$. Hence $\{y : x \in Q_y^{\mathbf{P}}\} \nsubseteq X$. Thus $\{y : \langle i \rangle(y) \subseteq \langle i \rangle(x)\} \nsubseteq X$. This means that there is $g \notin X$ such that $\langle i \rangle(g) \subseteq \langle i \rangle(x) \subseteq \langle i \rangle(X)$ so that $\Vdash^{\smile} \langle i \rangle(g) \subseteq \Vdash^{\smile} \langle i \rangle(x) \subseteq \Vdash^{\smile} \langle i \rangle(X)$. It follows that $X \subsetneq cl^{\mathbf{P}}(X)$.

Corollary 7. *Let* $\mathbf{Q}(\mathbf{S})$ *be an IQRS. Then,*

(a) $Q_{(\ldots)}^{\mathbf{S}} = \langle R_{\mathbf{S}}^{\smile} \rangle = cl$; (b) $Q_{(\ldots)}^{\mathbf{Q}(\mathbf{S})} = \langle R_{\mathbf{S}} \rangle = \mathcal{A}$; (c) $[R_{\mathbf{S}}^{\smile}] = int$; (d) $[R_{\mathbf{S}}] = \mathcal{C}$.

where the operators $cl, int, \mathcal{C}, \mathcal{A}$ *are intended over* $\mathbf{Q}(\mathbf{S})$.

Corollary 8. *Let* **S** *be an A-system or a P-system and* $A, B \subseteq G$. *Then,*

1. $Q^{\mathbf{S}}_{(\ldots)}, Q^{\mathbf{Q(S)}}_{(\ldots)}, \langle R_{\mathbf{S}} \rangle$ *and* $\langle R_{\widetilde{\mathbf{S}}} \rangle$ *are topological closure operators and their images are closed under intersections.*
2. $[R_{\mathbf{S}}]$ *and* $[R_{\widetilde{\mathbf{S}}}]$ *are topological interior operators and their images are closed under unions.*

Proof. (1) From *Corollary 7*, we have that $Q^{\mathbf{S}}_{(\ldots)}, Q^{\mathbf{Q(S)}}_{(\ldots)}, \langle R_{\mathbf{S}} \rangle$ and $\langle R_{\widetilde{\mathbf{S}}} \rangle$ are closure operators. Moreover, since they are lower adjoints in the category $\langle \wp(G), \subseteq \rangle$ they preserve unions. Finally, from *Proposition 7.2* they are normal and their images are closed under intersections. (2) From *Corollary 7*, $[R_{\mathbf{S}}]$ and $[R_{\widetilde{\mathbf{S}}}]$ are interior operators. Moreover as they are lower adjoints in the category $\langle \wp(G), \subseteq \rangle$ they preserve intersections. Finally from *Proposition 7.2* they are conormal and their images are closed under unions.

Corollary 9 (I-quantum systems). *Let* **S** *be an A-system or a P-system. Then,*

1. $\mathbf{Sat}_Q(\mathbf{S}) = \langle \Omega_Q(\mathbf{S}), \cup, \cap, \emptyset, G \rangle$ *is a distributive lattice, called the* I-quantum system - IQS *induced by* **S**.
2. $\mathbf{Sat}_Q(\mathbf{Q(S)}) = \langle \Omega_Q(\mathbf{Q(S)}), \cup, \cap, \emptyset, G \rangle$ *is a distributive lattice, called the co-I-quantum system - co-IQS* induced by **S**.
3. *The set theoretical complement is an antisomorphism between* $\mathbf{Sat}_Q(\mathbf{S})$ *and* $\mathbf{Sat}_Q(\mathbf{Q(S)})$.
4. $\mathbf{Sat}_{int}(\mathbf{Q(S)}), \mathbf{Sat}_{cl}(\mathbf{Q(S)}), \mathbf{Sat}_{\mathcal{A}}(\mathbf{Q(S)})$ *and* $\mathbf{Sat}_{\mathcal{C}}(\mathbf{Q(S)})$, *equipped with the set-theoretical operations, are distributive lattices.*
5. $\langle G, \Omega_Q(\mathbf{S}) \rangle$ *and* $\langle G, \Omega_Q(\mathbf{Q(S)}) \rangle$ *are topological spaces, where the interior operators are* $int^{\mathbf{Q}_2(\mathbf{S})}$ *and, respectively,* $int^{\mathbf{Q}_1(\mathbf{S})}$.

Proof. (1) We know that the operator $Q_{(\ldots)}$ is additive. Thus $\Omega_Q(\mathbf{S})$ is closed under unions. From *Corollary 8* it is closed under intersections too. Moreover, for $\Omega_Q(\mathbf{S})$ is a (finite) lattice of sets $\mathbf{Sat(S)}$ inherits distributivity from the corresponding property of unions and intersections. (2) Since $\mathbf{Q(S)}$ is a *P-system* the above considerations apply to this structure. (3) From *Corollary 5* we know that $-R_{\mathbf{S}} = R_{\widetilde{\mathbf{S}}}$, so that we obtain immediately the thesis. (4) From *Proposition 7.2* and *Corollary 8*. (5) Any family of open sets of a topological space enjoys distributivity of arbitrary unions over finite intersections and of intersection over arbitrary unions. Moreover, from *Proposition 11* and *Corollary 7* we obtain that, $\Omega_Q(\mathbf{S}) = \Gamma_{cl}(\mathbf{Q(S)}) = \Omega_{int}(\mathbf{Q}_2(\mathbf{S}))$ and $\Omega_Q(\mathbf{Q(S)}) = \Omega_{\mathcal{A}}(\mathbf{Q(S)}) = \Omega_{int}(\mathbf{Q(S)})$.

Now by means of the above mathematical machinery we prove a key statement in the theory of Approximation Space and Rough Sets, namely the well-known fact that the family of definable sets can be made into a Boolean algebra.

Proposition 12 (Quantum relations and Boolean algebras). *Let* **S** *be an Information system. If* $R_{\mathbf{S}}$ *is an equivalence relation, then* $\mathbf{Sat}_Q(\mathbf{S})$ *is a Boolean algebra.*

Proof. We show that if $R_{\mathbf{S}}$ is an equivalence relation then any element Q_X of $\Omega_Q(\mathbf{S})$ is complemented by $\overline{Q_X} = \bigcup_{z \notin Q_X} Q_z$. First, let us prove that $\overline{Q_X} \cup Q_X = G$. In fact for all $g \in G$ if $g \notin Q_X$ then $g \in \overline{Q_X}$ because $g \in Q_g$ (q-reflexivity). Now we prove that $\overline{Q_X} \cap Q_X = \emptyset$. Assume $z \notin Q_X$. We have just to prove that if $z' \in Q_z$ then $z' \notin Q_X$. So let $z' \in Q_z$. For q-symmetry $z \in Q_{z'}$. So, if there is an $x \in X$ such that $z' \in Q_x$ we have $z \in Q_x$ too (for q-transitivity), hence $z \in Q_X$. Contradiction.[19]

Corollary 10. *Let* $\mathbf{S}$ *be an Information system. Then, if* $\mathbf{S}$ *is an A-system, a dichotomic or a functional P-system, then* $\mathbf{Sat}_Q(\mathbf{S})$ *is a Boolean algebra.*

About the family of co-prime elements of $\mathbf{Sat}_Q(\mathbf{S})$ we have:

Lemma 6. *Let* $\mathbf{S}$ *be an A-system or a P-system. Then for any* $X \in \mathcal{J}(\mathbf{Sat}_Q (\mathbf{S}))$, $X = Q_g$ *for some* $g \in X$.

Proof. Trivial from the very additive definition of the operator Q and its increasing property.

Lemma 7. *Let* $\mathbf{P}$ *be a P-system and* $g \in G$. *Then* $Q_g = \bigcap\{\langle e \rangle(m) : m \in \langle i \rangle(g)\}$

Proof. Indeed, $x \in Q_g$ iff $\langle i \rangle(x) \supseteq \langle i \rangle(g)$ iff $x \in \langle e \rangle(m), \forall m \in \langle i \rangle(g)$.

Proposition 13. *Let* $\mathbf{P}$ *be a P-system such that cl (int) is topological. Then* $\mathbf{Sat}_Q(\mathbf{P}) = \mathbf{Sat}_{int}(\mathbf{P})$.

Proof. We have seen in *Lemma* 5 that $\Omega_{int}(\mathbf{P}) \subseteq \Omega_Q(\mathbf{P})$. Now we need just to show that if $X \in \mathcal{J}(\mathbf{Sat}_Q(\mathbf{P}))$ then $X = int(X)$. The proof is immediate. Indeed, the family $\{\langle e \rangle(m) : \langle e \rangle(m) \subseteq X\}$ is a *base* of $\Omega_{int}(\mathbf{S})$. Moreover, if *int* is topological then it is multiplicative and since for all $m \in M$, $int(\langle e \rangle(m)) = \langle e \rangle(m)$ (from *Lemma* 3), in view of the above *Lemma* 7 we have the result.

Corollary 11. *Let* $\mathbf{F}$ *be an FP-system. Then* $\Omega_Q(\mathbf{Q}_n(\mathbf{F})) = \Omega_{int}(\mathbf{Q}_n(\mathbf{F}))$, $n \geq 0$.

Proof. From *Corollary* 4 and an inductive extension of *Proposition* 13.

Hence we can note that *P-systems* such that *int* and *cl* are topological behave like functional systems.

Corollary 12. *If* $\mathbf{S}$ *is a preordered set (that is,* $G = M$ *and* $R \subseteq G \times G$ *is a preorder), then* $\Omega_Q(\mathbf{S}) = \Omega_{int}(\mathbf{S})$.

---

[19] There is another way to obtain this result. In fact, J. L. Bell proved that if $\mathbf{T} = \langle A, T \rangle$ is a relational structure with T a tolerance relation (that is, reflexive and symmetric) then the family $\Omega_{QL}(\mathbf{T})$ of all unions of principal order filters $\uparrow_T x$ (i. e. $\langle T^{\smile} \rangle(x)$ i. e., for symmetry of T, $\langle T \rangle(x)$) can be made into an ortholattice. But if $R_{\mathbf{S}}$ is an equivalence relation then it is a tolerance relation too and for any $x \in A$, $\uparrow_{R_{\mathbf{S}}} x = \downarrow_{R_{\mathbf{S}}} x = cl(x) = Q_x$ so that $\mathbf{Sat}_Q(\mathbf{S})$ can be made into a *distributive* ortholattice, that is, a Boolean algebra.

Proof. From *Corollary* 11 and *Corollary* 9 (3).

At this point we can end this subsection with an analogue of the duality between distributive lattices and preorders in the context of i-quantum relations and *P-systems*.

Proposition 14 (Duality between preorders and *P-systems*)

1. *Let* $\mathbf{O} = \langle G, R \rangle$ *be a preorder, then there is a P-system* $\mathbf{I}(\mathbf{O})$ *over* G *such that* $R_{\mathbf{I}(\mathbf{O})} = R$ *(hence,* $\mathbf{Q}(\mathbf{I}(\mathbf{O})) \cong_I \mathbf{O}$ *).*
2. *Let* $\mathbf{S}$ *be an A-system or a P-system. Then* $\mathbf{I}(\mathbf{Q}(\mathbf{S})) \cong_I \mathbf{S}$.

Proof. (1) Let $\mathcal{F}(\mathbf{O})$ be the set of order filters of $\mathbf{O}$. Thus $\mathcal{F}(\mathbf{O}) = \Omega_Q(\mathbf{Q}(\mathbf{O}))$ (i. e. $\Omega_Q(\mathbf{O}^\smile)$)), so that we know that $\mathbf{F}(\mathbf{O})$ can be made into the distributive lattice $\mathbf{Sat}_Q(\mathbf{Q}(\mathbf{O}))$. Then let $\mathcal{J}(\mathbf{Sat}_Q(\mathbf{Q}(\mathbf{O})))$ be the set of co-prime elements of $\mathbf{Sat}_Q(\mathbf{Q}(\mathbf{O}))$. Notice that co-prime elements have the form $\uparrow_R x$, i. e. $\langle R^\smile \rangle(x)$, for some element $x \in G$ and that they may be understood as properties fulfilled by the elements of G such that $g \Vdash x$ only if $\langle x, g \rangle \in R$. Let us then define $\mathbf{I}(\mathbf{O})$ as $\langle G, \mathcal{J}(\mathbf{Sat}_Q(\mathbf{Q}(\mathbf{O}))), \Vdash \rangle$. Thus, $\langle g, g' \rangle \in R_{\mathbf{I}(\mathbf{O})}$ iff $g' \in Q_g^{\mathbf{I}(\mathbf{O})}$, iff $\langle i \rangle(g) \subseteq \langle i \rangle(g')$, iff $g \in \langle R^\smile \rangle(x) \Longrightarrow g' \in \langle R^\smile \rangle(x)$ for all $\langle R^\smile \rangle(x) \in \mathcal{J}(\mathbf{Sat}_Q(\mathbf{Q}(\mathbf{O})))$. In particular, since R is reflexive, $g \in \langle R^\smile \rangle(g)$ so that $g' \in \langle R^\smile \rangle(g)$ holds, i. e. $\langle g, g' \rangle \in R$. Conversely, if $\langle g, g' \rangle \in R$ and $\langle x, g \rangle \in R$, for transitivity $\langle x, g' \rangle \in R$ too. It follows that $g \in \langle R^\smile \rangle(x) \Rightarrow g' \in \langle R^\smile \rangle(x)$, all $x \in G$. (2) For $\mathbf{Q}(\mathbf{S})$ is a preorder, from the previous point we have $\mathbf{Q}(\mathbf{I}(\mathbf{Q}(\mathbf{S}))) \cong_I \mathbf{Q}(\mathbf{S})$ so that trivially $\mathbf{I}(\mathbf{Q}(\mathbf{S})) \cong_I \mathbf{S}$.

EXAMPLE 3
Consider the *P-system* $\mathbf{P}$ and the *A-system* $\mathbf{A}$ of Example 2. Here below we display the lattices $\mathbf{Sat}_Q(\mathbf{P})$ and $\mathbf{Sat}_Q(\mathbf{A})$:

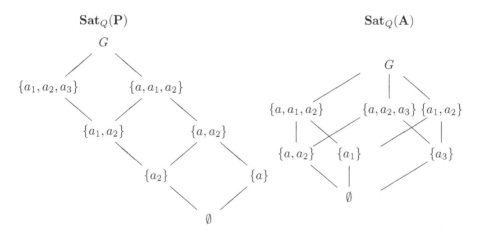

It is easy to verify that both of them are distributive lattices and that, moreover, $\mathbf{Sat}_Q(\mathbf{A})$ is a Boolean algebra. Indeed, for instance, the element $Q_{\{a,a_1\}}^{\mathbf{A}} = \{a, a_1, a_2\}$ is complemented by the element $\bigcup_{z \notin Q_{\{a,a_1\}}^{\mathbf{A}}} Q_z^{\mathbf{A}} = Q_{\{a_3\}}^{\mathbf{A}} = \{a_3\}$.

We can straightforwardly verify that $\mathbf{Sat}_{cl}(\mathbf{Q(P)})$ is $\mathbf{Sat}_{int}(\mathbf{P})$ plus some missed elements which are the difference between $\mathbf{Sat}_Q(\mathbf{P})$ and $\mathbf{Sat}_{int}(\mathbf{P})$. Indeed, the missed element is $\{a_1, a_2\}$ which equals $cl^{\mathbf{Q(P)}}(\{a_1, a_2\})$. On the contrary, $int^{\mathbf{P}}(\{a_1, a_2\}) = \emptyset$.

9 Generalised Topological Approximation Operators

In view of the duality between Information Systems and preorders, we can develop the rest of the theory from a more abstract point of view. Thus, from now on we shall deal with preordered structures and assume, intuitively, that they represent some information quantum relation system.

Corollary 13. *Let* $\mathbf{O} = \langle G, G, R \rangle$ *be a preordered set. Let* $X \subseteq G$. *Then:*

The application	is the least fixpoint of	including X
$\langle R \rangle(X)$	$\langle R \rangle\,[R^{\smile}]\,\mathcal{A}\,int$	
$\langle R^{\smile} \rangle(X)$	$\langle R^{\smile} \rangle\,[R]\,cl\,\mathcal{C}$	

The application	is the largest fixpoint of	included in X
$[R](X)$	$[R]\,\langle R^{\smile} \rangle\,\mathcal{C}\,cl$	
$[R^{\smile}](X)$	$[R^{\smile}]\,\langle R \rangle\,int\,\mathcal{A}$	

Proof. First notice that from *Proposition* 11 the listed fixpoints collapse. Nonetheless it is worthwhile proving the first two cases by means of two different approaches.
(a) Obviously $\langle R \rangle(X) \supseteq X$ and for idempotence $\langle R \rangle(X)$ is a fixed point of $\langle R \rangle$. Suppose Z is a fixed point of $\langle R \rangle$ such that $X \subseteq Z$. From monotonicity $\langle R \rangle(X) \subseteq \langle R \rangle(Z) = Z$. Hence $\langle R \rangle(X)$ is the least fixpoint of $\langle R \rangle$ including X.
(b) From *Proposition* 6.3 $[R]\langle R^{\smile} \rangle(X)$ is the smallest image of $[R]$ greater than or equal to X. Since $[R]$ is idempotent it is the least fixpoint of $[R]$ which includes X and from *Proposition* 11, it is also the least fixpoint of $\langle R^{\smile} \rangle(X)$ including X. The remaining cases are proved analogously.

Corollary 14. *Let* $\mathbf{O} = \langle G, G, R \rangle$ *be a preordered set. Then for all* $X \subseteq G$,
(i) $\langle R \rangle(X) = \bigcap\{Z : Z \in \Omega_{\mathcal{A}}(\mathbf{O})\ \&\ Z \supseteq X\}$; (ii) $[R](X) = \bigcup\{Z : Z \in \Gamma_{\mathcal{C}}(\mathbf{O})\ \&\ Z \subseteq X\}$;
(iii) $\langle R^{\smile} \rangle(X) = \bigcap\{Z : Z \in \Gamma_{cl}(\mathbf{O})\ \&\ Z \supseteq X\}$; (iv) $[R^{\smile}](X) = \bigcup\{Z : Z \in \Omega_{int}(\mathbf{O})\ \&\ Z \subseteq X\}$.

Henceforth, for obvious reasons we shall adopt the following terminology:

$\langle R \rangle(X)$	direct upper R-approximation of X, also denoted with $(uR)(X)$
$\langle R^{\smile} \rangle(X)$	inverse upper R-approximation of X, also denoted with $(uR^{\smile})(X)$
$[R](X)$	direct lower R-approximation of X, also denoted with $(lR)(X)$
$[R^{\smile}](X)$	inverse lower R-approximation of X, also denoted with $(lR^{\smile})(X)$

The information-oriented reading of the above operators is:

$\langle R \rangle (X)$	Set of the elements specialised by some member of X (or, which approximate some member of X)
$\langle R^\smile \rangle (X)$	Set of the elements approximated by some member of X (or, which specialise some member of X)
$[R](X)$	Set of the elements specialised only by members of X (or, which approximate only members of X)
$[R^\smile](X)$	Set of elements approximated only by members of X (or, which specialise only elements of X)

Particularly we can give an information-oriented interpretation to some combinations of operators:

$[R]\langle R \rangle (X)$	Set of the elements which are specialised just by elements specialised by some member of X ($x \in [R]\langle R \rangle (X)$ only if each element which specialises x is specialised by some member of X)
$[R^\smile]\langle R^\smile \rangle (X)$	Set of the elements which are approximated just by elements approximated by some member of X ($x \in [R^\smile]\langle R^\smile \rangle (X)$ only if each element which approximates x is approximated by some member of X)

Besides these operators we add also the interpretation of $[[R]]$ and $[[R^\smile]]$:

$[[R]](X)$	Set of the elements specialised by all the members of X (or, which approximate all the members of X)
$[[R^\smile]](X)$	Set of the elements approximated by all the members of X (or, which specialise all the members of X)

9.1 Topological Approximation Spaces

Eventually we define some interesting examples of topological Approximation Spaces.

Definition 14. *Let* $\mathbf{Q(S)} = \langle G, G, R_{\mathbf{S}} \rangle$ *be an IQRS. Then,*

1. $\langle G, [R_{\mathbf{S}}], \langle R_{\mathbf{S}} \rangle \rangle$ - *will be called a* Direct Intuitionistic Approximation Space.
2. $\langle G, [R_{\widetilde{\mathbf{S}}}], \langle R_{\widetilde{\mathbf{S}}} \rangle \rangle$ - *will be called an* Inverse Intuitionistic Approximation Space.
3. $\langle G, [R_{\widetilde{\mathbf{S}}}], \langle R_{\mathbf{S}} \rangle \rangle$ - *will be called a* Galois Intuitionistic Approximation Space.
4. $\langle G, [R_{\mathbf{S}}], \langle R_{\widetilde{\mathbf{S}}} \rangle \rangle$ - *will be called a* co-Galois Intuitionistic Approximation Space.

Definition 15. *Let* $\langle G, G, E \rangle$ *be a relational structure such that E is an equivalence relation. Let* $\mathbf{I}$ *and* $\mathbf{C}$ *be the interior and, respectively, topological operators of the topological space induced by taking* $\{[x]_E : x \in G\}$ *as a subbasis. Then* $\langle G, \mathbf{I}, \mathbf{C} \rangle$ *is called a* Pawlak Approximation Space

From the above discussion the following statement is obvious:

Proposition 15. *Let* $\mathbf{E} = \langle G, G, E \rangle$ *be a relational structure such that E is an equivalence relation. Then* $\langle G, int^{\mathbf{E}}, cl^{\mathbf{E}} \rangle$ *is a Pawlak Approximation Space.*

But we can prove a further fact. To this end we introduce the notion of an *Approximation Equivalence*, or *a-equivalence* between (pre) topological Approximation Spaces:

Definition 16. *Let* $\mathbf{A} = \langle G, \alpha, \beta \rangle$ *and* $\mathbf{A}' = \langle G, \gamma, \delta \rangle$ *two topological or pretopological Approximation Spaces. Then we say that* $\mathbf{A}$ *and* $\mathbf{A}'$ *are a-equivalent, in symbols,* $\mathbf{A} \cong_a \mathbf{A}'$ *if and only if* $\Omega_\alpha(G) = \Omega_\gamma(G)$ *and* $\Gamma_\beta(G) = \Gamma_\delta(G)$.

Clearly, by duality one equality implies the other. We use this definition in the following statement:

Proposition 16. *Let* $\mathbf{S}$ *be an A-system or FP-system or DP-system. Let us set* $\lozenge = \langle R_\mathbf{S} \rangle$ *and* $\square = [R_\mathbf{S}]$. *Then* $\lozenge = \langle R_{\widetilde{\mathbf{S}}} \rangle$, $\square = [R_{\widetilde{\mathbf{S}}}]$ *and* $\langle G, \square, \lozenge \rangle$ *is a Pawlak Approximation Space. Moreover, if* S *is an FP-system then* $\langle G, \square, \lozenge \rangle \cong_a \langle G, int^\mathbf{S}, cl^\mathbf{S} \rangle$.

Proof. Immediate, from the fact, proved in *Proposition* 9.(3), that $R_\mathbf{S}$ in this case is an equivalence relation. For the last part it is sufficient to use in addition *Proposition* 4.(3) together with *Proposition* 7, or the latter Proposition and *Proposition* 9.(3) which together with *Proposition* 8.(4) states that $cl^\mathbf{S}(g) = [g]_{k_f}$, any $g \in G$.

Note that the former system of the previous Proposition is, actually, $\langle G, G, \square, \lozenge \rangle$ while the latter is $\langle G, M, int^\mathbf{S}, cl^\mathbf{S} \rangle$. Therefore, we cannot put $\langle G, \square, \lozenge \rangle = \langle G, int^\mathbf{S}, cl^\mathbf{S} \rangle$. However, if $\mathbf{S}$ is an *FP-system*, then Single-agent (pre)topological Approximation Spaces, and Pawlak Approximation Spaces induce the same family of fixed points.

10 Comparing Information Systems

The notion of an i-quantum makes it possible to compare Information Systems. First of all we should ask whether it is possible to compare two quanta of information Q_g and $Q_{g'}$. At first sight we would say that Q_g is finer than $Q_{g'}$ if $Q_g \subseteq Q_{g'}$. However, this intuition works for *P-systems*, but not for *A-systems* because from *Proposition* 9.(2) if $Q_g \subseteq Q_{g'}$ then $Q_{g'} \subseteq Q_g$. Thus non trivial comparisons of quanta of information in *A-systems* require a specialised notion of an i-quantum, which, in any case, is useful for *P-systems* too.

Definition 17 (Relativised quanta of information)
  – *Let* $\mathbf{A}$ *be an A-system. The* quantum of information *of* g *relative to a subset* $A \subseteq At$ *is defined as:* $Q_g \upharpoonright A = \{g' \in G : \forall a \in A, \forall x \in V_a((a(g) = x) \implies (a(g') = x))\}$.
  – *Let* $\mathbf{P}$ *be a P-system. The* quantum of information *of* g *relative to a subset* $A \subseteq M$ *is defined as:* $Q_g \upharpoonright A = \{g' \in G : \forall a \in A(g \Vdash a \implies g' \Vdash a)\}$.

Definition 18 (I-quantum dependence)
Let $\mathbf{S}$ *be an A-system or a P-system. Let* $A, A' \subseteq At$ *(or* $A, A' \subseteq M$), $g \in G$.

1. *We say that A' functionally depends on A at g, in symbols $A \mapsto_g A'$, if for all $g' \in G$, $g' \in Q_g \restriction A \Longrightarrow g' \in Q_g \restriction A'$ (that is, if $Q_g \restriction A \subseteq Q_g \restriction A'$).*
2. *We say that A' functionally depends on A, in symbols $A \mapsto A'$, if for all $g \in G$, $A \mapsto_g A'$.*
3. *If $A \mapsto A'$ and $A' \mapsto A$, we say that A and A' are informationally equivalent, $A \cong_I A'$ (thus, $A \cong_I A'$ if for all $g \in G$, $Q_g \restriction A = Q_g \restriction A'$).*

So, a set of attributes (properties) A' functionally depends on a set of attributes (properties) A if A has a higher discriminatory capability than A'.

Clearly, if $\mathbf{S}$ is an A-*system* then the notion of an i-quantum dependence relation turns into the usual notion of a *functional dependence*.

From now on, if $\Vdash \restriction X$ denotes the relation $\Vdash$ with co-domain restricted to X then with $\mathbf{S} \restriction X$ we shall denote the subsystem $\langle G, X, \Vdash \restriction X \rangle$. If $\mathbf{S}$ is an A-*system* and $X \subseteq At$, with $\mathbf{S} \restriction X$ we shall denote the subsystem $\langle G, X, \{V_a\}_{a \in X} \rangle$.

The following statement formalises the above intuitions with respect to i-quantum relations:

Proposition 17. *Let $\mathbf{S}$ be an A-system or a P-system. Let $A, A' \subseteq At$ $(A, A' \subseteq M)$ such that $A \mapsto A'$. Then $R_{(\mathbf{A} \restriction A)} \subseteq R_{(\mathbf{A} \restriction A')}$.*

Proof. The proof is immediate. Suppose $A \mapsto A'$. Then for all $g \in G$, $Q_g \restriction A \subseteq Q_g \restriction A'$, so that $\langle g, g' \rangle \in R_{(\mathbf{A} \restriction A)}$ implies $\langle g, g' \rangle \in R_{(\mathbf{A} \restriction A')}$.

It follows that we can naturally extend the notion of a functional dependence in order to compare two sets X and X' of properties or attributes from two distinct (property or attribute) systems $\mathbf{S}$ and $\mathbf{S}'$ over the same set of points G. Thus, we can extend the notion of "informational equivalence" to entire systems:

Definition 19. *Let $\mathbf{S}$ and $\mathbf{S}'$ be A-systems or P-systems over the same set of points G. Let S and S' be the sets of attributes (properties) of $\mathbf{S}$ and, respectively, $\mathbf{S}'$. We say that $\mathbf{S}$ and $\mathbf{S}'$ are informationally equivalent, in symbols $\mathbf{S} \cong_I \mathbf{S}'$, if and only if for any $g \in G$, $Q_g \restriction S = Q_g \restriction S'$, that is, if and only if $Q_g^{\mathbf{S}} = Q_g^{\mathbf{S}'}$.*

Informational equivalence tells something about the behaviour of cl and int:

Proposition 18. *Let $\mathbf{P}$ and $\mathbf{P}'$ be P-systems and $\mathbf{P} \cong_I \mathbf{P}'$. Then for all $x \in G$, $cl^{\mathbf{P}}(x) = cl^{\mathbf{P}'}(x)$. If both $cl^{\mathbf{P}}$ and $cl^{\mathbf{P}'}$ are topological, then $cl^{\mathbf{P}}(X) = cl^{\mathbf{P}'}(X)$ and $int^{\mathbf{P}}(X) = int^{\mathbf{P}'}(X)$, for any $X \subseteq G$.*

Proof. Suppose $cl^{\mathbf{P}}(x) \neq cl^{\mathbf{P}'}(x)$. Then there is $g \in G$ such that, say, $g \in cl^{\mathbf{P}}(x)$ and $g \notin cl^{\mathbf{P}'}(x)$. It follows that $\langle \Vdash^\smile \rangle(g) \subseteq \langle \Vdash^\smile \rangle(x)$ but $\langle \Vdash'^\smile \rangle(g) \not\subseteq \langle \Vdash'^\smile \rangle(x)$. Thus $x \in Q^{\mathbf{P}}(g)$ and $x \notin Q^{\mathbf{P}'}(g)$, so that $\mathbf{P} \not\cong_I \mathbf{P}'$. If both closure operators are additive, then by easy induction we obtain that $cl^{\mathbf{P}}(X) = cl^{\mathbf{P}'}(X)$ for any $X \subseteq G$. Moreover, suppose $int^{\mathbf{P}}(X) \neq int^{\mathbf{P}'}(X)$. Then $-int^{\mathbf{P}}(X) \neq -int^{\mathbf{P}'}(X)$, so that $cl^{\mathbf{P}}(-X) \neq cl^{\mathbf{P}'}(-X)$ - contradiction.

Notice that if either $cl^{\mathbf{P}}$ or $cl^{\mathbf{P}'}$ is not topological then the equality between $cl^{\mathbf{P}}$ and $cl^{\mathbf{P}'}$ is guaranteed just for singletons so that $cl^{\mathbf{P}}(-X) \neq cl^{\mathbf{P}'}(-X)$ is not a contradiction. Moreover, we can have $\mathbf{P}$ and $\mathbf{P'}$ such that $int^{\mathbf{P}}(x) \neq int^{\mathbf{P'}}(x)$ but still $\mathbf{P} \cong_I \mathbf{P'}$. Therefore, the relation $\cong_I$ is far to be considered the "best" way to compare P-systems, though very useful for our purposes.

Now we want to stress the fact that we can compare not only the informational behaviour of the same point g with respect two different sets of properties (attributes) X and X', but we can also compare the behaviours of two different points g and g' with respect to the same set of properties (attributes) P.

Definition 20. *Let* $\mathbf{S}$ *be an A-system or a P-system,* $X \subseteq M$ *(or* $X \subseteq At$*) and* $g, g' \in G$.

1. *We say that* g *is an* X-*specialisation of* g' *(or that* g' *is an* X-*approximation of* g*), in symbols* $g' \preccurlyeq_X g$, *if and only if the following condition holds:*

$$\forall x \in G(g' \in Q_x \restriction X \Longrightarrow g \in Q_x \restriction X).$$

2. *We say that* g *is a* specialisation *of* g', $g' \preccurlyeq g$, *if and only if* $g' \preccurlyeq_M g$.

Since for q-reflexivity $x \in Q_x$, any $x \in G$, if $g' \preccurlyeq_X g$ then $g \in Q_{g'} \restriction X$, so that $g' \preccurlyeq_X g$ says that g fulfills at least all the properties from X that are fulfilled by g'. Therefore, $g' \preccurlyeq g$ implies $\langle g', g \rangle \in R_{\mathbf{S}}$. Conversely, if $\langle g', g \rangle \in R_{\mathbf{S}}$ then $g \in Q_{g'}$. Hence $g' \in Q_x$ implies $g \in Q_x$, any $x \in G$, from transitivity of $R_{\mathbf{S}}$. It follows that the two relations $\preccurlyeq$ and $R_{\mathbf{S}}$ coincide. In fact they are the same instance of the usual topological notion of a *specialisation preorder*. In view of *Proposition* 9.(1) we can construct a topological space $\langle G, Im_Q \rangle$ on G whose specialisation preorder is indeed $\preccurlyeq$ (that is, $R_{\mathbf{S}}$).

11   Transforming Perception Systems

Now we are equipped with a sufficient machinery in order to compare transformed systems.

Let $\mathbf{A}$ be an *A-system*. To get a *P-system* out of $\mathbf{A}$, the basic step derives from the observation that any attribute a is actually a set of properties, namely the possible attribute values for a. Thus we start associating each attribute a with the family $\mathcal{N}(a) = \{a_v\}_{v \in V_a}$. We set $\mathcal{N}(At) = \bigcup_{a \in At} \mathcal{N}(a)$. For each value v, a_v is the property "taking value v for attribute a". This transform is usually called a "scale nominalisation". Now let us set a relation $\Vdash^{\mathcal{N}}$ as:

$$g \Vdash^{\mathcal{N}} a_v \text{ if and only if } a(g) = v, \text{ all } g \in G, a \in At, v \in V_a.$$

We call the resulting system, $\mathcal{N}(\mathbf{A}) = \langle G, \mathcal{N}(At), \Vdash^{\mathcal{N}} \rangle$, the "*nominalisation of* $\mathbf{A}$". $\mathcal{N}(\mathbf{A})$ will be called a *nominal A-system* or *NA-system*.

Proposition 19. *Let* $\mathbf{A}$ *be an A-system. Then: (a)* $\mathcal{N}(\mathbf{A})$ *is a P-system; (b)* $\mathcal{N}(\mathbf{A}) \cong_I \mathbf{A}$.

Proof. (a) is obvious. (b) Let us prove that for any $g \in G, Q_g \restriction At = Q_g \restriction \mathcal{N}(At)$. Indeed, if $g' \in Q_g \restriction At$, then $a(g) = x$ if and only if $a(g') = x$, all $a \in At$.

Therefore for any $x \in \mathcal{N}(a)$, $g \Vdash a_x$ if and only if $g' \Vdash a_x$, whence $g' \in Q_g \upharpoonright \mathcal{N}(a)$. Finally, $g' \not\Vdash a_{x'}$ for any other $x' \neq x$, so we have the reverse implication.

Moreover, if we formally consider P-systems as binary A-systems, we can also nominalise P-systems. But in this case we have a further property:

Proposition 20. Let $\mathbf{P}$ be a P-system. Then $\mathcal{N}(\mathbf{P})$ is a dichotomic system.

Proof. This is obvious, because for any property p, the nominalisation $\mathcal{N}(p) = \{p_1, p_0\}$ forms a pair of complementary properties, since for all $g \in G$, $g \Vdash^{\mathcal{N}} p_1$ if and only if $g \Vdash p$ and $g \Vdash^{\mathcal{N}} p_0$ if and only if $g \not\Vdash p$.

Nominalisation of dichotomic or functional systems does not give rise to any further result.

Proposition 21. If $\mathbf{P}$ is a DP system or an FP system, then $\mathcal{N}(\mathbf{P}) \cong_I \mathbf{P}$.

Proof. If $\mathbf{P}$ is dichotomic let $\langle p, \overline{p} \rangle$ be a pair of complementary properties. After nominalisation we shall obtain two pairs $\mathcal{N}(p) = \{p_1, p_0\}$ and $\mathcal{N}(\overline{p}) = \{\overline{p_1}, \overline{p_0}\}$. Clearly, for any $g \in G$, $g \Vdash p$ in $\mathbf{P}$ if and only if $g \Vdash^{\mathcal{N}} p_1$ in $\mathcal{N}(\mathbf{P})$. But $g \Vdash^{\mathcal{N}} p_1$ if and only if $g \not\Vdash^{\mathcal{N}} p_0$ if and only if $g \Vdash^{\mathcal{N}} \overline{p_0}$. Conversely, $g \Vdash \overline{p}$ if and only if $g \Vdash^{\mathcal{N}} p_0$ if and only if $g \not\Vdash^{\mathcal{N}} p_1$ if and only if $g \Vdash^{\mathcal{N}} \overline{p_1}$. If $\mathbf{P}$ is functional and $g' \in Q_g \upharpoonright M$ then $g \Vdash m$ if and only if $g' \Vdash m$, since $\langle \Vdash^{\smile} \rangle(g) = \langle \Vdash^{\smile} \rangle(g') = m$. Thus the proof runs as in *Proposition 19.(b)*.

For $\mathcal{N}(\mathbf{A})$ is not only a P-system but it is still an A-system with $At = \{0, 1\}$, we obtain the following corollary:

Corollary 15. Let $\mathbf{S}$ be an A-system or a P-system. Then $\mathcal{N}(\mathbf{S}) \cong_I \mathcal{N}(\mathcal{N}(\mathbf{S}))$.

Proof. If $\mathbf{S}$ is a P-system then $\mathcal{N}(\mathbf{S})$ is a dichotomic systems so that from *Proposition 21* $\mathcal{N}(\mathcal{N}(\mathbf{S})) \cong_I \mathcal{N}(\mathbf{S})$. If $\mathbf{S}$ is an A-system then $\mathcal{N}(\mathbf{S})$ is a binary A-system and from *Proposition 19.(b)* $\mathcal{N}(\mathbf{S}) \cong_I \mathcal{N}(\mathcal{N}(\mathbf{S}))$.

Corollary 16. If $\mathbf{A}$ is an A-system then there is a dichotomic system $\mathbf{D}$ such that $\mathbf{D} \cong_I \mathbf{A}$.

Proof. Since $\mathcal{N}(\mathbf{A})$ is a P-system, from *Proposition 20* $\mathcal{N}(\mathcal{N}(\mathbf{A}))$ is dichotomic. But from *Proposition 19.(b)* and *Corollary 15* $\mathbf{A} \cong_I \mathcal{N}(\mathbf{A}) \cong_I \mathcal{N}(\mathcal{N}(\mathbf{A}))$.

As a side result we again obtain *Proposition 9.(2)*. Notice that this *Proposition*, as well as *Corollary 15*, relies on the fact that we are dealing with deterministic A-systems so that either two objects converge on the same attribute-value, or they diverge, but not both.

EXAMPLE 4

Here are some examples: a P-system $\mathbf{P} = \langle G, M, \Vdash \rangle$, an FP-system $\mathbf{F} = \langle G, M', \hat{f} \rangle$ and an A-system $\mathbf{A} = \langle G, At, V \rangle$ over the same set G:

⊩	b b' b'' b'''	$\hat{f}$	m m' m''		A A' A''
a	1 1 0 0	a	1 0 0	a	1 b α
a'	0 1 0 1	a'	0 1 0	a'	0 c α
a''	0 1 1 1	a''	1 0 0	a''	1 b α
a'''	0 0 0 1	a'''	0 0 1	a'''	3 f δ

Considering the system **P** let $A = \{b, b'\}$ and $B = \{b'', b'''\}$. Then $Q_{a''} \upharpoonright A = \{a, a', a''\}$ while $Q_{a''} \upharpoonright B = \{a''\}$. It follows that $B \mapsto_{a''} A$. On the contrary, $Q_{a'} \upharpoonright A = \{a, a', a''\}$ and $Q_{a'} \upharpoonright B = \{a', a'', a'''\}$ are not comparable. Hence $B \mapsto A$ does not hold. If we compare the above systems we notice what follows:
a) $\mathbf{A} \not\cong_I \mathbf{P}$ because $Q_a^{\mathbf{A}} = \{a, a''\}$ while $Q_a^{\mathbf{P}} = \{a\}$. Neither $\mathbf{P} \mapsto \mathbf{A}$ because $Q_{a'}^{\mathbf{P}} = \{a', a''\}$ while $Q_{a'}^{\mathbf{A}} = \{a'\}$. b) $\mathbf{F} \cong_I \mathbf{A}$, because for all $g \in G, Q_g^{\mathbf{A}} = Q_g^{\mathbf{F}}$.
Let us now nominalise the systems $\mathbf{A}$ and $\mathbf{P}$:

⊩$^{\mathcal{N}}_{\mathbf{A}}$	A_0 A_1 A_3 A'_b A'_d A'_f A''_α A''_δ	⊩$^{\mathcal{N}}_{\mathbf{P}}$	b_1 b_0 b'_1 b'_0 b''_1 b''_0 b'''_1 b'''_0
a	0 1 0 1 0 0 1 0	a	1 0 1 0 0 1 0 1
a'	1 0 0 0 1 0 1 0	a'	0 1 1 0 0 1 1 0
a''	0 1 0 1 0 0 1 0	a''	0 1 1 0 1 0 1 0
a'''	0 0 1 0 0 1 0 1	a'''	0 1 0 1 0 1 1 0

Thus $\mathcal{N}(A) = \{A_0, A_1, A_3\}, \mathcal{N}(b) = \{b_1, b_0\}$ and so on. It is evident that, for instance, $a \in Q_{a''}^{\mathcal{N}(\mathbf{A})}$ and $a'' \in Q_a^{\mathcal{N}(\mathbf{A})}$. But the same happens already in $\mathbf{A}$. Indeed, $Q_a^{\mathbf{A}} = \{a, a''\} = Q_{a''}^{\mathbf{A}}$. On the contrary, $Q_{a'}^{\mathbf{P}} = \{a', a''\}$ but $Q_{a'}^{\mathcal{N}(\mathbf{P})} = \{a'\}$. In fact $a'' \in Q_{a'}^{\mathbf{P}}$ because it fulfills all the properties fulfilled by a' (i. e. b' and b''') plus the additional property b''. But in $\mathcal{N}(\mathbf{P})$ this latter fact prevents a''' from belonging to $Q_{a'}^{\mathcal{N}(\mathbf{P})}$, because property b'' splits into the pair $\langle b''_0, b''_1 \rangle$ and $a' \Vdash^{\mathcal{N}}_{\mathbf{P}} b''_0$ while $a'' \Vdash^{\mathcal{N}}_{\mathbf{P}} b''_1$, what are mutually exclusive possibilities. If we further nominalise $\mathcal{N}(\mathbf{P})$ and split, for instance, $\langle b''_0, b''_1 \rangle$ into $\langle b''_{0_1}, b''_{0_0}, b''_{1_1}, b''_{1_0} \rangle$, it is obvious that the pairs $\langle b''_{0_1}, b''_{1_0} \rangle$ and $\langle b''_{0_0}, b''_{1_1} \rangle$ give the same information as b''_0 and, respectively, b''_1. It is not difficult to verify that $R_{\mathcal{N}(\mathbf{A})} = R_{\mathbf{A}}$ so that $\mathcal{N}(\mathbf{A}) \cong_I \mathbf{Q}(\mathbf{A})$.

11.1   Dichotomic, Functional and Nominal Systems

First notice that the reverse of *Proposition* 9.(2) does not hold. For instance, if $\mathbf{P'}$ is such that $G = \{1, 2, 3, 4\}, M = \{A, B, C\}$ and ⊩ (1) $= \{A, B\}$, ⊩ (2) $= \{A, B\}$, ⊩ (3) $= \{B, C\}$ and ⊩ (4) $= \{B, C\}$, Q_g is an equivalence class, any $g \in G$ though $\mathbf{P'}$ is neither dichotomic nor functional. Also, if $\mathbf{A}$ is an *A-system*, then $\mathcal{N}(\mathbf{A})$ is not necessarily dichotomic. However $\mathcal{N}(\mathbf{A}) \cong_I \mathcal{N}(\mathcal{N}(\mathbf{A}))$ which is dichotomic (see *Corollary* 15). Indeed, notice that $\mathcal{N}(\mathcal{N}(\mathbf{A}))$ is informationally equivalent to the system defined as follows:

1) For each a_v in $\mathcal{N}(\mathbf{A})$, if V_a is not a singleton set $\neg a_v = \{a_{v'}\}_{v' \neq v, v' \in V_a}$, while if $V_a = \{v\}$ then set $\neg a_v = \{a_{v'}\}$. We set $P = \{a_v\}_{v \in V_a} \cup \{\neg a_v\}_{v \in V_a}$.
2) For each $g \in G$ set $g \Vdash^* \neg a_v$ if and only if $g \not\Vdash a_v$ and $g \Vdash^* a_v$ if and only if $g \Vdash a_v$. Clearly $\neg a_v$ is the complementary copy of a_v. Thus, 3) set $\mathbf{S} = \langle G, P, \Vdash^* \rangle$. We can easily verify that $\mathbf{S}$ is a dichotomic system and that $\mathbf{S} \cong_I \mathcal{N}(\mathbf{A})$.

In reversal, since for any *P-system* $\mathbf{P}$, $\mathcal{N}(\mathbf{P})$ induces an equivalence relation, we can ask whether $\mathcal{N}(\mathbf{P})$ itself "is", in some form, an *A-system*. Indeed it is trivially an *A-system* with set of attributes values $V = \{0, 1\}$ and such that $m_1(g) = 1$ iff $g \Vdash m_1$ iff $g \nVdash m_0$ iff $m_0(g) = 0$ and $m_1(g) = 0$ iff $g \Vdash m_0$ iff $g \nVdash m_1$ iff $m_0(g) = 1$, all $m \in M$ and by trivial inspection one can verify that $\langle G, \mathcal{N}(M), \{0, 1\}\rangle \cong_I \mathcal{N}(\mathbf{P})$.

Finally we discuss another natural equivalence. We know that if $\mathbf{S}$ is an *A-system*, or a *DP* or a *FP system* then $R_{\mathbf{S}}$ is an equivalence relation (see *Proposition 9*). Thus a question arises as how to define a functional system $F(\mathbf{S})$ informationally equivalent to a given A or DP system $\mathbf{S}$. The answer is simple. If $\mathbf{S}$ is a *P-system* consider it as an *A-system*. Any tuple $t \in \prod_{a \in At} V_a$ is a combination of attribute-values and has the form $\langle a_{1_m}, \ldots, a_{j_n}\rangle$. We set $g \Vdash^* t$ only if $a_1(g) = a_{i_m}$ for any $a_i \in At$ and $a_{i_m} \in t$. The resulting system $\langle G, \prod_{a \in At} V_a, \Vdash^*\rangle$ is the required $F(\mathbf{S})$. Indeed $\Vdash^*$ is a map because no $g \in G$ can satisfy different tuples. Thus $R_{F(\mathbf{S})}$ is an equivalence relation such that $\langle g, g'\rangle \in R_{F(\mathbf{S})}$ only if $a(g) = a(g')$ for all $a \in At$ (or in M). It follows that $\mathcal{N}(\mathbf{S}) \cong_I F(\mathbf{S})$ so that if $\mathbf{S}$ is dichotomic or it is an *A-system* then $R_{\mathbf{S}} = R_{F(\mathbf{S})}$ and $\mathbf{S} \cong_I F(\mathbf{S})$.

12   Conclusions

We have seen how modal operators naturally arise from the satisfaction relation which links points and properties in a Property System. Combinations of two modal operators which fulfill a adjunction relations define pre-topological interior and closure operators, as studied in Formal Topology. Thus we have shown that approaching approximation problems by means of the mathematical machinery provided by Formal Topology and Galois Adjunction theory makes it possible to define well-founded generalization of the classical upper and lower approximation operators.

Moreover Galois Adjunction theory provides a set of results that can be immediately applied to these operators, so that we have a good understanding of the structure of the system of their fixed points (i. e. exact sets).

We have also seen how to define higher order information systems, namely Information Quantum Relation Systems, from property systems in order to define topological (that is, continuous) approximation operators, through the notion of a "quantum of information". And we have shown when these operators coincide with the lower and upper approximations defined in classical Rough Set Theory.

Eventually, we have seen how we can make different kinds of information systems, property systems and attribute systems, into a uniform theoretical framework, and control these manipulations by means of a particular notion of an "informational equivalence" induced by the concept of quanta of information. This has practical consequences too. Indeed, the relational modal or/and topological operators that we have defined over *P-systems* may be directly translated into extremely simple constructs of functional languages such as LISP or APL (see [16]), thus providing a sound implementation. Therefore, this approach directly links the logical interpretation of approximation operators to the manipulation of concrete data structures for it coherently embeds the concrete operations on Boolean matrices into a very general logical framework (the same relational interpretation of a modal operator applies to any sort of binary Kripke frame).

References

1. M. Banerjee & M. Chakraborty, "Rough Sets Through Algebraic Logic". *Fundamenta Informaticae*, XXVIII, 1996, pp. 211-221.
2. Banerjee, M., Chakraborty, M. K.: Foundations of Vagueness: a Category-theoretic Approach. In Electronic Notes in Theoretical Comp. Sc., 82 (4), 2003.
3. Barwise; J., Seligman, J.: Information Flow: the Logic of Distributed Systems. Cambridge University Press, Cambridge, 1997.
4. Bell, J. L.: Orthologic, Forcing, and the Manifestation of Attributes" In C. TR. Chong & M. J. Wiks (Eds.): Proc. Southeast Asian Conf. on Logic. North-Holland, 1983, pp. 13-36.
5. Blyth, T. S., Janowitz, M. F.: Residuation Theory. Pergamon Press, 1972.
6. Chakraborty, M. K., Banerjee, M.: Dialogue in Rough Context. Fourth International Conference on Rough Sets and Current Trends in Computing 2004 (RSCTC'2004), June 1-June 5, 2004, Uppsala, Sweden.
7. Düntsch, I., Gegida, G.: Modal-style operators in qualitative data analysis. Proc. of the 2002 IEEE Int. Conf. on Data Mining, 2002, pp. 155-162.
8. Düntsch, I., Orłowska, E.: Mixing modal and sufficiency operators. In Bulletin of the Section of Logic, Polish Academy of Sciences, 28, 1999, pp. 99-106.
9. Gierz, G., Hofmann, K. H., Keimel, K., Lawson, J. D., Mislove, M. and Scott, D. S.: A compendium of Continuous Lattices. Springer-Verlag, 1980.
10. Humberstone, I. L.: Inaccessible worlds. In Notre Dame Journal of Formal Logic, 24 (3), 1983, pp. 346-352.
11. Lin, T. Y.: Granular Computing on Binary Relations. I: Data Mining and Neighborhood Systems. II: Rough Set Representation and Belief Functions. In Polkowski L. & Skowron A. (Eds.): Rough Sets in Knowledge Discovery. 1: Methodology and Applications, Physica-Verlag, 1998, pp.107-121 and 122-140.
12. Orłowska, E.: Logic for nondeterministic information. In Studia Logica, XLIV, 1985, pp. 93-102.
13. Pagliani, P.: From Concept Lattices to Approximation spaces: Algebraic Structures of some Spaces of Partial Objects. In Fund. Informaticae, 18 (1), 1993, pp. 1-25.
14. P. Pagliani: A pure logic-algebraic analysis on rough top and rough bottom equalities. In W. P. Ziarko (Ed.): Rough Sets, Fuzzy Sets and Knowledge Discovery, Proc. of the Int. Workshop on Rough Sets and Knowledge Discovery, Banff, October 1993. Springer-Verlag, 1994, pp. 227-236.
15. Pagliani, P.: Rough Set Systems and Logic-algebraic Structures. In E. Orlowska (Ed.): Incomplete Information: Rough Set Analysis, Physica Verlag, 1997, pp. 109-190.
16. Pagliani, P.: Modalizing Relations by means of Relations: a general framework for two basic approaches to Knowledge Discovery in Database. In Proc. of the International Conference on Information Processing and Management of Uncertainty in Knowledge-Based Systems. IPMU 98, July, 6-10, 1998. "La Sorbonne", Paris, France, pp. 1175-1182.
17. Pagliani, P.: A practical introduction to the modal relational approach to Approximation Spaces. In A. Skowron (Ed.): Rough Sets in Knowledge Discovery. Physica-Verlag, 1998, pp. 209-232.

18. Pagliani, P.: Concrete neighbourhood systems and formal pretopological spaces (draft). Conference held at the Calcutta Logical Circle Conference on Logic and Artificial Intelligence. Calcutta October 13-16, 2003.
19. Pagliani, P.: Pretopology and Dynamic Spaces. In Proc. of RSFSGRC'03, Chongqing, R. P. China 2003. Extended version in Fundamenta Informaticae, 59(2-3), 2004, pp. 221-239.
20. Pagliani, P.: Transforming Information Systems. In Proc. of RSFDGrC 2005, Vol. I, pp. 660-670.
21. Pagliani, P., Chakraborty, M. K.: Information Quanta and Approximation Spaces. I: Non-classical approximation operators. In Proc. of the IEEE Int. Conf. on Granular Computing. Beijing, R. P. China 2005, pp. 605-610.
22. Pagliani, P., Chakraborty, M. K.: Information Quanta and Approximation Spaces. II: Generalised approximation operators. In Proc. of the IEEE Int. Conf. on Granular Computing. Beijing, R. P. China 2005, pp. 611-616.
23. Pawlak, Z.: Rough Sets: A Theoretical Approach to Reasoning about Data. Kluwer, 1991.
24. Polkowski, L.: Rough Sets: Mathematical Foundations. Advances in Soft Computing, Physica-Verlag, 2002.
25. Sambin, G.: Intuitionistic formal spaces and their neighbourhood. In Ferro, Bonotto, Valentini and Zanardo (Eds.) Logic Colloquium '88, Elsevier (North-Holland), 1989, pp. 261-285.
26. Sambin, G., Gebellato, S.: A Preview of the Basic Picture: A New Perspective on Formal Topology. In TYPES 1998, pp. 194-207.
27. Sambin, G.: Formal topology and domains. In Proc. of the Workshop on Domains, IV. Informatik-Bericht, Nr. 99-01, Universität GH Siegen, 1999.
28. Smyth, M.: Powerdomains and predicate transformers: a topological view. In J. Diaz (Ed.) Automata, languages and Programming. Springer LNCS, 154, 1983, pp. 662-675.
29. Skowron, A. & Stepaniuk, J.: Tolerance Approximation Spaces. Fundamenta Informaticae, 27 (2-3), IOS Press, 1996, pp. 245-253.
30. Skowron, A., Swiniarski, R. & Synak, P.: Approximation Spaces and Information Granulation. Transactions on Rough Sets III, LNCS 3400, Springer, 2005, pp. 175-189.
31. Skowron, A., Stepaniuk, J., Peters, J. F. & Swiniarski, R.: Calculi of approximation spaces. Fundamenta Informaticae, 72 (1-3), 2006, pp. 363–378.
32. Vakarelov, D.: Information systems, similarity relations and modal logics. In E. Orlowska (Ed.) Incomplete Information - Rough Set Analysis Physica-Verlag, 1997, pp. 492-550.
33. Vickers, S.: Topology via Logic. Cambridge University Press, 1989.
34. Wille, R.: Restructuring Lattice Theory. In I. Rival (Ed.): Ordered Sets, NATO ASI Series 83, Reidel, 1982, pp. 445-470.
35. Yao, Y. Y.: Granular computing using neighborhood systems. In R. Roy, T. Fumhashi, and P.K. Chawdhry (Eds.): Advances in Soft Computing: Engineering Design and Manufacturing, Springer-Verlag, London, U.K., 1999.
36. Yao, Y. Y.: A comparative study of formal concept analysis and rough set theory in data analysis. Manuscript 2004.
37. Zhang, G. -Q.: Chu spaces, concept lattices and domains. In Proc. of the 19th Conf. on the Mathematical Found. of Programming Semantics. March 2003, Montreal, Canada. Electronic Notes in Theor. Comp. Sc., Vol. 83, 2004.

On Conjugate Information Systems: A Proposition on How to Learn Concepts in Humane Sciences by Means of Rough Set Theory

Maria Semeniuk–Polkowska

Chair of Formal Linguistics, Warsaw University
Browarna 8/12, 00991 Warsaw, Poland
m_polkowski@hotmail.com

To the memory of Professor Zdzisław Pawlak

Abstract. Rough sets, the notion introduced by Zdzisław Pawlak in early 80's and developed subsequently by many researchers, have proved their usefulness in many problems of Approximate Reasoning, Data Mining, Decision Making. Inducing knowledge from data tables with data in either symbolic or numeric form, rests on computations of dependencies among groups of attributes, and it is a well–developed part of the rough set theory.

Recently, some works have been devoted to problems of concept learning in humane sciences via rough sets. This problem is distinct as to its nature from learning from data, as it does involve a dialogue between the teacher and the pupil in order to explain the meaning of a concept whose meaning is subjective, vague and often initially obscure, through a series of interchanges, corrections of inappropriate choices, explanations of reasons for corrections, finally reaching a point, where the pupil has mastered enough knowledge of the subject to be able in future to solve related problems fairly satisfactorily.

We propose here an approach to the problem of learning concepts in humane sciences based on the notion of a conjugate system; it is a family of information systems, organized by means of certain requirements in order to allow a group of students and a teacher to analyze a common universe ofobjects and to correct faulty choices of attribute value in order to reach a more correct understanding of the concept.

Keywords: learning of cognitive concepts, rough sets, information systems, conjugate information systems.

1 Introduction

In addition to a constant flux of research papers on inducing knowledge from data expressed in either symbolic or numerical form, there are recently papers on learning cognitive concepts by means of the rough set theory, see, e.g., [2], [12], [13], [14], [15].

We propose in this work an approach to the problem of learning/teaching of concepts in humane sciences that stems from an analysis of the process of

J.F. Peters et al. (Eds.): Transactions on Rough Sets VI, LNCS 4374, pp. 298–307, 2007.
© Springer-Verlag Berlin Heidelberg 2007

learning in humane sciences, and of learning approach in library sciences, in particular, that has been worked out during our seminars at Warsaw University, [12], [13], [14], [15].

In the process of learning of humane concepts, in general, a teacher, a tutor, is directing the pupil, the student, toward understanding of the problem, and toward its correct solutions, by means of a dialogue that involves expositions, responses, corrections, explanations etc., etc., aimed at developing a satisfactory understanding of the concept meaning by the student.

In order to formally render this mechanism and to study this problem, we recall here a notion of a *conjugate information system* introduced in [14] (under the name of SP–systems), and discussed shortly in [12].

The main idea underlying this approach can be introduced as follows:

1. Both the tutor and the student are equipped with information/decision systems (see Sect. 2, for all relevant notions of the rough set theory) that possess identical sets of attributes and the same universe of objects, and they differ from each other only in value assignment to attributes; the assumption is, that the values of attributes are correctly assigned in the tutor system whereas the student can initially assign those values incorrectly, which results in a faulty classification of objects to decision classes, and those values are gradually corrected during the interchange of messages with the tutor;

2. In order to learn a correct assignment of values to attributes, the student has also in his disposal a family of auxiliary decision systems, one for each attribute. Attributes in those decision systems are for simplicity (and, actually, in conformity with the practice of learning in many parts of humane sciences) assumed to be Boolean; this means that the value of an attribute on an object is selected on the basis of whether the object has/has not given Boolean features (for instance, when deciding whether romance books should be acquired for a library, one may look at the feature: *majority/minority in a poll opted for romance books in the library*).

In what follows, we present a formal framework for conjugate information systems. We refrain from discussing here the interface between the tutor and the pupil, being satisfied with presenting the formal apparatus of conjugate information systems.

Our methodology presented in what follows was tested in our courses given to the students in the Department of Library and Information Sciences at the University of Warsaw. The author does express gratitude to her students to whom she is indebted for many works on applying the ideas presented in this paper.

2   Auxiliary Notions of Rough Set Theory

All basic notions relevant to rough sets may be found in [5], [9], or in [3].

We recall here, for the reader's convenience, some basic notions that are used in the sequel.

2.1   Information Systems

An *information system* $\mathcal{A}$ is defined as a triple (U, A, h) where:

1. U is a finite set of *objects*;
2. A is a finite set of *conditional attributes*. In the sequel, we will use the term *attribute* instead of the term *conditional attribute*. Attributes act on objects, to each pair of the form (a, u), where a is an attribute and u is an object, a value $a(u)$ is assigned. In our setting, we wish to work with systems where U and A are fixed, but value assignments are distinct, hence the need for the component h, representing the value assignment in a given information system;
3. A mapping $h : U \times A \to V$, with $h(u, a) = a(u)$, is an $\mathcal{A}$–*value assignment*, where $V = \bigcup\{V_a : a \in A\}$ is the attribute value set.

2.2   Decision Systems

Decision systems are a variant of information systems in which a new attribute $d \notin A$, called *the decision*, is introduced; formally, a decision system is a quadruple (U, A, d, h) where U, A are as in sect. 2.1, $d : U \to V_d$ is the decision with values in the decision value set V_d, and the value assignment h does encompass d, i.e., $h : U \times (A \cup \{d\}) \to V \cup V_d$ with the obvious proviso that values of h on pairs of the form (d, u) belong in V_d.

2.3   Indiscernibility and Its Extensions

The crucial notion on which the idea of rough sets does hinge is that of the *indiscernibility relation* [5], [6]. For an information system $I = (U, A, h)$, the indiscernibility relation $IND^I(B)$, induced by a set $B \subseteq A$ of attributes, is defined as follows,

$$IND^I(B) = \{(u, v) : h(a, u) = h(a, v) \text{ for } a \in B\}, \tag{1}$$

and equivalence classes of $IND^I(B)$ generate by means of the set–theoretical operation of the union of sets the family of B–exact sets (or, concepts); concepts that are not B–exact are called B–rough. Rough set theory deals with constructs that are invariant with respect to indiscernibility relations hence they can be expressed in terms of indiscernibility classes.

Indiscernibility classes may be generalized to μ–granules, where μ is a *rough inclusion* [11]. Rough inclusions are relations of the form $\mu(u, v, r)$ read as: "u is a part of v to degree at least r". Examples of rough inclusions and deeper applications can be found, e.g., in [10]; let us quote from there an example of the rough inclusion μ_L induced from the Łukasiewicz t–norm $t_L(x, y) = max\{0, x+y-1\}$ by means of the formula $\mu_L(u, v, r) \Leftrightarrow g(\frac{|DIS(u,v)|}{|A|}) \geq r$ in which $DIS(u, v) = \{a \in A : h(a, u) \neq h(a, v)\}$, $|A|$ stands for the cardinality of A, and g is a function from the reals into $[0, 1]$ that figures in the representation : $t_L(x, y) = g(f(x) + f(y))$ (see,e.g., [7]). As $g(x) = 1 - x$, the formula for μ_L is finally:

$$\mu_L(u, v, r) \text{ iff } \frac{|IND(u, v)|}{|A|} \geq r. \tag{2}$$

In case $r = 1$, one obtains from (2) indiscernibility classes as sets (granules) of the form $g(u)_1 = \{v : \mu(u, v, 1)\}$; for $r < 1$, one obtains a collection of granules being unions of indiscernibility classes with respect to various sets of attributes.

2.4   Approximations to Rough Concepts

One more crucial notion due to Zdzisław Pawlak is the notion of an approximation. In the classical case, any set (concept) $X \subseteq U$ is approximated with indiscernibility classes $[u]_B$ of the relation $IND^I(B)$, where $B \subseteq A$, from below (the lower approximation) and from above (the upper approximation):

$$\underline{B}X = \bigcup\{[u]_B : [u]_B \subseteq X\}, \tag{3}$$

$$\overline{B}X = \bigcup\{[u]_B : [u]_B \cap X \neq \emptyset\}. \tag{4}$$

More generally, one can replace in above definitions classes $[u]_B$ with the μ–granules $g(u)_r$ of a fixed or subject to some conditions radius r.

3   Conjugate Information Systems

The notion of a conjugate information system reflects the mechanism of learning in the interaction between the tutor and the student (or, students). In this process, the correct evaluation scheme is transferred from the tutor to students, who initially may have incorrect evaluation schemes and gradually learn better evaluations in order to finally come up with schemes satisfactorily close to the correct one.

3.1   On Conjugate Information Systems

By a *conjugate information system*, we understand a triple,

$$\mathcal{C} = (\{\mathcal{A}_i = (U_i, A_i, h_i) : i \in I\}, \{F_{a,i}^d : a \in A, i \in I\}, i_0), \tag{5}$$

where I is a set of participants in the learning process, with i_0 denoting the tutor and $i \in I \setminus \{i_0\}$ denoting students, consisting of:

1. a family of information systems $\{\mathcal{A}_i = (U_i, A_i, h_i) : i \in I\}$ such that for some finite sets U, A we have $U_i = U, A_i = A$ for $i \in I$;
2. a family of decision systems $\{F_{a,i}^d : a \in A \text{ and } i \in I\}$. Thus the difference between information systems $\mathcal{A}_i, \mathcal{A}_j$ with $i \neq j, i, j \in I$ is in functional assignments h_i, h_j. The information system corresponding to i_0 is said to be the *tutor system*;

3. for each pair $(a, i), a \in A, i \in I$, the decision system $F_{a,i}^d$ is a decision system $(U, Feat_a, a, h_{a,i})$, where U is the $\mathcal{C}$–universe, $Feat_a$ is the set of a–*features*, each $f \in Feat_a$ being a binary attribute, a is the decision attribute of the system $F_{a,i}^d$, and $h_{a,i}$ is the value assignment.

The system F_{a,i_0}^d will be regarded as the system belonging to the tutor, while its realization by an agent $i \in I, i \neq i_0$ will be the system $F_{a,i}^d$ of the student i for the evaluation of the attribute $a \in A$.

An Assumption. We assume that each system $F_{a,i}^d$ is reduced in the sense that for each value v_a of the attribute a, there exists at most one object u with the property that $h_{a,i}(u, a) = v_a$. In the case when such an object u exists (and then, by our assumption, it is unique) we will denote by $h_{a,i}^{\leftarrow}(v_a)$ the information vector $(f(u) : f \in Feat_a)$, i.e.,

$$h_{a,i}^{\leftarrow}(v_a) = (f(u) : f \in Feat_a). \tag{6}$$

The symbol $(h_{a,i}^{\leftarrow}(v_a))_f$ will denote the $f-th$ coordinate of the vector $h_{a,i}^{\leftarrow}(v_a)$.

We deliberately omit the communication aspect of the process; formally, its presence could be marked with some mappings between the corresponding systems. However, we deem it unnecessarily complicating the picture for the purpose of this paper.

3.2   A Metric on Conjugate Systems

It is now important to introduce some means for organizing the unordered as of now set of participants in the learning process; to this end, we exploit the idea in [12] of some distance function on a conjugate system.

We introduce a distance function $dist$ on the conjugate system $\mathcal{C}$. To this end, we first introduce for an object $u \in U$ the set,

$$DIS_{i,j}(u) = \{a \in A : h_i(a, u) \neq h_j(a, u)\}, \tag{7}$$

of attributes discerning on u between systems $\mathcal{A}_i$ and $\mathcal{A}_j$. Thus, $DIS_{i,j}(u)$ collects attributes which are assigned distinct values on the object u by students $i \neq j$.

Now, we let,

$$dist(\mathcal{A}_i, \mathcal{A}_j) = max_u |DIS_{i,j}(u)|. \tag{8}$$

The function $dist(\mathcal{A}_i, \mathcal{A}_j)$ is a pseudo–metric, i.e., it has all properties of a metric (see, e.g., [1]) except for the fact that its value may be 0 whereas the arguments may be formally distinct as discerned by distinct indices i and j; we offer a simple argument showing that $dist(., .)$ is a pseudo–metric. Only the triangle inequality may need a proof.

Thus, assume that information systems $\mathcal{A}_i, \mathcal{A}_j$, and $\mathcal{A}_p$ are given. If $h_i(a, u) = h_j(a, u)$ and $h_j(a, u) = h_p(a, u)$ then $h_i(a, u) = h_p(a, u)$; thus, $h_i(a, u) \neq h_p(a, u)$ implies that either $h_i(a, u) \neq h_j(a, u)$ or $h_j(a, u) \neq h_p(a, u)$.

In consequence,

$$DIS_{i,p}(u) \subseteq DIS_{i,j}(u) \cup DIS_{j,p}(u), \tag{9}$$

and from (9) one gets that,

$$max_u|DIS_{i,p}(u)| \leq max_u|DIS_{i,j}(u)| + max_u|DIS_{j,p}(u)|. \tag{10}$$

The formula (10) is the required triangle equality.

Learning starts with the pupil(s) closest to the tutor, and continues in the decreasing order of the distance.

3.3   Basic Parameters for Learning

At the learning stage, each agent $\mathcal{A}_j$ (represented by the corresponding information system) learns to assign values of attributes in its set A from features in decision systems $\{F_{a,j}^d : a \in A\}$.

Parameters for Learning Feature Values and Attribute Values. First, at the training stage, each agent student learns to assign correct values to features in sets $Feat_a = \{f_{k,a} : k = 1, 2, ..., n_a\}$ for each attribute $a \in A$.

We assume that values at the tutor system are already established as correct.

The measure of learning quality is the *assurance–level–function* $m_j(k, a)$; for each triple (j, k, a), where $j \in I \setminus \{i_0\}$, $k \leq n_a$, and $a \in A$, it is defined as follows:

$$m_j(k, a) = \frac{pos(j, k, a)}{ex(j, k, a)}, \tag{11}$$

where $ex(j, k, a)$ is the number of examples for learning $f_{k,a}$ and $pos(j, k, a)$ is the number of positively classified examples in the set U by the agent $\mathcal{A}j$.

The process of learning, as mentioned above, proceeds in a dialogue between the tutor and the student, aimed at explaining the meaning of the attribute a, and its dependence on features in the set $Feat_a$; after that discussion, the $j - th$ student proceeds with assigning values to features for objects from the universe U, in the training sample. The assignment is evaluated by the tutor and on the basis of that evaluation, assurance levels are calculated, to judge the understanding of the pupil.

We order features in $F_{a,j}^d$ according to decreasing value of $m_j(k, a)$; the resulting linear order is denoted ρ_j^a, and the system $F_{a,j}^d$ with values assigned by the agent $\mathcal{A}_j$ is denoted with the symbol $F_{a,j}^d(\rho)$.

Metrics on Value Sets of Attributes. We set a distance function $\phi_a^j(v, w)$ on values of the attribute a for each $a \in A$, $v, w \in V_a$, and $j \in I$, estimated in the system $F_{a,j}^d$ by letting,

$$\phi_a^j(v, w) = |DIS_{a,j}(v, w)|, \tag{12}$$

where,

$$DIS_{a,j}(v, w) = \{f \in Feat_a : (h_{a,j}^{\leftarrow}(v))_f \neq (h_{a,j}^{\leftarrow}(w))_f\}. \tag{13}$$

This definitions are possible, due to our assumption about systems $F_{a,j}^d$ in Sect.3.1.

Thus, $\phi_a^j(v, w)$ does express the distance at the pair v, w of values of the attribute a measured as the number of differently classified features in the row defined by v, w, respectively.

4   Learning of Attribute Values

We now address the problem of learning from the tutor of the correct evaluation of attribute values. Objects $u \in U$ are sent to each agent $\mathcal{A}_i$ for $i \in I$ one-by-one.

Step 1. Assignment of attribute values based on training examples. At that stage the values $dist(i, i_0)$ of distances from agents $\mathcal{A}_i$ to the tutor $\mathcal{A}_{i_0}$ are calculated.

Step 2. The feedback information passed from the tutor to the agent $\mathcal{A}_i$ is the following:

$$Inf_i = (r, Error\_set\_i = \{a_{i_1}, ...a_{i_{p_r}}\}, Error\_vector\_i = [s_{i_1}, ..., s_{i_{p_r}}]), \tag{14}$$

where:

1. r is the value of the distance $dist$ $(\mathcal{A}_{i_0}, \mathcal{A}_i)$ from the student i to the tutor i_0;
2. p_r is the number of misclassified attributes in A between agents i, i_0. Clearly, $p(r) \leq |U| \cdot r$, depends on r;
3. a_{i_j} is the jth misclassified attribute;
4. for $j \in \{1, .., p_r\}$, the value s_{i_j} is the distance $\phi_{a_{i_j}}(v_j, w_j)$ where v_j is the correct (tutor) value of the attribute a_{i_j} and w_j is the value assigned to a_{i_j} by the agent $\mathcal{A}_i$.

Step 3. The given agent $\mathcal{A}_i$ begins with the attribute $a = a_{Error-set_i}^i$ for which the value of the assurance-level-function is maximal (eventually selected at random from attributes with this property).

For the attribute a, the value $s = s_a$ is given, meaning that $s \times 100$ percent of features has been assigned incorrect values by $\mathcal{A}_i$ in the process of determining the value of a.

Step 4. Features in $Feat_a$ are now ordered into a set $F_{a,i}^d(\rho)$ according to decreasing values of the assurance–level–function $m_i(k, a)$ i.e. by ρ_i^a: starting with the feature $f = f_{Feat_a}^{i,a}$ giving the minimal value of the function $m_i(k, a)$, the agent i goes along the ordered set changing the value at subsequent nodes. If the value of ϕ remains unchanged after the change at the node, the error counter remains unchanged, otherwise its value is decremented/incremented by one.

Step 5. When the error counter reaches the value 0, stop and go to the next feature.

Step 6. Go to the next attribute in the set A.

5   An Example

Our example is a simple case that concerns grading essays written by students in French, taken from [13], [12].

Grading is done on the basis of three attributes: a_1: *grammar*, a_2: *structure*, and a_3: *creativity*. We present below tables showing the tutor decision systems F_a for $a = a_1, a_2, a_3$.

Example 1. Decision systems $F_{a_1}, F_{a_2}, F_{a_3}$

Table 1. Decision systems $F_{a_1}, F_{a_2}, F_{a_3}$

$f_{a_1}^1$	$f_{a_1}^2$	$f_{a_1}^3$	a_1	$f_{a_2}^1$	$f_{a_2}^2$	$f_{a_2}^3$	a_2	$f_{a_3}^1$	$f_{a_3}^2$	$f_{a_3}^3$	a_3
+	+	+	3	-	-	+	3	-	-	-	3
-	-	+	2	-	+	-	2	+	+	-	2
-	-	-	1	+	-	-	1	+	+	+	1

where $f_{a_1}^1$ takes value +/- when the percent of declination errors is $\geq$ /< 20 ; $f_{a_1}^2$ is +/- when the percent of conjugation errors is $\geq$ / < 20, and $f_{a_1}^3$ is +/- when the percent of syntax errors is $\geq$ / < 20; $f_{a_2}^1$ takes value +/- when the structure is judged *rich/not rich*, $f_{a_2}^2$ is +/- when the structure is judged *medium/not medium*, and $f_{a_2}^3$ is +/- when the structure is judged to be *weak/ not weak*. $f_{a_3}^1$ takes value +/- when the lexicon is judged *rich/not rich*, $f_{a_3}^2$ is +/- when the source usage is judged *extensive/not extensive*, and $f_{a_3}^3$ is +/- when the analysis is judged to be *deep/ not deep*.

Consider a pupil $\mathcal{A}_1$ and a testing information system with $U = \{t_1, t_2, t_3\}$, $A = \{a_1, a_2, a_3\}$ which is completed with the following value assignments.

Example 2. Information systems $\mathcal{A}_0$ of the tutor and $\mathcal{A}_1$ of the pupil.

Table 2. Decision systems $F_{a_1}, F_{a_2}, F_{a_3}$

t	a_1	a_2	a_3	t	a_1	a_2	a_3
t_1	1	2	1	t_1	1	2	2
t_2	1	1	1	t_2	1	1	2
t_3	3	2	3	t_3	3	2	3

The distance $dist(\mathcal{A}_0, \mathcal{A}_1)$ is equal to 1 as $DIS_{0,1}(t_1) = \{a_3\} = DIS_{0,1}(t_2)$; $DIS_{0,1}(t_3) = \emptyset$.

Thus, the pupil misclassified the attribute a_3 due to a faulty selection of feature values: in case of t_1, the selection by the tutor is +,+,+ and by the pupil: +,+,-. The distance $\phi_{a_3,1}$ is equal to 1 and the information sent to the pupil in case of t_1 is $Inf_1 = (1, \{a_3\}, (1))$.

Assuming the values of assurance–level–function $m(1, k, a_3)$ are such that $f^{3,a_3} = f^3_{a_3}$, the pupil starts with $f^3_{a_3}$ and error–counter =1 and changing the value at that node reduces the error to 0. This procedure is repeated with t_2 etc.

6 Conclusion

We have presented a skeleton on which the mechanism of learning cognitive concepts can be developed. It has been the principal aim in this paper to show that the notion of a conjugate information system may be helpful in fulfilling this task as a model of dependence between the tutor and the student.

Acknowledgement

The topic of the paper has been discussed at seminars conducted by the author at the Institute of Library and Information Sciences at Warsaw University. The author wishes to thank the participants in those seminars. Thanks go also to Professor Lech Polkowski for useful discussions and valuable help with the preparation of this note.

References

1. Bourbaki, N.: Éléments de Mathématique. Topologie Générale. Hermann, Paris, France (1960).
2. Dubois, V., Quafafou, M.: Concept learning with approximations: rough version spaces. In: Lecture Notes in Artificial Intelligence vol. 2475, Springer–Verlag, Berlin, Germany. (2002) 239–246.
3. Komorowski, J., Pawlak, Z., Polkowski, L., Skowron, A.: Rough sets: A tutorial. In: Pal, S. K., Skowron, A. (Eds.): Rough–Fuzzy Hybridization: A New Trend in Decision Making. Springer–Verlag, Singapore Pte. Ltd. (1999) 3–98.
4. Pal, S. K., Polkowski, L., Skowron, A. (Eds.): Rough-Neural Computing. Techniques for Computing with Words. Springer–Verlag, Berlin, Germany (2004).
5. Pawlak, Z.: Rough Sets: Theoretical Aspects of Reasoning about Data. Kluwer, Dordrecht, the Netherlands (1991).
6. Pawlak, Z.: Rough sets. *International Journal of Computer and Information Science* 11 (1982) 341–356.
7. Polkowski, L.: Rough Sets. Mathematical Foundations. Physica–Verlag, Heidelberg, Germany (2002).
8. Polkowski, L., Tsumoto, S., Lin, T. Y. (Eds.): Rough Set Methods and Applications. Physica–Verlag, Heidelberg, Germany (2000).
9. Polkowski, L., Skowron, A. (Eds.): Rough Sets in Knowledge Discovery 1,2. Physica–Verlag, Heidelberg, Germany (1998).

10. Polkowski, L., Semeniuk–Polkowska, M.: On rough set logics based on similarity relations. *Fundamenta Informaticae* 64 (2005) 379–390.
11. Polkowski, L., Skowron, A.: Rough mereology: A new paradigm for approximate reasoning. *International Journal of Approximate Reasoning* 15 (1997) 333-365.
12. Semeniuk-Polkowska, M., Polkowski, L.: Conjugate information systems: Learning cognitive concepts in rough set theory. In: Lecture Notes in Artificial Intelligence 2639, Springer–Verlag, Berlin, Germany. (2003) 255–259.
13. Semeniuk–Polkowska, M.: Applications of Rough Set Theory. Seminar Notes (in Polish), Fasc. II, III, IV. Warsaw University Press, Warsaw, Poland (2000–2002).
14. Semeniuk–Polkowska, M.: On Some Applications of Rough Sets in Library Sciences (in Polish). Warsaw University Press, Warsaw, Poland (1997).
15. Stępień, E.: A study of functional aspects of a public library by rough set techniques.PhD Thesis, Warsaw University, Department of Library and Information Sciences, M. Semeniuk-Polkowska, supervisor (2002).

Discovering Association Rules in Incomplete Transactional Databases*

Grzegorz Protaziuk and Henryk Rybinski

Institute of Computer Science, Warsaw University of Technology
gprotazi@ii.pw.edu.pl, hrb@ii.pw.edu.pl

Abstract. The problem of incomplete data in the data mining is well known. In the literature many solutions to deal with missing values in various knowledge discovery tasks were presented and discussed. In the area of association rules the problem was presented mainly in the context of relational data. However, the methods proposed for incomplete relational database can not be easily adapted to incomplete transactional data. In this paper we introduce postulates of a statistically justified approach to discovering rules from incomplete transactional data and present the new approach to this problem, satisfying the postulates.

Keywords: association rules, frequent itemsets, incompleteness, transactional data.

1 Introduction

Very often one of the main restrictions in using data mining methodology is imperfection of data, which is a common fact in real-life databases, especially those exploited for a long period. Imperfection can be divided into several different categories: inconsistency, vagueness, uncertainty, imprecision and incompleteness [19]. In the paper we consider the problem of discovering knowledge from incomplete database. Within the knowledge discovery process the incompleteness of data can be taken into consideration at two stages, namely (1) at the preprocessing step, and (2) at the data mining step. The objective of (1) is to fill missing values in order to pass to the next steps of the process and process data as they were complete. Here, one can use simple approaches, such as replacing unknown values by special ones (e.g. average or dominant value), as well as more advanced methods, such as e.g. completing data methods based on classifiers or sets of rules [8]. In the case of (2), missing or unknown values are subject of processing by the data mining algorithms.

In the literature many such approaches for different data mining tasks were introduced. In particular, the problem of classifying incomplete objects has been addressed in the context the rough set theory [20,21]. The main idea of the approach is based on the indiscernibility relation and lower and upper approximation of a given set X. Originally proposed for complete information system,

* Research has been supported by the grant No 3 T11C 002 29 received from Polish Ministry of Education and Science.

J.F. Peters et al. (Eds.): Transactions on Rough Sets VI, LNCS 4374, pp. 308–328, 2007.
© Springer-Verlag Berlin Heidelberg 2007

the approach was successfully extended to deal with incomplete data. Various modifications have been proposed and discussed in the papers [10,16,28,29,30]. Yet another group of data mining algorithms dealing with incomplete data can be distinguished, the algorithms from this group are based on the methods for building decision tree. The modification of the C4.5 algorithm was introduced in [25]. In the CART algorithm [6] the surrogate tests are used for dealing with missing values. Different aspects of using decision trees for working with incomplete data have been presented in [9,17,32].

In the paper we concentrate on the algorithms discovering association rules from incomplete data sets. In [12] a notion of *legitimate approach* has been defined. It consists in satisfying a set of postulates resulting from statistical properties of the support and confidence parameters and being necessary conditions. We claim that any method dealing with incompleteness should satisfy the postulates of the approach in order to properly asses expected support and confidence. The original definition of the postulates referred to the relational database. Here we generalize it, so that it also covers discovering association rules from transactional data. In addition we define a novel data mining algorithm (DARIT), very well suited for discovering rules from transactional databases and satisfying the postulates of the statistically justified approach.

The rest of the paper is organized in the following manner. Section 2 formally defines association rules and their properties. Section 3 reviews the methods of discovering association rules from incomplete data. Section 4 presents the concept of support and confidence under incompleteness. In Section 5 we discuss details of the presented DARIT algorithm. The results of experiments are presented in Section 6, whereas Section 7 concludes this paper.

2   Association Rules and Their Properties

Below we introduce basic notions necessary for analyzing the process of discovering rules from incomplete data. We consider two types of databases, namely transactional and relational ones. A transactional database, denoted as DT, consists of finite set of transactions, $DT = \{t_1, t_2, t_3, \ldots, t_n\}$. Each transaction has a unique identifier and a non empty set of elements (items). Each element belongs to the finite set of items $I = \{elem_1, \ elem_2, \ldots, elem_m\}$. A relational database, denoted by DR, is a finite set of records $DR = \{r_1, r_2, \ldots, r_k\}$. Each record consists of n scalar values, belonging to the domains of n attributes respectively. The set of attributes is further denoted by A. By *k-itemset* we denote a set of k items from the database. In the sequel, if we do not distinguish between relational and transactional database, we denote it by D. Association rules are one of the simplest and the most comprehensive forms for representing discovered knowledge. The problem of discovering association rules was first defined in [1], in the context of market basket data with the goal to identify customers' buying habits. An exemplary association rule would state that 70% customers who buy bread also buy milk.

The basic property of an itemset is its support. It is defined as percentage of those transactions in the database D, which contain given itemset. It is referred to as a relative support, and is formally defined as:

$$support(X) =| \{t \in D \mid X \subseteq t\}|/ \mid D \mid . \tag{1}$$

where X – itemset, t – transaction or record.

Sometimes the notion of *absolute support* of an itemset is used, which is defined as a number of the transactions supporting given itemset.

If a given transaction (record) includes an itemset X we say that the transaction (record) supports the itemset. *Frequent itemset* is an itemset with support not less than a given minimal level called minimal support and denoted by *minSup*. *Association rule* is an expression in the form: $X \rightarrow Y$, where X, Y are itemsets over I and $X \neq \emptyset, Y \neq \emptyset$ and $X \cap Y = \emptyset$. X is called an antecedent of the rule, Y is called a consequent of the rule. The support of the rule $X \rightarrow Y$ is equal to $support(X \cup Y)$. *Confidence* of the rule $X \rightarrow Y$, denoted by confidence$(X \rightarrow Y)$, is defined as:

$$confidence(X \rightarrow Y) = support(X \rightarrow Y)/support(X). \tag{2}$$

The parameter *minConf* is defined by the user and indicates minimal confidence that the discovered rules should satisfy.

The basic task concerning association rules is to find all such rules which satisfy the minimal support and minimal confidence requirements.

3   Related Works

The most known algorithm for discovering all association rules is the *Apriori* algorithm, proposed in [2]. In the algorithm candidate sets (potentially frequent itemsets) of the length k are generated from frequent itemsets of the length $k-1$. Another approach proposed in [11] is based on the special data structure called *FP-tree* (frequent pattern tree).

Various aspects of discovering association rules are widely presented and discussed in the literature. In [3,15,24,27,31] the problem of generating only interesting rules with respect to additional measure is considered. In [4,5,23] the methods for discovering rules from dense data were introduced. The lossless concise representations of frequent patterns were proposed in [7,14,22]. However, the problem of discovering association rules from incomplete data is discussed relatively rarely, especially with respect to transactional data sets. In the case of missing values the fundamental problem is evaluating the real support of a given itemset, so that one can determine if the given itemset is frequent or not.

In [26] the new definitions of support and confidence of the rule were introduced for the relational databases with missing values. In the definitions, the notions of *disabled data* and *valid database* are used. A record r is disabled for a given set X if, it includes at least one unknown value for the attributes, for which there are elements in X. A set of disabled records for X is denoted by

$Dis(X)$. A valid database vdb for the set X consists only of those records from DR which are not disabled for X:

$$vdb = DR\backslash Dis(X).$$

Given these notions the authors define support as: $support(X) = |DX|/|vdb(X)|$, where D_X stands for a set of records containing the set X. The confidence was defined as follows: confidence$(X \rightarrow Y) = |D_{XY}|/|D_X| - |Dis(Y) \cap D_X|)$. It is worth mentioning that the proposed definition of support may give rise to the situations in which a multielement set has greater support than its subsets.

This drawback has been eliminated in [13], where the probabilistic approach was presented. The approach is dedicated to relational databases. It is based on the assumption that the missing values of an attribute do not depend on the values of other attributes. Given an attribute a, unknown values in all records are assigned a probability distribution over all known values from the domain of the attribute. Each value v has assigned a probability, denoted by $prob(v, a)$, which is equal to its frequency of occurring in all records having known values for the considered attribute. The main idea of the approach is based of the notion of *probable support*. Probable support, denoted by $probSup_r$, is calculated for the element $elem(v, a)$ of the value v from the domain of the attribute a for single record in the following manner:

$$probSup_r(elem(v, a)) = \begin{array}{l} 1 \text{ if } r.a = v \\ prob(v, a) \text{ if } r.a \text{ is unknown} \\ 0 \text{ otherwise} \end{array} \qquad (3)$$

where $r.a$ stands for the value for the attribute a in the record r.

The support $probSup_r$ of a set $X = \{elem(v_1, a_1), elem(v_2, a_2), \dots, elem(v_k, a_k)\}$ is computed by the following formula:

$$probSupr(X) = probSup_r(elem(v_1, a_1)) * probSup_r(elem(v_2, a_2)) * \dots * \qquad (4)$$
$$probSup_r(elem(v_k, a_k))$$

A similar approach has been applied for transactional databases in [18] where the algorithm $\sim AR$ has been proposed. The algorithm is a modification of the well known *Apriori* algorithm. Also here the main idea is based on partial support of itemsets, in this case by transactions. The following way of calculating support of itemsets for single transaction was introduced: given k-itemset Z^k, each element included in a transaction t and the set Z^k contributes in $1/k$ to total value of support of the Z^k set calculated for the transaction t. The total value of support of the Z^k set is computed by summing up all values contributed by elements included in the transaction. If the set Z^k is contained in the transaction t then the value of support of the set calculated for the transaction is equal to $k * (1/k) = 1$.

The $\sim AR$ algorithm starts from replacing each unknown element u_elem in the transactions by all known items k_elem corresponding to the unknown ones in other transactions. Each element k_elem replaces the u_elem with the probability evaluated based on its frequency of occurring in the transaction, of which the k_elem elements come form. The authors assumed that such replacing is possible because of presence of names of classes of elements or unified ordering of items in the transactions. A value contributed by a single k_elem to the total value of support of the Z_k set calculated for the transaction t is additionally multiplied by the probability of its occurring.

4   Discovering of Association Rules Under Incompleteness

Incompleteness of data creates problem in interpreting minimal support and minimal confidence thresholds given by the user. In the case of missing or unknown values it is not possible to calculate exact values of the support and confidence measures. Instead, we can provide the estimation of these values and the range, limited by optimistic and pessimistic support (confidence), in which the true value is placed.

4.1   Support and Confidence Under Incompleteness

In the order to express the properties of data incompleteness we will apply the following notions:

- by $minSet(X)$ we denote the maximal set of records (transactions) which certainly support the itemset X.
- by $maxSet(X)$ we denote the maximal set of records (transactions) which possibly support the itemset X.
- by nkD we denote the maximal set of records (transactions) which are incomplete (i.e. include at least one item with unknown or missing value).
- by kmD we denote the maximal set of records (transactions) which include only items with known values.

Definition 1. Minimal (pessimistic) support of an itemset X, denoted further as $pSup(X)$, is defined as the number of records (transactions) which certainly support the itemset X, i.e. $pSup(X) = |minSet(X)|$.

Definition 2. Maximal (optimistic) support of an itemset X, denoted further as $oSup(X)$, is defined as the number of records (transactions) which possibly support the itemset $X : oSup(X) = |maxSet(X)|$.

The estimated (probable) support of an itemset X is denoted further as $probSup(X)$.

Definition 3. Minimal (pessimistic) confidence of a rule $X \rightarrow Y$, denoted further as $pConf(X \rightarrow Y)$, is defined as:

$$pConf(X \rightarrow Y) = |minSet(X \cup Y)|/|maxSet(X)|.$$

Definition 4. Maximal (optimistic) confidence of a rule $X \rightarrow Y$, denoted further as $oConf(X \rightarrow Y)$, is defined as:

$$oConf(X \rightarrow Y) = |maxSet(X \cup Y)|/(|maxSet(X \cup Y)| + |minSet(X) \setminus minSet(Y)|).$$

Definition 5. Estimated (probable) confidence of a rule $X \rightarrow Y$, denoted further as $probConf(X \rightarrow Y)$, is defined as:
$probConf(X \rightarrow Y) = probSup(X \cup Y)/probSup(X)$.

4.2 Support Calculation for Single Record or Transaction

The standard method of calculating support for a complete record can be expressed by the formula:

$$Sup_r(X) = \begin{array}{l} 1 \text{ if } X \text{ is present in a record } r \\ 0 \text{ otherwise} \end{array} \qquad (5)$$

where $Sup_r(X)$ denotes support of the itemset X calculated for the record r, $r \in DR$. Instead, in the case of incomplete records we estimate probable support, for which we can use a generalization of (5) in the form:

$$probSup_r(X) = prob(sat(X, r)) \text{ if } X \text{ may be present in a record } r \qquad (6)$$

where $prob(sat(X, r))$ denotes probability that the itemset X is present in the record r. Obviously if X is certainly present in a record r (all needed values in r are present) then $probSup_r(X) = 1$. If X cannot be present in r then $probSup_r(X) = 0$.

For transactional data the method of calculating support in a single transaction is analogous.

4.3 Postulates for Relational Data

The simplest approach to the calculation of support and confidence based on optimistic or pessimistic estimation does not promise good results, especially if the incompleteness of data is significant and cannot be neglected. To obtain result of a higher quality more advanced techniques for the support and confidence estimation should be used. In the literature some proposals in this direction have been published. As mentioned above, we claim that any such method should satisfy the postulates defined in [12], in order to properly asses statistically justified expected support and confidence. The original postulates refer to the relational database. Let us recall them:

(P1) $probSup(X) \in [pSup(X), oSup(X)]$
(P2) $probSup(X) \geq probSup(Y)$ for $X \subset Y$
(P3) $probConf(X \rightarrow Y) = probSup(X \cup Y)/probSup(X)$
(P4) $probConf(X) \in [pConf(X \rightarrow Y), oConf(X \rightarrow Y)]$
(P5) $\sum_{X Instances(A)} probSup(X) = 1$ for any $A \subseteq AT$

where X, Y are itemsets, and $Instances(A)$ is the set of all possible tuples over the set of attributes A.

Postulate P1 assures the natural limitation of estimated support – it can not be greater than optimistic support and less than pessimistic support. The second postulate says that an itemset can not be present in database more often than its proper subset. The postulate 3 introduces the standard way for confidence calculation. The fourth one is analogous to P1 but refers to confidence. The last one states that for freely chosen set of attributes the sum of support of all itemsets consisting of items which belong to a domain of selected attributes is equal to the number of records included in a database. In the [16] it is shown that P4 is redundant.

With the satisfied condition of P2 the estimated support is consistent with the fact that any superset of an itemset X does not occur more often than X. However, satisfying these criteria does not necessary means that the sum of support of supersets of an itemset X, specified on the same set of attributes, is not greater than support of X.

Example: For the record $r8$ from the base DR_n1 (Table 1) the probabilities Pr_o of occurrences are defined for the following sets: $Pr\_w(z1 = \{atr3 = a\}) = 2/3$, $Pr\_w(z2 = \{atr3 = a1\}) = 1/3$, $Pr_w(z3 = \{atr4 = b\}) = 1/2$, $Pr_w(z4 = \{atr4 = c\}) = 1/2$, $Pr_w(z5 = \{atr3 = a, atr4 = b\}) = 1/2$, $Pr_w(z6 = \{atr3 = a, atr4 = c\}) = 1/2$.

The support of an itemset Y calculated for the record $r8$ is as follows:

$$probSup(Y) = \begin{matrix} Pr\_w(zN) & \text{if } Y = zN \cup r8.known_i \\ 1 & \text{if } Y \subset r8.known \\ 0 & \text{Otherwise} \end{matrix}$$

where: zN – is one of the sets z_i defined above, $r8.known$ – is a set of all sets which can be generated from known values included in the record $r8$, $r8.known_i$ – one of sets included in the $r8.known$ set.

The support defined in this way satisfies all the postulates, however after counting supports of the following set $[Y1 = \{atr2 = x, atr3 = a\}$, $probSup(Y1) = 42/3]$, $[Y2 = \{atr2 = x, atr3 = a, atr4 = b\}$, $probSup(Y2) = 2/1]$, $[Y3 = \{atr2 = x, atr3 = a, atr4 = c\}$, $probSup(Y3) = 2/1]$ one can conclude that the sets $Y2$ and $Y3$ occur in the database more often than the set $Y1$, which of course is not possible.

Table 1. Relational database DR_n1

id	atr2	atr3	atr4	atr5	atr6
r1	x	A	b	Y	v1
r2	x	A	b	Y	v2
r3	x	A	c	Y	v3
r4	x	A	c	Y	v4
r5	x	a1	b1	Z	v1
r6	x	a1	c1	P	v3
r7	k	a2	b2	P	v3
r8	x	*	*	Y	v2

* – the missing value

To exclude such situations the postulate P2 has to be modified. Below the new version of the postulate is presented.

$$(P2n) : probSup(X) = \sum_{Z \in Instances(A)} probSup(X \cup Z) \text{ for any } A \subseteq AT. \quad (7)$$

where $X \neq \emptyset, Z \neq \emptyset$ and $X \cap Z = \emptyset$ for any set Z.

Rationale: Support of a set X can be counted by testing values in records only for those attributes over which the set X is defined. If we consider larger set of attributes then of course the support of the set X does not change. In this case we can say that we examine whether in a given record the set X is present along with an additional set Z defined over attributes out of those in X. If in place of Z consecutively all the sets defined over the additional attributes will be considered then finally we will obtain valid value of the support of X.

4.4   Postulates for Transactional Data

For transactional data the postulates from [12] can be used directly, except for P2n and P5. In the definitions below we use the following notations:

- $superset(X, k)$ is a set of all supersets of X which include k more elements than X.
- t_{max} is a maximal transaction i.e. transaction belonging to the given DT which has the most known elements.
- $|t|$ is the length of transaction t – number of elements included in this transaction
- DT_k – set of transaction which include at least k elements
- $sets(I_k)$ denotes sets of k-itemsets
- t_i denotes the ith transaction.

The postulate 2 for transactional data is defined as follows:

$$probSup(X) \geq \sum_{k=1}^{t_{max}|-|X|} (-1)^{(k+1)} \sum_{Y \in superset(X,k)} probSup(Y). \quad (8)$$

The inequality in the formula results from taking into consideration the different lengths of transactions.

The postulate 5 for transactional data is defined as follows:

$$\frac{\sum_{X \in sets(I_k)} probSup(X)}{\sum^{|D_k|}(C^k_{|t_i|}, t_i \in DT_k} = 1 \quad (9)$$

where $C^k_{|t|}$ is a number of all combinations of k items from t.
In the definition of the postulate the absolute support is used.

Rationale for Postulate 2: Calculation of a support of an itemset X can be done by summing number of transactions which contain only all elements included in X with the number of transactions that include the set X and at least one more element i.e. which contain a set $z = \{X \cup \{y\} | \{y\} \cap X = \emptyset, y - \text{an item}\}$. There are up to n such sets z, where n is a number of items y, which occur in the database but not in X. A number of transaction in which at least one of the sets z occurs as the percentage of the number of all transaction can be calculated by adopting the following formula (the probability of the sum of events):

$$P(A_1 \cup A_2 \cup \ldots \cup A_n) =$$
$$P(A_1) + P(A_2) + \ldots + P(A_n) - P(A_1 \cap A_2) - P(A_1 \cap A_3) - \ldots -$$
$$P(A_{n-1} \cap A_n) + P(A_1 \cap A_2 \cap A_3) + P(A_1 \cap A_2 \cap A_4) + \ldots + \qquad (10)$$
$$P(A_{n-2} \cap A_{n-1} \cap A_n) + \ldots + (-1)^{n-1} P(A_1 \cap \ldots \cap A_n).$$

Assuming that an event A_i represents occurrence of sets z_i^1 in a transaction (where z_i^1 is an ith set composed from item y_i and elements included in X) we can write the following equation:

$$P(Z^1) = \sum_{k=1}^{n} (-1)^{k+1} \sum_{z_i^k \in Z^k} P(z_i^k) \qquad (11)$$

where $Z^k = \text{superset}(X, k)$, and $P(Z^1)$ is the probability that the transaction t contains X and at least one additional item. If we replace the probability by frequencies, and express the frequencies by relative support of the sets we obtain the following formula:

$$ptobSup(Z^1) = \sum_{k=1}^{n} (-1)^{(k+1)} \sum_{z_i^k \in Z^k} probSub(z_i^k) \qquad (12)$$

As we are interested in calculating support of itemsets, we should consider only such z^k sets for which $k + |X| \leq |t_{max}|$. All the more numerous sets have support equal to 0. Hence, we can rewrite our formula in the following manner:

$$probSup(Z^1) = \sum^{|t_{max}| - |X|} (-1)^{(k+1)} \sum_{z_1^k \in Z^k} probSup(z_i^k). \qquad (13)$$

To calculate support of the set X we have to sum the values: (i) resulting from the formula (13) and (ii) support of X for the transaction with exactly the same items as in X. This leads us finally to the formula (8) above.

Rationale for postulate 5: Each transaction supports n k-itemsets, where $k \leq |t|$ and n is the number of all different k-elements sets which can be created from the items belonging to the transaction. Hence for each transaction t: $\sum_{X \in sets(I_k)} probSup_t(X) = C_{|t|}^k$, where $probSup_t(X)$ denotes the support of the itemset X counted for the transaction t.

The fulfillment of the postulate 5 requires estimation of support for each possible itemset, even if the given itemset is certainly infrequent. In the tasks of discovering association rules only frequent sets are interested, so there is no need to take into consideration infrequent itemsets. In our opinion the postulate 5 can be weaken by replacing the equality relation (=) by the relation ≤ in the formula (9). This modification preserves natural definition of the postulate and makes it more practical.

The modified postulates will be denoted as SJATD (Statistically Justified Approach for Transactional Data).

4.5   Postulate for Single Record or a Transaction

The postulates SJATD presented above provide conditions for the support calculated for the entire database. On the other hand they say nothing about required properties of support counted for a single record or transaction. For the methods that estimate support of an itemset X by summing its support calculated for single record it seems that more practical is to define conditions that should be fulfilled by support calculated for single record or transaction rather than calculated for entire database. Below we present such requirements.

Postulates for single record
(P1r) $probSup_R(X) \in [0,1]$
(P2r) $probSup_R(X) \geq probSup_R(Y)$ for $X \subset Y$
(P3r) $probSup_R(X) \geq \sum_{Z \in sets(At)} probSup_R(X \cup Y) X \neq \emptyset, Z \neq \emptyset$ and $X \cap Z = \emptyset$
for any set Z
(P4r) $\sum_{X \in set(At)}, probSup_R(X) = 1$ for each set of attributes $At \subseteq A$.

For transactional data the postulates: (P3r) and (P4r) have to be redefined. The appropriate formulas are given below.

Postulate P3r for transactional data:

$$probSup_T(X) \geq \sum_{k=1}^{|T|-|X|} (-1)^{(k+1)} \sum_{Y \in superset(X,k)} probSup_T(Y) \tag{14}$$

Postulate P4r for transactional data

$$\frac{\sum_{X \in sets(I_k)} probSup_T(X)}{C_{|T|}^k} = 1 \tag{15}$$

where $probSup_T(X)$ is a support of an itemsets X counted for transaction T.

5   Algorithm DARIT

In this section we present the new algorithm for discovering association rules from incomplete transactional data, called in the sequel DARIT (Discovering Association Rules in Incomplete Transactions). In our approach we allow that incomplete transaction may have any number of missing element. We start from the description of the data structure, called $mT\text{-}tree$, which is used in the algorithm.

5.1 The mT-Tree

The tree-like *mT-tree* structure is used for storing information concerning sets tested by DARIT. Each node of this tree has assigned level, (the root is at level 1). Data associated with itemsets (supports, items) are stored in a node in a table of elements, denoted further as *tblElem*. Each field of the *tblElem* table in a node at level l may have pointer to another node at level $l+1$. With each element stored in the *tblElem* table there are three values associated: *val_pSup*, *val_probSup*, *val_oSup*, which are used for calculating pessimistic, estimated and optimistic supports respectively. On Figure 1 a simplified structure of *mT-tree* is presented. In the tables are only shown elements and name of a *val_pSup* parameter with itemsets to which this parameter concerns. Items belonging to a path in the *mT-tree*, beginning from the root and composed of the fields of the *tblElem* tables, form a frequent or potentially frequent itemset. The items stored in the root of the *mT-tree* are frequent 1 itemsets. For instance, the set {a, b, c, d} is represented by the path ($n1[a]$, $n2[b]$, $n5[c]$, $n8[d]$) on Figure 1, where $nN[x]$ stands for the field in *tblElem* where the item x is stored in a node of number N.

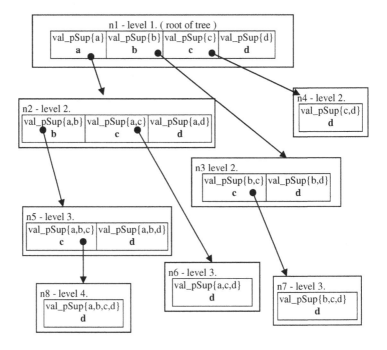

Fig. 1. mT-tree - simplified schema

5.2 Algorithm DARIT

The pseudo code of the algorithm DARIT is presented below.

```
Algorithm DARIT
1)    mTd: mT-tree;
2)      Apriori_Adapt(kmD, nkD, mTd)
3)      for each transaction t ∈ nkD
4)      begin
5)         Generate_Set_NZ(t, mTd);
6)         mTd.Modify_probSup(t,t.NZ);
7)      end
8)      Generate_Rules(mTd);
```

At the beginning the set of potentially frequent itemsets is generated (line 2) by calling the *Apriori_Adapt* procedure. In the procedure for determining potentially frequent itemsets the pessimistic support and minimal pessimistic supports thresholds are used. The minimal pessimistic support, denoted further as *min_minSup*, is an additional parameter of the algorithm. This parameter defines threshold which should be exceeded by the pessimistic support of each itemset in order to consider the itemset as potentially frequent. It allows for appropriate limitation of number of sets taken into consideration during execution of the procedure *Apriori_Adapt*, especially in the case of significant data incompleteness. In the next step of the DARIT algorithm for each incomplete transaction t, based on the sets stored in the *mT-tree* a set $t.NZ$ is generated; it consists of the sets which may occur in place of the special element *null*, which indicates missing elements in the transactions (line 5). For each element of the $t.NZ$ set the probability of occurring in the considered transaction is assigned. Based on the results obtained in this step, the values of the *probSup* support of itemsets represented in the *mT-tree* are modified (line 6). At the end of the algorithm the procedure *Generate_Rules* is called – it produces the association rules from the *mT-tree* using values of estimated support of itemsets.

Procedure Apriori_Adapt

```
Procedure Apriori_Adapt(Set of transaction kmD,nkD; mT-tree mTd)
1)     Add_Frequent_Items(kmD, nkD, mTd)
2)     p=2;
3)     while(Generate_Candidates(p, mTd)> 0)
4)     begin
5)        for each transaction t ∈ kmD
6)           Calculate_Support(t, mTd);
7)        for each transaction t ∈ nkD
8)           Calculate_Support_Incomplete(t, mTd);
9)        mTd.Remove_NotFrequent( minSup, min_minSup);
10)       p=p+1;
11)    end
```

The procedure *Apriori_Adapt* starts from adding 1-itemset potentially frequent to the root of the *mT-tree* (line 1). The function *Generate_Candidates* creates candidate sets and returns their number. Candidate sets in a node n at level p are generated by creating child nodes at level $p + 1$, for each field in the table $tblElem$, except for the last one. In the child node cn created for the jth field, the table of elements consists of all those elements from the table $tblElem$ of the parent node, which are stored in the fields of the index greater than j. The procedure $Calculate\_Support$ increases value of the optimistic support and of the pessimistic support for those of the candidate sets which are supported by a complete transaction t. The procedure $Calculate\_Support\_Incomplete$ (line 8) differs from the procedure $Calculate\_Support$ in that it increases values of the optimistic support for each candidate set. The method $Remove\_NotFrequent$ removes the candidate sets which certainly will not be frequent, and for the remaining candidate itemsets it sets value of $probSup$ support to the value of the pessimistic support.

Procedure Generate_SetNZ

```
Procedure Generate_Set_NZ(transaction t, mT-tree mTd)
1)    k = min(mTd.max_length_set -1, t.nb_known_items)
2)    while(Stop_condition = false and k > 0)
3)    begin
4)       t.NZ= t.NZ ∪ mTd.Find_set_nZ(t.known^k);
5)       k = k - 1;
6)       Stop_condidtion = Check_stop_condition();
7)    end
8)    Calculate_Probability(mTd);
```

At the beginning of procedure $Generate\_Set\_NZ$ the initial value of k is defined. It is calculated as a minimum of the 2 values (i) number of known items in the transaction t, and (ii) the number of elements in the most numerous potentially frequent set stored in $mT\text{-}tree$. Next, in the method $Find\_set\_nZ$ for each incomplete transaction t the $(k + j)$-itemsets, (denoted as zpc) are looked for in the $mT\text{-}tree$. Formally, $zpc = \{nZ \cup t.known_i^k\}$ and $t.known_i^k$ is a set consisting of k known elements from the transaction t. First, 1-item nZ sets are found, and then the $mT\text{-}tree$ is traversed deeper in order to find more numerous nZ sets. The nZ set is added to the $t.NZ$ set if it does not include known elements from the transaction t. With each set nZ the $parmProb$ parameter is associated. Further on, the parameter is used to estimate probability of occurrence of the given set nZ instead of the special element $null$ in the transaction, thus to estimate the value of the $probSup$ support. The value of the $parmProb$ is equal to the value of $minSup(zpc)$ of the currently examined set zpc, or if the given nZ set is already in the $t.NZ$ set, the value of the parameter is increased by $minSup(zpc)$. This procedure is repeated for $k = k - 1$ down to $k = 0$, or until the stop condition is met. The stop condition is fulfilled when the sum of the $parmProb$ parameters of 1-item nZ sets included in the $t.NZ$ set exceeds the following value:

$$max\_nb\_unknown = max\_len\_trans - t.nb\_known\_elem \tag{16}$$

where max_len_trans is the maximal number of the potentially frequent items included in the single transaction, and $t.nb\_known\_elem$ is the number of known items contained in the transaction t.

The Calculate_Probability procedure

The way in which the value of the parameter $parmProb$ of the nZ sets is computed causes that it can not be directly used as probability of occurrence of the given nZ set in a transaction. Generally, we have to deal with the following basic problems:

1) A value of the parameter for a single set or a sum of values for group of sets exceeds thresholds. In the former case it is 1 – the maximal value of probability, in the latter the threshold is associated with number of items included in a transaction, for instance, the sum of probabilities of occurrences of single items in a given transaction cannot be greater than the maximal length of the transaction. This condition may be expressed as: $\sum_{i=1..n} prob(item_i, t) \leq \max(|t|)$, where $prob(item_i, t)$ is the probability of occurring of ith item in the transaction t, $\max(|t|)$ is the maximal length of the transaction t, n is the number of items potentially frequent. The solution of this problem is a simple normalization of values to the required level.

2) Values of the parameter are very small. In this case the results obtained by applying the DARIT algorithm is comparable with the results obtained by using the methods in which the incompleteness of data is neglected (pessimistic support is used), but with much greater computational cost. To solve the problem these values are multiplied by certain ratio, which is calculated based of most probable number of items which should be present in the transaction in the place of the element $null$. The ratio is computed in such a way that the sum of probabilities of occurring of single items in a given transaction is not greater than the possible maximal length of the transaction.

The Modify_probSup method

The $Modify\_probSup$ method increases the value of the $probSup$ support for each such zpc set stored in the $mT\text{-}tree$, that $zpc = \{nZ \cup zZ\}$, where $nZ \in t.NZ$ and $zZ \in t.known$. Note that the set zZ may be empty. The value of the $probSup$ support is increased by the value of the $parmProb$ for the given nZ set.

The Generate_Rules procedure

The $Generate\_Rules$ procedure generates association rules from the sets stored in the $mT\text{-}tree$ in the following way: for each set cZ for which the minimal support requirements are fulfilled, all its supersets nZc are found, such that the value $probSup(nZc)/probSup(cZ)$ is not less than the minimal confidence threshold. Next, the rule $cZ \rightarrow \{nZc \backslash cZ\}$ is generated with support equal to $probSup(nZc)$ and confidence equal to $probSup(nZc)/probSup(cZ)$.

5.3   The SJATD Postulates

Theorem: DARIT satisfies the SJATD postulates.

Proof:

Postulate 1. Value of the *probSup* support for each set is of course not less than 0. The way of calculating *probSup* applied in the *Calculate_Probabilty* procedure ensures that the value of *probSup* support never exceeds 1.

Postulate 2. The method of computing the values of the *parmSup* parameters - summing the values of pessimistic support, ensures that the value of the parameter *parmSup* for any set will be not less than *parmSup* for its supersets. Multiplication of these values by the same factor does not influence this relation. Additionally, the way of adding the sets nZ to the set $t.NZ$ by traversing mT-*tree* in depth, and adding more and more numerous sets guarantees that for any given set all its subsets have been taken into consideration.

Postulate 3. Fulfilling this postulate follows from the method of computing the *probSup* support described in the proof for Postulate 2 and from the fact that the pessimistic support meets this postulate.

Postulate 5 (weakened). The postulate says that sum of the support of n-itemsets calculated for a single transaction cannot be greater than the number of k-elements sets which can be created from the items included in the transaction. According to the definition for an incomplete transaction, support of the set $dowZ$ is equal to:

- 1, if $dowZ \in t.known$,
- $nZ.parmSup$, if $dowZ = nZ$ or $dowZ = nZ \cup zZ$, $nZ \in t.NZ$, $zZ \in t.known$,
- 0 otherwise

where $t.known$ contains all the sets which can be formed from the known elements included in the transaction t.

Assuming that the number of elements nb_elem_t in the transaction t is equal to $\max(|t.NZ_k|) + t.nb\_known$, where $t.nb\_known$ denotes the number of known elements in the transaction t and $t.NZ_k$ denotes $k - itemsets$ from the set $t.NZ$, then number on n-elements sets, created from the items included in the transaction t is equal to $C^n_{nb\_elem\_t}$. The sum S_m of the support of n-itemsets counted for the transaction t can be split into tree parts:

$$S_m = S_{known} + S_{unknown} + S_{joined} \tag{17}$$

Of course, not all parts occur in all cases. The partial sums in the equation above means, respectively:

- S_{known} — the sum of the support of the sets $dowZ \in t.known$. The sum is equal to the number of n-elements sets created from the known items included in the transaction t.
- $S_{unknown}$ — the sum of supports of the sets $dowZ \in t.NZ$. In this case the sum is less than the number of the sets created from $\max(|t.NZ_k|)$ elements. It is ensured by the way of computing the normalization ratio in the method *Calculate_Probability*.
- S_{joined} — the sum of supports of the sets $dowZ = nZ \cup zZ$, $nZ \in t.NZ$, $zZ \in t.known$, $nZ \neq \emptyset, zZ \neq \emptyset$. The sum is smaller than the number of n-element sets created by joining the i-itemsets created from $t.nb\_known$

elements and j-itemsets created from the $\max(|t.NZ_k|)$ elements for $j, i > 0$ and $j + i = k$. The number of such sets can be computed from the following equation:

$$
\sum_{i=1}^{i=\min(n-1, t.nb\_known)} C_i^{t.nb\_known} * C_{n-1}^{\max(t.NZ_k)} \tag{18}
$$

for $i \leq t.nb\_known$ and $n - i \leq \max(|t.NZ_k|)$.

During the calculation of the sum S_{joined} the second factor is replaced by the sum of supports of the sets $nZ \in t.NZ_{n-i}$, which as it follows from explanation presented for S_{unknown} is smaller than the value of this factor in the original formula.

6   Experiments

The objective of the performed experiments was to evaluate the practical value of the proposed approach. In order to evaluate the quality of the results obtained by executing the DARIT algorithm the following procedure was applied. First, the sets of association rules and frequent itemsets, further denoted as the referential sets, were generated from a complete transactional database. Next, the incomplete database was created by random removing some elements from

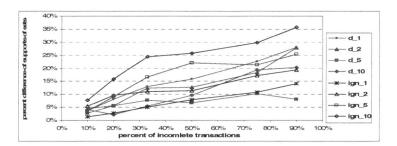

Fig. 2. Percent difference of supports of sets

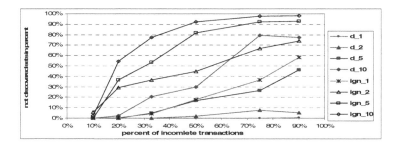

Fig. 3. Percent of non-discovered sets

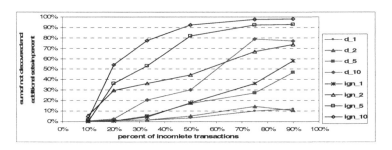

Fig. 4. Erroneous sets as percentage of the size of the referential set

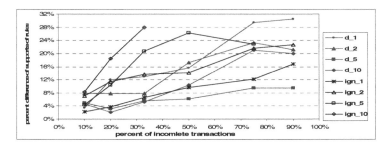

Fig. 5. Percent difference of supports of association rules

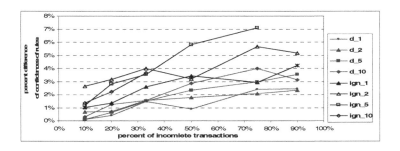

Fig. 6. Percent difference of confidences of association rules

some transactions. The number of incomplete transactions and the number of unknown elements in a single transaction vary in the experiments. The last step in the procedure was to use the DARIT algorithm to discover association rules in the incomplete data. The evaluation of quality of the results was based on the difference between the referential sets of association rules or frequent sets and the sets obtained from analyzing incomplete data. The difference was described by the following measures:

- percent difference between the supports of rules and frequent itemsets and between the confidence measures

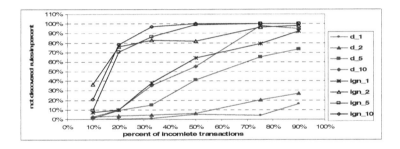

Fig. 7. Percent of non-discovered association rules

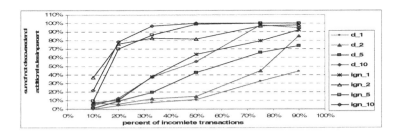

Fig. 8. Erroneous rules as a percentage of the size of the referential set

- percentage of the sets (rules) presented in the referential set that have not been discovered from incomplete data
- number of additional sets (rules) discovered from incomplete data but not present in the referential set, expressed as a percentage of the referential set.

The results obtained from DARIT were also compared with the results obtained from the other data mining algorithms, where the pessimistic support estimation is used. In the experiments the synthetic data were used. The set consisted of 592 transactions, the average length of the transactions was 20 elements. It was 1116 different items in the database. The test databases were incomplete in the degree between 10% up to 90% of transactions. For each case four situations with different number of missing elements in a single transaction were analyzed, namely the situation in which 1, 2, 5, or 10 elements in a single transaction were missing. In all experiments the minimal support threshold was set to 0.06 and the minimal confidence threshold was set 0.9. The tests were carried out for four values of minimal pessimistic support threshold calculated as a percentage of the minimal support value, namely 10%, 20%, 50%, 80% of this value. On the charts for Figure 2 to Figure 8 the average results for the minimal pessimistic support are presented. On the legend d denotes the results obtained from DARIT, *ign* stands for the *ignore* method, and the index at d or *ign* indicates the number of missing elements in a single transaction.

The results concerning frequent itemsets are presented on the figures 2, 3 and 4. The proposed approach is generally much better than the *ignore* method.

On Figure 4 the error of the algorithms DARIT and *ignore* is presented in the form of the sum of numbers of missing and erroneously discovered sets. Such a sum better presents differences between results obtained by the considered methods, since the *ignore* method does not produce erroneous sets. The results coming from those two methods are comparable only in the situations where the incompleteness of data is small. The presented results show that by applying the DARIT algorithm one can achieve good results also in the case of quite high incompleteness of data (up to 50% of incomplete transactions).

On Figures 5 – 8 we present the results concerning comparisons of the sets of discovered association rules. Also with respect to association rules DARIT is much better than the *ignore* method, though here the difference is smaller. The obtained results are generally worse than the results obtained for the frequent sets. The errors in estimating support rather cumulate then eliminate. In the case of association rules only the best configurations concerning the minimal pessimistic support thresholds allow to obtain good results when the incompleteness of data is higher.

7   Conclusions

In the paper we extended some postulates of legitimate approach to discovering association rules, as defined in [12], so that they may be applied also to transactional data. We have presented in detail a new approach for discovering association rules from incomplete transactional databases. In the presented DARIT algorithm we do not use any knowledge that is external to the dataset, but exploit only relations discovered in the investigated database. We have shown that the approach satisfies all the postulates of SJATD. We presented a number of experimental results using synthetic data. The performed experiments show that the proposed algorithm well foresees real values of the support and outperform the methods based on ignoring unknown values.

References

1. Agrawal R., Imielinski T., Swami A.: Mining Associations Rules between Sets of Items in Large Databases. In: Proc. of the ACM SIGMOD Conference on Management of Data, Washington, USA (1993) 207–216
2. Agrawal R., Srikant R.: Fast Algorithms for Mining Association Rules. In: Proc. of the 20th International Conference on Very Large Databases Conference (VLDB), Santiago, Chile, 1994. Morgan Kaufmann (1994) 487–499
3. Bayardo R.J., Agrawal R.: Mining the Most Interesting Rules. In: Proc. of the Fifth ACM SIGKDD International Conference on Knowledge Discovery and Data Mining (KDD), San Diego, CA, USA, 1999. ACM (1999) 145–154
4. Bayardo R.J., Agrawal R., Gunopulos D.: Constraint-Based Rule Mining in Large, Dense Databases. Data Mining and Knowledge Discovery, Vol. 4, No. 2/3 (2000) 217–240
5. Bayardo R. J. Jr.: Efficiently Mining Long Patterns from Databases, Proceedings of ACM SIGMOD International Conference on Management of Data, Seattle, (1998).

6. Breiman L., Friedman J. H., Olshen R. A. Stone C. J.: Classification and regression trees, Belmont Wadsworth, (1984).
7. Calders T., Goethals B.: Mining All Non-derivable Frequent Item Sets, Proc. of Principles of Data Mining and Knowledge Discovery, 6th European Conf., Helsinki, (2002)
8. Dardzińska-Głębocka A., Chase Method Based on Dynamic Knowledge Discovery for Prediction Values in Incomplete Information Systems, PhD thesis, Warsaw, 2004.
9. Friedman H. F., Kohavi R., Yun Y., Lazy decision trees, Proceedings of the 13th National Conference on Artificial Intelligence, Portland, Oregon, (1996)
10. Grzymala-Busse J. W.: Characteristic Relations for Incomplete Data: A Generalization of the Indiscernibility Relation, Proceedings Rough Sets and Current Trends in Computing, 4th International Conference, Uppsala, (2004).
11. Han J., Pei J., Yin Y.: Mining Frequent Patterns without Candidate Generation. In: Proc. Of the 2000 ACM SIGMOD International Conference on Management of Data, Dallas, Texas, USA, 2000. SIGMOD Record, Vol. 29, No. 2 (2000) 1–12
12. Kryszkiewicz M., Rybinski H.: Legitimate Approach to Association Rules under Incompleteness. In: Foundations of Intelligent Systems. Proc. of 12th International Symposium (ISMIS), Charlotte, USA, 2000. Lecture Notes in Artificial Intelligence, Vol. 1932. Springer-Verlag (2000) 505–514
13. Kryszkiewicz M.: Probabilistic Approach to Association Rules in Incomplete Databases, Proceedings of Web-Age Information Management, First International Conference, WAIM 2000, Shanghai, (2000).
14. Kryszkiewicz M.: Concise Representation of Frequent Patterns based on Disjunction-Free Generators. In: Proc. of the 2001 IEEE International Conference on Data Mining (ICDM), San Jose, California, USA, 2001. IEEE Computer Society (2001) 305–312
15. Kryszkiewicz M.: Representative Association Rules. In: Research and Development in Knowledge Discovery and Data Mining. Proc. of Second Pacific-Asia Conference (PAKDD). Melbourne, Australia, 1998. Lecture Notes in Computer Science, Vol. 1394. Springer (1998) 198–209
16. Kryszkiewicz M.: Concise Representations of Frequent Patterns and Association Rules Habilitation Thesis, Warsaw University of Technology, (2002)
17. Liu W. Z., White A. P.: Thompson S. G., Bramer M. A.: Techniques for Dealing with Missing Values in Classification, Proceedings of Advances in Intelligent Data Analysis, Reasoning about Data, Second International Symposium, London, (1997)
18. Nayak J. R., Cook D. J.: Approximate Association Rule Mining, Proceedings of the Fourteenth International Artificial Intelligence Research Society Conference, Key West, Florida, (2001)
19. Parsons S.: Current Approach to Handling Imperfect Information in Data and Knowledge Bases, IEEE Transaction on knowledge and data engineering Vol. 8, (1996)
20. Pawlak Z.: Rough Sets. International Journal of Information and Computer Sciences No. 11 (1982) 341–356
21. Pawlak Z.: Rough Sets: Theoretical Aspects of Reasoning about Data. Kluwer Academic Publishers, Vol. 9 (1991)
22. Pasquier N., Bastide Y., Taouil R., Lakhal L.: Discovering Frequent Closed Itemsets for Association Rules. In: Proc. of Database Theory - ICDT '99. Proc. of 7th International Conference (ICDT), Jerusalem, Israel, 1999. Lecture Notes in Computer Science, Vol. 1540. Springer (1999) 398–416

23. Protaziuk G., Sodacki P., Gancarz ., Discovering interesting rules in dense data, The Eleventh International Symposium on Intelligent Information Systems, Sopot, (2002).
24. Bastide Y., Pasquier N., Taouil R., Stumme G., Lakhal L.: Mining Minimal Non-redundant Association Rules Using Frequent Closed Itemsets. Comp. Logic (2000) 972–986
25. Quinlan J. R., C4.5 Programs for Machine Learning, San Mateo, California, (1993)
26. Ragel A., Cremilleux B.: Treatment of Missing Values for Association Rules. In: Research and Development in Knowledge Discovery and Data Mining. Proc. of Second Pacific-Asia Conference (PAKDD). Melbourne, Australia, 1998. Lecture Notes in Computer Science, Vol. 1394. Springer (1998) 258–270
27. Srikant R., Vu Q., Agrawal R.: Mining Association Rules with Item Constraints. In: Proc. Of the Third International Conference on Knowledge Discovery and Data Mining (KDD). Newport Beach, California, USA, 1997. AAAI Press (1997) 67–73
28. Stefanowski J., Tsoukias A.: Incomplete Information Tables and Rough Classification. Int. Journal of Computational Intelligence, Vol. 17, No 3 (2001) 545–566
29. Stefanowski J.: Algorytmy indukcji regu decyzyjnych w odkrywaniu wiedzy (Algorithms of Rule Induction for Knowledge Discovery). Habilitation Thesis, Poznan University of Technology, No. 361 (2001)
30. Wang G.: Extension of Rough Set under Incomplete Information Systems, Proceedings of the 2002 IEEE International Conf. on Fuzzy Systems, Honolulu, (2002)
31. Zaki M.J.: Generating Non-Redundant Association Rules. In Proc. of 6th ACM SIGKDD International Conference on Knowledge Discovery and Data Mining, Boston, MA, 2000. ACM Press (2000) 34–43
32. Zhang J., Honavar V.: Learning Decision Tree Classifiers from Attribute Value Taxonomies and Partially Specified Data, Proceedings of the Twentieth International Conference (ICML 2003), Washington, DC, (2003)

On Combined Classifiers, Rule Induction and Rough Sets

Jerzy Stefanowski

Institute of Computing Science
Poznań University of Technology,
60-965 Poznań, ul.Piotrowo 2, Poland
Jerzy.Stefanowski@cs.put.poznan.pl

Abstract. Problems of using elements of rough sets theory and rule induction to create efficient classifiers are discussed. In the last decade many researches attempted to increase a classification accuracy by combining several classifiers into integrated systems. The main aim of this paper is to summarize the author's own experience with applying one of his rule induction algorithm, called MODLEM, in the framework of different combined classifiers, namely, the bagging, n^2–classifier and the combiner aggregation. We also discuss how rough approximations are applied in rule induction. The results of carried out experiments have shown that the MODLEM algorithm can be efficiently used within the framework of considered combined classifiers.

1 Introduction

Rough sets theory has been introduced by Professor Zdzisław Pawlak to analyse granular information [25,26]. It is based on an observation that given information about objects described by attributes, a basic relation between objects could be established. In the original Pawlak's proposal [25] objects described by the same attribute values are considered to be *indiscernible*. Due to limitations of available information, its natural granulation or vagueness of a representation language some elementary classes of this relation may be *inconsistent*, i.e. objects having the same descriptions are assigned to different categories. As a consequence of the above inconsistency it is not possible, in general, to precisely specify a set of objects in terms of elementary sets of indiscernible objects. Therefore, Professor Zdzisław Pawlak introduced the concept of the *rough set* which is a set characterized by a pair of precise concepts – *lower* and *upper approximations* constructed from elementary sets of objects.

This quite simple, but smart, idea is the essence of the Pawlak's theory. It is a starting point to other problems, see e.g. [27,20,26,9]. In particular many research efforts have concerned *classification* of objects represented in data tables. Studying relationships between elementary sets and categories of objects (in other terms, target concepts or decision classes in the data table) leads to, e.g., evaluating dependency between attributes and objects classification, determining the level of this dependency, calculating importance of attributes for objects

J.F. Peters et al. (Eds.): Transactions on Rough Sets VI, LNCS 4374, pp. 329–350, 2007.
© Springer-Verlag Berlin Heidelberg 2007

classification, reducing the set of attributes or generating decision rules from data. It is also said that the aim is to synthesize reduced, approximate models of concepts from data [20]. The *transparency* and *explainability* of such models to human is an important property. Up to now rough sets based approaches were applied to many practical problems in different domains – see, e.g., their list presented in [20].

Besides "classical" rough sets, based on the indiscernibility relation, several generalizations have been introduced. Such data properties as, e.g., imprecise attribute values, incompleteness, preference orders, are handled by means of tolerance, similarity, fuzzy valued or dominance relations [9,20,37].

Looking into the previous research on rough sets theory and its applications, we could distinguish two main perspectives: *descriptive* and *predictive* ones.

The descriptive perspective includes extraction information patterns or regularities, which characterize some properties hidden in available data. Such patterns could facilitate understanding dependencies between data elements, explaining circumstances of previous decisions and generally gain insight into the structure of the acquired knowledge. In this context presentation of results in a human readable form allowing an interpretation is a crucial issue.

The other perspective concerns *predicting* unknown values of some attributes on the basis of an analysis of previous examples. In particular, it is a prediction of classes for new object. In this context rough sets and rules are used to construct a classifier that has to classify new objects. So, the main evaluation criterion is a *predictive classification accuracy*. Let us remind that the predictive classification has been intensively studied since many decades in such fields as machine learning, statistical learning, pattern recognition. Several efficient methods for creating classifiers have been introduced; for their review see, e.g., [16,19,23]. These classifiers are often constructed with using a search strategy optimizing criteria strongly related to predictive performance (which is not directly present in the original rough sets theory formulation). Requirements concerning interpretability are often neglected in favor of producing complex transformations of input data – an example is an idea of support vector machines.

Although in both perspectives we could use the same knowledge representation – rules, since motivation and objectives are distinct, algorithmic strategies as well as criteria for evaluating a set of rules are quite different. For instance, the prediction perspective directs an interest to classification ability of the complete rules, while in the descriptive perspective each rule is treated individually as a possible representative of an 'interesting' pattern evaluated by measures as confidence, support or coverage - for a more exhaustive discussion see, e.g., [42].

In my opinion, basic concepts of the rough sets theory have been rather considered in the way similar to a descriptive analysis of data tables. Nevertheless, several authors have developed their original approaches to construct decision rules from rough approximations of decision classes which joined together with classification strategies led to good classifiers, see e.g. [1,11,20,34,37]. It seems to me that many authors moved their interest to this direction in the 90's because of at least two reasons: (1) a research interest to verify whether knowledge derived from "closed world" of the data table could be efficiently applied to new objects coming from the "open world" – not seen in the analysed data table; (2) as a result of working with real life applications.

Let us also notice that the majority of research has been focused on developing *single classifiers* – i.e. based on the single set of rules. However, both empirical observations and theoretical works confirm that one cannot expect to find one single approach leading to the best results on overall problems [6]. Each learning algorithm has its *own area of superiority* and it may outperform others for a specific subset of classification problems while being worse for others. In the last decade many researches attempted to increase classification accuracy by combining several single classifiers into an *integrated system*. These are sets of learned classifiers, whose individual predictions are combined to produce the final decision. Such systems are known under names: *multiple classifiers, ensembles* or *committees* [6,45]. Experimental evaluations shown that these classifiers are quite effective techniques for improving classification accuracy. Such classifiers can be constructed in many ways, e.g., by changing the distributions of examples in the learning set, manipulating the input features, using different learning algorithms to the same data, see e.g. reviews [6,45,36]. Construction of integrated classifiers has also attracted the interest of some rough sets researchers, see e.g. [2,8,24]. The author and his co-operators have also carried out research, first on developing various rule induction algorithms and classification strategies (a review is given in [37]), and then on multiple classifiers [18,36,38,40,41].

The main aim of this paper is to summarize the author's experience with applying one of his rule induction algorithm, called MODLEM [35], in the framework of different multiple classifiers: the popular bagging approach [4], the n^2-classifier [18] – a specialized approach to solve multiple class learning problems, and the combiner approach to merge predictions of heterogeneous classifiers including also MODLEM [5]. The second aim is to briefly discuss the MODLEM rule induction algorithm and its experimental evaluation.

This paper is organized as follows. In the next section we shortly discuss rule induction using the rough sets theory. Section 3 is devoted to the MODLEM algorithm. In section 4 we briefly present different approaches to construct multiple classifiers. Then, in the successive three sections we summarize the experience of using rule classifiers induced by MODLEM in the framework of three different multiple classifiers. Conclusions are grouped in section 8.

2   Rules Generation and Rough Sets

2.1   Notation

Let us assume that objects – learning examples for rule generation – are represented in *decision table* $DT = (U, A \cup \{d\})$, where U is a set of objects, A is a set of *condition attributes* describing objects. The set V_a is a domain of a. Let $f_a(x)$ denotes the value of attribute $a \in A$ taken by $x \in U$; $d \notin A$ is a decision attribute that partitions examples into a set of decision classes $\{K_j : j = 1, \ldots, k\}$.

The *indiscernibility relation* is the basis of Pawlak's concept of the rough set theory. It is associated with every non-empty subset of attributes $C \subseteq A$ and $\forall x, y \in U$ is defined as $x I_C y \Leftrightarrow \{(x, y) \in U \times U \ f_a(x) = f_a(y) \ \forall a \in C \}$.

The family of all equivalence classes of relation $I(C)$ is denoted by $U/I(C)$. These classes are called *elementary sets*. An elementary equivalence class containing element x is denoted by $I_C(x)$.

If $C \subseteq A$ is a subset of attributes and $X \subseteq U$ is a subset of objects then the sets: $\{x \in U : I_C(x) \subseteq X\}$, $\{x \in U : I_C(x) \cap X \neq \emptyset\}$ are called *C-lower* and *C-upper* approximations of X, denoted by $\underline{C}X$ and $\overline{C}X$, respectively. The set $BN_C(X) = \underline{C}X - \overline{C}X$ is called the *C-boundary* of X.

A decision rule r describing class K_j is represented in the following form:

$$if \ P \ then \ Q,$$

where $P = w_1 \wedge w_2 \wedge \ldots w_p$ is a condition part of the rule and Q is decision part of the rule indicating that example satisfying P should be assigned to class K_j. The elementary condition of the rule r is defined as $(a_i(x) \ rel \ v_{a_i})$, where rel is a relational operator from the set $\{=, <, \leq, >, \geq\}$ and v_{a_i} is a constant being a value of attribute a_i.

Let us present some definitions of basic rule properties. $[P]$ is a *cover* of the condition part of rule r in DT, i.e. it is a set of examples, which description satisfy elementary conditions in P. Let B be a set of examples belonging to decision concept (class K_j or its appropriate rough approximation in case of inconsistencies). The rule r is discriminant if it distinguishes positive examples of B from its negative examples, i.e. $[P] = \bigcap[w_i] \subseteq B$. P should be a minimal conjunction of elementary conditions satisfying this requirement. The set of decision rules R completely describes examples of class K_j, if each example is covered by at least one decision rules.

Discriminant rules are typically considered in the rough sets literature. However, we can also construct *partially discriminant* rules that besides positive examples could cover a limited number of negative ones. Such rules are characterized by the *accuracy* measure being a ratio covered positive examples to all examples covered by the rule, i.e. $[P \cap B]/[P]$.

2.2   Rule Generation

If decision tables contain inconsistent examples, decision rules could be generated from rough approximations of decision classes. This special way of treating inconsistencies in the input data is the main point where the concept of the rough sets theory is used in the rules induction phase. As a consequence of using the approximations, induced decision rules are categorized into *certain* (discriminant in the sense of the previous definition) and *possible* ones, depending on the used lower and upper approximations, respectively.

Moreover, let us mention other rough sets approaches that use information on class distribution inside boundary and assign to lower approximation these inconsistent elementary sets where the majority of examples belong to the given class. This is handled in the *Variable Precision Model* introduced by Ziarko

[47] or *Variable Consistency Model* proposed by Greco *et al.* [10] – both are a subject of many extensions, see e.g. [31]. Rules induced from such variable lower approximations are not certain but partly discriminant ones.

A number of various algorithms have been already proposed to induce decision rules – for some reviews see e.g. [1,11,14,20,28,34,37]. In fact, there is no unique "rough set approach" to rule induction as elements of rough sets can be used on different stages of the process of induction and data pre-processing. In general, we can distinguish approaches producing *minimal set of rules* (i.e. covering input objects using the minimum number of necessary rules) and approaches generating more extensive rule sets.

A good example for the first category is LEM2, MODLEM and similar algorithms [11,35]. The second approaches are nicely exemplified by *Boolean reasoning* [28,29,1]. There are also specific algorithms inducing the set of decision rules which satisfy user's requirements given a priori, e.g. the threshold value for a minimum number of examples covered by a rule or its accuracy. An example of such algorithms is Explore described in [42]. Let us comment that this algorithm could be further extended to handle imbalanced data (i.e. data set where one class – being particularly important – is under-represented comparing to cardinalities of other classes), see e.g. studies in [15,43].

3   Exemplary Rule Classifier

In our study we will use the algorithm, called MODLEM, introduced by Stefanowski in [35]. We have chosen it because of several reasons. First of all, the union of rules induced by this algorithm with a classification strategy proved to provide efficient single classifiers, [14,41,37]. Next, it is designed to handle various data properties not included in the classical rough sets approach, as e.g. numerical attributes without its pre-discretization. Finally, it produces the set of rules with reasonable computational costs – what is important property for using it as a component inside combined classifiers.

3.1   MODLEM Algorithm

The general schema of the MODLEM algorithm is briefly presented below. More detailed description could be found in [14,35,37]. This algorithm is based on the idea of a *sequential covering* and it generates a *minimal set* of decision rules for every decision concept (decision class or its rough approximation in case of inconsistent examples). Such a minimal set of rules (also called *local covering* [11]) attempts to cover all *positive examples* of the given decision concept, further denoted as B, and not to cover any *negative examples* (i.e. $U \setminus B$). The main procedure for rule induction scheme starts from creating a first rule by choosing sequentially the 'best' elementary conditions according to chosen criteria (see the function *Find best condition*). When the rule is stored, all learning positive examples that match this rule are removed from consideration. The process is repeated while some positive examples of the decision concept remain still uncovered. Then, the procedure is sequentially repeated for each set of examples from a succeeding decision concept.

In the MODLEM algorithm numerical attributes are handled during rule induction while elementary conditions of rules are created. These conditions are represented as either $(a < v_a)$ or $(a \geq v_a)$, where a denotes an attribute and v_a is its value. If the same attribute is chosen twice while building a single rule, one may also obtain the condition $(a = [v_1, v_2))$ that results from an intersection of two conditions $(a < v_2)$ and $(a \geq v_1)$ such that $v_1 < v_2$. For nominal attributes, these conditions are $(a = v_a)$ or could be extended to the set of values.

Procedure MODLEM
(**input** B - a set of positive examples from a given decision concept;
      criterion - an evaluation measure;
output $\mathcal{T}$ – single local covering of B, treated here as rule condition parts)
begin
    $G := B$; {A temporary set of rules covered by generated rules}
    $\mathcal{T} := \emptyset$;
    while $G \neq \emptyset$ **do** {look for rules until some examples remain uncovered}
    begin
        $T := \emptyset$; {a candidate for a rule condition part}
        $S := U$; {a set of objects currently covered by T}
        while $(T = \emptyset)$ or $(\text{not}([T] \subseteq B))$ **do** {stop condition for accepting a rule}
        begin
            $t := \emptyset$; {a candidate for an elementary condition}
            for each attribute $q \in C$ **do** {looking for the best elementary condition}
            begin
                $new\_t :=$Find_best_condition(q, S);
                if Better$(new\_t, t, criterion)$ **then** $t := new\_t$;
                {evaluate if a new condition is better than previous one
                according to the chosen evaluation measure}
            end;
            $T := T \cup \{t\}$; {add the best condition to the candidate rule}
            $S := S \cap [t]$; {focus on examples covered by the candidate}
        end; { while not($[T] \subseteq B$ }
        for each elementary condition $t \in T$ **do**
            if $[T - t] \subseteq B$ then $T := T - \{t\}$; {test a rule minimality}
        $\mathcal{T} := \mathcal{T} \cup \{T\}$; {store a rule}
        $G := B - \bigcup_{T \in \mathcal{T}} [T]$; {remove already covered examples}
    end; { while $G \neq \emptyset$ }
    for each $T \in \mathcal{T}$ **do**
        if $\bigcup_{T' \in \mathcal{T}-T} [T'] = B$ **then** $\mathcal{T} := \mathcal{T} - T$ {test minimality of the rule set}
end {procedure}

function Find_best_condition
(**input** c - given attribute; S - set of examples; **output** $best\_t$ - bestcondition)
begin
    $best\_t := \emptyset$;
    if c is a numerical attribute **then**

begin
    H:=list of sorted values for attribute c and objects from S;
    { $H(i)$ - ith unique value in the list }
    for i:=1 to length(H)-1 **do**
    if object class assignments for $H(i)$ and $H(i + 1)$ are different **then**
    begin
        $v := (H(i) + H(i + 1))/2$;
        create a new_t as either $(c < v)$ or $(c \geq v)$;
        if Better($new\_t, best\_t, criterion$) **then** $best\_t := new\_t$;
    end
    end
    else { attribute is nominal }
    begin
        for each value v of attribute c **do**
        if Better($(c = v), best\_t, criterion$) **then** $best\_t := (c = v)$;
    end
end {function}.

For the evaluation measure (i.e. a function *Better*) indicating the best condition, one can use either *class entropy* measure or *Laplacian* accuracy. For their definitions see [14] or [23]. It is also possible to consider a lexicographic order of two criteria measuring the rule positive cover and, then, its conditional probability (originally considered by Grzymala in his LEM2 algorithm or its last, quite interesting modification called MLEM). In all experiments, presented further in this paper, we will use the entropy as an evaluation measure. Having the best cut-point we choose a condition $(a < v)$ or $(a \geq v)$ that covers more positive examples from the concept B.

In a case of nominal attributes it is also possible to use another option of *Find best condition* function, where a single attribute value in the elementary condition $(a = v_i)$ is extended to a multi-valued set $(a \in W_a)$, where W_a is a subset of values from the attribute domain. This set is constructed in the similar way as in techniques for inducing binary classification trees. Moreover, the author created MODLEM version with another version of rule stop condition. Let us notice that in the above schema the candidate T is accepted to become a rule if $[T] \subseteq B$, i.e. a rule should cover learning examples belonging to an appropriate approximation of the given class K_j. For some data sets – in particular noisy ones – using this stop condition may produce too specific rules (i.e. containing many elementary conditions and covering too few examples). In such situations the user may accept partially discriminating rules with high enough accuracy – this could be done by applying another stop condition ($[T \cap B]/[T] \geq \alpha$. An alternative is to induce all, even too specific rules and to post-process them – which is somehow similar to pruning of decision trees.

Finally we can illustrate the use of MODLEM by a simple example. The data table contains examples of 17 decision concerning classification of some customers into three classes coded as d, p, r. All examples are described by 5 qualitative and numerical attributes.

Table 1. A data table containing examples of customer classification

Age	Job	Period	Income	Purpose	Decision
m	u	0	500	K	r
sr	p	2	1400	S	r
m	p	4	2600	M	d
st	p	16	2300	D	d
sr	p	14	1600	M	p
m	u	0	700	W	r
sr	b	0	600	D	r
m	p	3	1400	D	p
sr	p	11	1600	W	d
st	e	0	1100	D	p
m	u	0	1500	D	p
m	b	0	1000	M	r
sr	p	17	2500	S	p
m	b	0	700	D	r
st	p	21	5000	S	d
m	p	5	3700	M	d
m	b	0	800	K	r

This data table is consistent, so lower and upper approximations are the same. The use of MODLEM results in the following set of certain rules (square brackets contain the number of learning examples covered by the rule):

rule 1. if $(Income < 1050)$ then $(Dec = r)$ [6]
rule 2. if $(Age = sr) \wedge (Period < 2.5)$ then $(Dec = r)$ [2]
rule 3. if $(Period \in [3.5, 12.5))$ then $(Dec = d)$ [2]
rule 4. if $(Age = st) \wedge (Job = p)$ then $(Dec = d)$ [3]
rule 5. if $(Age = m) \wedge (Income \in [1050, 2550))$ then $(Dec = p)$ [2]
rule 6. if $(Job = e)$ then $(Dec = p)$ [1]
rule 7. if $(Age = sr) \wedge (Period \geq 12.5)$ then $(Dec = p)$ [2]

Due to the purpose and page limits of this paper we do not show details of MODLEM working steps while looking for a single rule - the reader is referred to the earlier author's papers devoted to this topic only.

3.2   Classification Strategies

Using rule sets to predict class assignment for an unseen object is based on matching the object description to condition parts of decision rules. This may result in unique matching to rules from the single class. However two other ambiguous cases are possible: matching to more rules indicating different classes or the object description does not match any of the rules. In these cases, it is necessary to apply proper strategies to solve these conflict cases. Review of different strategies is given in [37]

In this paper we employ two classification strategies. The first was introduced by Grzymala in LERS [12]. The decision to which class an object belongs to is made on the basis of the following factors: strength and support. The *Strength* is the total number of learning examples correctly classified by the rule during training. The *support* is defined as the sum of scores of all matching rules from the class. The class K_j for which the support, i.e., the following expression

$$\sum_{matching\ rules\ R\ describing\ K_i} Strength\_factor(R)$$

is the largest is the winner and the object is assigned to K_j.

If complete matching is impossible, all partially matching rules are identified. These are rules with at least one elementary condition matching the corresponding object description. For any partially matching rule R, the factor, called *Matching factor* (R), defined as a ratio of matching conditions to all conditions in the rule, is computed. In partial matching, the concept K_j for which the following expression is the largest

$$\sum_{partially\ matching\ rules\ R} Matching\_factor(R) * Strength\_factor(R)$$

is the winner and the object is classified as being a member of K_j.

The other strategy was introduced in [32]. The main difference is in solving no matching case. It is proposed to consider, so called, *nearest rules* instead of partially matched ones. These are rules nearest to the object description in the sense of chosen distance measure. In [32] a weighted heterogeneous metric DR is used which aggregates a normalized distance measure for numerical attributes and $\{0;1\}$ differences for nominal attributes. Let r be a nearest matched rule, e denotes a classified object. Then $DR(r,e)$ is defined as:

$$Dr(r,e) = \frac{1}{m}(\sum_{a\in P} d_a^p)^{1/p}$$

where p is a coefficient equal to 1 or 2, m is the number of elementary conditions in P – a condition part of rule r. A distance d_a for numerical attributes is equal to $|a(e) - v_{ai}|/|v_{a-max} - v_{a-min}|$, where v_{ai} is the threshold value occurring in this elementary condition and v_{a-max}, v_{a-min} are maximal and minimal values in the domain of this attribute. For nominal attributes present in the elementary condition, distance d_a is equal to 0 if the description of the classified object e satisfies this condition or 1 otherwise. The coefficient expressing rule similarity (complement of the calculated distance, i.e. $1-DR(r,e)$) is used instead of matching factor in the above formula and again the strongest decision K_j wins. While computing this formula we can use also heuristic of choosing the first k nearest rules only. More details on this strategy the reader can find in papers [32,33,37].

Let us consider a simple example of classifying two objects $e_1 = \{(Age = m), (Job = p), (Period = 6), (Income = 3000), (Purpose = K)\}$ and $e_2 = \{(Age = m), (Job = p), (Period = 2), (Income = 2600), (Purpose = M)\}$. The first object is completely matched by to one rule no. 3. So, this object is be assigned to class d. The other object does not satisfy condition part of any rules. If we use the first strategy for solving no matching case, we can notice that object e_2 is partially matched to rules no. 2, 4 and 5. The support for class r is equal to $0.5 \cdot 2 = 1$. The support for class d is equal to $0.5 \cdot 2 + 0.5 \cdot 2 = 2$. So, the object is assigned to class d.

3.3   Summarizing Experience with Single MODLEM Classifiers

Let us shortly summarize the results of studies, where we evaluated the classification performance of the single rule classifier induced by MODLEM. There are some options of using this algorithm. First of all one can choose as decision concepts either lower or upper approximations. We have carried out several experimental studies on benchmark data sets from ML Irvine repository [3]. Due to the limited size of this paper, we do not give precise tables but conclude that generally none of approximations was better. The differences of classification accuracies were usually not significant or depended on the particular data at hand. This observation is consistent with previous experiments on using certain or possible rules in the framework of LEM2 algorithm [13]. We also noticed that using classification strategies while solving ambiguous matching was necessary for all data sets. Again the difference of applied strategies in case of non-matching (either Grzymala's proposal or nearest rules) were not significant. Moreover, in [14] we performed a comparative study of using MODLEM and LEM2 algorithms on numerical data. LEM2 was used with preprocessing phase with the good discretization algorithm. The results showed that MODLEM can achieved good classification accuracy comparable to best pre-discretization and LEM2 rules.

Here, we could comment that elements of rough sets are mainly used in MODLEM as a kind of preprocessing, i.e. approximations are decision concepts. Then, the main procedure of this algorithm follows rather the general inductive principle which is common aspect with many machine learning algorithms – see e.g. a discussion of rule induction presented in [23]. Moreover, the idea of handling numerical attributes is somehow consistent with solutions also already present in classification tree generation. In this sense, other rule generation algorithms popular in rough sets community, as e.g. based on Boolean reasoning, are more connected with rough sets theory.

It is natural to compare performance of MODLEM induced rules against standard machine learning systems. Such a comparative study was carried out in [37,41] and showed that generally the results obtained by MODLEM (with nearest rules strategies) were very similar to ones obtained by C4.5 decision tree.

4   Combined Classifiers – General Issues

In the next sections we will study the use of MOLDEM in the framework of the combined classifiers. Previous theoretical research (see, e.g., their summary in [6,45]) indicated that combining several classifiers is effective only if there is a substantial *level of disagreement* among them, i.e. if they make errors independently with respect to one another. In other words, if they make errors for a given object they should indicate different class assignments. Diversified base classifiers can be generated in many ways, for some review see, e.g. [6,36,45]. In general, either *homogeneous* or *heterogeneous classifiers* are constructed.

In the first category, the same learning algorithm is used over different samples of the data set. The best-known examples are either *bagging* and *boosting* techniques which manipulate set of examples by including or weighting particular examples, or methods that manipulate set of attributes, e.g. randomly choosing several attribute subsets. Moreover, multiple classifiers could be trained over different samples or partitions of data sets.

In the second category, different learning *algorithms* are applied to the same data set, and the diversity of results comes from heterogeneous knowledge representations or different evaluation criteria used to construct them. The *stacked generalization* or *meta-learning* belong to this category. In section 7 we study the *combiner* as one of these methods.

Combining classification predictions from single classifiers is usually done by *group* or *specialized decision making*. In the first method all base classifiers are consulted to classify a new object while the other method chooses only these classifiers whose are expertised for this object. *Voting* is the most common method used to combine single classifiers. The vote of each classifier may be weighted, e.g., by an evaluation of its classification performance.

Moreover, looking into the rough sets literature one can notice a growing research interest in constructing more complex classification system. First works concerned rather an intelligent integration of different algorithms into *hybrid* system. For instance, some researchers tried to refine rule classifiers by analysing relationships with neural networks [44]. More related works included an integration of k - nearest neighbor with rough sets rule generation, see e.g. RIONA system, which offered good classification performance [8]. Yet another approach comprises two level knowledge representation: rules induced by Explore representing general patterns in data and case base representing exceptions [36], which worked quite well for the difficult task of credit risk prediction [43]. Recently Skowron and his co-operators have been developing *hierarchical classifiers* which attempt at approximating more complex concepts [2]. Classifiers on different hierarchy level correspond to different levels of pattern generalization and seems to be a specific combination of multiple models, which could be obtained in various ways, e.g. using a special lattice theory [46] or leveled rule generation. Nguyen et al. described in [24] an application concerning detecting sunspots where hierarchical classifier is constructed with a domain knowledge containing an ontology of considered concepts.

5   Using MODLEM Inside the Bagging

Firstly, we consider the use of MODLEM induced classifier inside the most popular homogeneous multiple classifiers [38].

This approach was originally introduced by Breiman [4]. It aggregates classifiers generated from different bootstrap samples. The *bootstrap sample* is obtained by uniformly *sampling with replacement* objects from the training set. Each sample has the same size as the original set, however, some examples do not appear in it, while others may appear more than once. For a training set with m examples, the probability of an example being selected at least once is $1 - (1 - 1/m)^m$. For a large m, this is about $1 - 1/e$. Given the parameter R which is the number of repetitions, R bootstrap samples $S_1, S_2, \ldots, S_R$ are generated. From each sample S_i a classifier C_i is induced by the same learning algorithm and the final classifier C^* is formed by aggregating these R classifiers. A final classification of object x is built by a uniform voting scheme on $C_1, C_2, \ldots, C_R$, i.e. is assigned to the class predicted most often by these sub-classifiers, with ties broken arbitrarily. For more details and theoretical justification see e.g. [4].

Table 2. Comparison of classification accuracies [%] obtained by the single MODLEM based classifier and the bagging approach; R denotes the number of component classifiers inside bagging

Name of data set	Single classifier	Bagging	R
bank	93.81 ± 0.94	95.22 ± 1.02	7
buses	97.20 ± 0.94	99.54 ± 1.09	5
zoo	94.64 ± 0.67	93.89* ± 0.71	7
hepatitis	78.62 ± 0.93	84.05 ± 1.1	5
hsv	54.52 ± 1.05	64.78 ± 0.57	7
iris	94.93 ± 0.5	95.06* ± 0.53	5
automobile	85.23 ± 1.1	83.00 ±0.99	5
segmentation	85.71 ± 0.71	87.62 ± 0.55	7
glass	72.41 ± 1.23	76.09 ± 0.68	10
bricks	90.32* ± 0.82	91.21* ± 0.48	7
vote	92.67 ± 0.38	96.01 ± 0.29	10
bupa	65.77 ± 0.6	76.28 ± 0.44	5
election	88.96± 0.54	91.66 ± 0.34	7
urology	63.80 ± 0.73	67.40 ± 0.46	7
german	72.16 ± 0.27	76.2 ± 0.34	5
crx	84.64 ± 0.35	89.42 ± 0.44	10
pima	73.57 ± 0.67	77.87 ± 0.39	7

In this paper we shortly summarize main results obtained in the extensive computational study [38]. The MODLEM algorithm was applied to generate base classifiers in the bagging combined classifier. In table 2 we present the comparison of the classification accuracy obtained for the best variant of the bagging against the single rule classifier (also induced by MODLEM). The experiments were carried out on several data sets coming mainly from ML Irvine repository [3]. For each data set, we show the classification accuracy obtained by a single classifier over the 10 cross-validation loops. A standard deviation is also given. An asterisk indicates that the difference for these compared classifiers and a given data set is not statistically significant (according to two-paired t-Student test with $\alpha=0.05$). The last column presents the number of R component classifiers inside the bagging - more details on tuning this value are described in [38].

We conclude that results of this experiment showed that the bagging significantly outperformed the single classifier on 14 data sets of total 18 ones. The difference between classifiers were non-significant on 3 data sets (those which were rather easy to learn as, e.g. *iris* and *bricks* - which were characterized by a linear separation between classes). Moreover, we noticed the slightly worse performance of the bagging for quite small data (e.g. *buses*, *zoo* - which seemed to be too small for sampling), and significantly better for data sets containing a higher number of examples. For some of these data sets we observed an substantial increase of predictive accuracy, e.g. for *hsv* – over 10%, *bupa* – around 10% and *hepatitis* – 5.43%.

However, we should admit that this good performance was expected as we know that there are many previous reports on successful use of decision trees in bagging or boosting.

6   On Solving Multiclass Problems with the n^2-Classifier

One can say the bagging experiment has been just a variant of a standard approach. Now we will move to more original approach, called the n^2-classifier, which was introduced by Jelonek and author in [18,36]. This kind of a multiple classifier is a specialized approach to solve *multiple class learning problems*.

The n^2-classifier is composed of $(n^2 - n)/2$ base binary classifiers (where n is a number of decision classes; $n > 2$). The main idea is to discriminate each pair of the classes: (i, j), $i, j \in [1..n], i \neq j$, by an independent binary classifier C_{ij}. Each base binary classifier C_{ij} corresponds to a pair of two classes i and j only. Therefore, the specificity of the training of each base classifier C_{ij} consists in presenting to it a subset of the entire learning set that contains only examples coming from classes i and j. The classifier C_{ij} yields a binary classification indicating whether a new example $\mathbf{x}$ belongs to class i or to class j. Let us denote by $C_{ij}(\mathbf{x})$ the classification of an example $\mathbf{x}$ by the base classifier C_{ij}.

The complementary classifiers: C_{ij} and C_{ji} (where $i, j \in\; < 1 \ldots n >$; $i \neq j$) solve the same classification problem – a discrimination between class i-th and j-th. So, they are equivalent ($C_{ij} \equiv C_{ji}$) and it is sufficient to use only $(n^2 - n)/2$ classifiers $C_{ij}(i < j)$, which correspond to all combinations of pairs of n classes.

An algorithm providing the final classification assumes that a new example $\mathbf{x}$ is applied to all base classifiers C_{ij}. As a result, their binary predictions $C_{ij}(\mathbf{x})$ are computed. The final classification is obtained by an aggregation rule, which is based on finding a class that wins the most pairwise comparisons. The more sophisticated approach includes a *weighted* majority voting rules, where the vote of each classifier is modified by its credibility, which is calculated as its classification performance during learning phase; more details in [18].

We have to remark that the similar approach was independently studied by
Friedman [7] and by Hastie and Tibshirani [17] – they called it *classification by
pairwise coupling*. The experimental studies, e.g. [7,17,18], have shown that such
multiple classifiers performed usually better than the standard classifiers. Previously the author and J.Jelonek have also examined the influence of a learning
algorithm on the classification performance of the n^2-classifier.

Table 3. Comparison of classification accuracies [%] and computation times [s] for
the single MODLEM based classifier and the n^2-classifier also based on decision rules
induced by MODLEM algorithm

Name of data set	Accuracy of single MODLEM (%)	Accuracy of n^2_{MODLEM} (%)	Time of comput. MODLEM	Time of comput. n^2_{MODLEM}
automobile	85.25 ± 1.3	87.96 ± 1.5	15.88 ± 0.4	5.22 ± 0.3
cooc	55.57 ± 2.0	59.30 ± 1.4	4148,7 ± 48.8	431.51 ± 1.6
ecoli	79.63 ± 0.8	81.34 ± 1.7	27.53 ± 0.5	11.25 ± 0.7
glass	72.07 ± 1.2	74.82 ± 1.4	45.29 ± 1.1	13.88 ± 0.4
hist	69.36 ± 1.1	73.10 ± 1.4	3563.79 ± 116.1	333.96 ± 0.8
meta-data	47.2 ± 1.3	49.83 ± 1.9	252.59 ± 78.9	276.71 ± 5.21
iris	94.2 ± 0.6	95.53* ± 1.2	0.71 ± 0.04	0.39 ± 0.04
soybean-large	91.09 ± 0.9	91.99* ± 0.8	26.38 ± 0.3	107.5 ± 5.7
vowel	81.81 ± 0.5	83.79 ± 1.2	3750.57 ± 30.4	250.63 ± 0.7
yeast	54.12 ± 0.7	55.74 ± 0.9	1544.3 ± 13.2	673.82 ± 9.4
zoo	94.64 ± 0.5	94.46* ± 0.8	0.30 ± 0.02	0.34 ± 0.12

Here, we summarize these of our previous results, where the MODLEM was
applied to generate base classifiers inside the n^2-classifier [38]. In table 3 we
present classification accuracies obtained by the n^2-classifier and compare them
against the single rule classifier induced by MODLEM on 11 data sets, all concerning multiple-class learning problems, with a number of classes varied from
3 up to 14. The second and third columns are presented in a similar way as in
Table 2. These results showed that the n^2-classifier significantly (again in the
sense of paired t test with a significance level $\alpha = 0.05$) outperformed the single
classifier on 7 out of 11 problems, e.g. for *hist* – over 3.7%, *glass* – around 2.7%,
automobile – 2.5% and *meta-data* – 2.6%. These improvements were not so high
as in the bagging but still they occurred for many difficult multi-class problems.
Again, the multiple classifier was not useful for easier problems (e.g. *iris*). Moreover, we noticed that its performance was better for data sets with a higher
number of examples. Coming back to our previous results for the n^2-classifier
[18] we can again remark that the comparable classification improvements were
observed for the case of using decision trees.

Then, let us focus our attention on interesting phenomena concerning computation costs of using the MODLEM in a construction of the n^2-classifier. Table 3
(two last columns) contains computation times (in seconds calculated as average

values over 10 folds with standard deviations). We can notice that generally constructing a combined classifiers does not increase the computation time. What is even more astonishing, for some data sets constructing the n^2-classifier requires even less time than training the standard single classifier. Here, we have to stress that in our previous works [18,37] we noticed that the increase of classification accuracy (for other learning algorithms as e.g. decision trees, k-nearest neighbor or neural networks) was burden with increasing the computational costs (sometimes quite high). In [38] we attempted to explain the good performance of MODLEM inside the n^2-classifier. Shortly speaking, the n^2-classifier should be rather applied to solving difficult ("complex") classification tasks, where examples of decision classes are separated by non-linear decision borders – these are often difficult concepts to be learned by standard classifiers, while pairwise decision boundaries between each pair of classes may be simpler and easier to be learned with using a *smaller number* of attributes. Here, MODLEM could gain its performance thanks to his sequential covering and greedy heuristic search. It generates rules distinguishing smaller number of learning examples (from two classes only) than in the multiple class case and, above all, testing a smaller number of elementary conditions. To verify hypothesis we inspect syntax of rule sets induced by the single classifier and the n^2-classifier. Rules for binary classifiers were using less attributes and covered more learning example on average than rules from the single set generated in the standard way [38].

7   Combining Predictions of Heterogeneous Classifiers

In two previous sections we described the use of MODLEM based classifiers inside the architecture of homogeneous classifiers. In these solutions, the MODLEM was the only algorithm applied to create base classifiers inside multiple classifiers and could directly influence their final performance. Diversification of base classifiers is one of the conditions for improving classification performance of the final system. Let us repeat that in previously considered solutions it was achieved by changing the distribution of examples in the input data.

Another method to obtain component classifier diversity is constructing, so called, *heterogeneous* classifiers. They are generated from the same input data by different learning algorithms which use different representation language and search strategies. These base classifiers could be put inside a layered architecture. At the first level base classifiers receive the original data as input. Their predictions are then aggregated at the second level into the final prediction of the system. This could be done in various ways. In one of our studies we used a solution coming from Chan & Stolfo [5], called a *combiner*.

The combiner is based on an idea of merging predictions of base classifiers by an additional classifier, called *meta-classifier*. This is constructed in an *extra meta-learning* step, i.e. first base classifiers are learned, then their predictions made on a set of extra validation examples, together with correct decision labels, form a meta-level training set. An extra learning algorithm is applied to this set to discover how to merge base classifier predictions into a final decision.

Table 4. Classification accuracies [%] for different multiple classifiers

Data set	Bagging	n^2-classifier	Combiner
Automobile	83.00	87.90	84.90
Bank	95.22	–	95.45
Bupa	76.28	–	69.12
Ecoli	85.70	81.34	85.42
Glass	74.82	74.82	71.50
HSV	64.75	–	59.02
Meta-data	48.11	49.80	51.33
Pima	75.78	–	74.78
Voting	93.33	–	94.67
Yeast	58.18	55.74	58.36
Zoo	93.89	94.46	95.05

In [41] we performed a comparative study of using a combiner approach against the single classifiers learned by these algorithms which were applied to create its component classifiers. In this study base classifiers were induced by k-NN, C4.5 and MODLEM. The meta-classifier was a Naive Bayes. This comparative study was performed on 15 data sets. However, the obtained results showed that the combiner did not improve classification accuracy in so many cases as previously studied homogeneous classifiers. Only in 33% data we observed a significant improvement comparing against single classifiers. In table 4 we present only some of these results concerning the final evaluation of the combiner compared also against the previous multiple classifiers. However, while comparing these classifiers we should be cautious as the number of the results on common data sets was limited. Moreover, MODLEM is only one of three component classifiers inside the combiner that influences the final result.

We could also ask a question about other elements of the architecture of heterogeneous classifier, e.g. number of component classifiers or the aggregation techniques. In recent experiments we focus our interest on testing two other techniques instead of the meta-combiner:

- a simple aggregation performed by means of a majority voting rule (denoted as MV in table 4),
- using a quite sophisticated approach – SCANN; It was introduced by Merz [22] and uses a mechanism of the correspondence analysis to discover hidden relationships between the learning examples and the classification done by the component classifiers.

Results from ongoing experiments are given in Table 5. There is also a difference to previous architecture, i.e. adding an additional, forth component classifiers Naive Bayesian at the first level. We can remark that the more advanced aggregation technique could slightly increase the classification accuracy comparing to simpler one. On the other hand they are much time consuming.

Table 5. Comparison of different methods producing the final decision inside the heterogeneous classifiers - classification accuracies [%]

Data set	MV	SCANN	Combiner
credit-a	86.2 ± 0.6	87 ± 0.7	86.6 ± 0.4
glass	68.5 ± 0.3	70.1 ± 0.2	70.5 ± 0.6
ecoli	86.1 ± 0.9	81.5 ± 0.8	84.5 ± 0.5
zoo	95 ± 0.9	92.2 ± 0.7	95.1 ± 0.4

8 Discussion of Results and Final Remarks

As Professor Zdzisław Pawlak wrote in the introductory chapter of his book on rough sets [26] knowledge of human beings and other species is strictly connected with their ability to classify objects. Finding classification patterns of sensor signals or data form fundamental mechanisms for very living being. In his point of view it was then connected with a partition (classification) operation leading to basic blocks for constructing knowledge. Many researchers followed the Pawlak's idea. One of the main research directions includes constructing approximations of knowledge from tables containing examples of decisions on object classification. Rules were often induced as the most popular knowledge representation. They could be used either to describe the characteristics of available data or as the basis for supporting classification decisions concerning new objects. Up to now several efficient rule classifiers have been introduced.

In this study we have attempted to briefly describe the current experience with using the author's rule induction algorithm MODLEM, which induces either certain or possible rules from appropriate rough approximations. This is the main point where elements of the rough sets theory is applied in this algorithm. Given as an input learning examples from approximations, the rule generation phase follows the general idea of sequential covering, which is somehow in common with machine learning paradigms. The MODLEM produces a minimal set of rules covering examples from rough approximations. This rule sets should be joined with classification strategies for solving ambiguous matching of the new object description to condition parts of rules. An extra property of this algorithm is it ability to handle directly numerical attributes without prior discretization. The current experience with comparative studies on benchmark data sets and real life applications showed that the classification performance of this approach was comparable to other symbolic classifiers, in particular to decision trees.

Although the MODLEM classifier and other machine learning approaches are efficient for many classification problems, they do not always lead to satisfactory classification accuracy for more complex and difficult problems. This is our motivation to consider new approaches for increasing classification accuracy by combining several classifiers into an integrated system. Several proposals of constructing such multiple classifiers are already proposed. Most of them are general approaches, where many different algorithms could be applied to induce the component classifiers.

Thus, our main research interest in this study is to summarize our experiments with using MODLEM induced rule classifiers inside the framework of three different multiple classifiers, namely the bagging, the n^2-classifier and the combiner. A classification accuracy for the multiple classifier has been compared against the standard classifiers – also induced by MODLEM. These results and their detailed discussion has been given in the previous sections.

Firstly we could notice that using MODLEM inside the bagging was quite effective. However, it was a kind of standard approach and we could expect such good performance as MODLEM performs similarly to decision trees (which have been extensively studied in the bagging) and could be seen as *unstable* learning algorithm - i.e. an algorithm whose output classifier undergoes changes in response to small changes in the training data. This kind of algorithm may produce base classifiers diversified enough (but not too much, see e.g. discussion of experimental study by Kuncheva and Whitaker [21]) which is a necessary condition for their effective aggregation. Following the same arguments we also suspect that MODLEM should nicely work inside the boosting classifier. Further on, we could hypothesize that slightly worse improvements of the classification accuracy in the combiner approach may result from insufficient diversification of component heterogeneous classifiers. This has been verified by analysing distributions of wrong decisions for base classifiers, presented in [41]. It showed the correlation of errors for some data sets, where finally we did not notice the improvement of the classification accuracy.

The most original methodological approach is Jelonek and author's proposal of the n^2-classifier which is in fact a specialized approach to learning multiple class problems. The n^2-classifier is particularly well suited for multiple class data where exist "simpler" pairwise decision boundaries between pair of classes. MODLEM seems to be a good choice to be used inside this framework as it leads to an improvement of classification performance and does not increase computational costs - reasons for this have been discussed in section 7. Let us notice that using other learning algorithms inside the n^2-classifier and applying MODLEM in two other multiple classifier requires an extra computation efforts comparing to learning the single, standard classifier [38].

Comparing results of all together multiple classifiers "head to head" we should be cautious as we had a limited number of common data sets. It seems that the n^2-classifier is slightly better for these data. While the standard multiple classifiers, as bagging or combiner, are quite efficient for simpler data and are easier to be implemented.

To sum up, the results of our experiments have shown that the MODLEM algorithm can be efficiently used within the framework of three multiple classifiers for data sets concerning more "complex" decision concepts. However, the relative merits of these new approaches depends on the specifies of particular problems and a training sample size.

Let us notice that there is a disadvantage of the multiple classifiers - loosing a simple and easy interpretable structure of knowledge represented in a form decision rules. These are ensembles of diversified rule sets specialized for predictive aims not one set of rules in a form for a human inspection.

As to future research directions we could consider yet another way of obtaining diversified data – i.e. selecting different subsets of attributes for each component classifiers. The author has already started research on extending bootstrap samples inside the bagging by applying additionally attribute selection [39,40]. In this way each bootstrap is replicated few times, each of them using different subset of attributes. We have considered the use of different selection techniques and observed that besides random choice or wrapper model, techniques which use either entropy based measures or correlation merits are quite useful. The results of comparative experiments carried out in [40] have showed that the classification accuracy of such a new extended bagging is higher than for standard one. In this context one could come back to the classical rough sets topic of reducts, which relates to finding an ensemble of few attribute subsets covering different data properties and constructing in this way a set of diversified examples for an integrated system. However, we are not limited to "classical" meaning of pure rough sets reducts but rather to approximate ones, where the entropy measure is also considered [30].

Acknowledgment. The author would like to thank his colleagues Jacek Jelonek, Sławomir Nowaczyk and his M.Sc. students Michał Brończyk, Ryszard Gizelski, Maciej Łuszczyński who have worked with him on the software implementations of the classifiers or took part in some experiments.

References

1. Bazan J.: A comparison of dynamic and non-dynamic rough set methods for extracting laws from decision tables. In Polkowski L., Skowron A. (eds.), *Rough Sets in Data Mining and Knowledge Discovery* vol. 1, Physica-Verlag, 1998, 321–365.
2. Bazan J., Nguyen Hung Son, Skowron A.: Rough sets methods in approximation of hierarchical concepts. In *Proc. of the Conference on Rough Sets and New Trends in Computing, RSCTC –* 2004, LNAI 2066, Springer Verlag, 2004, 346–355.
3. Blake C., Koegh E., Mertz C.J.: Repository of Machine Learning, University of California at Irvine (1999).
4. Breiman L.: Bagging predictors. *Machine Learning*, 24 (2), 1996, 123–140.
5. Chan P.K., Stolfo S.: On the accuracy of meta-learning for scalable data mining. *Journal of Intelligent Information Systems*, **8**, (1), 1997, 5-28.
6. Dietrich T.G.: Ensemble methods in machine learning. In *Proc. of 1st Int. Workshop on Multiple Classifier Systems*, 2000, 1–15.
7. Friedman J.: Another approach to polychotomous classification, Technical Report, Stanford University, 1996.
8. Góra G., Wojna A.: RIONA: a new classification system combining rule induction and instance based learning. *Fundamenta Informaticae* **51** (4), 2002, 369-390.
9. Greco S., Matarazzo B., Słowiński R.: The use of rough sets and fuzzy sets in MCDM. In Gal T., Stewart T., Hanne T. (eds), *Advances in Multiple Criteria Decision Making*, Kluwer, chapter 14, 1999, pp. 14.1-14.59.

10. Greco S., Matarazzo B., Słowiński R., Stefanowski J.: Variable consistency model of dominance-based rough set approach. In *Proc. 2nd Int. Conference on Rough Sets and New Trends in Computing, RSCTC –* 2000, LNAI 2005, Springer Verlag, 2001,170–181.

11. Grzymala-Busse J.W. LERS - a system for learning from examples based on rough sets. In Slowinski R. (ed.), *Intelligent Decision Support*, Kluwer Academic Publishers, 1992, 3–18.

12. Grzymala-Busse J.W.: Managing uncertainty in machine learning from examples. In *Proc. 3rd Int. Symp. in Intelligent Systems*, Wigry, Poland, IPI PAN Press, 1994, 70–84.

13. Grzymala-Busse J.W. Zou X.: Classification strategies using certain and possible rules. In *Proceedings of the 1th Rough Sets and Current Trends in Computing Conference, RSCTC–98* , LNAI 1424, Springer Verlag, 1998, 37-44.

14. Grzymala-Busse J.W., Stefanowski J.: Three approaches to numerical attribute discretization for rule induction. *International Journal of Intelligent Systems*, 16 (1), (2001) 29–38.

15. Grzymala-Busse J.W., Stefanowski J. Wilk Sz.: *A comparison of two approaches to data mining from imbalanced data*. In Proc. of the KES 2004 - 8-th Int. Conf. on Knowledge-based Intelligent Information & Engineering Systems, Springer LNCS vol. **3213**, 2004, 757-763.

16. Han J., Kamber M.: *Data mining: Concepts and techniques*, San Francisco, Morgan Kaufmann, 2000.

17. Hastie T., Tibshirani R.: Classification by pairwise coupling. In Jordan M.I. (ed.) *Advances in Neural Information Processing Systems*: 10 (NIPS-97), MIT Press, 1998, 507-513.

18. Jelonek J., Stefanowski J.: Experiments on solving multiclass learning problems by the n^2-classifier. In *Proceedings of 10th European Conference on Machine Learning ECML 98*, Springer LNAI no. 1398, 1998, 172–177.

19. Klosgen W., Żytkow J.M. (eds.): *Handbook of Data Mining and Knowledge Discovery*, Oxford Press, 2002.

20. Komorowski J., Pawlak Z., Polkowski L. Skowron A.: Rough Sets: tutorial. In Pal S.K., Skowron A. (eds) *Rough Fuzzy Hybridization. A new trend in decision making*, Springer Verlag, Singapore, 1999, 3–98.

21. Kuncheva L., Whitaker C.J.: Measures of diversity in classifier ensembles and their relationship with the ensemble accuracy. *Machine Learning*, 51, 2003, 181–207.

22. Merz C.: Using correspondence analysis to combine classifiers. *Machine Learning*, 36 (1/2), 1999, 33–58.

23. Mitchell Tom M.: *Machine learning*, McGraw Hill, 1997.

24. Nguyen Sinh Hoa, Trung Tham Nguyen, Nguyen Hung Son: Rough sets approach to sunspot classification problem. In *Proc. of the Conference RSFDGrC –* 2005, vol 2, LNAI 3642, Springer Verlag, 2005, 263-272.

25. Pawlak Z.: Rough sets. *Int. J. Computer and Information Sci.*, 11, 1982, 341–356.

26. Pawlak Z.: *Rough sets. Theoretical aspects of reasoning about data.* Kluwer Academic Publishers, Dordrecht, 1991.

27. Pawlak Z., Grzymala-Busse J., Slowinski R., Ziarko W.: Rough sets. *Communications of the ACM*, vol. 38, no. 11, 1995, 89-95.

28. Skowron A.: Boolean reasoning for decision rules generation. In Komorowski J., Ras Z. (des.) *Methodologies for Intelligent Systems*, LNAI 689, Springer-Verlag, 1993, 295–305.

29. Skowron A., Rauszer C.: The discernibility matrices and functions in information systems. In Slowinski R. (ed.), *Intelligent Decision Support. Handbook of Applications and Advances of Rough Set Theory.* Kluwer Academic Publishers, 1992, 331–362.

30. Slezak D.: Approximate entropy reducts. *Fundamenta Informaticae* **53** (3/4), 2002, 365-387.

31. Slowinski R., Greco S.: Inducing Robust Decision Rules from Rough Approximations of a Preference Relation. In Rutkowski L. et al. (eds): Artiffcial Intelligence and Soft Computing, LNAI 3070, Springer-Verlag, 2004, 118-132.

32. Stefanowski J.: Classification support based on the rough sets. *Foundations of Computing and Decision Sciences*, vol. 18, no. 3-4, 1993, 371-380.

33. Stefanowski J.: Using valued closeness relation in classification support of new objects. In Lin T. Y., Wildberger (eds) *Soft computing: rough sets, fuzzy logic, neural networks uncertainty management, knowledge discovery*, Simulation Councils Inc., San Diego CA, 1995, 324–327.

34. Stefanowski J.: On rough set based approaches to induction of decision rules. In Polkowski L., Skowron A. (eds), *Rough Sets in Data Mining and Knowledge Discovery*, vol. 1, Physica-Verlag, 1998, 500–529.

35. Stefanowski J.: The rough set based rule induction technique for classification problems. In *Proceedings of 6th European Conference on Intelligent Techniques and Soft Computing* EUFIT 98, Aachen 7-10 Sept., 1998, 109–113.

36. Stefanowski J.: Multiple and hybrid classifiers. In Polkowski L. (ed.) *Formal Methods and Intelligent Techniques in Control, Decision Making, Multimedia and Robotics*, Post-Proceedings of 2nd Int. Conference, Warszawa, 2001, 174–188.

37. Stefanowski J.: Algorithims of rule induction for knowledge discovery. (In Polish), Habilitation Thesis published as Series Rozprawy no. 361, Poznan Univeristy of Technology Press, Poznan (2001).

38. Stefanowski J.: The bagging and n2-classifiers based on rules induced by MODLEM. In *Proceedings of the 4th Int. Conference Rough Sets and Current Trends in Computing*, RSCTC – 2004, LNAI 3066, Springer-Verlag, 2004, 488-497.

39. Stefanowski J.: An experimental study of methods combining multiple classifiers - diversified both by feature selection and bootstrap sampling. In K.T. Atanassov, J. Kacprzyk, M. Krawczak, E. Szmidt (eds), *Issues in the Representation and Processing of Uncertain and Imprecise Information*, Akademicka Oficyna Wydawnicza EXIT, Warszawa, 2005, 337-354.

40. Stefanowski J., Kaczmarek M.: Integrating attribute selection to improve accuracy of bagging classifiers. In *Proc. of the AI-METH 2004. Recent Developments in Artificial Intelligence Methods*, Gliwice, 2004, 263-268.

41. Stefanowski J., Nowaczyk S.: On using rule induction in multiple classifiers with a combiner aggregation strategy. In *Proc. of the 5th Int. Conference on Intelligent Systems Design and Applications* - ISDA 2005, IEEE Press, 432-437.

42. Stefanowski J., Vanderpooten D.: Induction of decision rules in classification and discovery-oriented perspectives. *International Journal of Intelligent Systems* **16** (1), 2001, 13–28.

43. Stefanowski J., Wilk S.: Evaluating business credit risk by means of approach integrating decision rules and case based learning. *International Journal of Intelligent Systems in Accounting, Finance and Management* **10** (2001) 97–114.

44. Szczuka M: Refining classifiers with neural networks. *International Journal of Intelligent Systems* **16** (1), 2001, 39–56.

45. Valentini G., Masuli F.: Ensambles of learning machines. In R. Tagliaferri, M. Marinaro (eds), *Neural Nets WIRN Vietri-2002*, Springer-Verlag LNCS, vol. 2486, 2002 , 3–19.
46. Wang H., Duntsch I., Gediga G., Skowron A.: Hyperrelations in version space. *International Journal of Approximate Reasoning*, **23**, 2000, 111–136.
47. Ziarko W.: Variable precision rough sets model. *Journal of Computer and Systems Sciences*, vol. 46. no. 1, 1993, 39–59.

Approximation Spaces in Multi Relational Knowledge Discovery

Jarosław Stepaniuk

Department of Computer Science, Białystok University of Technology
Wiejska 45a, 15-351 Białystok, Poland
jstepan@ii.pb.bialystok.pl

Abstract. Pawlak introduced approximation spaces in his seminal work on rough sets more than two decades ago. In this paper, we show that approximation spaces are basic structures for knowledge discovery from multi-relational data. The utility of approximation spaces as fundamental objects constructed for concept approximation is emphasized. Examples of basic concepts are given throughout this paper to illustrate how approximation spaces can be beneficially used in many settings. The contribution of this paper is the presentation of an approximation space-based framework for doing research in various forms of knowledge discovery in multi relational data.

Keywords: rough sets, approximation spaces, multi-relational data mining, rough inclusion, uncertainty function.

1 Introduction

Approximation spaces are fundamental structures for the rough set approach [7,8,10]. In this paper we present a generalization of the original approximation space model. Using such approximation spaces we show how the rough set approach can be used for approximation of concepts assuming that only partial information on approximation spaces is available. Hence, searching for concept approximation, i.e., the basic task in machine learning and pattern recognition can be formulated as searching for relevant approximation spaces.

Rough set approach has been used in a lot of applications aimed at description of concepts. In most cases, only approximate descriptions of concepts can be constructed because of incomplete information about them. In learning approximations of concepts, there is a need to choose a description language. This choice may limit the domains to which a given algorithm can be applied. There are at least two basic types of objects: structured and unstructured. An unstructured object is usually described by attribute-value pairs. For objects having an internal structure first order logic language is often used. Attribute-value languages have the expressive power of propositional logic. These languages sometimes do not allow for proper representation of complex structured objects and relations among objects or their components. The background knowledge that can be

J.F. Peters et al. (Eds.): Transactions on Rough Sets VI, LNCS 4374, pp. 351–365, 2007.
© Springer-Verlag Berlin Heidelberg 2007

used in the discovery process is of a restricted form and other relations from the database cannot be used in the discovery process. Using first-order logic (or FOL for short) has some advantages over propositional logic [1,2,4]. First order logic provides a uniform and very expressive means of representation. The background knowledge and the examples, as well as the induced patterns, can all be represented as formulas in a first order language. Unlike propositional learning systems, the first order approaches do not require that the relevant data be composed into a single relation but, rather they can take into account data organized in several database relations with various connections existing among them.

The paper is organized as follows. In Section 2 we recall the definition of approximation spaces. Next, we describe a constructive approach for computing values of uncertainty and rough inclusion functions. These functions are the basic components of approximation spaces. Parameters of the uncertainty and rough inclusion functions are tuned in searching for relevant approximation spaces. Among such parameters we distinguish sensory environments and their extensions. These parameters are used for constructive definition of uncertainty and rough inclusion functions. In Section 3 we discuss notions of relational learning. In Sections 4 and 5 we consider application of rough set methods to discovery of interesting patterns expressed in a first order language. In Section 4 rough set methodology is used in the process of translating first–order data into attribute–value data. Some properties of this algorithm were presented in [13]. In Section 5 rough set methodology is used in the process of selecting literals which may be a part of a rule. The criterion of selecting a literal is as follows: only such a literal is selected which added to the rule makes the rule discerning most of the examples which were indiscernible so far. Some properties of this algorithm were presented in [14,15].

2   Approximation Spaces

In this section we recall the definition of an approximation space from [10,13,11].

Definition 1. *A parameterized approximation space is a system*
$AS_{\#,\$} = (U, I_{\#}, \nu_{\$})$, *where*

- *U is a non-empty set of objects,*
- *$I_{\#} : U \to P(U)$ is an uncertainty function, where $P(U)$ denotes the power set of U,*
- *$\nu_{\$} : P(U) \times P(U) \to [0,1]$ is a rough inclusion function,*

and $\#, \$$ denote vectors of parameters (the indexes $\#, \$$ will be omitted if it does not lead to misunderstanding).

2.1   Uncertainty Function

The uncertainty function defines for every object x, a set of objects described similarly to x. The set $I(x)$ is called the neighborhood of x (see, e.g., [8,10]).

We assume that the values of the uncertainty function are defined using a *sensory environment* [11], i.e., a pair $(\Sigma, \|\cdot\|_U)$, where Σ is a set of formulas, called the *sensory formulas*, and $\|\cdot\|_U : \Sigma \longrightarrow P(U)$ is the *sensory semantics*. We assume that for any sensory formula α and any object $x \in U$ the information whether $x \in \|\alpha\|_U$ holds is available. The set $\{\alpha : x \in \|\alpha\|_U\}$ is called the *signature of x* in AS and is denoted by $Inf_{AS}(x)$. For any $x \in U$, the set $\mathcal{N}_{AS}(x)$ of *neighborhoods of x* in AS is defined by $\{\|\alpha\|_U : x \in \|\alpha\|_U\}$ and from this set the neighborhood $I(x)$ is constructed. For example, $I(x)$ is defined by selecting an element from the set $\{\|\alpha\|_U : x \in \|\alpha\|_U\}$ or by $I(x) = \bigcap \mathcal{N}_{AS}(x)$. Observe that any sensory environment $(\Sigma, \|\cdot\|_U)$ can be treated as a parameter of I from the vector $\#$ (see Definition 1).

Let us consider two examples. Any decision table $DT = (U, A, d)$ [8] defines an approximation space $AS_{DT} = (U, I_A, \nu_{SRI})$, where, as we will see, $I_A(x) = \{y \in U : a(y) = a(x) \text{ for all } a \in A\}$. Any sensory formula is a descriptor, i.e., a formula of the form $a = v$ where $a \in A$ and $v \in V_a$ with the standard semantics $\|a = v\|_U = \{x \in U : a(x) = v\}$. Then, for any $x \in U$ its signature $Inf_{AS_{DT}}(x)$ is equal to $\{a = a(x) : a \in A\}$ and the neighborhood $I_A(x)$ is equal to $\bigcap \mathcal{N}_{AS_{DT}}(x)$. Another example can be obtained assuming that for any $a \in A$ there is given a tolerance relation $\tau_a \subseteq V_a \times V_a$ (see, e.g., [10]). Let $\tau = \{\tau_a\}_{a \in A}$. Then, one can consider a tolerance decision table $DT_\tau = (U, A, d, \tau)$ with tolerance descriptors $a =_{\tau_a} v$ and their semantics $\|a =_{\tau_a} v\|_U = \{x \in U : v\tau_a a(x)\}$. Any such tolerance decision table $DT_\tau = (U, A, d, \tau)$ defines the approximation space AS_{DT_τ} with the signature $Inf_{AS_{DT_\tau}}(x) = \{a =_{\tau_a} a(x) : a \in A\}$ and the neighborhood $I_A(x) = \bigcap \mathcal{N}_{AS_{DT_\tau}}(x)$ for any $x \in U$.

The fusion of $\mathcal{N}_{AS_{DT_\tau}}(x)$ for computing the neighborhood of x can have many different forms, the intersection is only an example. For example, to compute the value of $I(x)$ some subfamilies of $\mathcal{N}_{AS}(x)$ may first be selected and the family consisting of intersection of each such a subfamily is next taken as the value of $I(x)$.

2.2 Rough Inclusion Function

One can consider general constraints which the rough inclusion functions should satisfy. Searching for such constraints initiated investigations resulting in creation and development of rough mereology (see, the bibliography in [9]). In this subsection, we present some examples of rough inclusion functions only.

The rough inclusion function $\nu_\$: P(U) \times P(U) \to [0, 1]$ defines the degree of inclusion of X in Y, where $X, Y \subseteq U$.

In the simplest case it can be defined by (see, e.g., [10,8]):

$$\nu_{SRI}(X, Y) = \begin{cases} \frac{card(X \cap Y)}{card(X)} & \text{if } X \neq \emptyset \\ 1 & \text{if } X = \emptyset. \end{cases}$$

This measure is widely used by the data mining and rough set communities. It is worth mentioning that Jan Łukasiewicz [3] was the first one who used this idea to estimate the probability of implications. However, rough inclusion can have a much more general form than inclusion of sets to a degree (see, e.g., [9]).

Another example of rough inclusion is used for relation approximation [12] and in the variable precision rough set approach [16].

2.3 Lower and Upper Approximations

The lower and the upper approximations of subsets of U are defined as follows.

Definition 2. *For any approximation space $AS_{\#,\$} = (U, I_\#, \nu_\$)$ and any subset $X \subseteq U$, the lower and upper approximations are defined by*
$$LOW\left(AS_{\#,\$}, X\right) = \{x \in U : \nu_\$\left(I_\#\left(x\right), X\right) = 1\},$$
$$UPP\left(AS_{\#,\$}, X\right) = \{x \in U : \nu_\$\left(I_\#\left(x\right), X\right) > 0\}, \text{ respectively.}$$

The lower approximation of a set X with respect to the approximation space $AS_{\#,\$}$ is the set of all objects, which can be classified with certainty as objects of X with respect to $AS_{\#,\$}$. The upper approximation of a set X with respect to the approximation space $AS_{\#,\$}$ is the set of all objects which can possibly be classified as objects of X with respect to $AS_{\#,\$}$.

Several known approaches to concept approximation can be covered using the approximation spaces discussed here, e.g., the approach given in [8] or tolerance (similarity) rough set approximations (see, e.g., references in [10]).

We recall the notions of the positive region and the quality of approximation of classification in the case of generalized approximation spaces [13].

Definition 3. *Let $AS_{\#,\$} = (U, I_\#, \nu_\$)$ be an approximation space and let $r > 1$ be a given natural number and let $\{X_1, \ldots, X_r\}$ be a classification of objects (i.e. $X_1, \ldots, X_r \subseteq U$, $\bigcup_{i=1}^r X_i = U$ and $X_i \cap X_j = \emptyset$ for $i \neq j$, where $i, j = 1, \ldots, r$).*

1. *The positive region of the classification $\{X_1, \ldots, X_r\}$ with respect to the approximation space $AS_{\#,\$}$ is defined by*
   $$POS\left(AS_{\#,\$}, \{X_1, \ldots, X_r\}\right) = \bigcup_{i=1}^r LOW\left(AS_{\#,\$}, X_i\right).$$
2. *The quality of approximation of the classification $\{X_1, \ldots, X_r\}$ in the approximation space $AS_{\#,\$}$ is defined by*
   $$\gamma\left(AS_{\#,\$}, \{X_1, \ldots, X_r\}\right) = \frac{card\left(POS\left(AS_{\#,\$}, \{X_1,\ldots,X_r\}\right)\right)}{card(U)}.$$

The quality of approximation of the classification coefficient expresses the ratio of the number of all $AS_{\#,\$}$-correctly classified objects to the number of all objects in the data table.

3   Relational Data Mining

Knowledge discovery is the process of discovering particular patterns over data. In this context data is typically stored in a database. Approaches using first order logic (FOL, for short) languages for the description of such patterns offer data mining the opportunity of discovering more complex regularities which may be out of reach for attribute-value languages.

3.1 Didactic Example

In this section we present an example inspired by [2].

Example 1. There are two information systems:

$$IS_{Customer} = (U_{Customer}, A_{Customer})$$

where the set of objects $U_{Customer} = \{x_1, \ldots, x_7\}$, and the set of attributes $A_{Customer} = \{Name, Gender, Income, BigSpender\}$ (see Table 1) and

$$IS_{MarriedTo} = (U_{MarriedTo}, A_{MarriedTo})$$

where $U_{MarriedTo} = \{y_1, y_2, y_3\}$, and $A_{MarriedTo} = \{Spouse1, Spouse2\}$ (see Table 2).

Table 1. An Information System $IS_{Customer}$

$U_{Customer}$	Name	Gender	Income	BigSpender
x_1	Mary	Female	70000	yes
x_2	Eve	Female	120000	yes
x_3	Kate	Female	80000	no
x_4	Meg	Female	80000	yes
x_5	Jim	Male	100000	yes
x_6	Tom	Male	100000	yes
x_7	Henry	Male	60000	no

Table 2. An Information System $IS_{MarriedTo}$

$U_{MarriedTo}$	Spouse1	Spouse2
y_1	Mary	Jim
y_2	Meg	Tom
y_3	Kate	Henry

Using attribute–value language we obtain for example the following decision rules:

if $Income \geq 100000$ **then** $BigSpender = yes$
if $Income \leq 75000$ **then** $BigSpender = yes$ (May be this rule is not intuitive.)
if $Name = Meg$ **then** $BigSpender = yes$ (This rule is generally not applicable to new objects.)

Using first order language one can obtain the following two rules:

$BigSpender(var_1, var_2, var_3) \leftarrow var_3 \geq 100000$
$BigSpender(var_1, var_3, var_3) \leftarrow MarriedTo(var_1, var_1')$ **and**
$Customer(var_1', var_2', var_3', var_4')$ **and** $var_3' \geq 100000$

which involve the predicates $Customer$ and $MarriedTo$. It predicts a person to be a big spender if the person is married to somebody with high income (compare this to the rule that states a person is a big spender if he/she has high income, listed above the relational rules). Note that the two persons var_1 and var'_1 are connected through the relation MarriedTo. Relational patterns are typically expressed in subsets of first-order logic (also called predicate or relational logic). Essentials of predicate logic include predicates ($MarriedTo$) and variables (var_1, var'_1), which are not present in propositional logic (attribute–value language). Relational patterns are thus more expressive than the propositional ones.

Knowledge discovery based on FOL has other advantages as well. Complex background knowledge provided by experts can be encoded as first order formulas and be used in the discovery task. The expressiveness of FOL enables the discovered patterns to be described in a concise way, which in most cases increases readability of the output. Multiple relations can be naturally handled without explicit (and expensive) joins.

3.2 Relational Learning

Before moving on to the algorithm for learning of a set of rules, let us introduce some basic terminology from relational learning.

Relational learning algorithms learn classification rules for a concept [2] (for relational methods and their applications in computer science see also [5]). The program typically receives a large collection of positive and negative examples from real-world databases as well as background knowledge in the form of relations. Let p be a target predicate of arity m and $r_1, \ldots, r_l$ be background predicates, where $m, l > 0$ are given natural numbers. We denote the constants by $con_1, \ldots, con_n$, where $n > 0$. A term is either a variable or a constant. An atomic formula is of the form $p(t_1, \ldots, t_m)$ or $r_i(t_1, \ldots)$ where the t's are terms and $i = 1, \ldots, l$. A literal is an atomic formula or its negation. If a literal contains a negation symbol ($\neg$), we call it a negative literal, otherwise it is a positive literal. A clause is any disjunction of literals, where all variables are assumed to be universally quantified. The learning task for relational learning systems is as follows:

Input
a set X^+_{target} of positive and a set X^-_{target} of negative training examples (expressed by literals without variables) for the target relation, background knowledge (or BK for short) expressed by literals without variables and not including the target predicate.

Output
a set of $\xi \leftarrow \lambda$ rules, where ξ is an atomic formula of the form $p(var^p_1, \ldots, var^p_m)$ with the target predicate p and λ is a conjunction of literals over background predicates $r_1, \ldots, r_l$, such that the set of rules satisfies the positive examples relatively to background knowledge.

Example 2. Let us consider the data set related to document understanding. The learning task involves identifying the purposes served by components of single-page letters such as that in Figure 1.

Background predicates describe properties of components such as their width and height, and relationships such as horizontal and vertical alignment with other components. Target predicates describe whether a block is one of the five predetermined types: sender, receiver, logo, reference, and date. For example, for letter presented in Figure 1, we obtain the following predicate data:

$$date\,(c_8),\ logo\,(c_3),\ receiver\,(c_{21}),\ on\_top\,(c_8, c_{21}),\ on\_top\,(c_{21}, c_{14}),$$
$$on\_top\,(c_5, c_{24}),\ on\_top\,(c_3, c_5),\ aligned\_only\_left\_col\,(c_1, c_3),$$
$$aligned\_only\_right\_col\,(c_5, c_{21}),\ \ldots$$

We consider generation of rules of the form:
$$sender\,(var_1) \leftarrow on\_top\,(var_1, var_2)\ \textbf{and}\ logo\,(var_2).$$

We will adopt the lower and the upper approximations for subsets of the set of target examples. First, we define the coverage of a rule.

Definition 4. *The coverage of Rule, written Coverage(Rule), is the set of examples such that there exists a substitution giving values to all variables appearing in the rule and all literals of the rule are satisfied for this substitution.*

The set of the positive (negative) examples covered by *Rule* is denoted by $Coverage^+(Rule)$, $Coverage^-(Rule)$, respectively.

Remark 1. For any literal L, we obtain

$$Coverage(h \leftarrow b) = Coverage(h \leftarrow b \wedge L) \cup Coverage(h \leftarrow b \wedge \neg L).$$

Let $U = X_{target}^+ \cup X_{target}^-$ and $Rule\_Set = \{Rule_1, \ldots, Rule_n\}$.

Definition 5. *For the set of rules Rule_Set and any example $x \in U$ the uncertainty function is defined by*

$$I_{Rule\_Set}(x) = \{x\} \cup \bigcup_{i=1}^{n} \{Coverage(Rule_i) : x \in Coverage(Rule_i)\}.$$

The lower and upper approximations may be defined as earlier but in this case they are equal to the forms presented in Remark 2.

Remark 2. For an approximation space $AS_{Rule\_Set} = (U, I_{Rule\_Set}, \nu_{SRI})$ and any subset $X \subseteq U$ the lower and the upper approximations are defined by

$$LOW\,(AS_{Rule\_Set}, X) = \{x \in U : I_{Rule\_Set}(x) \subseteq X\},$$

$$UPP\,(AS_{Rule\_Set}, X) = \{x \in U : I_{Rule\_Set}(x) \cap X \neq \emptyset\},$$

respectively.

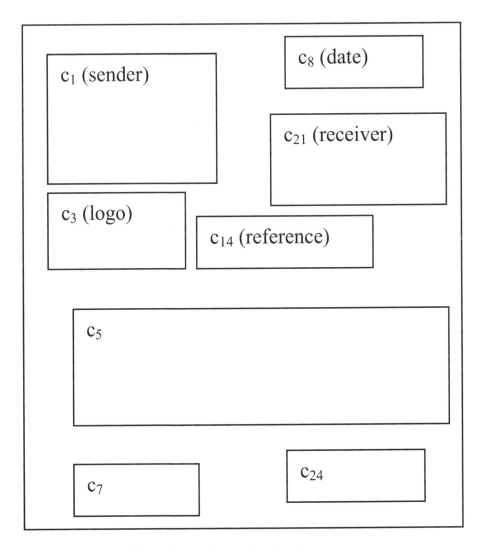

Fig. 1. Sample Letter Showing Components

4   Translating First–Order Data into Attribute–Value Form

In this section we discuss the approach based on two steps. First, the data is transformed from first-order logic into decision table format by the iterative checking whether a new attribute adds any relevant information to the decision table. Next, the reducts and rules from reducts [8,10,13] are computed from the decision table obtained.

Data represented as a set of formulas can be transformed into attribute–value form. The idea of translation was inspired by LINUS and DINUS systems (see, e.g., [2]). We start with a decision table directly derived from the positive and negative examples of the target relation. Assuming that we have m-ary target predicate, the set U of objects in the decision table is a subset of $\{con_1, \ldots, con_n\}^m$. Decision attribute $d_p : U \rightarrow \{+, -\}$ is defined by the target predicate with possible values " $+$ " or " $-$ ". All positive and negative examples of the target predicate are now put into the decision table. Each example forms a separate row in the table. Then background knowledge is applied to the decision table. We determine all the possible applications of the background predicates to the arguments of the target relation. Each such application introduces a new Boolean attribute.

To analyze the complexity of the obtained data table, let us consider the number of condition attributes. Let A_{r_i} be a set of attributes constructed for every predicate symbol r_i, where $i = 1, \ldots, l$. The number of condition attributes in constructed data table is equal to $\sum_{i=1}^{l} card\,(A_{r_i})$ resulting from the possible applications of the l background predicates on the variables of the target relation. The cardinality of A_{r_i} depends on the number of arguments of target predicate p (denoted by m) and the arity of r_i. Namely, $card\,(A_{r_i})$ is equal to $m^{ar(r_i)}$, where $ar\,(r_i)$ is the arity of the predicate r_i. The number of condition attributes in obtained data table is polynomial in the arity m of the target predicate p and the number l of background knowledge predicates, but its size is usually so large that its processing is unfeasible. Therefore, one can check interactively if a new attribute is relevant, i.e., if it adds any information to the decision table and, next we add to the decision table only relevant attributes.

Two conditions for testing if a new attribute a is relevant are proposed:

1. $\gamma\left(AS_{B \cup \{a\}}, \{X_+, X_-\}\right) > \gamma\left(AS_B, \{X_+, X_-\}\right)$,
   where X_+ and X_- denote the decision classes corresponding to the target concept. An attribute a is added to the decision table if this results in a growth of the positive region with respect to the attributes selected previously.
2. $Q_{DIS}(a) = \nu_{SRI}\left(X_+ \times X_-, \{(x,y) \in X_+ \times X_- : a\,(x) \neq a\,(y)\}\right) \geq \theta$,
   where $\theta \in [0, 1]$ is a given real number. An attribute a is added to the decision table if it introduces some discernibility between objects belonging to different non-empty classes X_+ and X_-.

Each of these conditions can be applied to a single attribute before it is introduced to the decision table. If this attribute does not meet a condition, it should not be included into the decision table. The received data table is then analyzed by a rough set based systems. First, reducts are computed. Next, decision rules are generated.

Example 3. The problem with three binary predicates r_1, r_3, p and one unary predicate r_2 can be used to demonstrate the transformation of relational learning problem into attribute–value form. Suppose that there are the following positive and negative examples of a target predicate p :

$$X_{target}^+ = \{p(1,2), p(4,1), p(4,2)\}, \quad X_{target}^- = \{\neg p(6,2), \neg p(3,5), \neg p(1,4)\}.$$

Consider the background knowledge about relations, r_1, r_2, and r_3 :

$r_1(5,1), r_1(1,2), r_1(1,4), r_1(4,1), r_1(3,1), r_1(2,6), r_1(3,5), r_1(4,2),$
$r_2(1), r_2(2), r_2(3), r_2(4), r_2(6), r_3(2,1), r_3(1,4), r_3(2,4),$
$r_3(2,5), r_3(3,2), r_3(3,5), r_3(5,1), r_3(5,3), r_3(2,6), r_3(4,2).$

We then transform the data into attribute–value form (decision table). In Table 3, a quality index Q_{DIS} of potential attributes is presented.

Table 3. Quality Q_{DIS} of Potential Attributes

Symbol	Attribute	$Q_{DIS}(\bullet)$
a_1	$r_2(var_1)$	0
a_2	$r_2(var_2)$	0.33
a_3	$r_1(var_1, var_1)$	0
a_4	$r_1(var_1, var_2)$	0.33
a_5	$r_1(var_2, var_1)$	0.56
a_6	$r_1(var_2, var_2)$	0
a_7	$r_3(var_1, var_1)$	0
a_8	$r_3(var_1, var_2)$	0.56
a_9	$r_3(var_2, var_1)$	0.33
a_{10}	$r_3(var_2, var_2)$	0

Using conditions introduced in this section some attributes will not be included in the resulting decision table. For example, the second condition with $Q_{DIS}(\bullet) \geq \theta = 0.3$ would permit the following attribute set into the decision table: $A_{0.3} = \{a_2, a_4, a_5, a_8, a_9\}$.

Therefore, $DT_{0.3} = (U, A_{0.3} \cup \{d\})$ finally. We obtain two decision classes: $X_+ = \{(1,2), (4,1), (4,2)\}$ and $X_- = \{(6,2), (3,5), (1,4)\}$. For the obtained decision table we construct an approximation space $AS_{A_{0.3}} = (U, I_{A_{0.3}}, \nu_{SRI})$ such that the uncertainty function and the rough inclusion are defined in Table 4. Then, we can compute reducts and decision rules.

5   The Rough Set Relational Learning Algorithm

In this section we introduce and investigate the RSRL (**R**ough **S**et **R**elational **L**earning) algorithm. Some preliminary versions of this algorithm were presented in [14,15].

5.1   RSRL Algorithm

To select the most promising literal from the candidates generated at each step, RSRL considers the performance of the rule over the training data. The evaluation function $card(R(L, NewRule))$ used by RSRL to estimate the utility of adding a new literal is based on the numbers of discernible positive and negative examples before and after adding the new literal (see, Figure 2).

Table 4. Resulting Decision Table $DT_{0.3}$, Uncertainty Function and Rough Inclusion

(var_1, var_2)	a_2	a_4	a_5	a_8	a_9	d_p	$I_{A_{0.3}}(\bullet)$	$\nu_{SRI}(\bullet, X_+)$	$\nu_{SRI}(\bullet, X_-)$
$(1,2)$	true	true	false	false	true	+	$\{(1,2)\}$	1	0
$(4,1)$	true	true	true	false	true	+	$\{(4,1)\}$	1	0
$(4,2)$	true	true	false	true	true	+	$\{(4,2)\}$	1	0
$(6,2)$	true	false	true	false	true	-	$\{(6,2)\}$	0	1
$(3,5)$	false	true	false	true	true	-	$\{(3,5)\}$	0	1
$(1,4)$	true	true	true	true	false	-	$\{(1,4)\}$	0	1

Some modification of the algorithm RSRL were presented in [15]. The modified algorithm generates rules as the original RSRL but its complexity is lower because it performs operations on the cardinalities of sets without computing the sets.

5.2 Illustrative Example

Let us illustrate the RSRL algorithm on a simple problem of learning a relation.

Example 4. The task is to define the target relation $p(var_1, var_2)$ in terms of the background knowledge relations r_1 and r_3. Let $BK = \{r_1(1,2), r_1(1,3), r_1(2,4),$ $r_3(5,2), r_3(5,3), r_3(4,6), r_3(4,7)\}$. There are two positive and three negative examples of the target relation:

$$X_{target}^+ = \{e_1, e_2\} \text{ and } X_{target}^- = \{e_3, e_4, e_5\}, \text{ where}$$

$$e_1 = p(1,4), e_2 = p(2,6), e_3 = \neg p(5,4), e_4 = \neg p(5,3) \text{ and } e_5 = \neg p(1,2).$$

Let us see how the algorithm generates rules for $h = p(var_1, var_2), app = lower$. The successive steps of the algorithm:

$Pos = \{e_1, e_2\}, Learned\_rules = \emptyset$.
$Pos \neq \emptyset$.
$R = \{(e_1, e_3), (e_1, e_4), (e_1, e_5), (e_2, e_3), (e_2, e_4), (e_2, e_5)\}$.
$R \neq \emptyset$.
We obtain the following candidates:

$r_i(var_1, var_1), r_i(var_1, var_2), r_i(var_1, var_3), r_i(var_2, var_1), r_i(var_2, var_2),$
$r_i(var_2, var_3), r_i(var_3, var_1), r_i(var_3, var_2),$ where $i = 1, 3$.

For ever1. candidate, we compute $R(L, NewRule)$ and we obtain the best result for $r_3(var_1, var_3)$.

In the first step, every example is covered either by the rule $p(var_1, var_2) \leftarrow r_3(var_1, var_3)$ or by $p(var_1, var_2) \leftarrow \neg r_3(var_1, var_3)$.

We obtain: $e_1, e_2, e_5 \in Coverage^+(h \leftarrow \neg Best\_literal) \cup Coverage^-(h \leftarrow \neg Best\_literal)$,
$e_3, e_4 \in Coverage^+(h \leftarrow Best\_literal) \cup Coverage^-(h \leftarrow Best\_literal)$.
From the intersection of R and the set
$(Coverage^+(h \leftarrow Best\_literal) \times Coverage^-(h \leftarrow \neg Best\_literal)) \cup$

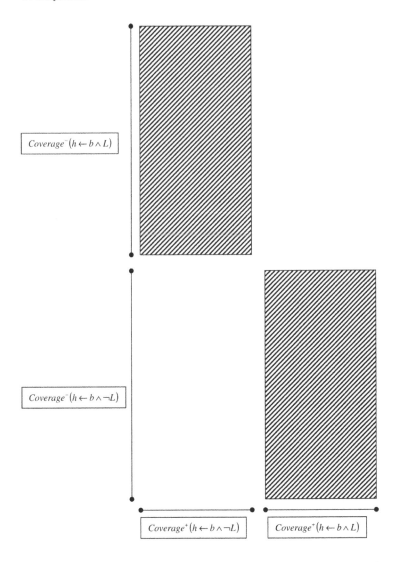

Fig. 2. The set $R(L, h \leftarrow b)$ is equal to the union of two Cartesian products

$(Coverage^+(h \leftarrow \neg Best\_literal) \times Coverage^-(h \leftarrow Best\_literal))$,
we obtain
$R(Best\_literal, NewRule) = \{(e_1, e_3), (e_1, e_4), (e_2, e_3), (e_2, e_4)\} \neq \emptyset$.
Since the value of the coverage of $p(var_1, var_2) \leftarrow \neg r_3(var_1, var_3)$
is greater than the value of the coverage of $p(var_1, var_2) \leftarrow r_3(var_1, var_3)$,
$Best\_literal = \neg r_3(var_1, var_3)$.
$app \neq upper$.
$b = \neg r_3(var_1, var_3)$.
$Coverage^-(NewRule) \neq \emptyset$.

We find new R considering a general case:
$b = b_1 = \neg r_3(var_1, var_3), L_1 = \neg r_3(var_1, var_3)$.
$R_1 = Coverage^+(h \leftarrow b_1) \times Coverage^-(h \leftarrow b_1) \cup S_1$.
$Coverage^+(h \leftarrow b_1) = \{e_1, e_2\}, Coverage^-(h \leftarrow b_1) = \{e_5\}$,
$Coverage^+(h \leftarrow b_0 \wedge \neg L_1) = \emptyset, Coverage^-(h \leftarrow b_0 \wedge \neg L_1) = \{e_3, e_4\}$,
$S_1 = Coverage^+(h \leftarrow b_0 \wedge \neg L_1) \times Coverage^-(h \leftarrow b_0 \wedge \neg L_1) = \emptyset$.
Hence, we obtain $R = R_1 = \{(e_1, e_5), (e_2, e_5)\}$.
The second step in the second loop:
$R \neq \emptyset$.

Algorithm 1. RSRL Algorithm

---

input : $Target\_predicate, BK, X^+_{target} \cup X^-_{target}, app$ //where $Target\_predicate$
is a target predicate with a set X^+_{target} of positive examples and a set
X^-_{target} of negative examples, BK is a background knowledge, app is a
type of approximation ($app \in \{lower, upper\}$).

output: $Learned\_rules$ //where $Learned\_rules$ is a set of rules for "positive
decision class".

$Pos \longleftarrow X^+_{target}$;
$Learned\_rules \longleftarrow \emptyset$;
while $Pos \neq \emptyset$ **do**

> $Learn\ a\ NewRule$;
> $NewRule \longleftarrow$ most general rule possible;
> $R \longleftarrow Pos \times X^-_{target}$;
> **while** $R \neq \emptyset$ **do**
>
> > $Candidate\_literals \longleftarrow$ generated candidates; // RSRL generates
> > candidate specializations of $NewRule$ by considering a new literal L that
> > fits one of the following forms:
> >
> > > - $r(var_1, \ldots, var_s)$, where at least one of the variable var_i in the created
> > >   literal must already exist in the positive literals of the rule;
> > > - the negation of the above form of literal;
> >
> > $Best\_literal \longleftarrow$ arg max $_{L \in Candidate\_literals} card(R(L, NewRule))$; //
> > (the explanation of $R(L, Rule)$ is in Figure 2 given)
> > **if** $R(Best\_literal, NewRule) = \emptyset$ or ($app = upper$ and ($NewRule \neq most$
> > general rule possible)
> > and $Coverage^+(NewRule) \neq Coverage^+(NewRule \wedge Best\_literal)$))
> > **then**
> >
> > > | exit while;
> >
> > **end**
> > Add $Best\_literal$ to $NewRule$ preconditions; //Add a new literal to
> > specialize $NewRule$;
> > **if** $Coverage^-(NewRule) = \emptyset$ **then**
> >
> > > | exit while;
> >
> > **end**
> > $R := R \setminus R(Best\_literal, NewRule)$;
>
> **end**
> $Learned\_rules \longleftarrow Learned\_rules \cup \{NewRule\}$;
> $Pos \longleftarrow Pos \setminus Coverage^+(NewRule)$;

end

We generate new candidates. We obtain the best result for the candidate $r_1(var_1, var_2)$ thus $Best\_literal = r_1(var_1, var_2)$.

Now $b = b_2 = \neg r_3(var_1, var_3) \wedge r_1(var_1, var_2), L_2 = r_1(var_1, var_2)$.

We compute the following sets:

$Coverage^+(h \leftarrow b_2) = \emptyset, Coverage^-(h \leftarrow b_2) = \{e_5\}$,

$Coverage^+(h \leftarrow b_1 \wedge \neg L_2) = \{e_1, e_2\}, Coverage^-(h \leftarrow b_1 \wedge \neg L_2) = \emptyset$.

$Coverage^+(h \leftarrow b_2) \times Coverage^-(h \leftarrow b_1 \wedge \neg L_2) \cup$

$Coverage^+(h \leftarrow b_1 \wedge \neg L_2) \times Coverage^-(h \leftarrow b_2) = \{(e_1, e_5), (e_2, e_5)\}$.

We obtain $R(Best\_literal, NewRule) = \{(e_1, e_5), (e_2, e_5)\} \neq \emptyset$.

Since the value of the coverage of

$p(var_1, var_2) \leftarrow \neg r_3(var_1, var_3) \wedge \neg r_1(var_1, var_2)$

is greater than the value of the coverage of

$p(var_1, var_2) \leftarrow \neg r_3(var_1, var_3) \wedge r_1(var_1, var_2)$

then $Best\_literal = \neg r_1(var_1, var_2)$.

$app \neq upper$.

$b = \neg r_3(var_1, var_3) \wedge \neg r_1(var_1, var_2)$.

$Coverage^-(NewRule) = \emptyset$. The end of the second loop.

$Learned\_rules = \{p(var_1, var_2) \leftarrow \neg r_3(var_1, var_3) \wedge \neg r_1(var_1, var_2)\}$.

$Coverage^+(p(var_1, var_2) \leftarrow \neg r_3(var_1, var_3) \wedge \neg r_1(var_1, var_2)) = \{e_1, e_2\} = Pos$, hence $Pos = Pos \setminus Coverage^+(NewRule) = \emptyset$.

The end of the algorithm proceeding.

In each step of the algorithm we obtain $Coverage^+(NewRule) = Coverage^+(NewRule \wedge Best\_literal)$. Hence, if $app = upper$ then we obtain the same rules as for $app = lower$. Hence, the lower and the upper approximations of X_{target}^+ are equal in our example. Let us compute the above sets to compare them. We have $Rule\_Set = \{p(var_1, var_2) \leftarrow \neg r_3(var_1, var_3) \wedge r_1(var_1, var_2), p(var_1, var_2) \leftarrow \neg r_3(var_1, var_3) \wedge \neg r_1(var_1, var_2)\}$ and $X_{target}^+ = \{e_1, e_2\}$.

We obtain the uncertainty function $I_{Rule\_Set}(e_1) = I_{Rule\_Set}(e_2) = \{e_1, e_2\}$, $I_{Rule\_Set}(e_3) = \{e_3\}, I_{Rule\_Set}(e_4) = \{e_4\}$ and $I_{Rule\_Set}(e_5) = \{e_5\}$.

Hence, $LOW\left(AS_{Rule\_Set}, X_{target}^+\right) = \{e_1, e_2\} = UPP\left(AS_{Rule\_Set}, X_{target}^+\right)$.

6   Conclusions

The first approach presented in this paper transforms input first-order logic formulas into decision table form, then uses reducts to select only meaningful data. The second approach is based on the algorithm RSRL for the first order rules generation. We showed that approximation spaces are basic structures for knowledge discovery from multi-relational data. Furthermore, our approach can be treated as a step towards the understanding of rough set methods in the first order rules generation.

Acknowledgements

The author wishes to thank the anonymous reviewers for their many helpful comments.

References

1. Bonchi, F., Boulicaut, J. F. (Eds.): Knowledge Discovery in Inductive Databases, Lecture Notes in Computer Science 3933, Springer–Verlag, Berlin Heidelberg, 2006.
2. Dzeroski, S., Lavrac, N. (Eds.): Relational Data Mining, Springer-Verlag, Berlin, 2001.
3. Łukasiewicz, J.: Die logischen Grundlagen der Wahrscheinlichkeitsrechnung, Kraków 1913. In: Borkowski, L. (ed.), *Jan Łukasiewicz - Selected Works*. North Holland, Amstardam, Polish Scientific Publishers, Warsaw, 1970.
4. Milton, R. S., Maheswari V. U., Siromoney A.: Rough Sets and Relational Learning, Transactions on Rough Sets I, Lecture Notes in Computer Science 3100, Springer, 2004, 321–337.
5. Orłowska, E., Szałas, A. (Eds.): Relational Methods for Computer Science Applications, Physica–Verlag, Heidelberg, 2001.
6. Pal, S.K., Polkowski, L., Skowron, A. (Eds.): Rough-Neural Computing: Techniques for Computing with Words. Springer-Verlag, Berlin, 2004.
7. Pawlak, Z.: Rough sets, International J. Comp. Inform. Science 11, 1982, 341–356.
8. Pawlak, Z.: Rough Sets. Theoretical Aspects of Reasoning about Data, Kluwer Academic Publishers, Dordrecht, 1991.
9. Polkowski, L., Skowron, A. (Eds.): Rough Sets in Knowledge Discovery 1 and 2. Physica-Verlag, Heidelberg, 1998.
10. Skowron, A., Stepaniuk, J.: Tolerance Approximation Spaces, Fundamenta Informaticae, 27, 1996, 245–253.
11. Skowron A., Stepaniuk J., Peters J. F., Swiniarski R.: Calculi of Approximation Spaces, Fundamenta Informaticae vol. 72(1–3), 2006, 363–378.
12. Stepaniuk, J.: Rough relations and logics. In: L. Polkowski, A. Skowron (Eds.), Rough Sets in Knowledge Discovery 1. Methodology and Applications, Physica Verlag, Heidelberg, 1998, 248–260.
13. Stepaniuk, J.: Knowledge Discovery by Application of Rough Set Models, L. Polkowski, S. Tsumoto, T.Y. Lin, (Eds.) Rough Set Methods and Applications. New Developments in Knowledge Discovery in Information Systems, Physica– Verlag, Heidelberg, 2000, 137–233.
14. Stepaniuk, J., Góralczuk, L.: An Algorithm Generating First Order Rules Based on Rough Set Methods, (ed.) J. Stepaniuk, Zeszyty Naukowe Politechniki Białostockiej Informatyka nr 1, 2002, 235–250. [in Polish]
15. Stepaniuk, J., Honko, P.: Learning First–Order Rules: A Rough Set Approach Fundamenta Informaticae, 61(2), 2004, 139–157.
16. Ziarko, W., Variable precision rough set model, Journal of Computer and System Sciences 46, 1993, 39–59.

Finding Relevant Attributes in High Dimensional Data: A Distributed Computing Hybrid Data Mining Strategy

Julio J. Valdés and Alan J. Barton

National Research Council Canada, M50, 1200 Montreal Rd., Ottawa, ON K1A 0R6
julio.valdes@nrc-cnrc.gc.ca,
alan.barton@nrc-cnrc.gc.ca
http://iit-iti.nrc-cnrc.gc.ca

Abstract. In many domains the data objects are described in terms of a large number of features (e.g. microarray experiments, or spectral characterizations of organic and inorganic samples). A pipelined approach using two clustering algorithms in combination with Rough Sets is investigated for the purpose of discovering important combinations of attributes in high dimensional data. The Leader and several k-means algorithms are used as fast procedures for attribute set simplification of the information systems presented to the rough sets algorithms. The data described in terms of these fewer features are then discretized with respect to the decision attribute according to different rough set based schemes. From them, the reducts and their derived rules are extracted, which are applied to test data in order to evaluate the resulting classification accuracy in crossvalidation experiments. The data mining process is implemented within a high throughput distributed computing environment. Nonlinear transformation of attribute subsets preserving the similarity structure of the data were also investigated. Their classification ability, and that of subsets of attributes obtained after the mining process were described in terms of analytic functions obtained by genetic programming (gene expression programming), and simplified using computer algebra systems. Visual data mining techniques using virtual reality were used for inspecting results. An exploration of this approach (using Leukemia, Colon cancer and Breast cancer gene expression data) was conducted in a series of experiments. They led to small subsets of genes with high discrimination power.

1 Introduction

As a consequence of the information explosion and the development of sensor and observation technologies, it is now common in many domains to have data objects characterized by an increasingly larger number of attributes, leading to high dimensional databases in terms of the set of fields. A typical example is a gene expression experiment, where the genetic content of samples of tissues are obtained with high throughput technologies (microchips). Usually, thousands of genes are investigated in such experiments. In other bio-medical research contexts, the samples are characterized by infrared, ultraviolet, and other kinds of spectra, where the absorption properties, with respect to a large number of wavelengths, are investigated. The same situation occurs in other domains, and the common denominator is to have a set of data objects of a very high dimensional nature.

J.F. Peters et al. (Eds.): Transactions on Rough Sets VI, LNCS 4374, pp. 366–396, 2007.
© Springer-Verlag Berlin Heidelberg 2007

This paper investigates one, of the possibly many approaches to the problem of finding relevant attributes in high dimensional datasets. The approach is based on a combination of clustering and rough sets techniques in a high throughput distributed computing environment, with low dimensional virtual reality data representations aiding data analysis understanding. The goals are:

i) to investigate the behavior of the combination of these techniques in a knowledge discovery process
ii) to perform preliminary comparisons of the experimental results from the point of view of the discovered relevant attributes, applied to the example problem of finding relevant genes

2   Datasets

In this study publicly available datasets were considered. They result from gene expression experiments in genomics, and appear in numerous studies about data mining and machine learning in bioinformatics. All of them share a feature typical of that kind of information: the data consist of a relatively small number of samples, described in terms of a large collection of attributes. Besides genomics, this situation is found in other fields as well, like experimental physics and astronomy. When infrared, ultraviolet or other spectral properties are used to describe the sampled objects, hundreds or thousands of energy intensity values for radiation emission or absorption at different wavelengths are used as sample attributes. The techniques investigated here are of a general nature, that is, not specific or tailored to any particular domain. The datasets considered for this study were:

- Leukemia ALL/AML dataset: (72 samples described in terms of 7129 genes [15]).
- Breast Cancer (24 samples described in terms of 12, 625 genes [7]).
- Colon Cancer: (62 samples described in terms of 2000 genes [1]).

The Leukemia dataset is that of [15], and consists of 7129 genes, where patients are separated into *i)* a training set containing 38 bone marrow samples: 27 acute lymphoblastic leukemia (ALL) and 11 acute myeloid leukemia (AML), obtained from patients at the time of diagnosis, and *ii)* a testing set containing 34 samples (24 bone marrow and 10 peripheral blood samples), where 20 are ALL and 14 AML. The test set contains a much broader range of biological samples, including those from peripheral blood rather than bone marrow, from childhood AML patients, and from different reference laboratories that used different sample preparation protocols. In the present study, however, the dataset will not be divided into training and test samples, because crossvalidation is used, as explained below.

The breast cancer data selected [7] was that provided by the Gene Expression Omnibus (GEO) (See www.ncbi.nlm.nih.gov/projects/geo/gds/gds_browse.cgi?gds=360). It consists of 24 core biopsies taken from patients found to be *resistant* (greater than 25% residual tumor volume, of which there are 14) or *sensitive* (less than 25% residual tumor volume, of which there are 10) to docetaxel treatment. The number of genes (probes) placed onto (and measured from) the microarray is 12, 625, and two classes are recognized: *resistant* and *sensitive*.

The Colon cancer data correspond to *tumor* and *normal* colon tissues probed by oligonucleotide arrays [1].

3   Foundational Concepts

3.1   Clustering Methods

Clustering with classical partition methods constructs crisp (non overlapping) subpopulations of objects or attributes. Two such classical algorithms were used in this study: the Leader algorithm [17], and several variants of k-means [2].

Leader Algorithm. The leader algorithm operates with a dissimilarity or similarity measure and a preset threshold. A single pass is made through the data objects, assigning each object to the first cluster whose leader (i.e. representative) is close enough (or similar enough) to the current object w.r.t. the specified measure and threshold. If no such matching leader is found, then the algorithm will set the current object to be a new leader; forming a new cluster. This technique is very fast; however, it has several negative properties:

i)   the first data object always defines a cluster and therefore, appears as a leader.
ii)   the partition formed is not invariant under a permutation of the data objects.
iii) the algorithm is biased, as the first clusters tend to be larger than the later ones since they get first chance at "absorbing" each object as it is allocated.

Variants of this algorithm with the purpose of reducing bias include:

a)   reversing the order of presentation of a data object to the list of currently formed leaders.
b)   selecting the absolute best leader found (thus making the object presentation order irrelevant).

The highest quality is obtained using *b)*, but at a higher computational cost because the set of leaders (whose cardinality increases as the process progresses), has to be completely explored for every data object. Nevertheless, even with this extra computational overhead, the technique is still very fast, and large datasets can be clustered very quickly. Usually the partitions generated by this method are used as initial approximations to more elaborated methods.

K-Means. The k-means algorithm is actually a family of techniques based on the concept of data reallocation. A dissimilarity or similarity measure is supplied, together with an initial partition of the data, and the goal is to alter cluster membership so as to obtain a better partition w.r.t. the chosen measure. The modification of membership is performed by reallocating the data objects to a different group w.r.t. the one in which it was a member. Different variants very often give different partition results. However, in papers dealing with gene expression analysis, very seldom are the specificities of the k-means algorithm described. For the purposes of this study, the following k-means variants were used: Forgy's, Jancey's, convergent, and MacQueen's [13], [20], [24], [2].

Let n_c be the number of clusters desired. The definition of an initial partition follows basically two schemes: *i)* direct specification of a set of n_c initial centroids (seed points), or *ii)* specification of n_c initial disjoint groups such that they cover the entire dataset, and compute from them initial centroids to start the process. There are many variations of these two schemes, and the following variants for defining an initial partition were considered in this paper:

1. Select n_c data objects and use them as initial centroids.
2. Divide the total number of objects into n_c consecutive clusters, compute the centroid of each, and use them as initial centroids.
3. Arbitrary n_c centroids are given externally.
4. Take the first n_c data objects and use them as initial centroids.

The classical k-means clustering is a simple algorithm with the following sequence of steps.

1. Allocate each data unit to the cluster with the nearest seed point (if a dissimilarity measure is used), or to the cluster with the most similar seed point (if a similarity measure is used).
2. Compute new seed points as the centroids of the newly formed clusters
3. *If* (termination criteria = true) *then* stop *else* goto 2

Several termination criteria (or a combination of them) can be established, which provide better control on the conditions under which a k-means process concludes. Among them are the folowing:

1. A preset number of object reallocations have been performed.
2. A preset number of iterations has been reached.
3. A partition quality measure has been reached.
4. The partition quality measure does not change in subsequent steps.

There are several variants of the general k-means scheme. That is why it is necessary to specify explicitly the specific variant applied. In this paper, several of them were used.

K-Means: Forgy's Variant. The classical Forgy's k-means algorithm [13] consists of the following steps:

i) Begin with any desired initial configuration. Go to *(ii)* if beginning with a set of seed objects, or go to *(iii)* if beginning with a partition of the dataset.
ii) Allocate each object to the cluster with the nearest (most similar) seed object (centroid). The seed objects remain fixed for a full cycle through the entire dataset.
iii) Compute new centroids of the clusters.
iv) Alternate *(ii)* and *(iii)* until the process converges (that is, until no objects change their cluster membership).

K-Means: Jancey's Variant. In Jancey's variant [20], the process is similar to Forgy's, but the first set of cluster seed objects is either given, or computed as the centroids of clusters in the initial partition. Then, at all succeeding stages, each new seed point is found by reflecting the old one through the new centroid for the cluster (a heuristic which tries to approximate the direction of the gradient of the error function).

K-Means: MacQueen's Variant. MacQueen's method [24] is another popular member of the k-means family, and is composed of the following steps:

i)   Take the first k data units as clusters of one member each.
ii) Assign each of the remaining objects to the cluster with the nearest (most similar) centroid. After each assignment, recompute the centroid of the gaining cluster.
iii) After all objects have been assigned in step *ii)*, take the existing cluster centroids as fixed points and make one more pass through the dataset assigning each object to the nearest (most similar) seed object.

K-Means: Convergent Variant. The so called convergent k-means [2] is a variant defined by the following steps:

i)   Begin with an initial partition like in Forgy's and Jancey's methods (or the output of MacQueen's method).
ii) Take each object in sequence and compute the distances (similarities) to all cluster centroids; if the nearest (most similar) is not that of the object's parent cluster, reassign the object and update the centroids of the losing and gaining clusters.
iii) Repeat steps *ii)* and *iii)* until convergence is achieved (that is, until there is no change in cluster membership).

Similarity Measure. The Leader and the k-means algorithms were used with a similarity measure rather than with a distance. In particular Gower's general coefficient was used [16], where the similarity between objects i and j is given by Eq-1:

$$S_{ij} = \sum_{k=1}^{p} s_{ijk} / \sum_{k=1}^{p} w_{ijk} \ , \tag{1}$$

where the weight of the attribute (w_{ijk}) is set equal to 0 or 1 depending on whether the comparison is considered valid for attribute k. If $v_k(i), v_k(j)$ are the values of attribute k for objects i and j respectively, an invalid comparison occurs when at least one them is missing. In this situation w_{ijk} is set to 0.

For quantitative attributes (like the ones in the datasets used in this paper), the scores s_{ijk} are assigned as in Eq-2:

$$s_{ijk} = 1 - |X_{ik} - X_{jk}| / R_k \ , \tag{2}$$

where X_{ik} is the value of attribute k for object i (similarly for object j), and R_k is the range of attribute k.

For symbolic attributes (nominal), the scores s_{ijk} are assigned as in Eq-3

$$s_{ijk} = \begin{cases} 1 \text{ if } X_{ik} = X_{jk} \\ 0 \ otherwise \ . \end{cases} \tag{3}$$

3.2   Rough Sets

The Rough Set Theory [31] bears on the assumption that in order to define a set, some knowledge about the elements of the dataset is needed. This is in contrast to the classical approach where a set is uniquely defined by its elements. In the Rough Set Theory, some elements may be indiscernible from the point of view of the available information and it turns out that vagueness and uncertainty are strongly related to indiscernibility. Within this theory, knowledge is understood to be the ability of characterizing all classes of the classification. More specifically, an information system is a pair $\mathbf{A} = (U, A)$ where U is a non-empty finite set called the universe and A is a non-empty finite set of attributes such that $a : U \rightarrow V_a$ for every $a \in A$. The set V_a is called the value set of a. For example, a decision table is any information system of the form $\mathbf{A} = (U, A \cup \{d\})$, where $d \in A$ is the decision attribute and the elements of A are the condition attributes.

Implicants. It has been described [28] that an m-variable function $f : \mathbf{B}^m \rightarrow \mathbf{B}$ is called a *Boolean function* if and only if it can be expressed by a Boolean formula. An *implicant* of a Boolean function f is a term p such that $p \leqslant f$, where $\leqslant$ is a partial order called the inclusion relation. A *prime implicant* is an implicant of f that ceases to be so if any of its literals are removed. An implicant p of f is a prime implicant of f in case, for any term q, the implication of Eq-4 holds.

$$p \leqslant q \leqslant f \Rightarrow p = q \ . \tag{4}$$

General Boolean Reasoning Solution Scheme. It has been described [28] that following the presentation of earlier work, the general scheme of applying Boolean reasoning to solve a problem P can be formulated as follows:

1. Encode problem P as a system of simultaneously-asserted Boolean equations as in Eq-5, where the g_i and h_i are Boolean functions on $\mathbf{B}$.

$$P = \begin{cases} g_1 = h_1 \\ \quad \vdots \\ g_k = h_k \end{cases} . \tag{5}$$

2. Reduce the system to a single Boolean equation (e.g. $f_p = 0$) as in Eq-6.

$$f_p = \sum_{i=1}^{k} \left(g_i' \cdot h_i + g_i \cdot h_i' \right) \ . \tag{6}$$

3. Compute Blake's Canonical Form ($BCF(f_p)$), the prime implicants of f_p.
4. Solutions to P are then obtained by interpreting the prime implicants of f_p.

Discernibility Matrices. An information system $\mathcal{A}$ defines a matrix M_A called a discernibility matrix. Each entry $M_A(x, y) \subseteq A$ consists of the set of attributes that can be used to discern between objects $x, y \in U$ according to Eq-7.

$$M_A(x, y) = \{a \in A : discerns\,(a, x, y)\} \ . \tag{7}$$

Where, $discerns\,(a, x, y)$ may be tailored to the application at hand.

Indiscernibility Relations and Graphs. A discernibility matrix M_A defines a binary relation $R_A \subseteq U^2$. The relation R_A is called an *indiscernibility relation* [28] (See Eq-8) with respect to A, and expresses which pairs of objects that we cannot discern between.

$$xR_Ay \Leftrightarrow M_A(x, y) = \emptyset \ . \tag{8}$$

An alternative way to represent R_A is via an *indiscernibility graph* (IDG), which is a graph $G_A = (U, R_A)$ with vertex set U and edge set R_A. It has been stated [28] that G_A is normally only interesting to consider when R_A is a tolerance relation, in which case G_A may be used for the purpose of clustering or unsupervised learning.

Discernibility Functions. A *discernibility function* [28] is a function that expresses how an object or a set of objects can be discerned from a certain subset of the full universe of objects. It can be constructed relative to an object $x \in U$ from a discernibility matrix M_A according to Eq-9.

$$f_A(x) = \prod_{y \in U} \left\{ \sum a^* : a \in M_A(x, y) \text{ and } M_A(x, y) \neq \emptyset \right\} \ . \tag{9}$$

The function $f_A(x)$ contains $|A|$ Boolean variables, where variable a^* corresponds to attribute a. Each conjunction of $f_A(x)$ stems from an object $y \in U$ from which x can be discerned and each term within that conjunction represents an attribute that discerns between those objects. The prime implicants of $f_A(x)$ reveal the minimal subsets of A that are needed to discern object x from the objects in U that are not members of $R_A(x)$.

In addition to defining discernibility relative to a particular object, discernibility can also be defined for the information system $\mathcal{A}$ as a whole. The full discernibility function $g_A(U)$ (See Eq-10) expresses how all objects in U can be discerned from each other. The prime implicants of $g_A(U)$ reveal the minimal subsets of A we need to discern all distinct objects in U from each other.

$$g_A(U) = \prod_{x \in U} f_A(x) \ . \tag{10}$$

Reducts. If an attribute subset $B \subseteq A$ preserves the indiscernibility relation R_A then the attributes $A \backslash B$ are said to be *dispensable*. An information system may have many such attribute subsets B. All such subsets that are minimal (i.e. that do not contain any dispensable attributes) are called reducts. The set of all reducts of an information system $\mathcal{A}$ is denoted $RED(\mathcal{A})$.

In particular, minimum reducts (those with a small number of attributes), are extremely important, as decision rules can be constructed from them [4]. However, the problem of reduct computation is NP-hard, and several heuristics have been proposed [43].

Rough Clustering. Based on the concept of a rough set, modifications to the classical family of k-means algorithms have been introduced in [22] and [23] observing that in data mining it is not possible to provide an exact representation of each class in the partition. For example, an approximation image classification method has been reported in [32]. Rough sets enable such representation using upper and lower bounds. In the case of rough k-means clustering, the centroids of the clusters have to be modified to include the effects of lower and upper bounds. The modified centroid calculations for a distance-based clustering would be as shown in Eq-11 [23]:

$$
x = \begin{cases} w_{lower} \times \frac{\sum_{v \in \underline{A}(x)} v_j}{|\underline{A}(x)|} + w_{upper} \times \frac{\sum_{v \in (\overline{A}(x) - \underline{A}(x))} v_j}{|\overline{A}(x) - \underline{A}(x)|} & if \overline{A}(x) - \underline{A}(x) \neq \phi \\ w_{lower} \times \frac{\sum_{v \in \underline{A}(x)} v_j}{|\underline{A}(x)|} & otherwise \end{cases} ,
$$

$$(11)$$

where $1 \leq j \leq m$ (the number of clusters). The parameters w_{lower} and w_{upper} control the importance of the lower and upper bonds. Equation 11 generalizes the corresponding k-means centroids update. If the lower and upper bounds are equal, conventional crisp clusters would be obtained (the boundary region $\overline{A}(x) - \underline{A}(x)$ is empty). The object membership w.r.t. the lower or upper bound of a cluster is determined in the following way: Let v be an object and x_i, x_j the centroids of clusters X_i, X_j respectively, where x_i is the closest centroid to object v, and x_j an arbitrary other centroid. Let $d(v, x_i), d(v, x_j)$ be the distances from object v to the corresponding centroids, and let T be a threshold value. If $d(v, x_i) - d(v, x_j) \leq T$, then $v \in \overline{A}(x_i)$, and $v \in \overline{A}(x_j)$ (i.e. v is not part of any lower bound). Otherwise, $v \in \underline{A}(x_i)$ and clearly $v \in \overline{A}(x_i)$. This algorithm depends on three parameters w_{lower}, w_{upper}, and T.

3.3   Virtual Reality Representation of Relational Structures

The role of visualization techniques in the knowledge discovery process is well known. Several reasons make Virtual Reality (VR) a suitable paradigm: Virtual Reality is *flexible*, in the sense that it allows the choice of different representation models to better accommodate different human perception preferences. In other words, allows the construction of different virtual worlds representing *the same* underlying information, but with a different look and feel. Thus, the user can choose the particular representation that is most appealing. VR allows *immersion*. VR creates a *living* experience. The user is not merely a passive observer or an outsider, but an actor in the world. VR is *broad and deep*. The user may see the VR world as a whole, and/or concentrate the focus of attention on specific details of the world. Of no less importance is the fact that in order to interact with a Virtual World, no mathematical knowledge is required, but only minimal computer skills.

A virtual reality based visual data mining technique, extending the concept of 3D modeling to relational structures, was introduced [40], [41] (see also http://www.hybridstrategies.com). It is oriented to the understanding of large heterogeneous, incomplete and imprecise data, as well as symbolic knowledge. The notion of data is not restricted to databases, but includes logical relations and other forms of both structured and non-structured knowledge. In this approach, the data objects are considered as tuples from a heterogeneous space [39]. Different information sources are associated with the attributes, relations and functions, and these sources are associated with the nature of what is observed (e.g. point measurements, signals, documents, images, directed graphs, etc). They are described by mathematical sets of the appropriate kind called source sets (Ψ_i), constructed according to the nature of the information source to represent. Source sets also account for incomplete information. A heterogeneous domain is a Cartesian product of a collection of source sets: $\hat{\mathcal{H}}^n = \Psi_1 \times \cdots \times \Psi_n$, where $n > 0$ is the number of information sources.

A *virtual reality space* is the tuple $\Upsilon = < \underline{O}, G, B, \Re^m, g_o, l, g_r, b, r >$, where $\underline{O}$ is a relational structure ($\underline{O} = < O, \Gamma^v >$, the O is a finite set of objects, and Γ^v is a set of relations), G is a non-empty set of *geometries* representing the different objects and relations. B is a non-empty set of *behaviors* of the objects in the virtual world. $\Re$ is the set of real numbers and $\Re^m \subset \mathbb{R}^m$ is a *metric space* of dimension m (Euclidean or not) which will be the actual virtual reality geometric space. The other elements are mappings: $g_o : O \to G, l : O \to \Re^m, g_r : \Gamma^v \to G, b : O \to B$.

Of particular importance is the mapping l. If the objects are in a heterogeneous space, $l : \hat{\mathcal{H}}^n \to \Re^m$. Several desiderata can be considered for building a VR-space. One may be to preserve one or more properties from the original space as much as possible (for example, the similarity structure of the data [6]). From an unsupervised perspective, the role of l could be to maximize some metric/non-metric structure preservation criteria [5], or to minimize some measure of information loss. From a supervised point of view l could be chosen as to emphasize some measure of class separability over the objects in O [41]. Hybrid requirements are also possible.

For example, if δ_{ij} is a dissimilarity measure between any two $i, j \in U$ ($i, j \in [1, N]$, where n is the number of objects), and $\zeta_{i^v j^v}$ is another dissimilarity measure defined on objects $i^v, j^v \in O$ from Υ ($i^v = \xi(i), j^v = \xi(j)$, they are in one-to-one correspondence). An error measure frequently used is shown in Eq-12 [35]:

$$Sammon\ error = \frac{1}{\sum_{i<j} \delta_{ij}} \frac{\sum_{i<j} (\delta_{ij} - \zeta_{ij})^2}{\delta_{ij}} . \tag{12}$$

Typically, classical algorithms have been used for directly optimizing measures of this type, like Steepest descent, Conjugate gradient, Fletcher-Reeves, Powell, Levenberg-Marquardt, and others. The l mappings within this paper were obtained using the method of Fletcher-Reeves [33]. The new nonlinear features are a form of dimensionality reduction and new attribute creation.

3.4   Gene Expression Programming

Pattern matching and function approximation are very important operations within data mining and data analysis. Typical examples of general function approximators are neural networks and fuzzy systems. While their performance is unquestioned, their interpretation is still awkward, sometimes extremely difficult in human terms. In the case of a neural network, the understanding of its performance is obscured by the intricacies of its architecture and its weights, some times very many. In the case of a fuzzy system, the set of fuzzy rules might be large in number and complexity. Moreover, the number of linguistic variables required and the collection of membership functions, might be large as well. Therefore, either a neural network or a fuzzy model may have an excellent performance, but interpretability issues might make a human user reluctant to use them. Analytic functions, have a relation with physical systems in general, which has a long history in science. They are easier to understand by humans, the preferred building blocks of modeling, and a highly condensed form of knowledge. Regression is an example where the family of functions is restricted to a few (typically just one), and the problem reduces to finding a set of parameters or coefficients which makes the function fulfill some desirable approximation property (for example, minimizing a least square error or other model quality measure). However, direct discovery of general analytic functions poses enormous challenges because of the (in principle) infinite size of the search space.

This important knowledge discovery problem can be approached from a computational intelligence perspective via evolutionary computation, and the solutions obtained are relevant to a large number of disciplines and domains. In particular genetic programming techniques aim at evolving computer programs, which ultimately are functions. Among this subfield of evolutionary computation, gene expression programming (GEP) is appealing [12]. Gene expression programming (GEP), like genetic algorithms (GAs), evolution strategies (ES) and genetic programming (GP), is an evolutionary algorithm as it uses populations of individuals, selects them according to fitness, and introduces genetic variation using one or more genetic operators. The fundamental difference between these techniques resides in the nature of the individuals. Different from GA, ES and GP, GEP individuals are nonlinear entities of different sizes and shapes (expression trees) encoded as strings of fixed length. For the interplay of the GEP chromosomes and the expression trees (ET), GEP uses a translation system to transfer the chromosomes into expression trees and vice versa [12]. The set of operators applied to GEP chromosomes always produces valid ETs.

The chromosomes in GEP itself are composed of genes structurally organized into a head and a tail [11]. The head contains symbols that represent both functions (from a function set F) and terminals (from a terminal set T), whereas the tail contains only terminals. Two different alphabets occur at different regions within a gene. For each problem, the length of the head h is chosen, whereas the length of the tail t is a function of h and the number of arguments of the function with the largest arity.

As an example, consider a gene composed of the function set F=$\{Q, +, -, *, /\}$, where Q represents the square root function, and the terminal set T=$\{a, b\}$. Such a gene (the tail is shown in **bold**) is: *Q-b++a/-b**baabaaabaab**, and encodes the ET which corresponds to the mathematical equation $f(a, b) = \sqrt{b} \cdot \left(\left(a + \frac{b}{a}\right) - ((a - b) + b) \right)$, which simplifies to $f(a, b) = \frac{b \cdot \sqrt{b}}{a}$.

GEP chromosomes are usually composed of more than one gene of equal length. For each problem the number of genes as well as the length of the head has to be chosen. Each gene encodes a sub-ET and the sub-ETs interact with one another forming more complex multi-subunit ETs through a connection function. To evaluate GEP chromosomes, different fitness functions can be used.

3.5   Distributed Computing and the Grid

Distributed computing can be defined in different ways, and there is no universally accepted formulation of the concept. It can be understood as an environment where idle CPU cycles and storage space of tens, hundreds, or thousands of networked systems can be harnessed to work together on a particular processing-intensive problem. The growth of such processing models has been limited, however, due to a lack of compelling applications and by bandwidth bottlenecks, combined with significant security, management, and standardization challenges. However, in the last years the interest has grown to the extent of making the technology an emergent fact. Increasing desktop CPU power and communications bandwidth have also helped to make distributed computing a more practical approach. The numbers of real applications are still somewhat limited, and the challenges (particularly standardization) are significant.

Grid computing is a form of distributed computing that involves coordinating and sharing computing, application, data, storage, or network resources across dynamic and geographically dispersed organizations. As previously stated, there is no universally accepted definition, but a consensus exists in that a Grid is a type of parallel and distributed system that enables the sharing, selection, and aggregation of geographically distributed "autonomous" resources dynamically at runtime depending on their availability, capability, performance, cost, and users' quality-of-service requirements. Grid technologies promise to change the way complex computational problems are approached and solved. However, the vision of large scale resource sharing is not yet a reality in many areas. Grid computing is an evolving area of computing, where standards and technology are still being developed to enable this new paradigm.

The grid computing concept aims to promote the development and advancement of technologies that provide seamless and scalable access to wide-area distributed resources. Computational Grids enable the sharing, selection, and aggregation of a wide variety of geographically distributed computational resources (such as supercomputers, compute clusters, storage systems, data sources, instruments, people) and presents them as a single, unified resource for solving large-scale compute and data intensive computing applications (e.g, molecular modelling for drug design, brain activity analysis, and high energy physics). The idea is analogous to electric power networks (grids) where power generators are distributed, but the users are able to access electric power without bothering about the source of energy and its location. Grids aim at exploiting synergies that result from cooperation–ablity to share and agreegrate distributed computational capabilities and deliver them as service. The use of grid technologies for data mining is an obvious choice for many exploratory data analysis tasks within the knowledge discovery process.

The identification of the research issues and their potential priorities for the years 2003-2010, as well as the formulation of proposal of suitable means for implementation, has been addressed by several groups of experts [14], [3].

Among distributed computing systems for delivering high throughput computing, the Condor system stands out [9], [36], [37], [38], (http://www.cs.wisc.edu/condor/). Condor is a specialized workload management system for compute-intensive jobs in a distributed computing environment, developed by the Condor

Research Project at the University of Wisconsin-Madison (UW-Madison). Like other full-featured batch systems, Condor provides a job queueing mechanism, scheduling policy, priority scheme, resource monitoring, and resource management. Users submit their serial or parallel jobs to Condor, Condor places them into a queue, chooses when and where to run the jobs based upon a policy, carefully monitors their progress, and ultimately informs the user upon completion.

While providing functionality similar to that of a more traditional batch queueing system, Condor's novel architecture allows it to succeed in areas where traditional scheduling systems fail. Condor can be used to manage a cluster of dedicated compute nodes (such as a "Beowulf" cluster), possibly mixed with individual nodes. In addition, unique mechanisms enable Condor to effectively harness wasted CPU power from otherwise idle desktop workstations. For instance, Condor can be configured to only use desktop machines where the keyboard and mouse are idle. Should Condor detect that a machine is no longer available (such as a key press detected), in many circumstances Condor is able to transparently produce a checkpoint and migrate a job to a different machine which would otherwise be idle. Condor does not require a shared file system across machines - if no shared file system is available, Condor can transfer the job's data files on behalf of the user, or Condor may be able to transparently redirect all of the job's I/O requests back to the submit machine. As a result, Condor can be used to seamlessly combine all of an organization's computational power into one resource.

3.6 Implementation

A detailed perspective of data mining procedures provides insight into additional important issues to consider (e.g. storage/memory/communication/management/time/etc) when evaluating a computational methodology consisting of combined techniques. This study presents one possible implementation, from which more software development may occur in order to integrate better and/or different tools. In addition, all of these issues become even more pronounced when, as in this study, a complex problem is investigated.

The implementation of the distributed pipeline is shown in Alg.1. It consists of two pieces; a sequential portion, and a distributed portion. For the sequential portion, a specific machine (usually the local host) is used to perform some preliminary processing on the data (as it only needs to be performed once) and then distributes the data via a specific distribution mechanism to a set of waiting computing nodes, which may include the distributing machine. Once all of the computations have completed, the sequential portion of the pipeline may then proceed to collect the results from all of the files that have been placed onto the distributing machine (again via the distribution mechanism, but this time from compute node (e.g. remote host) to distributing host). The resultant databases may then be queried for the purpose of analysis.

The specific distribution mechanism used, is a high throughput pipeline (Fig. 2) consisting of many co-operating programs. Such a pipeline structure is generated automatically in order to ease the proper configuration of each participating program within the

---

Algorithm 1. Abstract Conceptualization of the Distributed Pipeline

---

Input : A Data Matrix, D^{Input}
Output: A Set of Relevant Attributes.
From a Specific Host, Sequentially do
   | GenerateAndConfigurePipeline();
   | D^{Random} ⟵ ShuffleObjects$_{(i)}$($\mathsf{Opt}_{(i)}$, D^{Input}) ;
   | DistributeToComputeNodes$_{(j)}$($\mathsf{Opt}_{(j)}$, D^{Random}) ;
   | StartPipelineExecution();
   | BlockedWaitForResults();   // Monitor Each Job's Progress
   | // Store All Completed Job Results Locally
   | $\mathsf{Results}^{Rules}$ ⟵ ReformatAndCollectRules$_{(k)}$() ;
   | $\mathsf{Results}^{Statistics}$ ⟵ ReformatAndCollectStats$_{(l)}$() ;
   | AnalyzeResults(D^{Random}, $\mathsf{Results}^{Rules}$, $\mathsf{Results}^{Statistics}$) ;
end
On Each Compute Node Run A Job And do
   | $D^{Leaders}$ ⟵ ConstructLeaders$_{(m)}$($\mathsf{Opt}_{(m)}$, D^{Random}) ;
   | $D^{Subsets}$ ⟵ SubsetSelection$_{(n)}$($\mathsf{Opt}_{(n)}$, $D^{Leaders}$) ;
   | // e.g. create $D^{Subsets}$ by 10-fold cross-validation
   | **forall** $((D_i^{Tr}, D_i^{Te}) \in D^{Subsets})$ **do**
      | $(D_{i,discr}^{Tr}, \mathsf{Cuts}_i^{Tr})$ ⟵ Discretize$_{(o)}$($\mathsf{Opt}_{(o)}^{Tr}$, D_i^{Tr}) ;
      | $\mathsf{Reducts}_i^{Tr}$ ⟵ FormReducts$_{(p)}$($\mathsf{Opt}_{(p)}^{Tr}$, $D_{i,discr}^{Tr}$) ;
      | Rules_i^{Tr} ⟵ GenerateRules$_{(q)}$($\mathsf{Opt}_{(q)}^{Tr}$, $\mathsf{Reducts}_i^{Tr}$) ;
      | $D_{i,discr}^{Te}$ ⟵ Discretize$_{(o)}$($\mathsf{Opt}_{(o)}^{Te}$, D_i^{Te}, Cuts_i^{Tr}) ;
      | $\mathsf{RuleSetMerit}_i$ ⟵ Classify$_{(r)}$($\mathsf{Opt}_{(r)}$, $D_{i,discr}^{Te}$) ;
      | Record($\mathsf{RuleSetMerit}_i$)
   | **end**
end

---

pipeline. In this paper, the automatically generated pipeline was facilitated via *i)* a file generation program (written in Python and running on the local host) and *ii)* the Condor tool described in section 3.5.

The initial preprocessing stage of the pipeline, occurring on the distributing host after generation of files, involves shuffling the input data records as described previously and in Fig. 1. The shuffled data is stored on the distributing host's disk, in order to provide the same randomized data to the next stage of processing, which occurs on the computing hosts (Fig. 2).

A Condor submission program, which was also automatically generated, is used to specify all of the data and configuration files for the programs that will execute on the remote host. The submission process enables Condor to:

i)   schedule jobs for execution
ii) check point them (put a job on hold)
iii) transfer all data to the remote host
iv) transfer all generated data back to the local host (submitting machine)

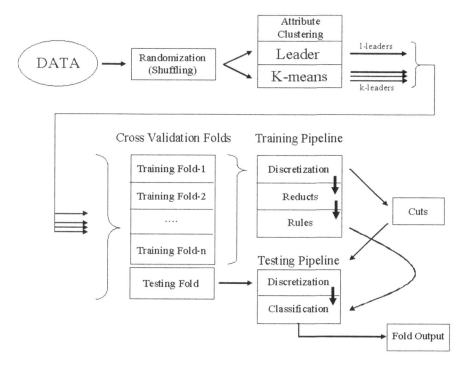

Fig. 1. Data processing strategy combining clustering, Rough Sets analysis and crossvalidation

The final postprocessing stage of the pipeline involves collecting all of the results (parsing the files) and reporting them in a database.

4 Experimental Methodology

The datasets consist of information systems with an attribute set composed of ratio and interval variables, and a nominal or ordinal decision attribute. More general information systems have been described in [39]. The general idea is to construct subsets of relatively similar attributes, such that a simplified representation of the data objects is obtained by using the corresponding attribute subset representatives. The attributes of these simplified information systems are explored from the point of view of their reducts. From them, rules are learned and applied systematically to testing data subsets not involved in the learning process (Fig.1). The whole procedure can be seen as a pipeline.

   In a first step, the objects in the dataset are shuffled using a randomized approach in order to reduce the possible biases introduced within the learning process by data chunks sharing the same decision attribute. Then, the attributes of the shuffled dataset are clustered using the two families of fast clustering algorithms described in previous sections (the leader, and k-means). Each of the formed clusters of attributes is

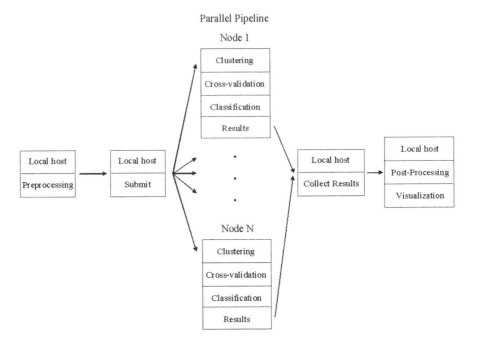

Fig. 2. Automatically generated high throughput pipeline oriented towards the Condor distributed computing environment

represented by exactly one of the original data attributes. By the nature of the leader algorithm, the representative is the leader (called an *l-leader*), whereas for a k-means algorithm, a cluster is represented by the most similar object w.r.t. the centroid of the corresponding cluster (called a *k-leader*). This operation can be seen as a filtering of the attribute set of the original information system. As a next step, the filtered information system undergoes a segmentation with the purpose of learning classification rules, and testing their generalization ability in a cross-validation framework. N-folds are used as training sets; where the numeric attributes present are converted into nominal attributes via a discretization process (many possibilities exist), and from them, reducts are constructed. Finally, classification rules are built from the reducts, and applied to a discretized version of the test fold (according to the cuts obtained previously), from which the generalization ability of the generated rules can be evaluated. Cross-validation is used in order to create a statistically meaningful estimate of the classification accuracy of the rules generated from the reducts for a particular experiment. The final database of experimental results was then sorted by minimum mean classification accuracy. The best mean accuracy experiments were then selected in order to extract the attributes from within the computed reducts for further analysis.

Besides the numeric descriptors associated with the application of classification rules to data, use of visual data mining techniques, like the virtual reality representation (section 3.3), enables structural understanding of the data described in terms of the selected subset of attributes and/or the rules learned from them. This technique can be applied at a pre- and/or post-processing stage. In this paper, all of the applications were made in the unsupervised mode. That is, the existing class information was not used during the process, and is incorporated within the resulting visualization only to enable thecomparison between the data structure provided by the predictor variables with the class distribution that is known to exist. Each stage of the process feeds its results to the next stage of processing, yielding a pipelined data analysis stream, with partial outputs that can be used for other kinds of analysis.

4.1 ROSETTA

The ROSETTA Software [28], [30] was used within this study with the algorithms that are described within the following sections.

Discretization: NaiveScaler. The heuristic implemented by Rosetta [29] was used and was described under the assumption that all condition attributes A are numerical. For each condition attribute a, sort its value set V_a to obtain the ordering indicated by Eq-13.

$$v_a^1 < \cdots v_a^i < \cdots < v_a^{|V_a|} \; . \tag{13}$$

Then let C_a denote the set of all cuts for attribute a generated in a naive fashion according to equations Eq-14, Eq-15, and Eq-16.

$$X_a^i = \left\{ x \in U : a\left(x\right) = v_a^i \right\} \; , \tag{14}$$

$$\Delta_a^i = \left\{ v \in V_d : \exists x \in X_a^i \text{ such that } d\left(x\right) = v \right\} \; , \text{and} \tag{15}$$

$$C_a = \left\{ \frac{v_a^i + v_a^{i+1}}{2} : \left|\Delta_a^i\right| > 1 \text{ or } \left|\Delta_a^{i+1}\right| > 1 \text{ or } \Delta_a^i \neq \Delta_a^{i+1} \right\} \; . \tag{16}$$

The set C_a consists of all cuts midway between two observed attribute values, except for the cuts that are clearly not needed due to the fact that the objects have the same decision value. Hence, such a cut would not discern the objects. If no cuts are found for an attribute, NaiveScaler leaves the attribute unprocessed. Missing values are ignored in the search for cuts. In the worst case, each observed value is assigned its own interval.

Discretization: SemiNaiveScaler. The discretization algorithm as implemented within Rosetta [29] is similar to the NaiveScaler but has more logic to handle the case where value-neighboring objects belong to different decision classes. This algorithm typically results in fewer cuts than the simpler NaiveScaler, but may still produce more cuts than are desired. In Eq-17, the set D_a^i collects the dominating decision values for the objects in X_a^i. If there are no ties, D_a^i is a singleton. The rationale used within Rosetta for not adding a cut if the sets of dominating decisions define an inclusion is that it is hoped (although it is stated that the implementation does not check) that a cut will be added for another attribute (different from a) such that the objects in X_a^i and X_a^{i+1} can be discerned.

$$D_a^i = \left\{ v \in V_d : v = \underset{v'}{argmax} \left| \left\{ x \in X_a^i : d(x) = v' \right\} \right| \right\} . \tag{17}$$

$$C_a = \left\{ \frac{v_a^i + v_a^{i+1}}{2} : D_a^i \not\subseteq D_a^{i+1} \text{ and } D_a^{i+1} \not\subseteq D_a^i \right\} . \tag{18}$$

Discretization: RSESOrthogonalScaler. This algorithm is an efficient implementation [25] of the Boolean reasoning algorithm [27] within the Rough Set Exploration System (RSES) (See http://logic.mimuw.edu.pl/~rses/). It is mentioned [29] that this algorithm is functionally similar to BROrthogonalScaler but much faster. Approximate solutions are not supported. If $a(x)$ is missing, object x is not excluded from consideration when processing attribute a, but is instead treated as an infinitely large positive value. If no cuts are found for an attribute, all entries for that attribute are set to 0.

Discretization: BROrthogonalScaler. The Rosetta implementation [29] of a previously outlined algorithm [27] was used, which is based on the combination of the NaiveScaler algorithm previously presented and a Boolean reasoning procedure for discarding all but a small subset of the generated cuts. Construct set of candidate cuts C_a according to Eq-16. Then construct a boolean function f from the set of candidate cuts according to Eq-19.

$$f = \prod_{(x,y)} \sum_a \left\{ \sum c^* : c \in C_a \text{ and } a(x) < c < a(y) \text{ and } \partial_A(x) \neq \partial_A(y) \right\} . \tag{19}$$

Then compute the prime implicant of f using a greedy algorithm [21] (see JohnsonReducer). This Boolean reasoning approach to discretization may result in no cuts being deemed necessary (because they do not aid discernibility) for some attributes. The Rosetta implementation does not alter such attributes.

Discretization: EntropyScaler. The Rosetta implementation [29] of the algorithm [8] is based on recursively partitioning the value set of each attribute so that a local measure of entropy is optimized. The minimum description length principle defines a stopping criterion for the partitioning process. Rosetta ignores missing values in the search for cuts and Rosetta does not alter attributes for which no cuts were found.

Reduct Computation: RSESExhaustiveReducer. The RSES algorithm included within Rosetta [29] computes all reducts by brute force. Computing reducts is NP-hard, so information systems of moderate size are suggested to be used within Rosetta.

Reduct Computation: Holte1RReducer. Rosetta's [29] algorithm creates all singleton attribute sets, which was inspired by a paper in Machine Learning [18]. The set of all 1R rules, (i.e. univariate decision rules) are thus directly constructed from the attribute sets.

Reduct Computation: `RSESJohnsonReducer`. Rosetta [29] invokes the RSES implementation of the greedy algorithm [21] for reduct computation. No support is provided for IDGs, boundary region thinning or approximate solutions.

Reduct Computation: `JohnsonReducer`. Rosetta [29] invokes a variation of a greedy algorithm to compute a single reduct [21]. The algorithm (See Alg.2) has a natural bias towards finding a single prime implicant of minimal length. The reduct R is found by executing the following algorithm, where $\sum w(X)$ denotes a weight for set $X \in S$ that is computed from the data. Support for computing approximate solutions is provided by aborting the loop when enough sets have been removed from S, instead of requiring that S has to be fully emptied.

Algorithm 2. Johnson Reducer

Input : A Data Matrix, D^{Input}
Output: One Reduct
$R \leftarrow \emptyset$;   // Reduct has no attributes within it
$S \leftarrow \{S_1, S_2, \ldots, S_n\}$;
repeat
  // A contains all attributes that maximizes $\sum w(X)$,
  // where the sum is taken over all sets $X \in S$ that
  // contain a.
  $A \leftarrow \{a : \text{maximal} \{ \sum_{\{X \in S: a \in X\}} w(X)\}\}$;
  // The Rosetta implementation resolves ties arbitrarily
  $a \leftarrow \text{RandomElementFromSet}(A)$;
  $R \leftarrow R \cup \{a\}$;   // Add attribute to growing reduct
  $S \leftarrow \{X \in S : a \notin X\}$;   // Stop considering sets containing a
until $S = \emptyset$;   // No more attributes left for consideration
return R ;

Rule Generation: `RSESRuleGenerator`. Rosetta [29] invokes the RSES implementation of an algorithm to generate rules from a set of reducts. Conceptually performed by overlaying each reduct in the reduct set over the reduct set's parent decision table and reading off the values.

5   Results

5.1   Leukemia Gene Expression Data

The example high dimensional dataset selected is that of [15], and consists of 7129 genes where patients are separated into i) a training set containing 38 bone marrow samples: 27 acute lymphoblastic leukemia (ALL) and 11 acute myeloid leukemia (AML), obtained from patients at the time of diagnosis, and ii) a testing set containing 34 samples (24 bone marrow and 10 peripheral blood samples), where 20 are ALL and 14 AML. The test set contains a much broader range of biological samples, including those from peripheral blood rather than bone marrow, from childhood AML patients,and from different reference laboratories that used different sample preparation protocols. Further, the dataset is known to have two types of ALL, namely B-cell and T-cell. For the purposes of investigation, only the AML and ALL distinction was made. The dataset distributed by [15] contains preprocessed intensity values, which were obtained by rescaling such that overall intensities for each chip are equivalent (A linear regression model using all genes was fit to the data). In this paper no explicit preprocessing of the data was performed, in order to not introduce bias and to be able to expose the behavior of the data processing strategy, the methods used, and their robustness. That is, no background subtraction, deletions, filtering, or averaging of samples/genes were applied, as is typically done in gene expression experiments.

In a preprocessing stage, a virtual reality representation of the opriginal dataset in a 3-dimensional space as described in section 3.3 was computed. Gower similarity was used for the original space, and normalized Euclidean distance for the target space. Steepest descent was used for optimizing Sammon's error. The purpose was to appreciate the relationship of the structure of the existing classes and the collection of original attributes. As shown in (Fig.3, the two Leukemia classes appear completely mixed, as approximated with the original set of attributes. Noisy attributes do not allow a resolution of the classes.

The pipeline (Fig.1) was investigated through the generation of 480 k-leader and 160 l-leader for a total of 640 experiments (Table-1). The discretization, reduct computation and rule generation algorithms are those included in the Rosetta system [30]. This approach leads to the generation of 74 files per experiment, with 10-fold cross-validation. From the experiments completed so far, one was chosen which illustrates the kind of results obtained with the explored methodology. It corresponds to a leader clustering algorithm with a similarity threshold of 0.99 (leading to 766 l-leader attributes), used as input to the data processing pipeline containing 38 samples. The results of the best 10 fold cross-validated experiment has a mean accuracy of 0.925 and a standard deviation

Table 1. The set of parameters and values used in the experiments using the distributed pipeline environment

Algorithm/Parameter	Values
Leader	ReverseSearch, ClosestSearch
Leader Similarity Threshold	0.7, 0.8, 0.9, 0.95, 0.99,
	0.999, 0.9999, 0.99999
K-Means	Forgy, Jancey, Convergent, MacQueen
Cross-validation	10 folds
Discretization	BROrthogonalScaler, EntropyScaler,
	NaiveScaler, SemiNaiveScaler
Reduct Computation	JohnsonReducer, Holte1RReducer
Rule Generation	RSESRuleGenerator

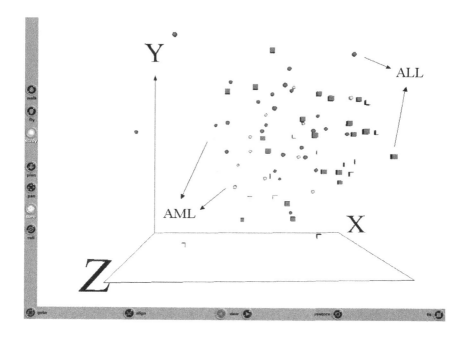

Fig. 3. Snapshot of the Virtual Reality representation of the original Leukemia data (training set with 38 samples + test set with 34, both with 7129 genes). Dark objects= ALL class, Light objects=AML class. Spheres = training, Cubes = test. Representation error = 0.143, Sammon error = $3.56e - 6$.

of 0.168. This experiment led to 766 reducts (all of them singleton attributes), which was consistent across each of the 10 folds. The obtained classification accuracy represents a slight improvement over those results reported in [42] (0.912). It was conjectured in that study that the introduction of a cross-validated methodology could improve the obtained classification accuracies, which is indeed the case. It is interesting to observe that all of the 7 relevant attributes (genes) reported in [42] are contained (subsumed) within the single experiment mentioned above. Moreover, they were collectively found using both the leader and k-means algorithms, with different dissimilarity thresholds and number of clusters, whereas with the present approach, a single leader clustering input was required to get the better result. Among the relevant attributes (genes) obtained, many coincide with those reported by [15], [10], and [42].

At a post-processing stage, a virtual reality space representation of the above mentioned experiment is shown in Fig.4. Due to the limitations of representing an interactive virtual world on static media, a snapshot from an appropriate perspective is presented. Sammon's error [35] was used as criteria for computing the virtual reality space, and Gower's similarity was used for characterizing the data in the space of the 766 selected genes. After 200 iterations a satisfactory error level of 0.0998 was obtained. It is interesting to see that the ALL and AML classes can be clearly differentiated.

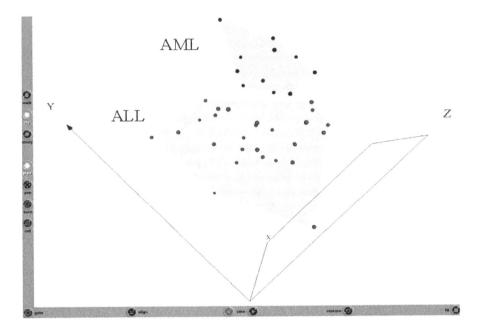

Fig. 4. Snapshot of the Virtual Reality representation of the union of all of the reducts obtained from 10 fold cross-validation input (38 samples with 766 genes) for the Leukemia data. The leader clustering algorithm was used with a similarity threshold of 0.99. The ALL and the AML classes are perfectly separated. Representation error = 0.0998.

5.2   Breast Cancer Data

A visual representation of the original data in terms of the 12, 625 genes obtained using Gower similarity in the original space, Sammon error, and steepest descent optimization [34] is shown in Fig.5. The *sensitive* and *resistant* classes are shown for comparison, and semi-transparent convex hulls wrap the objects from the corresponding classes. There is a little overlap between the two sets, indicating the classification potential of the whole set of attributes, but complete class resolution is not obtained with the nonlinear coordinates of the VR space.

Rough k-means [23] was used to cluster the 24 samples into 2 groups using the whole set of original attributes (genes) in order to illustrate the difficulties involved in using all of the original attributes. The particular algorithm parameters ($w_{lower} = 0.9$, $w_{upper} = 0.1$, $distanceThreshold = 1$). The rough k-means result for the 24 samples using the 12, 625 original attributes, and requesting 2 classes is shown in Fig.6. In the VR space, 2 classes are clearly well differentiated, but one of them contains 5 objects and the other contains the rest. Moreover, when the smaller class is investigated, it contains a mixture of samples from the *resistant* and *sensitive* class. Therefore, even the more elaborated rough k-means clustering can not resolve the two known classes from the point of view of all of the original attributes used at the same time. It is also interesting, that for this dataset that no boundary cases were obtained with the clustering parameters used.

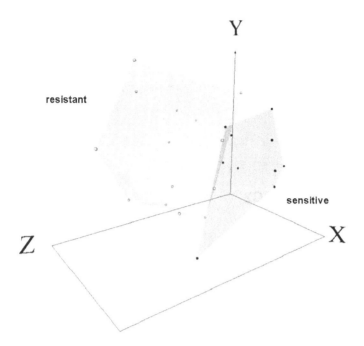

Fig. 5. Visual representation of 24 breast cancer samples with $12,625$ genes. Convex hulls wrap the *resistant*(size= 14) and *sensitive*(size= 10) classes. Absolute Error = $7.33 \cdot 10^{-2}$. Relative Mapping Error = $1.22 \cdot 10^{-4}$.

The experimental settings used in the investigation of breast cancer data with the distributed pipeline are reported in Table 2. For each experiment, the discretization, reduct computation and rule generation algorithms are those included in the Rosetta system [30]. The leader algorithm variants were described in section 3.1. This approach leads to the generation of 84 files per experiment, with 10-fold cross-validation.

From the series of l-leader Breast Cancer experiments performed, 4 experiments (Exp-81, Exp-82, Exp-145, Exp-146) were found to be equivalent when analyzing the mean (0.73), median (0.67), standard deviation (0.25), minimum (0.5) and maximum (1.0) of the 10-fold cross validated classification accuracy of the produced rules. For the l-leader algorithm a similarity threshold of 0.7 was used by all experiments, with Exp-81 and Exp-145 using closest placement criteria and Exp-82 and Exp-146 using reverse search criteria. The discretization algorithm as provided by the Rosetta system was the RSESOrthogonalScaler for Exp-81 and Exp-82 and the BROrthogonalScaler for Exp-145 and Exp-146. The reduct algorithm (RSESExhaustiveReducer) was the same for all 4 experiments with full discernibility and all selection. The rule generation algorithm (RSESRuleGenerator), was also the same for all 4 experiments.

In a postprocessing stage, the gene expression programming was applied to selected pipeline results. The idea was to try a simple function set ($F = \{+, -, *\}$) without the use of numeric constants, in order to reduce the complexity of the assembled functions as much as possible. In particular, experiments 81, 82, 145 and 146 all found a subset

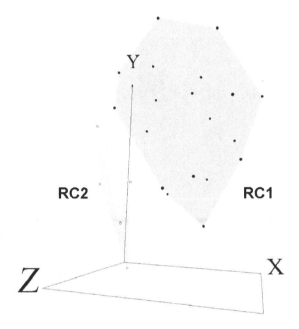

Fig. 6. Visual representation of 24 breast cancer samples with $12,625$ genes. Convex hulls wrap the *RC1*(size= 19) and *RC2*(size= 5) classes built by rough set based k-means. Absolute Error $= 7.06 \cdot 10^{-2}$. Relative Mapping Error $= 5.21 \cdot 10^{-5}$. *RC1, RC2* stand for the two rough clusters obtained.

of only 3 l-leader attributes with a mean crossvalidation error of 0.73. The subset of attributes was $\{2, 139, 222\}$, corresponding to genes $\{31307\_at, 31444\_s\_at, 31527\_at\}$. They represent $\{12359, 27, 239\}$ data objects respectively. Accordingly, the terminal set defined by the attributes found by the pipeline was set to $T = \{v_2, v_{139}, v_{222}\}$. The resulting class membership function emerging from the GEP process is as shown in Eq-20:

$$f(v_2, v_{139}, v_{222}) = ((v_{222} + v_{139}) + v_{139}) + \tag{20}$$
$$(((v_{222} - v_{139}) * (v_2 * v_2)) * ((v_{139} + v_{139}) - v_{139})) +$$
$$(v_{222} - (((v_2 - v_{222}) * (v_{139} - v_2)) * ((v_{222} + v_{139}) + v_2))) \ .$$

This analytic expression was simplified with the Yacas computer algebra system `http://www.xs4all.nl/~apinkus/yacas.html` which resulted in the expression in Eq-21:

$$f(v_2, v_{139}, v_{222}) = v_{222}^2 * v_{139} - v_{222}^2 * v_2 + v_{222} * v_{139}^2 +$$
$$v_{222} * v_{139} * v_2^2 - v_{222} * v_{139} * v_2 + 2 * v_{222} +$$
$$2 * v_{139} - (v_{139}^2 * v_2 + v_{139}^2 * v_2^2) + v_2^3 \ . \tag{21}$$

Table 2. The set of parameters and values used in the experiments with the Breast Cancer dataset using the distributed pipeline environment

Algorithm/Parameter	Values
Leader	ReverseSearch, ClosestSearch
Leader Similarity Threshold	0.7, 0.8, 0.9, 0.95, 0.99,
	0.999, 0.9999, 0.99999
Cross-validation	10 folds
Discretization	BROrthogonalScaler, EntropyScaler,
	NaiveScaler, RSESOrthogonalScaler, SemiNaiveScaler
Reduct Computation	RSESExhaustiveReducer, RSESJohnsonReducer
Rule Generation	RSESRuleGenerator

The classification rule associated with Eq-21 is shown in Eq-22. The classification accuracy on the original dataset was 91.67%, and it should be noted that only 3 genes out of the 12, 625 original ones are used. Moreover, the resulting model is relatively simple.

$$IF \quad f(v_2, v_{139}, v_{222}) \geq 0.5) \longrightarrow class = sensitive \qquad (22)$$
$$otherwise \longrightarrow class = resistant \ .$$

5.3 Colon Cancer Data

A virtual reality space representation of the dataset in terms of the original 2000 attributes was computed for an initial assesment of the structure of the data. Sammon error was used as structure measure, with normalized Euclidean distance as dissimilarity, and Powell's method for error optimization [34]. In 50 iterations an extremely low mapping error obtained $(1.067 \text{x} 10^{-6})$ is shown in Fig.7.

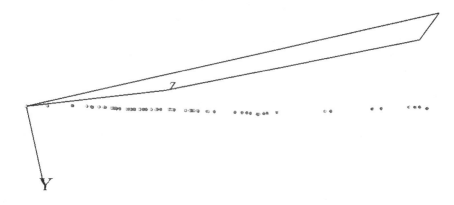

Fig. 7. Visual representation (3 dimensions) of 62 colon cancer samples with 2000 genes. Darker objects belong to the tumor class, and lighter objects to the normal class. After 50 iterations: Absolute Error $= 1.067 \cdot 10^{-6}$. Relative Mapping Error $= 0.0488$.

Table 3. Two Breast Cancer dataset experiments and their associated reducts for each of the 10 cross-validated folds. The GEP encodings are also reported.

Experiment	Fold	Reducts	GEP Encoding
	1	$\{31307\_at, 31444\_s\_at, 31527\_at\}$	$\{v_2, v_{139}, v_{222}\}$
	2	$\{31307\_at, 31527\_at\}$	$\{v_2, v_{222}\}$
	3	$\{31307\_at, 31527\_at\}$	$\{v_2, v_{222}\}$
	4	$\{31307\_at, 31444\_s\_at, 31527\_at\}$	$\{v_2, v_{139}, v_{222}\}$
81	5	$\{31307\_at, 31444\_s\_at, 31527\_at\}$	$\{v_2, v_{139}, v_{222}\}$
	6	$\{31307\_at, 31444\_s\_at, 31527\_at\}$	$\{v_2, v_{139}, v_{222}\}$
	7	$\{31307\_at, 31444\_s\_at, 31527\_at\}$	$\{v_2, v_{139}, v_{222}\}$
	8	$\{31307\_at, 31444\_s\_at, 31527\_at\}$	$\{v_2, v_{139}, v_{222}\}$
	9	$\{31307\_at, 31444\_s\_at, 31527\_at\}$	$\{v_2, v_{139}, v_{222}\}$
	10	$\{31307\_at, 31527\_at\}$	$\{v_2, v_{222}\}$
	1	$\{31307\_at, 31444\_s\_at, 31527\_at\}$	$\{v_2, v_{139}, v_{222}\}$
	2	$\{31444\_s\_at, 31527\_at\}$	$\{v_{139}, v_{222}\}$
	3	$\{31307\_at, 31527\_at\}$	$\{v_2, v_{222}\}$
	4	$\{31307\_at, 31444\_s\_at, 31527\_at\}$	$\{v_2, v_{139}, v_{222}\}$
82	5	$\{31307\_at, 31444\_s\_at, 31527\_at\}$	$\{v_2, v_{139}, v_{222}\}$
	6	$\{31307\_at, 31444\_s\_at, 31527\_at\}$	$\{v_2, v_{139}, v_{222}\}$
	7	$\{31307\_at, 31527\_at\}$	$\{v_2, v_{222}\}$
	8	$\{31307\_at, 31444\_s\_at, 31527\_at\}$	$\{v_2, v_{139}, v_{222}\}$
	9	$\{31307\_at, 31527\_at\}$	$\{v_2, v_{222}\}$
	10	$\{31307\_at, 31527\_at\}$	$\{v_2, v_{222}\}$

Such a small mapping error indicates that the VR space is a very accurate portrait of the 2000 dimensional original space. The most interesting feature in the VR space is the existence of an intrinsic unidimensionality in the data from the point of view of preserving the distance structure. Although the right hand side of the projected data line in Fig.7 predominantly contains objects of the tumor class, and the left half objects of the normal classes, they are mixed, and therefore, the space does not resolve the classes. Nevertheless, this result is an indication about the potential of finding relatively small subsets of attributes with reasonable classification power and about the large redundancy within the original attributes.

The experimental settings used in the investigation of colon cancer data with the distributed pipeline are reported in Table 4.For each experiment, the discretization, reduct computation and rule generation algorithms are those included in the Rosetta system [30]. The leader algorithm variants were described in section 3.1. This approach leads to the generation of 84 files per experiment, with 10-fold cross-validation. From the series of 320 l-leader Colon Cancer experiments, 5 are selected for illustration: Exp-113, Exp-304, Exp-195, Exp-180, and Exp-178. They were found to be equivalent when analyzing the mean (0.73), median (0.67), standard deviation (0.25), minimum (0.5), and maximum (1.0) of the 10-fold cross validated classification accuracy.From the series of l-leader experiments, 2 were selected for illustrating the number of created rules per fold. The *i)* rules and their respective reducts from which they were generated are shown in Table 6 and *ii)* the original attribute names as well as related genes as found

Table 4. The set of parameters and values used in the experiments with the Colon Cancer dataset using the distributed pipeline environment

Algorithm/Parameter	Values
Leader	ReverseSearch, ClosestSearch
Leader Similarity Threshold	0.7, 0.8, 0.9, 0.95, 0.99,
	0.999, 0.9999, 0.99999
Cross-validation	10 folds
Discretization	BROrthogonalScaler, EntropyScaler,
	NaiveScaler, RSESOrthogonalScaler, SemiNaiveScaler
Reduct Computation	JohnsonReducer, Holte1RReducer,
	RSESExhaustiveReducer, RSESJohnsonReducer
Rule Generation	RSESRuleGenerator

Table 5. Selected l-leader Colon Cancer experiments sorted by minimum 10-fold cross-validated classification accuracy. The last column shows the resultant number of pipeline selected attributes (from 2, 000) for each experiment.

Experiment	Mean	Median	Standard Deviation	Min.	Max.	Sim.	No. Attr.
178	0.562	0.500	0.261	0.167	1.0	0.7	3
180	0.629	0.646	0.219	0.333	1.0	0.8	9

within the source reference [1] are shown in Table 7. In these cases, all reducts were composed of singleton attributes found from the original 2000. They are presented due to their cross-validated minimum and maximum accuracies of [0.167-1.0] and [0.333-1.0] respectively. Experiment 178 contains the same 3 singleton reducts in each of the 10 folds, from which [59 − 73] rules were obtained. Whereas, Experiment 180 contains 9 singleton reducts in each of the 10 folds, from which [209 − 241] rules were obtained.

For the found l-leaders in these experiments, it can be seen that perfect classification has been made for some of the folds which is a sign of interestingness.

In a postprocessing stage, gene expression programming was applied to selected pipeline results. The idea was to try a simple function set ($F = \{+, -, *, sin, cos, log\}$) in order to reduce the complexity of the assembled nonlinear functions as much as possible. In particular, experiment 178 found a subset of only 3 l-leader attributes. The subset of attributes was $\{1, 2, 12\}$, corresponding to genes $\{H55933, R39465, H86060\}$. They represent $\{1, 11, 1988\}$ data objects respectively. Accordingly, the terminal set defined by the attributes found by the pipeline was set to $T = \{v_1, v_2, v_{12}\}$. The resulting class membership function emerging from the GEP process is shown in Eq-23:

$$f(v_1, v_2, v_{12}) = ((v_1 * (v_1 * cos((v_{12} * v_{12})))) - v_2) +$$
$$(v_1 * (cos(log(v_1)) * v_2)) +$$
$$((v_2 * (cos((v_2 + v_2)) * v_2)) + v_1) \ . \tag{23}$$

When Eq-23 is simplified, the resultant equation is that as shown in Eq-24:

$$f(v_1, v_2, v_{12}) = cos(v_{12}^2) * v_1^2 + v_1 * v_2 * cos(log(v_1)) +$$
$$v_1 + v_2^2 * cos(2 * v_2) - v_2 \ . \tag{24}$$

The classification rule associated to Eq-24 is shown in Eq-25, which has a classification accuracy on the original dataset of 88.7%, and it should be noted that only 3 genes out of the 2000 original ones were used.

$$IF \ (f(v_1, v_2, v_{12}) \geq 0.5) \longrightarrow class = normal \tag{25}$$
$$otherwise \longrightarrow class = tumor \ .$$

In a second application of the gene expression programming method, the attributes found by experiment 180 $\{1, 2, 4, 5, 6, 22, 27, 119, 878\}$ were used (again 1-leaders). The terminal set was allowed to have numeric constants. The resulting class membership function emerging from the GEP process is as shown in Eq-26:

$$f(v_1, v_2, v_4, v_5, v_6, v_{22}, v_{27}, v_{119}, v_{878}) =$$
$$(((((v_{878} - v_{22}) * v_5) * (v_4 * v_{878})) * ((v_6 + v_5) + v_5)) + v_1) +$$
$$(((((v_{119} - v_{878}) * v_{119}) * v_1) * ((v_{27} + v_{878}) * (v_1 + v_{22}))) + v_1) +$$
$$(v_{27} - (v_5 * (k_1 - (v_{119} + ((v_{119} + v_1) + v_5)))))) +$$
$$(k_2 * (((v_1 - (v_{878} + v_5)) + v_4) * v_6)) \ , \tag{26}$$

where $k_1 = 3.55777, k_2 = -7.828919$. After simplification, the resulting function is as shown in Eq-27:

$$f(v_1, \mathbf{v_2}, v_4, v_5, v_6, v_{22}, v_{27}, v_{119}, v_{878}) =$$
$$2 * v_{878}^2 * v_5^2 * v_4 +$$
$$v_{878}^2 * v_5 * v_4 * v_6 - v_{878}^2 * v_1^2 * v_{119} - v_{878}^2 * v_{22} * v_1 * v_{119} +$$
$$(-2) * v_{878} * v_{22} * v_5^2 * v_4 - v_{878} * v_{22} * v_5 * v_4 * v_6 +$$
$$v_{878} * v_{22} * v_1 * v_{119}^2 - v_{878} * v_{22} * v_1 * v_{119} * v_{27} +$$
$$v_{878} * v_1^2 * v_{119}^2 - v_{878} * v_1^2 * v_{119} * v_{27} - v_{878} * v_6 * k_2 +$$
$$v_{22} * v_1 * v_{119}^2 * v_{27} +$$
$$v_5^2 +$$
$$v_5 * v_1 +$$
$$2 * v_5 * v_{119} - v_5 * k_1 - v_5 * v_6 * k_2 +$$
$$v_4 * v_6 * k_2 +$$
$$v_6 * v_1 * k_2 +$$
$$v_1^2 * v_{119}^2 * v_{27} +$$
$$2 * v_1 +$$
$$v_{27} \ . \tag{27}$$

The classification rule associated to Eq-27 is as shown in Eq-28:

$$IF \ (f(v_1, v_4, v_5, v_6, v_{22}, v_{27}, v_{119}, v_{878}) \geq 0.5) \longrightarrow class = normal$$
$$otherwise \longrightarrow class = tumor \ , \quad (28)$$

Table 6. Two Colon Cancer dataset experiments. Exp. 178 has 3 reducts that are the same in all 10 folds. Exp. 180 has 9 reducts that are the same in all 10 folds.

Cross Validation Fold	1	2	3	4	5	6	7	8	9	10
No. Rules in Exp. 178	67	72	73	72	68	59	66	72	68	68
No. Rules in Exp. 180	230	236	233	234	227	209	216	241	228	227

Table 7. Discovered attributes for 2 Colon Cancer dataset experiments. Exp. 180 found one attribute (v_6 =R02593) that was also previously reported.

Exp. 178	Encoding:	v_1	v_2	v_{12}		
	Original:	H55933	R39465	H86060		
	Encoding:	v_1	v_2	v_{22}	v_{27}	v_{119}
	Original:	H55933	R39465	J02763	H86060	T72175
Exp. 180	Encoding	v_4	v_5	v_6	v_{878}	
	Original:	R85482	U14973	R02593	M87789	
	Compared to [1]:	(R85464)	(U14971)	(*Same*)		

and has a classification accuracy on the original dataset of 91.9%. From the point of view of classification accuracy, it is only slightly better than the one obtained with only 3 attributes. On the other hand, despite the fact that most of the individual terms are relatively simple (addition is the root of the expression tree), the expression as a whole is very complex, and certainly much more than the previous model. Likely such an expression is not an arguable replacement for a neural network or a set of fuzzy relations in terms of simplicity or understandability, and moreover, the situation can be even worse if the function set is extended with other nonlinear functions like e^x, $ln(x)$, transcendental functions, numeric constants, etc., as is required in complex function approximation tasks. However, despite these difficulties, genetic programming, and particularly GEP allows an explicit assessment of the role of predictor variables. It also provides analytic means to perform sensitivity analysis directly, through the study of the partial derivatives and the multidimensional gradient of the generated functions. The approach is promising, and new developments in genetic programming including meta-function approximation and the incorporation of more intelligent techniques may overcome the above mentioned difficulties.

It is interesting to observe that gene 2 is not found in Eq-28, indicating that it was irrelevant (boldfaced in Eq-27). However, despite the increased complexity, this independent technique showed that the set of genes suggested by the data mining pipeline has an important classification power.

6 Conclusions

Good results were obtained with the proposed high throughput pipeline based on the combination of clustering and rough sets techniques for the discovery of relevant attributes in high dimensional data. The use of several clustering and rough set analysis techniques, and their combination as a virtual data mining machine implemented in a grid and high throughput computing environment, proved to be a promising way to address complex knowledge discovery tasks. In particular, the introduction of a fast attribute reduction procedure aided rough set reduct discovery in terms of computational time, of which the former is further improvable via its amenability for parallel and distributed computing. Cross-validated experiments using three different sets of gene expression data demonstrate the possibilities of the proposed approach. With the hybrid methodology presented, in all cases it was possible to find subsets of attributes of size much smaller than the original set, retaining a high classification accuracy.

The pre- and post-processing stages of visual data mining using multidimensional space mappings and genetic programming techniques like gene expression programming (in combination with computer algebra systems), are effective elements within the data processing strategy proposed. The analytic functional models obtained for evaluating class memberships and ultimately for classification via gene expression programming, allowed a better understanding of the role of the different attributes in the classification process, as well as an explicit explanation of their influence.

More thorough studies are required to correctly evaluate the impact of the experimental settings on the data mining effectiveness, and further experiments with this approach are necessary.

Acknowledgements

This research was conducted within the scope of the BioMine project (National Research Council Canada (NRC), Institute for Information Technology (IIT)). The authors would like to thank Robert Orchard and Marc Leveille from the Integrated Reasoning Group (NRC-IIT), and to Ratilal Haria and Roger Impey from the High Performance Computing Group (NRC-IIT).

References

1. Alon, U., Barkai, N., Notterman, D.A., Gish, K., Ybarra, S., Mack, D., Levine, A.J.: Broad patterns of gene expression revealed by clustering analysis of tumor and normal colon tissues probed by oligonucleotide arrays. In: Proceedings National Academy of Science. USA v96 (1999) 67456750
2. Anderberg, M.: Cluster Analysis for Applications. Academic Press (1973)
3. Bal, H., de Laat, C., Haridi, S., Labarta, J., Laforenza, D., Maccallum, P., Mass, J., Matyska, L., Priol, T., Reinefeld, A., Reuter, A., Riguidel, M., Snelling, D., van Steen, M.: Next Generation Grid(s) European Grid Research 2005 - 2010 Expert Group Report, (2003)
4. Bazan, J.G., Skowron, A., Synak, P.: Dynamic Reducts as a Tool for Extracting Laws from Decision Tables. In: Proceedings of the Symp. on Methodologies for Intelligent Systems. Charlotte, NC, Oct. 16-19 1994. Lecture Notes in Artificial Intelligence 869, Springer-Verlag (1994) 346-355

5. Borg, I., Lingoes, J.: Multidimensional similarity structure analysis. Springer-Verlag, New York, NY (1987)
6. Chandon, J.L., Pinson, S.: Analyse typologique. Thorie et applications. Masson, Paris (1981)
7. Chang, J.C. et al.: Gene expression profiling for the prediction of therapeutic response to docetaxel in patients with breast cancer. Mechanisms of Disease. The Lancet, vol 362 (2003)
8. Dougherty, J., Kohavi, R., Sahami,M.: Supervised and unsupervised discretization of continuous features. In A. Prieditis and S. Russell, editors, Proc. Twelfth International Conference on Machine Learning, Morgan Kaufmann (1995) 194-202
9. Epema, D.H.J., Livny, M., van Dantzig, R., Evers, X., Pruyne, J.: A worldwide flock of Condors: Load sharing among workstation clusters. Journal of Future Generation Computer Systems, (1996) 53-65
10. Famili, F., Ouyang, J.: Data mining: understanding data and disease modeling. In: Proceedings of the 21st IASTED International Conference, Applied Informatics, Innsbruck, Austria, Feb. 10-13, (2003) 32-37
11. Ferreira C.: Gene Expression Programming: A New Adaptive Algorithm for Problem Solving. Journal of Complex Systems ˇ13, 2, (2001) 87-129
12. Ferreira C.: Gene Expression Programming: Mathematical Modeling by an Artificial Intelligence. Angra do Heroismo, Portugal (2002)
13. Forgy, E.W.: Cluster analysis of multivariate data: Efficiency versus interpretability of classifications. Biometric Soc. Meetings, Riverside, California (abstract in Biometrics, v.21, no. 3 (1965) 768
14. Foster, I., Kesselman, C., Tuecke, S.: The Anatomy of the Grid: Enabling Scalable Virtual Organizations. International. Journal of Supercomp. App., v.15(3):20 (2001) 222-237
15. Golub, T.R., et al.: Molecular classification of cancer: class discovery and class prediction by gene expression monitoring. Science, vol. 286 (1999) 531-537
16. Gower, J.C.: A general coefficient of similarity and some of its properties. Biometrics, v.1, no. 27, (1973) 857-871
17. Hartigan, J.: Clustering Algorithms. John Wiley & Sons (1975)
18. Holte, R.C.: Very simple classification rules perform well on most commonly used datasets. Machine Learning, 11(1) April (1993) 63-91
19. Jain, A.K., Mao, J.: Artificial Neural Networks for Nonlinear Projection of Multivariate Data. In: Proceedings 1992 IEEE Joint Conf. on Neural Networks (1992) 335-340
20. Jancey, R.C.: Multidimensional group analysis. Australian Journal of Botany, v.14, no. 1 (1966) 127-130
21. Johnson, D.S.: Approximation algorithms for combinatorial problems. Journal of Computer and System Sciences, 9 (1974) 256-278
22. Lingras, P.: Unsupervised Rough Classification using GAs. Journal of Intelligent Information Systems v16, 3 Springer-Verlag (2001) 215-228
23. Lingras, P., Yao, Y.: Time Complexity of Rough Clustering: GAs versus K-Means. Third. Int. Conf. on Rough Sets and Current Trends in Computing RSCTC 2002. Alpigini, Peters, Skowron, Zhong (Eds.) Lecture Notes in Computer Science (Lecture Notes in Artificial Intelligence Series) LNCS 2475. Springer-Verlag (2002) 279-288
24. MacQueen, J.B.: Some methods for classification and analysis of multivariate observations. In: Proceedings of the 5-th Symposium on Math. Statist. and Probability. Berkeley. AD669871 Univ. of California Press, Berkeley. v.1 (1967) 281–297
25. Nguyen, H.S., Nguyen, S.H.: Some efficient algorithms for rough set methods. In: Proceedings Fifth Conference on Information Processing and Management of Uncertainty in Knowledge-Based Systems (IPMU'96), Granada, Spain, July (1996) 1451–1456
26. Nguyen, H.S., Nguyen, S.H.: Discretization Methods in Data Mining. In: L. Polkowski, A. Skowron (eds.): Rough Sets in Knowledge Discovery. Physica-Verlag, Heidelberg (1998) 451-482

27. Nguyen, H.S., Skowron, A.: Quantization of real-valued attributes. In: Proceedings Second International Joint Conference on Information Sciences, Wrightsville Beach, NC, September (1995) 34-37
28. Øhrn, A.: Discernibility and Rough Sets in Medicine: Tools and Applications. PhD thesis, Norwegian University of Science and Technology, Department of Computer and Information Science, December NTNU report 1999:133. [http://www.idi.ntnu.no/~aleks/thesis/] (1999)
29. Øhrn, A.: Rosetta Technical Reference Manual. Department of Computer and Information Science, Norwegian University of Science and Technology, Trondheim, Norway (2001)
30. Øhrn, A., Komorowski, J.: Rosetta- A Rough Set Toolkit for the Analysis of Data. In: Proceedings of Third Int. Join Conf. on Information Sciences (JCIS97), Durham, NC, USA, March 1-5 (1997) 403-407
31. Pawlak, Z.: Rough sets: Theoretical aspects of reasoning about data. Kluwer Academic Publishers, Dordrecht, Netherlands (1991)
32. Peters, J.F., Borkowski, M.: K-means Indiscernibility Relation over Pixels, Fourth. Int. Conf. on Rough Sets and Current Trends in Computing RSCTC 2004. Tsumoto, Slowinski, Komorowki, Grzymala-Busse (Eds.) Lecture Notes in Computer Science (Lecture Notes in Artificial Intelligence Series) LNAI 3066, Springer-Verlag (2004) 580-585
33. Press, W.H., Flannery, B.P., Teukolsky, S.A., Vetterling, W.T.: Numerical Recipes in C, Cambridge University Press, New York (1986)
34. Press, W.H., Teukolsky, S.A., Vetterling, W.T., Flannery, B.P.: Numerical Recipes in C. The Art of Scientific Computing. Cambridge Univ. Press (1992)
35. Sammon, J.W.: A non-linear mapping for data structure analysis. IEEE Trans. on Computers C18 (1969) 401-409
36. Tannenbaum, T., Wright, D., Miller, K., Livny, M.: Condor – A Distributed Job Scheduler. In Thomas Sterling (Ed.) Beowulf Cluster Computing with Linux. MIT Press (2001)
37. Thain, D., Tannenbaum, T., Livny, M.: Condor and the Grid. In Fran Berman and Geoffrey Fox and Tony Hey (Eds.) Grid Computing: Making the Global Infrastructure a Reality. John Wiley & Sons (2002)
38. Thain, D., Tannenbaum, T., Livny, M.: Distributed Computing in Practice: The Condor Experience. Journal of Concurrency and Computation: Practice and Experience (2004)
39. Valdés, J.J.: Similarity-Based Heterogeneous Neurons in the Context of General Observational Models. Neural Network World. Vol **12**, No. 5 (2002) 499-508
40. Valdés, J.J.: Virtual Reality Representation of Relational Systems and Decision Rules: An exploratory Tool for understanding Data Structure. In: Theory and Application of Relational Structures as Knowledge Instruments. Meeting of the COST Action 274 (P. Hajek. Ed). Prague, November 14-16 (2002)
41. Valdés, J.J.: Virtual Reality Representation of Information Systems and Decision Rules: An Exploratory Tool for Understanding Data and Knowledge. Lecture Notes in Artificial Intelligence LNAI 2639, Springer-Verlag, 15–618 (2003)
42. Valdés, J.J., Barton, A.J.: Gene Discovery in Leukemia Revisited: A Computational Intelligence Perspective. In: Proceedings of the 17th International Conference on Industrial & Engineering Applications of Artificial Intelligence & Expert Systems May 17-20, 2004, Ottawa, Canada. Lecture Notes in Artificial Intelligence LNAI 3029, Springer-Verlag (2004) 118-127
43. Wróblewski, J.: Ensembles of Classifiers Based on Approximate Reducts. Fundamenta Informaticae 47 IOS Press (2001) 351–360

A Model PM for Preprocessing and Data Mining Proper Process

Anita Wasilewska[1], Ernestina Menasalvas[2], and Christelle Scharff[3]

[1] Department of Computer Science, Stony Brook University, NY, USA
anita@cs.sunysb.edu
[2] Departamento de Lenguajes y Sistemas Informaticos Facultad de Informatica,
U.P.M, Madrid, Spain
ernes@fi.upm.es
[3] Computer Science Department, Pace University, New York, NY, USA
cscharff@pace.edu

Extended Abstract

Data Mining, as defined in 1996 by Piatetsky-Shapiro ([1]) is a step (crucial, but a step nevertheless) in a KDD (Knowledge Discovery in Data Bases) process. The Piatetsky-Shapiro's definition states that the KDD process consists of the following steps: developing an understanding of the application domain, creating a target data set, choosing the data mining task i.e. deciding whether the goal of the KDD process is classification, regression, clustering, etc..., choosing the data mining algorithm(s), data preprocessing, data mining (DM), interpreting mined patterns, deciding if a re-iteration is needed, and consolidating discovered knowledge.

Since then the Data Mining (DM) term has evolved to become a name for all of the KDD process, or some parts of it, or even to be used as a name of an application of a data mining (or learning) algorithm.

For example, in 1997 a Cross-Industry Standard Process for Data Mining (CRISP-DM) was proposed ([5]) to establish a standard for what they called, and others adopted, a data mining process. CRISP-DM standard was developed for business purposes and it included all of KDD process steps plus some extra steps such as a business understanding, business goal understanding followed by the KDD standard steps. Hence the KDD process became Data Mining process for industrial applications and was and is more and more often called just by the name of Data Mining.

To clarify these naming confusions we follow the standard terminology developed by data mining researches in which we understand by Data Mining (DM) a KDD process in which its original data mining phase is now called *data mining proper phase*. For short we say that

Data Mining (DM) is a process that includes between the others the following phases: creating the target data, data preprocessing, data mining proper, pattern evaluation, and knowledge presentation.

We present here formal models DP and DMP for two essential phases of the Data Mining: preprocessing and data mining proper. They are defined in

J.F. Peters et al. (Eds.): Transactions on Rough Sets VI, LNCS 4374, pp. 397–399, 2007.
© Springer-Verlag Berlin Heidelberg 2007

such a way that put together they form a Process Model PM for the sequence of preprocessing and data mining proper processes, and hence for the most essential part of the KDD (Data Mining) process.

The main components of our models are: a Data Mining System DMS and preprocessing and data mining proper operators that form together a set of all process operators of our PM model.

The process operators reflect some ideas presented in [6] and [7], where some operators, called generalization operators were defined. The generalization operators were very abstract in nature and their definitions reflected the author's efforts to find a formal model for Data Mining viewed as the process of information generalization. The process operators defined here do not address the generalization issue and are specifically defined, one by one, and in a great detail in an effort to cover all known preprocessing and data mining proper techniques. We discuss the relationship of our new operators and the generalization operators of [6], [7] in the last section of the paper.

The Data Mining System DMS is a crucial component of all of our models and is defined as an extension of Pawlak's Information System ([3]).

Following the Rough Set tradition stated in the statement: *knowledge is an ability to classify objects* ([3], [4]) we observe that this is what not only Rough Sets algorithms do, but it is (as it should be) a common property of all of data mining algorithms, methods, models. We hence model here the data mining proper process as a process of grouping objects (records) into sets of objects. To be able to do so we need to define an extension of the notion of the information system where the information function acts on the sets of objects. We call such function, in the definition 1 of our data mining system DMS a *a knowledge function*. The name reflects the fact we are modelling data mining process as a transformation of an information (set of records as described by the information function) into a higher level *knowledge*. This knowledge obtained in the process (by algorithms, methods, models) comes in two forms: *semantic* and *syntactic*. The syntactic knowledge is always defined in terms of attributes and values of attributes of the initial data table, i.e. initial information system. It has different forms, depending on the goal of the data mining process and methods used. While modelling the semantic knowledge, i.e. the grouping objects (records) into sets of objects we want to model as well its syntactic descriptions. We want, at the end of the process be able to characterize these groups of sets (semantics) in terms of attributes and values of attributes of the initial data base (syntax) and moreover, do do so, as it often happens in terms of some accuracy parameters. Our extension of the notion of the information system accommodates all these demands and is defined formally as follows.

Definition 1. A Data Mining System DMS *based on* $I = (U, A, V_A, f)$ *is a system*

$$K_I = (\mathcal{P}(U), A, E, V_A, V_E, g)$$

where:

E is a finite set of **knowledge attributes** *(k-attributes) such that* $A \cap E = \emptyset$;

V_E *is a finite set of* **values of k- attributes***;*
g is a partial function called **a knowledge function** *(k-function)*

$$g : \mathcal{P}(U) \times (A \cup E) \longrightarrow (V_A \cup V_E)$$

such that:
(i) $g \mid (\bigcup_{x \in U}\{x\} \times A) = f;$
(ii) $\forall S \in \mathcal{P}(U),\ \forall a \in A\ ((S,a) \in dom(g)\ \Rightarrow\ g(S,a) \in V_A);$
(iii) $\forall S \in \mathcal{P}(U),\ \forall e \in E\ ((S,e) \in dom(g)\ \Rightarrow\ g(S,e) \in V_E);$

The models presented here generalize many ideas developed during years of investigations. First they appeared as a part of development of Rough Sets Theory (to include only few recent publications) [3], [10], [11], [2]; then in building Rough Sets inspired foundations of information generalization and Foundations of Data Mining in [6], [7], [9], [8].

References

1. Fayyad, Piatetsky-Shapiro, Smyth, *From Data Mining to Knowledge Discovery: An Overview*, Advances in Knowledge Discovery and Data Mining, Fayyad, Piatetsky-Shapiro, Smyth, Uthurusamy, editors, AAAI Press / The MIT Press, Menlo Park, CA, 1996, pp.1-34.
2. M. Inuiguchi, T. Tanino. *Classification versus Approximation oriented Generalization of Rough Sets* Bulletin of International Rough Set Society, Volume 7, No. 1/2.2003
3. Pawlak, Z. *Rough Sets- theoretical Aspects Reasoning About Data* Kluwer Academic Publishers 1991
4. Polkowski Lech. *Rough Sets- Mathematical Foundations* Physica- Verlag, A Springer- Verlag Company, 2002
5. Colin Shearer, *The CRISP-DM Model: The New Blueprint for Data Mining* Journal of Data Warehousing. Volume 5 Number 4 Fall 2000, pp 13-22.
6. Anita Wasilewska, Ernestina Menasalvas. *Data Preprocessing and Data Mining as Generalization Process* Proceedings of ICDM'04, The Fourth IEEE International Conference on Data Mining, Brighton, UK, Nov 1-4, 2004, pp 133-137.
7. Anita Wasilewska, Ernestina Menasalvas. *Data Mining Operators* Proceedings of ICDM'04, The Fourth IEEE International Conference on Data Mining, Brighton, UK, Nov 1-4, 2004, pp 209-214.
8. Anita Wasilewska, Ernestina Menasalvas, Christelle Scharff. *Uniform Model for Data Mining* Proceedings of FDM05 (Foundations of Data Mining), in ICDM2005, Fifth IEEE International Conference on Data Mining, Austin, Texas, Nov 27-29, 2005, pp 19-27.
9. A. Wasilewska, Ernestina Menasalvas Ruiz *Data Mining as Generalization: A Formal Model*, book chapter in Foundations and Novel Approaches in Data Mining, T.Y. Lin, S. Ohsuga, C. J. Liau, and X. Hu , editors. Springer 2006, Studies in Computational intelligence 9, pp 99-126.
10. Wojciech Ziarko. *Variable Precision Rough Set Model* Journal of Computer and Systen Sciences, Vol.46. No.1, pp. 39-59, 1993.
11. J.T. Yao, Y.Y. Yao. *Induction of Classification Rules by Granular Computing* Proceedings of Third International RSCTC'02 Conference, Malvern, PA, USA, October 2002, pp. 331-338. Springer Lecture Notes in Artificial Intelligence.

Lattice Theory for Rough Sets

Jouni Järvinen

Turku Centre for Computer Science (TUCS)
FI-20014 University of Turku, Finland
`jouni.jarvinen@utu.fi`

Abstract. This work focuses on lattice-theoretical foundations of rough set theory. It consist of the following sections: 1: Introduction 2: Basic Notions and Notation, 3: Orders and Lattices, 4: Distributive, Boolean, and Stone Lattices, 5: Closure Systems and Topologies, 6: Fixpoints and Closure Operators on Ordered Sets, 7: Galois Connections and Their Fixpoints, 8: Information Systems, 9: Rough Set Approximations, and 10: Lattices of Rough Sets. At the end of each section, brief bibliographic remarks are presented.

1 Introduction

The present work is written for readers interested in the lattice-theoretical background of rough sets. It contains the necessary part of lattice theory and shows how to formulate in an elegant way various concepts and facts about rough sets and Pawlak's information systems. Prerequisites are minimal and the work is self-contained.

Rough set theory consists of two key notions which both are introduced by Zdzisław Pawlak: rough set approximations and information systems. Rough set approximations are defined by means of indiscernibility relations which are equivalences interpreted so that two objects are equivalent if we cannot distinguish them by using our information. This means that our ability to discern objects is limited – we cannot observe individual objects, only their equivalence classes. Since we perceive just blocks of objects, three kinds of situations will occur: an equivalence class may be included in a given set, it may intersect with the set, or it can entirely lie outside the set in question. A consequence of this is that characteristic functions of sets become three-valued – the third value represents the possibility of belonging to a set. In an information system an indiscernibility relation arises naturally when one considers a given set of attributes: two objects are equivalent when their values of all attributes in the set are the same.

Lattices are relatively simple structures since the basic concepts of the theory include only orders, least upper bounds, and greatest lower bounds. Lattice theory has turned to be very useful in dealing with different structures in theoretical computer science. In this work particularly Galois connections are in a central role. They are pairs of maps which enable us to move back and forth between two ordered sets. Galois connections tie different structures firmly and when a Galois connection is found between two structures, we immediately know

J.F. Peters et al. (Eds.): Transactions on Rough Sets VI, LNCS 4374, pp. 400–498, 2007.
© Springer-Verlag Berlin Heidelberg 2007

that they have much in common. We will find out that the pair of upper and lower approximation mappings forms a Galois connection and several properties of rough approximations follow from this observation. It is also interesting to notice that in information systems the map attaching to each attribute set its indiscernibility relation has an adjoint and therefore it determines a Galois connection. We will show how Galois connections between complete lattices define dependency relations and this lets us to obtain the essential properties of dependencies easily. In particular, fixpoints of Galois connections are important because definable sets may be viewed as fixed points of the rough approximation mappings.

Section 2 presents the elemental notions and facts of sets, relations, and functions. In Section 3, the fundamental theory of lattices is developed. Section 4 is devoted to distributive, Boolean, and Stone lattices. It will turn out that topologies have an important role in the study of definable sets, and in Section 5 we deal with closure operators and topological spaces. In Section 6 we study closure operators in a more general setting of ordered sets. Since fixpoints of functions are closely related to closure operators, basic facts about fixpoints are presented here. Fixpoints of Galois connections deserve a special attention and the seventh section considers them. Section 8 begins with introducing Armstrong systems which are closed sets of dependencies and we show how Galois connections induce Armstrong systems. Pawlak's model for information systems is presented here and we examine in information systems their Galois connections, dependencies, and attribute reduction. Section 9 is devoted to rough set approximations and definable sets. The section begins with approximations determined by equivalences, but also approximations of other types of relations are studied. In the final section we investigate the lattice structures of the ordered set of all rough sets determined by different kinds of indiscernibility relations.

2   Basic Notions and Notation

In this section we consider the following preliminary subjects:

2.1 Sets
2.2 Relations
2.3 Functions

2.1   Sets

A *set* can be viewed as a collection of distinguishable objects, called its *members* or *elements*. If an object x is a member of a set A, we write $x \in A$; the notation $x \notin A$ denotes that x is not in A. The set which has no elements is called the *empty set* and is denoted by $\emptyset$.

If there are n distinct elements in a set A, where $n \in \mathbb{N} = \{0, 1, 2, 3, \ldots\}$, we say that A is a *finite set*. The number n is the *cardinality* of A and it is denoted by $|A|$. When $|A| = n$ it is often convenient to put $A = \{a_1, \ldots, a_n\}$. A set is *infinite* if it is not finite.

Two sets A and B are *equal*, denoted by $A = B$, if they contain the same elements. A set A is said to be a *subset* of B if every element of A is also an element of B. We write $A \subseteq B$ to indicate that A is a subset of B. A set A is a *proper subset* of B if $A \subseteq B$, but $A \neq B$; this is denoted by $A \subset B$.

We can form new sets from existing ones by applying the following *set operations*. The *intersection* of A and B is $A \cap B = \{x \mid x \in A \text{ and } x \in B\}$, the *union* of A and B is $A \cup B = \{x \mid x \in A \text{ or } x \in B\}$, and the set-theoretical *difference* of A and B (or *relative complement* of B in A) is $A - B = \{x \mid x \in A \text{ and } x \notin B\}$. Given a universe U, we define the *complement* of $A \subseteq U$ as $A^c = U - A$. The set of all subsets of A is denoted by $\wp(A)$ and is called the *power set* of A.

A set of which elements are sets is called a *family of sets*. For a family of sets $\mathcal{F}$, we may define its union and intersection by generalizing the notions of the union and the intersection of two sets. The *intersection* of $\mathcal{F}$ is $\bigcap \mathcal{F} = \{x \mid x \in A \text{ for all } A \in \mathcal{F}\}$ and the *union* of $\mathcal{F}$ is $\bigcup \mathcal{F} = \{x \mid x \in A \text{ for some } A \in \mathcal{F}\}$.

It is usually assumed that we consider subsets of some given universe U, and in such a case it is natural to define $\bigcap \emptyset = U$ and $\bigcup \emptyset = \emptyset$. The equality $\bigcap \emptyset = U$ can be interpreted so that every element of U belongs to all sets in $\emptyset$ because the empty family $\emptyset$ contains no sets. The equality $\bigcup \emptyset = \emptyset$ is more obvious since $\emptyset$ has no elements. An *indexed family of sets* is $\mathcal{F} = \{A_i \mid i \in I\}$, where I is a set, referred to as the *index set*. Note that we may also denote the indexed family of sets $\mathcal{F}$ simply by $\{A_i\}_{i \in I}$.

In our first proposition, we present some essential properties of set operations.

Proposition 1. *Let A, B, C be subsets of a universe U.*

(a) $A \cup A = A \cap A = A$;
(b) $A \cup B = B \cup A$ *and* $A \cap B = B \cap A$;
(c) $A \cup (B \cup C) = (A \cup B) \cup C$ *and* $A \cap (B \cap C) = (A \cap B) \cap C$;
(d) $A \cup (A \cap B) = A$ *and* $A \cap (A \cup B) = A$;
(e) $A \cup (B \cap C) = (A \cup B) \cap (A \cup C)$ *and* $A \cap (B \cup C) = (A \cap B) \cup (A \cap C)$;
(f) $A \cup A^c = U$ *and* $A \cap A^c = \emptyset$;
(g) $(A \cup B)^c = A^c \cap B^c$ *and* $(A \cap B)^c = A^c \cup B^c$;
(h) $A \cup B = B \iff A \subseteq B \iff A \cap B = A$;
(i) $A \subseteq B \iff B^c \subseteq A^c$.

Proof. We prove the first part of (e) as an example. Now, $x \in A \cup (B \cap C) \iff x \in A \text{ or } x \in (B \cap C) \iff x \in A \text{ or } (x \in B \text{ and } x \in C) \iff (x \in A \text{ or } x \in B) \text{ and } (x \in A \text{ or } x \in C) \iff x \in (A \cup B) \text{ and } x \in (A \cup C) \iff x \in (A \cup B) \cap (A \cup C)$. □

2.2 Relations

An *ordered pair* of elements a and b is a pair (a, b) arranged in a fixed order. For two sets A and B, the *Cartesian product* $A \times B$ of A and B is the set of all ordered pairs (a, b), where $a \in A$ and $b \in B$, that is, $A \times B = \{(a, b) \mid a \in A \text{ and } b \in B\}$.

A *binary relation* R from A to B is a subset of $A \times B$. If R is a binary relation from A to B, then an element $a \in A$ is said to be *related to* $b \in B$, when

$(a, b) \in R$. We often write $a\,R\,b$ for $(a, b) \in R$. For any binary relation R, we denote by $R^{-1} = \{(b, a) \mid a\,R\,b\}$ the *inverse relation* of R. A *binary relation on A* is a relation from A to A. We denote by $\mathrm{Rel}(A)$ the set of all binary relations on A.

Example 2. Binary relations on a set can be represented by diagrams such that the elements of the set are represented with circles, and if an element x is related to an element y, there is an arrow from the circle representing x to the circle representing y.

For example, if $A = \{a, b, c, d\}$ and a binary relation R on A consists of the pairs (a, a), (a, b), (a, c), (b, d), (c, b), (c, d), and (d, c), then R can be represented by Fig 1. Clearly, we obtain the diagram of the inverse relation simply by reversing the arrows.

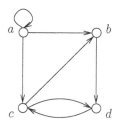

Fig. 1.

If R and S are binary relations from A to B, we can form new relations $R \cup S$, $R \cap S$, $R - S$, and R^c by the usual set-theoretical operations. Further, if R is a relation from A to B and S is a relation from B to C, then the *composition* $R \circ S$ is the relation from A to C defined by $(x, z) \in R \circ S$ if and only if there exists $y \in B$ such that $(x, y) \in R$ and $(y, z) \in S$.

In the following we introduce some properties of binary relations. Let R be a binary relation on a set A. The relation R is

(a) *connected*, if for all $x \in A$, there exists $y \in A$ such that $x\,R\,y$;
(b) *reflexive*, if $x\,R\,x$ for all $x \in A$;
(c) *symmetric*, if $x\,R\,y$ implies $y\,R\,x$ for all $x, y \in A$;
(d) *antisymmetric*, if $x\,R\,y$ and $y\,R\,x$ imply $x = y$ for all $x, y \in A$;
(e) *transitive*, if $x\,R\,y$ and $y\,R\,z$ imply $x\,R\,z$ for all $x, y, z \in A$.

Note that every reflexive relation is connected, and that a relation R is antisymmetric if and only if $(x, y) \in R$ and $x \neq y$ imply $(y, x) \notin R$ for all $x, y \in A$.

If a relation is reflexive and symmetric, it is called a *tolerance relation*, and if a relation is reflexive and transitive, it is called a *preorder* (or a *quasi-order*). A relation which is both a tolerance and a preorder is an *equivalence relation*. Notice that if a relation is connected, symmetric, and transitive, it is necessarily an equivalence.

Two elements that are related by an equivalence relation E are said to be *equivalent*. The set of all elements that are related to an element $x \in A$ is called the *equivalence class* of x and is denoted by $[x]_E = \{y \in A \mid x\,E\,y\}$. The *quotient set of A modulo E* is the family of the equivalence classes $A/E = \{\,[x]_E \mid x \in A\}$.

Two sets are called *disjoint* if their intersection equals the empty set. A *partition Π* of a set A is a family of nonempty subsets of A such that each element of A belongs to one, and only one, set of Π. It is thus obvious that the sets of a partition are disjoint.

Proposition 3. *The equivalence classes of an equivalence relation E on a set A forms a partition A/E of A, and for any partition Π of A, there exists an equivalence relation on A of which quotient set is Π.*

Proof. First we will show that A/E is a partition. Because E is reflexive, $a \in [a]_E$, and so the equivalence classes are nonempty and each element belongs to at least one equivalence class. We have to show that equivalence classes are disjoint. If equivalence classes $[a]_E$ and $[b]_E$ have a common element c, then $a\,E\,c$ and $b\,E\,c$, imply $a\,E\,b$ by symmetry and transitivity. This gives $[a]_E = [b]_E$ and thus each element can belong to only one equivalence class.

Let us define for a partition Π an equivalence E such that $a\,E\,b$ if and only if there exists a set X in Π such that $a, b \in X$. The relation E is reflexive, since each element belongs to one set of Π. The relation E is also symmetric, because $a\,E\,b$ means that a and b are in the same set, and hence $b\,E\,a$. The transitivity holds, because if $a\,E\,b$ and $b\,E\,c$, then a, b, and c are necessarily in the same set, and thus $a\,E\,c$ holds. Clearly, the equivalence classes of E consist of the sets in Π.   □

Example 4. For representing equivalences by diagrams, we may agree on two simplifications. Since equivalences are reflexive, there should be an arrow from each circle to the circle itself. Such loops can be omitted. Furthermore, if there is an arrow from x to y, there must be also an arrow from y to x by symmetry. Therefore, the situation that x and y are equivalent can be represented just by a line connecting x and y.

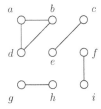

Fig. 2.

Let us consider the equivalence depicted in Fig. 2. The corresponding partition consists of the equivalence classes $\{a, b, d\}$, $\{c, e\}$, $\{f, i\}$, and $\{g, h\}$.

2.3 Functions

A *function* f from a set A to a set B, denoted by $f\colon A \to B$, is a relation from A to B such that for each $a \in A$, there exists exactly one $b \in B$ with $(a, b) \in f$, in which case we write $f(a) = b$ or $f\colon a \mapsto b$. The terms *map* and *mapping* are often used instead of function. The set of all functions from A to B is denoted by B^A. For a function $f\colon A \to B$, we write for all $S \subseteq A$, $f(S) = \{f(x) \mid x \in S\}$. The set $f(A)$ is called the *range* of f. The *preimage set* of $Y \subseteq B$ is $f^{-1}(Y) = \{x \in A \mid f(x) \in Y\}$.

The map $f\colon A \to B$ is *injective* (or *one-to-one*) if $f(a_1) = f(a_2)$ implies $a_1 = a_2$, and f is *surjective* (or *onto*) if for every $b \in B$, there exists an element $a \in A$ with $f(a) = b$; that is, $f(A) = B$. Furthermore, f is *bijective* if it is both injective and surjective. A map f from a set A to the same set A is called a *self-map*. A self-map $f\colon A \to A$ is *idempotent* if $f(f(a)) = f(a)$ for all $a \in A$.

For two maps $f\colon A \to B$ and $g\colon B \to C$, let $g \circ f\colon A \to C$ be the map defined by $(g \circ f)(a) = g(f(a))$. The map $g \circ f$ is called the *composition* (or *product*) of f and g. The map $1_A\colon A \to A, a \mapsto a$, is called the *identity map* of A. A map $g\colon B \to A$ is the *inverse map* of $f\colon A \to B$ if $g \circ f = 1_A$ and $f \circ g = 1_B$.

Lemma 5. *A function $f\colon A \to B$ has an inverse map if and only if f is a bijection.*

Proof. If $f\colon A \to B$ is a bijection, then for each $y \in B$ there exists exactly one $x \in A$ such that $f(x) = y$. The rule $g(y) = x$ defines a function $B \to A$ which is the inverse of f.

Conversely, suppose that f has the inverse f^{-1}. Given $y \in B$, we know that $f(f^{-1}(y)) = y$, and by setting $x = f^{-1}(y)$ we obtain $f(x) = y$. Thus, f is a surjection. If $f(x) = f(y)$, then $x = f^{-1}(f(x)) = f^{-1}(f(y)) = y$, that is, f is injective. □

The inverse of a bijection f is denoted by f^{-1}. It is clear that if functions are considered as relations, f^{-1} is the inverse relation of f. Note that for any function $f\colon A \to B$ we can form the preimage set $f^{-1}(Y)$ of every $Y \subseteq B$ even though f is not a bijection.

Example 6. Let us consider Fig. 3.

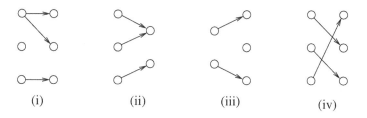

(i)   (ii)   (iii)   (iv)

Fig. 3.

The relation in part (i) is not a function because one element is related to two elements and there is also an element which is not related to anyone. The mapping (ii) is not an injection, because two elements have the same image. The map in (iii) is not a surjection since there exists an element which is not an image. The function in part (iv) is a bijection.

Notice that for a bijection, we can get the diagram of its inverse just by reversing the arrows.

Bibliographical Notes

Basic notions and notation concerning sets, relations and functions can be found in almost any elementary mathematical textbook. For instance, the books [5,16] provide introductions to discrete mathematics.

3   Orders and Lattices

This section consists of the following two subsections:

3.1 Orders
3.2 Lattices and Complete Lattices

3.1   Orders

Let P be a set. An *order* $\leq$ on P is a reflexive, antisymmetric, and transitive binary relation, that is, for all $a, b, c \in P$,

(a) $a \leq a$,
(b) $a \leq b$ and $b \leq a$ imply $a = b$, and
(c) $a \leq b$ and $b \leq c$ imply $a \leq c$.

An *ordered set* $(P, \leq)$ consists of a nonempty set P and an order $\leq$ on P.

The relation $\leq$ is read as usual: 'is less than or equal to'. Many authors use the term *partially ordered set* – and even the shorthand *poset* – for an ordered set. We denote by $\geq$ the inverse relation of $\leq$. Usually we say simply that 'P is an ordered set'. Where it is necessary to specify the order relation, we write $(P, \leq)$. An order $\leq$ gives rise to relation $<$ of *strict order*: $a < b$ if and only if $a \leq b$ and $a \neq b$.

Let P be an ordered set and let $a, b \in P$. We say that a is *covered by* b (or b *covers* a) and write $a \prec b$, if $a < b$ and there is no element c in P with $a < c < b$.

Every finite ordered set $(P, \leq)$ can be represented by a *Hasse diagram* that is determined by the covering relation. As before, the elements of P are represented with circles, and the circles representing two elements a and b are connected by a straight line if $a \prec b$ or $b \prec a$. Moreover, if a is covered by b, the circle representing a is lower than the circle representing b. It is also clear that the Hasse diagram of a finite ordered set determines uniquely the partial ordering: $a \leq b$ if and only if $a = b$ or the circle representing b can be reached from the circle representing a by moving upward along the lines.

Example 7. For any set A, the pair $(\wp(A), \subseteq)$ is an ordered set. For $A = \{a, b, c\}$, the ordered set $(\wp(A), \subseteq)$ is depicted in Fig. 4.

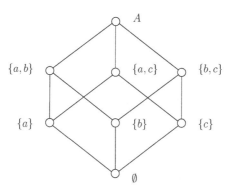

Fig. 4.

Next we consider some structure-preserving mappings. Let P and Q be two ordered sets. A map $f : P \to Q$ is

(a) *order-preserving*, if $a \le b$ in P implies $f(a) \le f(b)$ in Q;
(b) *order-reversing*, if $a \le b$ in P implies $f(a) \ge f(b)$ in Q;
(c) an *order-embedding*, if $a \le b$ in P is equivalent to $f(a) \le f(b)$ in Q;
(d) an *order-isomorphism* between P and Q if f is an order-embedding onto Q.

When there exists an order-isomorphism between P and Q, we say that P and Q are *order-isomorphic* and write $P \cong Q$. Notice that an order-embedding is always an injection, and that two finite ordered sets are order-isomorphic if and only if they can be represented by a same Hasse diagram.

Example 8. Let us consider Fig 5.

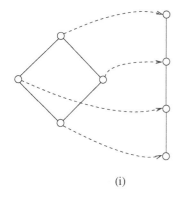

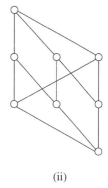

(i)   (ii)

Fig. 5.

The mapping in (i) is a bijective order-preserving map, but it is not an embedding. The ordered set of part (ii) is order-isomorphic to the powerset in Fig. 4.

For an ordered set P, we can form a new ordered set P^{op} by defining $x \leq y$ to hold in P^{op} if and only if $y \leq x$ holds in P. The ordered set P^{op} is called the *dual* (or the *opposite*) of P. The Hasse diagram of the dual P^{op} of a poset P is obtained from that of P by turning the Hasse diagram of P upside down. Trivially, $P = (P^{\mathrm{op}})^{\mathrm{op}}$.

If two ordered sets P and Q satisfy $P \cong Q^{\mathrm{op}}$, we say that P and Q are *dually order-isomorphic*. Many ordered sets are dually order-isomorphic with themselves, that is, $P \cong P^{\mathrm{op}}$. In such a case we say that P is a *self-dual*.

Example 9. Let A be a set. Then the ordered set $(\wp(A), \subseteq)$ is a self-dual. The map $\phi \colon X \mapsto X^c$ is onto $\wp(A)$, because $\phi(X^c) = (X^c)^c = X$ for all $X \subseteq A$, and ϕ is an order-embedding since by Proposition 1, $X \subseteq Y$ if and only if $\phi(Y) = Y^c \subseteq X^c = \phi(X)$.

If Φ is a statement about ordered sets, we get its dual statement Φ^{op} by replacing every occurrence of $\leq$ by $\geq$ and vice versa. We may now present the following 'meta-theorem', which in many cases can save a lot of work. Its proof is obvious, because if a statement Φ is true for an ordered set P, the dual statement Φ^{op} is true for P^{op}.

Duality Principle. If a statement Φ is true in all ordered sets, then its dual Φ^{op} is also true in all ordered sets.

Let P be an ordered set and let $S \subseteq P$. Then $x \in S$ is a *maximal element* of S, if $x \leq a \in S$ implies $a = x$. Further, $x \in S$ is the *greatest element* of S, if $x \geq a$ for all $a \in S$. A *minimal element* of S and the *least element* of S are defined dually. Notice that if S has a greatest element, it is unique by the antisymmetry of $\leq$. Similarly, the least element of S is unique.

The greatest element of P, if such exists, is called the *top element* of P and it is denoted by $\top$. Similarly, the least element of P, if it exists, is called the *bottom element* and is denoted by $\bot$. For example, the set $\mathbb{N}$ does not have a greatest element.

Lemma 10. *Any finite nonempty subset of an ordered set has maximal and minimal elements.*

Proof. Suppose that $S = \{x_1, x_2, \ldots, x_n\}$. Let us define elements $m_1, m_2, \ldots, m_n$ inductively in such a way that $m_1 = x_1$ and

$$m_k = \begin{cases} x_k & \text{if } x_k < m_{k-1} \\ m_{k-1} & \text{otherwise.} \end{cases}$$

Then m_n will be minimal in S. Similarly, S has a maximal element. $\qquad\square$

3.2   Lattices and Complete Lattices

Let P be an ordered set and let $S \subseteq P$. An element $x \in P$ is an *upper bound* of S if $a \leq x$ for all $a \in S$. A *lower bound* of S is defined dually. If there is a least element in the set of all upper bounds of S, it is called the *supremum* of S and is denoted by $\sup S$ or $\bigvee S$; dually a greatest lower bound is called *infimum* and written $\inf S$ or $\bigwedge S$. We also write $a \vee b$ for $\sup\{a, b\}$ and $a \wedge b$ for $\inf\{a, b\}$. Supremum and infimum are frequently called *join* and *meet*.

It is sometimes necessary to indicate that a join or a meet is being found in a certain ordered set P. In such cases we write $\bigvee_P S$ or $\bigwedge_P S$. If I is an index set and $S = \{x_i \mid i \in I\}$ is a subset of P, instead of $\bigvee S$ we also write $\bigvee_{i \in I} x_i$ and in place of $\bigwedge S$ we write $\bigwedge_{i \in I} x_i$.

It is clear that if $S = \emptyset$, then P is the set of upper bounds of S. This means that $\bigvee \emptyset$ exists in P if and only if P has a least element $\bot$, and then $\bigvee \emptyset = \bot$. By duality, $\bigwedge \emptyset = \top$ whenever P has a greatest element $\top$. Furthermore, if P has a greatest element $\top$, then the set of upper bounds of P is $\{\top\}$, and thus $\bigvee P = \top$. By duality, $\bigwedge P = \bot$ whenever P has a least element $\bot$. It is also obvious that if P has a greatest element $\top$, then $x \vee \top = \top$ and $x \wedge \top = x$ for all $x \in P$. Similarly, if P has a least element $\bot$, then for all $x \in P$, $x \vee \bot = x$ and $x \wedge \bot = \bot$.

Example 11. Let us consider the ordered set of Fig. 6. The pair of elements marked with filled circles does not have a supremum. These elements have two mutual minimal upper bounds, but not a least one.

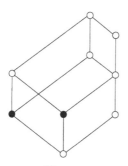

Fig. 6.

In the next lemma is given some simple but useful properties of joins and meets.

Lemma 12. *Let P be an ordered set and assume that S and T are subsets of P such that their joins and meets in P exist.*

(a) *If $a \in S$, then $\bigwedge S \leq a \leq \bigvee S$.*
(b) *If $x \in P$, then $x \leq \bigwedge S$ if and only if $x \leq a$ for all $a \in S$.*
(c) *If $x \in P$, then $x \geq \bigvee S$ if and only if $x \geq a$ for all $a \in S$.*
(d) *If $S \subseteq T$, then $\bigvee S \leq \bigvee T$ and $\bigwedge S \geq \bigwedge T$.*

Proof. Claim (a) is obvious by the definition of $\bigvee S$ and $\bigwedge S$.

(b) If $x \leq \bigwedge S$, then $x \leq \bigwedge S \leq a$ for all $a \in S$. If $x \leq a$ for all $a \in S$, then x is a lower bound of S, which yields $x \leq \bigwedge S$. Claim (c) can be proved in a similar way.

(d) If $S \subseteq T$, then $b \leq \bigvee T$ for all $b \in T$. Because $S \subseteq T$, $a \leq \bigvee T$ holds also for all $a \in S$. This implies $\bigvee S \leq \bigvee T$. The other part can be proved dually.   □

Let P be a nonempty ordered set. If $x \vee y$ and $x \wedge y$ exist for all $x, y \in P$, then P is called a *lattice*. If $\bigvee S$ and $\bigwedge S$ exist for all $S \subseteq P$, then P is called a *complete lattice*. By the Duality Principle, if L is a lattice or a complete lattice, so is its dual L^{op} in which joins and meets determined in L are mutually interchanged.

Lemma 13. *Every finite lattice is complete.*

Proof. As will be shown in the proof of Proposition 19, it is possible to write $\bigvee \{a, b, c\} = a \vee b \vee c$ without parenthesis. If $\emptyset \neq S = \{a_1, a_2, \ldots, a_n\}$, then by simple induction $\bigvee S = a_1 \vee a_2 \vee \cdots \vee a_n$, and by duality $\bigwedge S = a_1 \wedge a_2 \wedge \cdots \wedge a_n$.   □

In many cases, the following theorem makes it much easier to show that certain ordered set is a complete lattice.

Theorem 14. *If P is an ordered set such that $\bigwedge S$ exists for all $S \subseteq P$, then P is a complete lattice in which*

$$\bigvee S = \bigwedge \{x \in P \mid (\forall a \in S)\, a \leq x\}.$$

Proof. Any element of S is a lower bound of $\{x \in P \mid (\forall a \in S)\, a \leq x\}$ and thus $\bigwedge \{x \in P \mid (\forall a \in S)\, a \leq x\}$ is an upper bound of S. If z is an upper bound of S, then necessarily $\bigwedge \{x \in P \mid (\forall a \in S)\, a \leq x\} \leq z$.   □

Example 15. (a) An ordered set P is a *chain* if, for all $x, y \in P$, either $x \leq y$ or $y \leq x$. Every chain is a lattice in which $a \vee b = \min\{a, b\}$ and $a \wedge b = \max\{a, b\}$, that is, the *minimum* and *maximum* of a and b, respectively. In particular, the set of natural numbers $\mathbb{N}$ is a chain and a lattice under its usual order. Note that $\mathbb{N}$ is not a complete lattice since it lacks a top element.

(b) Let A be a set. Then $(\wp(A), \subseteq)$ is a lattice such that $X \vee Y = X \cup Y$ for all $X, Y \subseteq A$. Trivially, $X, Y \subseteq X \cup Y$ and if Z is an upper bound of X and Y, then $X \cup Y \subseteq Z$. Similarly, we can show that $X \wedge Y = X \cap Y$.

(c) For any set A, the ordered set $(\wp(A), \subseteq)$ is also a complete lattice in which $\bigvee \mathcal{H} = \bigcup \mathcal{H}$ and $\bigwedge \mathcal{H} = \bigcap \mathcal{H}$ for any $\mathcal{H} \subseteq \wp(A)$.

(d) Let $\emptyset \neq \mathcal{L} \subseteq \wp(A)$. Then $\mathcal{L}$ is a *ring of sets* if it is closed under finite unions and intersections, and a *complete ring of sets* if it is closed under arbitrary unions and intersections. If $\mathcal{L}$ is a ring of sets, then $(\mathcal{L}, \subseteq)$ is a lattice such that $A \vee B = A \cup B$ and $A \wedge B = A \cap B$. Similarly, if $\mathcal{L}$ is a complete ring of sets, then $(\mathcal{L}, \subseteq)$ is a complete lattice with join given by set union and meet given by set intersection.

Let x be an element of and ordered set P. The set

$$(x] = \{a \in P \mid a \leq x\}$$

is called the *principal ideal* of x and

$$[x) = \{a \in P \mid a \geq x\}$$

is called the *principal filter* of x. If P is a complete lattice, also $(x]$ and $[x)$ are complete lattices.

In the following lemma we will show that each ordered set can be embedded into a complete ring of sets. In general, for any set P and a complete lattice L, we say that L is a *completion* of P, if P can be embedded into L. Let P be an ordered set and $Q \subseteq P$. Then Q is a *down-set* if for all $x \in Q$ and $y \in P$, $x \geq y$ implies $y \in Q$. The family of all down-sets is denoted by $\mathcal{O}(P)$.

Lemma 16. *Let P be an ordered set.*

(a) $\mathcal{O}(P)$ *is a complete ring of sets.*
(b) P *can be embedded into* $\mathcal{O}(P)$.

Proof. (a) Suppose $\mathcal{H}$ is a subfamily of $\mathcal{O}(P)$. If $x \in \bigcap \mathcal{H}$ and $x \geq y$, then $y \in X$ for all $X \in \mathcal{H}$ because each X is a down-set. Hence, $y \in \bigcap \mathcal{H}$ and $\bigcap \mathcal{H} \in \mathcal{O}(P)$. That $\bigcup \mathcal{H}$ is in $\mathcal{O}(P)$ can be shown in a similar way.

(b) We show that $x \mapsto (x]$ is an order-embedding. Clearly, $(x]$ is a down-set. If $x \leq y$, then $a \in (x]$ implies $a \leq x \leq y$ and $a \in (y]$, that is, $(x] \subseteq (y]$. Conversely, if $(x] \subseteq (y]$, then $x \leq x$ implies $x \in (y]$, that is, $x \leq y$. $\qquad \square$

Example 17. In Fig. 7 is depicted the order-embedding $x \mapsto (x]$ from P to $\mathcal{O}(P)$.

The next lemma presents connections between the order $\leq$ and the 'operations' $\vee$ and $\wedge$.

Lemma 18. *If L is a lattice and $a, b, x \in L$, then*

(a) $a \leq b$ *if and only if $a \wedge b = a$ if and only if $a \vee b = b$;*
(b) $a \leq b$ *implies $a \vee x \leq b \vee x$ and $a \wedge x \leq b \wedge x$.*

Proof. (a) If $a \leq b$, then a is a lower bound of a and b. If z is a lower bound of a and b, then trivially $z \leq a$, which means that $a = a \wedge b$. On the other hand, if $a = a \wedge b$, then $a = a \wedge b \leq b$. We may prove the rest analogously.

(b) If $a \leq b$, then obviously $a \leq b \leq b \vee x$ and $x \leq b \vee x$. This gives $a \vee x \leq b \vee x$. The other part can be proved in a similar manner. $\qquad \square$

Next we give some important properties of $\vee$ and $\wedge$.

Proposition 19. *If L is a lattice, then for all $a, b, c \in L$,*

(L1) $a \vee a = a$ *and* $a \wedge a = a$;
(L2) $a \vee b = b \vee a$ *and* $a \wedge b = b \wedge a$;

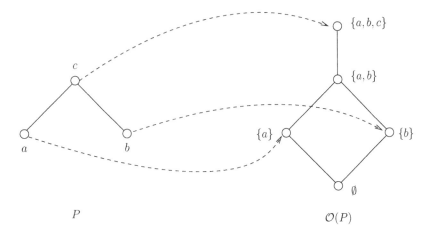

Fig. 7.

(L3) $a \vee (b \vee c) = (a \vee b) \vee c$ *and* $a \wedge (b \wedge c) = (a \wedge b) \wedge c$;
(L4) $a \vee (a \wedge b) = a$ *and* $a \wedge (a \vee b) = a$.

Proof. Claims (L1) and (L2) are obvious properties of $\vee$ and $\wedge$. For (L3) and (L4) we prove only the first claims, the rest follows then from the Duality Principle.

(L3) We prove $\bigvee \{a, b, c\} = ((a \vee b) \vee c)$ which by (L2) is in fact enough to prove the claim. Let $d = a \vee b$ and $e = c \vee d$. Clearly, $a \le d$, $b \le d$, $c \le e$, and $d \le e$. By transitivity, $a, b, c \le e$. If f is an upper bound of $\{a, b, c\}$, then $a \le f$, $b \le f$, and $c \le f$, which implies $d = a \vee b \le f$ and $e = c \vee d \le f$. This gives $\bigvee \{a, b, c\} = e = ((a \vee b) \vee c)$.

(L4) Since $a \le a \vee b$ and $a \wedge b \le a$, we obtain $a \wedge (a \vee b) = a$ and $a \vee (a \wedge b) = a$. □

Proposition 19 presents the characteristic properties of the operations $\vee$ and $\wedge$ as we see in the following theorem.

Theorem 20. *Let L be a non-empty set equipped with two binary operations $\vee$ and $\wedge$ that satisfy (L1)–(L4) of Proposition 19. If we define $\le$ on L by $a \le b$ if and only if $a \vee b = b$, then $(L, \le)$ is a lattice in which the original operations agree with the induced ones, that is, for all $a, b \in L$,*

$$a \vee b = \sup \{a, b\} \quad and \quad a \wedge b = \inf \{a, b\}.$$

Proof. First we show that $\le$ is an order. By (L1), $a \vee a = a$ which gives $a \le a$ for all $a \in L$. The relation $\le$ is antisymmetric since $a \le b$ and $b \le a$ mean that $a \vee b = b$ and $b \vee a = a$. This implies $a = b \vee a = a \vee b = b$ by (L2). If $a \le b$ and $b \le c$, then $a \vee b = b$ and $b \vee c = c$. Thus, $a \vee c = a \vee (b \vee c) = (a \vee b) \vee c = b \vee c = c$ by (L3), and so $a \le c$. This means that $\le$ is also transitive.

Next we show that $a \vee b = \sup \{a, b\}$; the proof for meets is similar. Now, $a \le a \vee b$, since

$$a \vee (a \vee b) = (a \vee a) \vee b = a \vee b.$$

Similarly, $a \vee b$ is an upper bound of b. Let c be an upper bound for $\{a, b\}$. Then $a \vee c = c$ and $b \vee c = c$. This gives

$$(a \vee b) \vee c = a \vee (b \vee c) = a \vee c = c,$$

and thus $a \vee b = \sup \{a, b\}$. □

Theorem 20 reveals the elegant feature letting lattices be regarded either as ordered sets $(L, \leq)$ or as algebras $(L, \vee, \wedge)$. We may thus say 'let L be a lattice' and replace L by $(L, \leq)$ or by $(L, \vee, \wedge)$. Notice that the powerset $\wp(A)$ of any set A is a lattice also because the 'powerset algebra' $(\wp(A), \cup, \cap)$ satisfies conditions (a)–(d) of Proposition 1.

Next we consider how we can obtain new ordered sets and lattices from given ones. Let L and K be ordered sets. Let us order $L \times K$ *coordinatewise* by setting

$$(x_1, y_1) \leq (x_2, y_2) \iff x_1 \leq x_2 \text{ and } y_1 \leq y_2.$$

If L and K are complete lattices, then $L \times K$ is a complete lattice such that

$$\bigvee_{i \in I} (x_i, y_i) = (\bigvee_{i \in I} x_i, \bigvee_{i \in I} y_i) \text{ and } \bigwedge_{i \in I} (x_i, y_i) = (\bigwedge_{i \in I} x_i, \bigwedge_{i \in I} y_i),$$

because clearly $(\bigvee_{i \in I} x_i, \bigvee_{i \in I} y_i)$ is an upper bound of $\{(x_i, y_i) \mid i \in I\}$, and if (u, v) is an upper bound of $\{(x_i, y_i) \mid i \in I\}$, then $x_i \leq u$ and $y_i \leq v$ for all $i \in I$, which implies $\bigvee_{i \in I} x_i \leq u$ and $\bigvee_{i \in I} y_i \leq v$, that is, $(\bigvee_{i \in I} x_i, \bigvee_{i \in I} y_i) \leq (u, v)$. Similar observations hold also for meets. It is also clear that if L and K are lattices, then $L \times K$ is a lattice.

If X is any set and P is an ordered set, we may order the set P^X of all maps from X to P by the *pointwise order*:

$$f \leq g \text{ in } P^X \text{ if and only if for all } x \in X, \ f(x) \leq g(x) \text{ in } P.$$

If L if a complete lattice, then L^X is a complete lattice in which for all $\{f_i\}_{i \in I} \subseteq L^X$ and $x \in X$,

$$(\bigvee_{i \in I} f_i)(x) = \bigvee_{i \in I} f_i(x) \text{ and } (\bigwedge_{i \in I} f_i)(x) = \bigwedge_{i \in I} f_i(x).$$

This is easy to see since obviously $\bigvee_{i \in I} f_i$ is an upper bound of $\{f_i\}_{i \in I}$ and if g is an upper bound of $\{f_i\}_{i \in I}$, then for all $x \in X$, $f_i(x) \leq g(x)$ and $\bigvee_{i \in I} f_i(x) \leq g(x)$. The equality for meets can be shown analogously. Further, it is clear that if L is a lattice, then L^X is a lattice in which joins and meets are formed pointwise.

Let L be a lattice and $\emptyset \neq H \subseteq L$. Then H is a *sublattice* of L if

$$a, b \in H \text{ implies } a \vee b \in H \text{ and } a \wedge b \in H.$$

Similarly, if L is a complete lattice and $\emptyset \neq H \subseteq L$, then H is a *complete sublattice* of L if

$$S \subseteq H \text{ implies } \bigvee S \in H \text{ and } \bigwedge S \in H.$$

We may also define *join-sublattices*, *meet-sublattices*, *complete join-sublattices*, and *complete meet-sublattices* in a similar manner.

Let P and Q be two ordered sets. A map $f: P \to Q$ is

(a) a *join-morphism* if whenever $a, b \in P$ and $a \vee b$ exists in P, then $f(a) \vee f(b)$ exists in Q and $f(a \vee b) = f(a) \vee f(b)$.

(b) a *complete join-morphism* if whenever $S \subseteq P$ and $\bigvee S$ exists in P, then $\bigvee f(S)$ exists in Q and $f(\bigvee S) = \bigvee f(S)$.

The notions of a *meet-morphism* and a *complete meet-morphism* are defined dually. Further, a map is called a *morphism* if it is a join-morphism and a meet-morphism. *Complete morphisms* are defined analogously. If P and Q are bounded, then $f: P \to Q$ is *bottom-preserving* if $f(\perp_P) = \perp_Q$, and it is *top-preserving* if $f(\top_P) = \top_Q$. Notice that between bounded ordered sets, any complete join-morphism is bottom-preserving and every complete meet-morphism is top-preserving.

Every order-isomorphism is a complete morphism, and every complete join-morphism, as well as every complete meet-morphism, is order-preserving. In case both P and Q are (complete) lattices and f is a (complete) join-morphism, $f(P)$ is a (complete) join-sublattice of Q. Analogous observations hold for meet-morphisms.

Example 21. The map f Fig. 8 is a complete morphism between L and K.

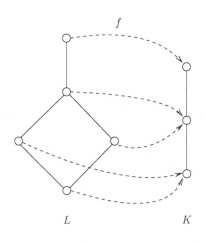

Fig. 8.

The next lemma states that to show that two lattices are isomorphic it suffices to find a bijective join-morphism – sometimes called a join-isomorphism – between them. This means that either the join or the meet operation completely determines lattice's ordering structure.

Lemma 22. *If L and K are lattices and $f\colon L \to K$ is a bijection, then the following assertions are equivalent:*

(a) *f is an order-isomorphism;*
(b) *f is a join-morphism;*
(c) *f is a meet-morphism.*

Proof. Suppose that (a) holds and let $a, b \in L$. Then $f(a) \vee f(b) \le f(a \vee b)$, because f is order-preserving. Assume x is an upper bound of $f(a)$ and $f(b)$. Because f is a bijection, $x = f(c)$ for some $c \in L$, and we must have $a, b \le c$. This means $a \vee b \le c$ and $f(a \vee b) \le f(c) = x$. Hence, $f(a \vee b) = f(a) \vee f(b)$, and (a) implies (b).

If (b) holds, then $a \le b$ implies trivially $f(a) \le f(b)$. On the other hand, if $f(a) \le f(b)$, then $f(a \vee b) = f(a) \vee f(b) = f(b)$, which gives $a \vee b = b$ and $a \le b$. Thus, also (b) implies (a).

We have now shown that (a) and (b) are equivalent. The equivalence of (a) and (c) can be proved dually. $\qquad\square$

We end this section by considering dense sets, which are subsets capable of 'generating' ordered sets and lattices. Let P be an ordered set and let $S \subseteq P$. Then S is called *join-dense* in P if for every element $a \in P$, there exists a subset A of S such that $a = \bigvee A$. The dual of join-dense is *meet-dense*.

Lemma 23. *Let L be a complete lattice. If $S \subseteq L$ is join-dense, then for any $x \in L$,*

$$x = \bigvee \{a \in S \mid a \le x\}.$$

Proof. If S is join-dense, then there exists $A \subseteq S$ such that $x = \bigvee A$. For all $a \in A$, $a \le x$ holds. Thus, $A \subseteq \{a \in S \mid a \le x\}$ and hence $x = \bigvee A \le \bigvee \{a \in S \mid a \le x\} \le x$. $\qquad\square$

Let L be a lattice. An element $x \in L$ is *join-irreducible* if

(a) $x \ne 0$ (in case L has a least element);
(b) $x = a \vee b$ implies $x = a$ or $x = b$ for all $a, b \in L$.

A *meet-irreducible* element is defined dually. We denote the set of join-irreducible elements of L by $\mathcal{J}(L)$ and the set of meet-irreducible elements by $\mathcal{M}(L)$.

In a finite lattice L, an element is clearly join-irreducible if and only if it covers precisely one element. Dually, an element is meet-irreducible if and only if it is covered by exactly one element. Notice also that in $\mathbb{N}$, each nonzero element is join-irreducible.

For finite lattices we can write the following lemma.

Lemma 24. *Let L be a finite lattice.*

(a) *Suppose that $x, y \in L$ and $x \not\le y$. Then there exists $a \in \mathcal{J}(L)$ such that $a \le x$ and $a \not\le y$.*
(b) *For all $x \in L$, $x = \bigvee \{a \in \mathcal{J}(L) \mid a \le x\}$.*

Proof. (a) Let $x \not\leq y$ and $S = \{a \in L \mid a \leq x \text{ and } a \not\leq y\}$. Let a be a minimal element of S. Note that $S \neq \emptyset$ since $x \in S$ and S is finite by assumption. This implies by Lemma 10 that S has at least one minimal element. We claim that a is join-irreducible. Suppose that $a = b \vee c$ for some $b < a$ and $c < a$. Since a is minimal in S, $b \notin S$ and $c \notin S$. However, $b \leq x$ and $c \leq x$ imply $b \leq y$ and $c \leq y$ because $b, c \notin S$. Hence, $a = b \vee c \leq y$, a contradiction!

(b) Let $x \in L$ and $S = \{a \in \mathcal{J}(L) \mid a \leq x\}$. Obviously, x is an upper bound of S. Let $y \in L$ be an upper bound of S and assume that $x \not\leq y$. Then, by (a), there exists $a \in \mathcal{J}(L)$ such that $a \leq x$ and $a \not\leq y$. This gives $a \in S$, and hence $a \leq y$ because y is an upper bound of S, a contradiction! Therefore, $x \leq y$ and $x = \bigvee S$. $\qquad\square$

The next simple proposition characterizes join-dense sets for finite lattices.

Proposition 25. *Let L be a finite lattice. Then $S \subseteq L$ is join-dense in L if and only if $\mathcal{J}(L) \subseteq S$.*

Proof. By Lemma 24(b), $\mathcal{J}(L)$ is join-dense. Trivially, any superset $S \subseteq L$ of $\mathcal{J}(L)$ is also join-dense.

Conversely, let $S \subseteq L$ be join-dense. Assume that $a \in \mathcal{J}(L)$. Because S is join-dense, there exists a subset A of S such that $a = \bigvee A$. Since A is finite and a is join-irreducible, we must have $a \in A \subseteq S$. Thus, $\mathcal{J}(L) \subseteq S$. $\qquad\square$

Example 26. In the lattice depicted in Fig. 9 the join-irreducible elements are marked with filled circles. Clearly, each element of the lattice can be represented as a join of some (or none) of marked elements.

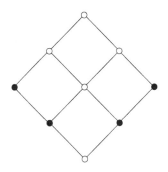

Fig. 9.

For complete rings of sets we may present stronger results. Let L be any complete lattice. An element $x \in L$ is *completely join-irreducible* if for every subset S of L, $x = \bigvee S$ implies that $x \in S$. Note that completely join-irreducible elements must be nonzero, because $0 = \bigvee \emptyset$ and $0 \in \emptyset$ cannot hold. Furthermore, every completely join-irreducible element is trivially join-irreducible.

Let $\mathcal{F} \subseteq \wp(U)$ be a complete ring of sets. For any $x \in U$, we denote

$$N_{\mathcal{F}}(x) = \bigcap \{X \in \mathcal{F} \mid x \in X\}.$$

Clearly, $x \in N_{\mathcal{F}}(x) \in \mathcal{F}$ for any $x \in U$.

Proposition 27. *Let $\mathcal{F} \subseteq \wp(U)$ be a complete ring of sets.*

(a) *The family of completely join-irreducible elements of $\mathcal{F}$ is $\{N_{\mathcal{F}}(x) \mid x \in U\}$.*
(b) *The family $\{N_{\mathcal{F}}(x) \mid x \in U\}$ is the smallest join-dense set in $\mathcal{F}$.*

Proof. (a) Let $x \in U$. If $N_{\mathcal{F}}(x) = \bigcup \mathcal{H}$ for some $\mathcal{H} \subseteq \mathcal{F}$, then $x \in X$ for some $X \in \mathcal{H}$. This implies $N_{\mathcal{F}}(x) \subseteq X$. The inclusion $X \subseteq N_{\mathcal{F}}(x)$ is trivial. Hence, $N_{\mathcal{F}}(x) = X \in \mathcal{H}$ and $N_{\mathcal{F}}(x)$ is completely join-irreducible. Assume that $X \in \mathcal{F}$ is completely join-irreducible. It is easy to see that $X = \bigcup \{N_{\mathcal{F}}(x) \mid x \in X\}$, because for all $x \in X$, $x \in N_{\mathcal{F}}(x) \subseteq X$. Since X is completely join-irreducible, $X = N_{\mathcal{F}}(x)$ for some, or in fact, all $x \in X$.

By the proof of (a), the set $\{N_{\mathcal{F}}(x) \mid x \in U\}$ is join-dense in $\mathcal{F}$. Assume that $\mathcal{H}$ is join-dense and $x \in U$. Then there exists $\mathcal{S} \subseteq \mathcal{H}$ such that $\bigcup \mathcal{S} = N_{\mathcal{F}}(x)$, which implies $N_{\mathcal{F}}(x) \in \mathcal{S} \subseteq \mathcal{H}$ because $N_{\mathcal{F}}(x)$ is completely join-irreducible. $\square$

Bibliographical Notes

Most lattice-theoretical notions and results presented in this section can be found in [3,7,11,21]. It should be noted that completely join-irreducible elements were originally introduced in [48] by defining that an element x is completely join-irreducible if for every subset S of L, $x \leq \bigvee S$ implies that there exists $y \in S$ such that $x \leq y$.

4   Distributive, Boolean, and Stone Lattices

As the title of the section suggests, we consider here the following topics:

4.1 Distributive Lattices
4.2 Boolean Lattices
4.3 Stone Lattices

4.1   Distributive Lattices

A nice property of unions and intersections is that they distribute over each other. Therefore, it is natural to consider lattices for which joins and meets have analogous properties.

A *distributive lattice* is a lattice L satisfying the *distributive laws*

(D1) $(\forall x, y, z \in L)\ x \wedge (y \vee z) = (x \wedge y) \vee (x \wedge y)$;
(D2) $(\forall x, y, z \in L)\ x \vee (y \wedge z) = (x \vee y) \wedge (x \vee z)$.

By the above definition, the dual L^{op} is distributive whenever L is.

Lemma 28. *A lattice L satisfies (D1) if and only if it satisfies (D2).*

Proof. Suppose that (D1) holds. Let $x, y, z \in L$ and let us denote $a = x \vee y$, $b = x$, and $c = z$. Then

$$(x \vee y) \wedge (x \vee z) = a \wedge (b \vee c) = (a \wedge b) \vee (a \wedge c)$$

and

$$\begin{aligned}
(a \wedge b) \vee (a \wedge c) &= ((x \vee y) \wedge x) \vee ((x \vee y) \wedge z) \\
&= x \vee ((x \vee y) \wedge z) \\
&= x \vee ((x \wedge z) \vee (y \wedge z)) \\
&= (x \vee (x \wedge z)) \vee (y \wedge z) \\
&= x \vee (y \wedge z).
\end{aligned}$$

Thus, (D1) implies (D2). By duality, (D2) implies (D1), too. □

The previous lemma means that to show that a lattice is distributive, we have to prove only (D1) or (D2). Also a 'part' of (D1) and (D2) is always true, as we see in the following lemma.

Lemma 29. *If L is a lattice, then for all $a, b, c \in L$,*

(a) $a \wedge (b \vee c) \geq (a \wedge b) \vee (a \wedge c)$;
(b) $a \vee (b \wedge c) \leq (a \vee b) \wedge (a \vee c)$.

Proof. (a) Since $a \wedge (b \vee c) \geq a \wedge b$ and $a \wedge (b \vee c) \geq a \wedge c$ by Lemma 18, we have that $a \wedge (b \vee c) \geq (a \wedge b) \vee (a \wedge c)$. Claim (b) can be proved dually. □

By combining Lemmas 28 and 29 it suffices to check either of the inequalities

$$a \wedge (b \vee c) \leq (a \wedge b) \vee (a \wedge c)$$

or

$$a \vee (b \wedge c) \geq (a \vee b) \wedge (a \vee c)$$

to show that a lattice is distributive.

Example 30. Next we consider some examples of distributive lattices.

(a) Any ring of sets is a distributive lattice, since for all sets X, Y, and Z, $X \cap (Y \cup Z) = (X \cap Y) \cup (X \cap Z)$ by Proposition 1. In particular, the lattice $(\wp(A), \subseteq)$ is distributive for every A.
(b) By Lemma 16, $\mathcal{O}(P)$ is a ring of set for any ordered set P. This gives that each ordered set can be embedded into a distributive lattice.
(c) Every chain is a distributive lattice. To verify (D1), we need only consider the three cases: (i) $x \leq y \leq z$, (ii) $y \leq x \leq z$, and (iii) $y \leq z \leq x$. For instance in case (ii),

$$x \wedge (y \vee z) = x \wedge z = x$$

and

$$(x \wedge y) \vee (x \wedge z) = y \vee x = x.$$

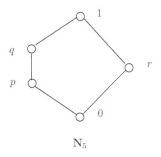
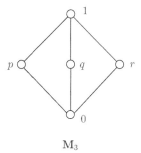

N_5 M_3

Fig. 10.

The two usually mentioned non-distributive lattices are N_5 and M_3, whose Hasse diagrams are presented in Fig. 10. The lattice N_5 is usually called the *pentagon* and M_3 is commonly referred to as the *diamond*.

It is easy to verify that N_5 is not distributive, because

$$p \vee (q \wedge r) = p \vee 0 = p \text{ and } (p \vee q) \wedge (p \vee r) = q \vee 1 = q.$$

Similarly, M_3 is not distributive, since

$$p \vee (q \wedge r) = p \vee 0 = p \text{ and } (p \vee q) \wedge (p \vee r) = 1 \vee 1 = 1.$$

We have seen that new lattices can be obtained by forming sublattices or products of lattices, as well as taking all functions from a set to a lattice. These constructions preserve distributivity.

Lemma 31. *Let L and K be distributive lattices.*

(a) *Any sublattice of L is distributive.*
(b) *The product $L \times K$ is distributive.*
(c) *For any set X, L^X is distributive.*

Proof. Claim (a) is trivial and (b) holds because joins and meets are defined in $L \times K$ coordinatewise. Similarly, in L^X operations are defined pointwise, which implies, for example, that

$$\begin{aligned}
(\varphi_1 \wedge (\varphi_2 \vee \varphi_3))(x) &= \varphi_1(x) \wedge (\varphi_2 \vee \varphi_3)(x) \\
&= \varphi_1(x) \wedge (\varphi_2(x) \vee \varphi_3(x)) \\
&= (\varphi_1(x) \wedge \varphi_2(x)) \vee (\varphi_1(x) \wedge \varphi_3(x)) \\
&= (\varphi_1 \wedge \varphi_2)(x) \vee (\varphi_1 \wedge \varphi_3)(x) \\
&= ((\varphi_1 \wedge \varphi_2) \vee (\varphi_1 \wedge \varphi_3))(x)
\end{aligned}$$

for all $x \in X$.   □

By the previous lemma, if a lattice has a sublattice isomorphic to N_5 or M_3, it cannot be distributive. In fact, we could also prove the converse stating that if a lattice does not have a sublattice isomorphic to $\mathbf{N}_5$ or $\mathbf{M}_3$, it is distributive. However, the proof is too long to be presented here and it can be found in almost any introductory book on lattice theory.

Example 32. By Lemma 31(c), the set of mappings $\mathbf{n}^X$ from a X to the n-element chain $\mathbf{n}$ is distributive.

4.2   Boolean Lattices

Let L be a bounded lattice with a least element 0 and a greatest element 1. For an element $a \in L$, we say that an element $b \in L$ is a *complement* of a if

$$a \vee b = 1 \quad \text{and} \quad a \wedge b = 0.$$

If the element a has a unique complement, we denote it by a'.

If a bounded lattice is not distributive, it is possible that some elements have several complements. For example, consider the lattices $\mathbf{N}_5$ and $\mathbf{M}_3$ of Fig. 10.

Lemma 33. *In a bounded distributive lattice any element can have at most one complement.*

Proof. Let L be a bounded distributive lattice and $a \in L$. If a has complements b_1 and b_2, then

$$b_1 = b_1 \wedge 1 = b_1 \wedge (a \vee b_2) = (b_1 \wedge a) \vee (b_1 \wedge b_2) = 0 \vee (b_1 \wedge b_2) = b_1 \wedge b_2.$$

This implies that $b_1 \leq b_2$. By symmetry, $b_2 \leq b_1$, and hence $b_1 = b_2$.   □

A lattice L is a *Boolean lattice* if it is distributive, bounded, and its every element a has a unique complement $a' \in L$. The next lemma gives some useful properties of complements in Boolean lattices.

Lemma 34. *Let B be a Boolean lattice and $a, b, c \in B$.*

(a) $0' = 1$ *and* $1' = 0$;
(b) $a'' = a$;
(c) $(a \vee b)' = a' \wedge b'$ *and* $(a \wedge b)' = a' \vee b'$;
(d) $a \wedge b = 0 \iff a \leq b'$;
(e) $a \leq b \implies b' \leq a'$.

Proof. Claims (a) and (b) follow directly from the definition of complements.
     (c) By the distributive laws,

$$(a \vee b) \vee (a' \wedge b') = ((a \vee b) \vee a') \wedge ((a \vee b) \vee b') = 1 \wedge 1 = 1$$

and

$$(a \vee b) \wedge (a' \wedge b') = (a \wedge (a' \wedge b')) \vee (b \wedge (a' \wedge b')) = 0 \vee 0 = 0.$$

The other equality follows by duality.
     (d) If $a \wedge b = 0$, then

$$a = a \wedge (b \vee b') = (a \wedge b) \vee (a \wedge b') = a \wedge b',$$

that is, $a \leq b'$. On the other hand, $a \leq b'$ implies $a \wedge b \leq b' \wedge b = 0$.
     (e) If $a \leq b$, then $a \wedge b' = 0$ and $b' \wedge a = 0$ by (d). This gives $b' \leq a'$.   □

Example 35. (a) By Proposition 1, the powerset $\wp(A)$ of A forms with respect to the inclusion relation $\subseteq$ a Boolean lattice such that the complement of any $X \subseteq A$ is $X^c = A - X$.

(b) The lattice depicted in Fig. 11(i) is a Boolean lattice such that $0' = 1$, $a' = b$, $b' = a$ and $1' = 0$.

(c) Let us consider the 2-element chain $\mathbf{2} = \{0, 1\}$. For any X, the distributive lattice $\mathbf{2}^X$ is a complete Boolean lattice. Recall that joins and meets are defined pointwise, and similarly the complement f' of a map $f \colon X \to \mathbf{2}$ is defined by

$$f'(x) = \begin{cases} 1 & \text{if } f(x) = 0 \\ 0 & \text{if } f(x) = 1, \end{cases}$$

that is, $f'(x) = f(x)'$. If X is a finite set with n elements, $\mathbf{2}^X$ can be identified with the set of all ordered n-tuples $\{(x_1, \ldots, x_n) \mid x_i \in \mathbf{2}\}$. The diagram of the ordered set $\mathbf{2}^3$ is in Fig. 11(ii).

Lemma 36. *For any set A, $\wp(A) \cong \mathbf{2}^A$.*

Proof. We define for any $X \subseteq A$ the so-called *characteristic function* $\mu_X \colon A \to \mathbf{2}$ of X by setting

$$\mu_X(x) = \begin{cases} 1 & \text{if } x \in A \\ 0 & \text{if } x \notin A. \end{cases}$$

The map $\varphi \colon X \mapsto \mu_X$ is clearly from $\wp(A)$ onto $\mathbf{2}^A$. Further, for all $X, Y \subseteq A$,

$$\begin{aligned} X \subseteq Y &\iff (\forall a \in A)\, a \in X \implies a \in Y \\ &\iff (\forall a \in A)\, \mu_X(a) = 1 \implies \mu_Y(a) = 1 \\ &\iff \mu_X \leq \mu_Y. \qquad \square \end{aligned}$$

Let L be a lattice with a least element 0. Then $a \in L$ is called an *atom* of L, if $0 \prec a$. The set of atoms of L is denoted by $\mathcal{A}(L)$. In Boolean lattices the atoms are exactly the join-irreducible elements, as we see in the next lemma. Note that in Fig. 11 the atoms are marked with filled circles.

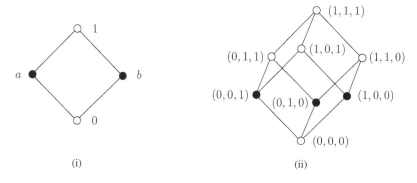

(i)

(ii)

Fig. 11.

Lemma 37. *Let L be a lattice with a least element 0.*

(a) *Then $\mathcal{A}(L) \subseteq \mathcal{J}(L)$.*
(b) *If L is Boolean lattice, then $\mathcal{A}(L) = \mathcal{J}(L)$.*

Proof. (a) Suppose that $0 \prec x$ and $x = a \vee b$ with $a < x$ and $b < x$. Because $0 \prec x$, we have $a = b = 0$, from which we get $x = 0$, a contradiction!

(b) Let L be a Boolean lattice. Assume that $x \in \mathcal{J}(L)$. If $0 \le y < x$, then

$$x = x \vee y = (x \vee y) \wedge (y' \vee y) = (x \wedge y') \vee y.$$

Because x is join-irreducible and $y < x$, we must have $x = x \wedge y'$. This implies $x \le y'$ and thus $y = x \wedge y \le y' \wedge y = 0$. So, x is an atom and $\mathcal{J}(L) \subseteq \mathcal{A}(L)$. $\square$

The lattice L is *atomic* if every element x of L is the supremum of the atoms below it, that is, $x = \bigvee \{a \in \mathcal{A}(L) \mid a \le x\}$. If L is an atomic lattice, then for all $x \ne 0$, there exists an atom $a \in \mathcal{A}(L)$ such that $a \le x$. Namely, if $\{a \in \mathcal{A}(L) \mid a \le x\} = \emptyset$, then $x = \bigvee \{a \in \mathcal{A}(L) \mid a \le x\} = \bigvee \emptyset = 0$.

Lemma 38. *Any finite Boolean lattice is atomic.*

Proof. By Lemma 37, $\mathcal{A}(B) = \mathcal{J}(B)$ for a Boolean lattice B. Since B is finite,

$$x = \bigvee \{a \in \mathcal{J}(B) \mid a \le x\} = \bigvee \{a \in \mathcal{A}(B) \mid a \le x\}$$

for all $x \in B$ by Lemma 24(b). $\square$

Example 39. (a) For any set U, $\wp(U)$ is a complete atomic Boolean lattice in which the set of atoms is $\{\{x\} \mid x \in U\}$.
(b) The Cartesian product $\wp(U) \times \wp(U)$ ordered with coordinatewise order is an atomic complete Boolean lattice in which the complement of an element (X, Y) is (X^c, Y^c) and the atoms are the pairs $(\{a\}, \emptyset)$ and $(\emptyset, \{a\})$.
(c) In general, if B is a complete atomic Boolean lattice, then $B \times B$ is a complete atomic Boolean lattice, in which joins and meets are defined coordinatewise. The lattice is distributive by Lemma 31, $(0,0)$ and $(1,1)$ are the least and the greatest elements, and the complement of (x, y) is (x', y'). The atoms are the pairs $(0, a)$ and $(a, 0)$, where a is any atom of B.

A ring of sets $\mathcal{F} \subseteq \wp(U)$ is called a *field of sets*, if $X \in \mathcal{F}$ implies $X^c \in \mathcal{F}$. A complete ring of sets which is also a field of sets is called a *complete field of sets*.

Example 40. (a) For any set U, the power set $\wp(U)$ of U is complete field of sets. Furthermore, the set $\mathrm{Rel}(U)$ of all binary relations on U is a complete field of sets, because $\mathrm{Rel}(U)$ is equal to $\wp(U \times U)$.
(b) In Example 9 we showed that $(\wp(U), \subseteq)$ is self-dual. Here we note that every complete field of sets $\mathcal{F} \subseteq \wp(U)$ is self-dual with respect to the set-inclusion relation. The map $\phi \colon X \mapsto X^c$ is clearly the required dual order-isomorphism.

Proposition 41. *Every complete field of sets $\mathcal{F} \subseteq \wp(U)$ is a complete atomic Boolean lattice with respect to the set-inclusion relation such that the set of atoms is $\{N_{\mathcal{F}}(x) \mid x \in U\}$.*

Proof. Clearly, $(\mathcal{F}, \subseteq)$ is a complete Boolean lattice such that $\bigwedge \mathcal{H} = \bigcap \mathcal{H}$ and $\bigvee \mathcal{H} = \bigcup \mathcal{H}$ for all $\mathcal{H} \subseteq \mathcal{F}$, and X^c is the complement of any $X \in \mathcal{F}$. We have to still show that $\mathcal{F}$ is atomic.

We know by Proposition 27 that for each $X \in \mathcal{F}$,

$$X = \bigcup \{N_{\mathcal{F}}(x) \mid x \in X\}.$$

It is enough to prove that $\{N_{\mathcal{F}}(x) \mid x \in X\}$ is the set of atoms. Let $x \in U$. Since $x \in N_{\mathcal{F}}(x)$, $\emptyset \subset N_{\mathcal{F}}(x)$. Assume that $\emptyset \subset X \subseteq N_{\mathcal{F}}(x)$. If $x \notin X$, then $x \in X^c$ and $\emptyset \neq X \subseteq N_{\mathcal{F}}(x) \subseteq X^c$, a contradiction! This implies $x \in X$ and $N_{\mathcal{F}}(x) \subseteq X$. Hence, each $N_{\mathcal{F}}(x)$ is an atom. It is also obvious that atoms must be of the form $N_{\mathcal{F}}(x)$, because for any $X \subseteq U$, $x \in X$ implies $N_{\mathcal{F}}(x) \subseteq X$. □

4.3   Pseudocomplements and Stone Lattices

Here we introduce a weaker type of complement which may exists in lattices that are not complemented in the usual sense.

Suppose that L is a lattice with a least element 0. An element x^* is a *pseudocomplement* of $x \in L$, if $x \wedge x^* = 0$ and for all $a \in L$, $x \wedge a = 0$ implies $a \leq x^*$. An element can have at most one pseudocomplement. A lattice is *pseudocomplemented* if each element has a pseudocomplement.

Lemma 42. *If L is a pseudocomplemented lattice, then for all $a, b \in L$,*

(a) $a \leq a^{**}$;
(b) $a \leq b$ implies $a^* \geq b^*$;
(c) $a^* = a^{***}$.

Proof. (a) By definition, $a \wedge a^* = 0$ and thus $a \leq a^{**}$.
(b) If $a \leq b$, then $a \wedge b^* \leq b \wedge b^* = 0$ and hence $b^* \leq a^*$.
(c) By (a) and (b), $a \leq a^{**}$ and $a^* \geq a^{***}$. Further, $a^* \leq a^{***}$ by (a). □

Example 43. (a) Every Boolean lattice is a pseudocomplemented lattice in which the pseudocomplements are the usual complements.
(b) Every finite distributive lattice L is pseudocomplemented. Obviously, L is bounded. Let us define for any $x \in L$,

$$x^* = \bigvee \{y \in L \mid x \wedge y = 0\}.$$

Let $\{y \in L \mid x \wedge y = 0\} = \{y_1, y_2, \dots, y_n\}$. Then

$$\begin{aligned}
x \wedge x^* &= x \wedge (y_1 \vee y_2 \vee \cdots \vee y_n) \\
&= (x \wedge y_1) \vee (x \wedge y_2) \vee \cdots \vee (x \wedge y_n) \\
&= 0 \vee 0 \vee \cdots \vee 0 \\
&= 0.
\end{aligned}$$

Further, if $x \wedge a = 0$, then $a = y_i$ for some i, which gives $a \leq x^*$.

(c) In every pseudocomplemented lattice, $0^* = 1$ and $1^* = 0$. Trivially, $0 \wedge 1 = 0$. For any $a \in L$, $0 \wedge a = 0$ and $a \leq 1$. If $1 \wedge a = 0$, then necessary $a = 0$.

A bounded pseudocomplemented distributive lattice L satisfying the identity

$$a^* \vee a^{**} = 1$$

is called a *Stone lattice*. For a Stone lattice L, the set

$$S(L) = \{a^* \mid a \in L\}$$

is called the *skeleton* of L.

Lemma 44. *Let L be a Stone lattice.*

(a) $a \in S(L)$ *if and only if* $a = a^{**}$.
(b) $(a \wedge b)^* = a^* \vee b^*$ *for all* $a, b \in L$.

Proof. (a) If $a \in S(L)$, then $a = b^*$ for some $b \in L$. So, $a^{**} = b^{***} = b^* = a$. Conversely, $a = a^{**}$ implies trivially $a \in S(L)$.
  (b) We show that $a^* \vee b^*$ is the pseudocomplement of $a \wedge b$. For all $a, b \in L$,

$$(a \wedge b) \wedge (a^* \vee b^*) = (a \wedge b \wedge a^*) \vee (a \wedge b \wedge b^*) = 0 \vee 0 = 0.$$

If $(a \wedge b) \wedge x = 0$, then $(b \wedge x) \wedge a = 0$ and $b \wedge x \leq a^*$. Hence $b \wedge x \wedge a^{**} \leq a^* \wedge a^{**} = 0$. Thus, $x \wedge a^{**} \leq b^*$ and

$$x = x \wedge 1 = x \wedge (a^* \vee a^{**}) = (x \wedge a^*) \vee (x \wedge a^{**}) \leq a^* \vee b^*. \qquad \square$$

Proposition 45. *If L is a Stone lattice, then the skeleton $S(L)$ is a sublattice of L such that $0, 1 \in S(L)$. Further, $S(L)$ is a Boolean lattice in which the complement of any $a \in S(L)$ is a^*.*

Proof. Let $a, b \in S(L)$. Then by Lemma 44, $a \vee b = a^{**} \vee b^{**} = (a^* \wedge b^*)^* \in S(L)$. Further, $a = a^{**} \geq (a \wedge b)^{**}$ and $b = b^{**} \geq (a \wedge b)^{**}$. Hence, $a \wedge b \geq (a \wedge b)^{**}$. By Lemma 42, $a \wedge b \leq (a \wedge b)^{**}$. Thus, $a \wedge b \in S(L)$.
  By Example 43, $0^* = 1$ and $1^* = 0$. This gives $0^{**} = 1^* = 0$ and $1^{**} = 0^* = 1$. Hence, $0, 1 \in S(L)$.
  Because $S(L)$ is a sublattice of a distributive lattice, it is distributive. Let $a \in S(L)$, then

$$a \vee a^* = a^{**} \vee a^* = 1 \quad \text{and} \quad a \wedge a^* = 0. \qquad \square$$

For a Stone lattice L, let us define the set

$$D(L) = \{a \mid a^* = 0\}.$$

The members of $D(L)$ are called *dense*. Dense elements should not be confused with join- or meet-dense subsets of ordered sets. The set $D(L)$ is a sublattice of L, since for all $a, b \in D(L)$, $(a \vee b)^* \leq a^* \vee b^* = 0 \vee 0 = 0$ and $(a \wedge b)^* = a^* \vee b^* =$

$0 \vee 0 = 0$. This implies that also $D(L)$ is distributive. It is easy to see that for any $a \in L$,

$$a = a^{**} \wedge (a \vee a^*),$$

$a^{**} \in S(L)$, and $a \vee a^* \in D(L)$. This can be interpreted so that any $a \in L$ can represented in the form

$$a = b \wedge c,$$

where $b \in S(L)$ and $c \in D(L)$.

Lemma 46. *For any complete Boolean lattice B, the set*

$$B^{[2]} = \{(a,b) \in B \times B \mid a \leq b\}$$

is a complete Stone lattice, in which joins and meets are given by

$$\bigvee_{i \in I} (a_i, b_i) = \left(\bigvee_{i \in I} a_i, \bigvee_{i \in I} b_i \right) \quad \text{and} \quad \bigwedge_{i \in I} (a_i, b_i) = \left(\bigwedge_{i \in I} a_i, \bigwedge_{i \in I} b_i \right)$$

and $(a,b)^* = (b', b')$ for all $(a,b) \in B^{[2]}$. Further, $S(B^{[2]}) = \{(a,a) \mid a \in B\} \cong B$ and $D(B^{[2]}) = \{(a,1) \mid a \in B\}$.

Proof. Suppose $\{(a_i, b_i)\}_{i \in I}$ is a subset of $B^{[2]}$. Then, for all $i \in I$, $a_i \leq b_i$ which gives $a_i \leq \bigvee_{i \in I} b_i$ and hence $\bigvee_{i \in I} a_i \leq \bigvee_{i \in I} b_i$. The analogous fact holds also for meets. So, $B^{[2]}$ is a sublattice of $B \times B$.

If $a \leq b$, then $b' \leq a'$. Thus, $(a,b) \wedge (b', b') = (a \wedge b', b \wedge b') \leq (a \wedge a', b \wedge b') = (0,0)$. Further, if $(a,b) \wedge (x,y) = 0$ for some $x \leq y$, then $b \wedge y = 0$ implies $x \leq y \leq b'$, which gives $(x,y) \leq (b', b')$. Thus, $(a,b)^* = (b', b')$.

By definition,

$$S(B^{[2]}) = \{(a,b)^* \mid (a,b) \in B^{[2]}\} = \{(b', b') \mid b \in B\} = \{(a,a) \mid a \in B\}$$

and

$$\begin{aligned} D(B^{[2]}) &= \{(a,b) \in B^{[2]} \mid (a,b)^* = (0,0)\} \\ &= \{(a,b) \in B^{[2]} \mid (b', b') = (0,0)\} \\ &= \{(a,1) \mid a \in B\}. \end{aligned} \qquad \square$$

Example 47. (a) Let B be the 4-element Boolean lattice in Fig 11. Then $B^{[2]} = \{(0,0), (0,a), (0,b), (0,1), (a,a), (a,1), (b,b), (b,1), (1,1)\}$ is the Stone lattice depicted in Fig. 12. The set $S(L)$ is denoted by filled circles and the elements of $D(L)$ are boxed.

(b) Let us consider the 3-element chain $\mathbf{3} = \{0, u, 1\}$. Then clearly $\mathbf{3}$ is a Stone lattice in which $0^* = 1$, $u^* = 0$, and $1^* = 0$. Further, for any set X, $\mathbf{3}^X$ is a Stone lattice in which the pseudocomplement f^* of f is defined by

$$f^*(x) = \begin{cases} 1 & \text{if } f(x) = 0 \\ 0 & \text{otherwise,} \end{cases}$$

that is, $f^*(x) = f(x)^*$. Notice that $\mathbf{3}^X$ is isomorphic to the lattice in Fig. 12 for any two-element X.

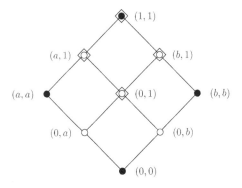

Fig. 12.

Bibliographical Notes

Basic definitions and results concerning distributive and Boolean lattices can be found in [3,7,11,21]. The facts about pseudocomplemented lattices and Stone lattices presented here can be found in [3,21].

5   Closure Systems and Topologies

This section has the following subsections:

5.1 Closure Systems and Closure Operators
5.2 Topological Spaces
5.3 Alexandrov Spaces

5.1   Closure Systems and Closure Operators

A family $\mathcal{L}$ of subsets of a set U is said to be a *closure system* if $\mathcal{L}$ is closed under intersections, which means that for all $\mathcal{H} \subseteq \mathcal{L}$, we have $\bigcap \mathcal{H} \in \mathcal{L}$. If $\mathcal{L}$ is a closure system on U, then the ordered set $(\mathcal{L}, \subseteq)$ is a complete lattice according to Theorem 14. The meet is just set intersection, but the join not need to be the union. Note that $U = \bigcap \emptyset$ belongs to every closure system on U.

A map $C \colon \wp(U) \to \wp(U)$ is a *closure operator* on U if, for all $X, Y \subseteq U$, it satisfies the conditions:

(CO1) $X \subseteq C(X)$ (extensive)
(CO2) $X \subseteq Y$ implies $C(X) \subseteq C(Y)$ (order-preserving)
(CO3) $C(C(X)) = C(X)$ (idempotent)

A subset X of U is *closed* with respect to C if $C(X) = X$. We denote by $\mathcal{L}_C$ the set of C-closed subsets of U.

Lemma 48. *Let C be a closure operator on U.*

(a) $\mathcal{L}_C = \{ C(X) \mid X \subseteq U \}$;
(b) $C(X) = \bigcap \{ B \in \mathcal{L}_C \mid X \subseteq B \}$ *for all $X \subseteq U$.*

Proof. (a) If $X \in \mathcal{L}_C$, then $X = C(X)$, that is, $X \in \{C(X) \mid X \subseteq U\}$. Conversely, if $Y \in \{C(X) \mid X \subseteq U\}$, then $Y = C(Z)$ for some $Z \subseteq U$ which gives $C(Y) = C(C(Z)) = C(Z) = Y$, that is, $Y \in \mathcal{L}_C$.

(b) Clearly $C(X)$ is the least element of $\{B \in \mathcal{L}_C \mid X \subseteq B\}$.   □

The following theorem revels a bijective correspondence between closure operators and closure systems.

Theorem 49. *Let U be a set.*

(a) *If C is a closure operator on U, then the family $\mathcal{L}_C$ of closed subsets of U is a closure system and so it forms a complete lattice with respect to inclusion such that for all $\mathcal{H} \subseteq \mathcal{L}$,*

$$\bigwedge \mathcal{H} = \bigcap \mathcal{H} \quad \text{and} \quad \bigvee \mathcal{H} = C(\bigcup \mathcal{H}).$$

(b) *If $\mathcal{L}$ is a closure system on U, the formula*

$$C_{\mathcal{L}}(X) = \bigcap \{B \in \mathcal{L} \mid X \subseteq B\}$$

defines a closure operator $C_{\mathcal{L}}$ on U.

Proof. (a) Assume that $\mathcal{H} \subseteq \mathcal{L}_C$. Then $\bigcap \mathcal{H} \subseteq C(\bigcap \mathcal{H}) \subseteq C(X) = X$ for all $X \in \mathcal{H}$. This means that $C(\bigcap \mathcal{H}) = \bigcap \mathcal{H}$ and $\bigcap \mathcal{H} \in \mathcal{L}_C$. Hence, $\mathcal{L}_C$ is a complete lattice such that $\bigwedge \mathcal{H} = \bigcap \mathcal{H}$. By Theorem 14,

$$\bigvee \mathcal{H} = \bigcap \{B \in \mathcal{L} \mid X \subseteq B \text{ for all } X \in \mathcal{H}\}$$
$$= \bigcap \{B \in \mathcal{L} \mid \bigcup \mathcal{H} \subseteq B\}$$
$$= C(\bigcup \mathcal{H}).$$

(b) It is obvious that $C_{\mathcal{L}}$ is extensive. If $X \subseteq Y$, then $\{B \in \mathcal{L} \mid X \subseteq B\} \supseteq \{B \in \mathcal{L} \mid Y \subseteq B\}$, which implies $C_{\mathcal{L}}(X) = \bigcap \{B \in \mathcal{L} \mid X \subseteq B\} \subseteq \bigcap \{B \in \mathcal{L} \mid Y \subseteq B\} = C_{\mathcal{L}}(Y)$, that is, $C_{\mathcal{L}}$ is order-preserving. By definition, $C_{\mathcal{L}}(C_{\mathcal{L}}(X)) = \bigcap \{B \in \mathcal{L} \mid C_{\mathcal{L}}(X) \subseteq B\}$. Since $C_{\mathcal{L}}(X) \in \{B \in \mathcal{L} \mid C_{\mathcal{L}}(X) \subseteq B\}$, we have $C_{\mathcal{L}}(C_{\mathcal{L}}(X)) \subseteq C_{\mathcal{L}}(X)$. The inclusion $C_{\mathcal{L}}(X) \subseteq C_{\mathcal{L}}(C_{\mathcal{L}}(X))$ is obvious.   □

The relationship between closure systems and closure operators is bijective. The closure operator induced by the closure system $\mathcal{L}_C$ is C itself, and similarly the closure system induced by the closure operator $C_{\mathcal{L}}$ is $\mathcal{L}$. In symbols,

$$C_{(\mathcal{L}_c)} = C \quad \text{and} \quad \mathcal{L}_{(C_{\mathcal{L}})} = \mathcal{L}.$$

Note that if $\mathcal{L}$ is a closure system on U, then in the complete lattice $\mathcal{L}$, $\bigvee \mathcal{H} = C_{\mathcal{L}}(\bigcup H)$ for all $\mathcal{H} \subseteq \mathcal{L}$.

In Section 3.2 we saw that every ordered set can be embedded into a complete lattice of sets. Here we show that closure systems are important also because each complete lattice is isomorphic to some closure system.

Proposition 50. *Every complete lattice is isomorphic to some closure system.*

Proof. Let L be a complete lattice. We define a family $\mathcal{L}$ of subsets of L by setting

$$\mathcal{L} = \{\,(x] \mid x \in L\,\}.$$

We know by the proof of Lemma 16 that the map $x \mapsto (x]$ is an order-isomorphism between L and $\mathcal{L}$. If $\{\,(x] \mid x \in S\}$ is a subfamily of $\mathcal{L}$, then for all $a \in L$,

$$a \in \bigcap_{x \in S}(x] \iff (\forall x \in S)\,a \le x \iff a \le \bigwedge S \iff a \in \left(\bigwedge S\right].$$

Thus, $\mathcal{L}$ is a closure system. $\qquad\square$

Notice that the previous proposition implies directly that closure systems are not necessarily distributive lattices.

In Section 9 we will consider rough set approximations which are determined by equivalences called indiscernibility relations. Here we consider the set of all equivalences on a set U, which is denoted by $\mathrm{Eq}(U)$.

Lemma 51. *If $\mathcal{H} \subseteq \mathrm{Eq}(U)$, then $\bigcap \mathcal{H}$ is an equivalence on U.*

Proof. We show that $\bigcap \mathcal{H}$ is transitive. The rest can be proved in an analogous way. If $(x,y) \in \bigcap \mathcal{H}$ and $(y,z) \in \bigcap \mathcal{H}$, then $(x,y) \in E$ and $(y,z) \in E$ for every $E \in \mathcal{H}$. This implies $(x,z) \in E$ for all $E \in \mathcal{H}$ and hence $(x,z) \in \bigcap \mathcal{H}$. $\qquad\square$

By the previous lemma, the family $\mathrm{Eq}(U)$ of all equivalences on U is a closure system on $\mathrm{Rel}(U)$. The corresponding closure operator is

$$^{E}\colon \mathrm{Rel}(U) \to \mathrm{Rel}(U),\, R \mapsto \bigcap \{E \in \mathrm{Eq}(U) \mid R \subseteq E\}.$$

Hence, $(\mathrm{Eq}(U), \subseteq)$ is a complete lattice in which

$$\bigwedge \mathcal{H} = \bigcap \mathcal{H} \quad \text{and} \quad \bigvee \mathcal{H} = \left(\bigcup \mathcal{H}\right)^{E}.$$

Next we determine the number of equivalence relations for a finite set. We define the number $\left\{{n \atop k}\right\}$ for any $n, k \ge 1$ by setting

$$\left\{{n \atop 1}\right\} = \left\{{n \atop n}\right\} = 1 \quad \text{and} \quad \left\{{n \atop k}\right\} = \left\{{n-1 \atop k-1}\right\} + k \cdot \left\{{n-1 \atop k}\right\},$$

for $2 \le k \le n - 1$. The numbers $\left\{{n \atop k}\right\}$ are called the *Stirling's numbers of the second kind*.

Proposition 52. *If A is a set with n elements, then $\left\{{n \atop k}\right\}$ is the number of partitions of the cardinality k of A.*

Proof. There is only one partition with one block, namely A itself, and the only partition into n parts is the family of singletons $\{x\}$.

Let $x \in A$. For every partition Π of A either (i) the singleton $\{x\} \in \Pi$ or (ii) $\{x\} \notin \Pi$. When the set $\{x\}$ is removed from a partition of type (i), we obtain a partition of the $n - 1$-element set $A - \{x\}$ into $k - 1$ parts, and there are $\left\{{n-1 \atop k-1}\right\}$ of those. Conversely, if we are given such a partition, we can restore the set $\{x\}$, so that the correspondence is a bijection.

Suppose that a partition Π of type (ii) consists of the sets $X_1, X_2, \ldots, X_k$. Now this situation determines a pair (x, Π_x) such that $x \in X_i$ and Π_x is a partition on the $n - 1$-element set $A - \{x\}$ with parts $X_1, \ldots, X_{i-1}, X_i - \{x\}, X_{i+1}, \ldots, X_k$. There are k possible values of i and $\left\{{n-1 \atop k}\right\}$ possible partitions Π_x, so we have $k\left\{{n-1 \atop k}\right\}$ such pairs. Furthermore, if we are given such a pair, we can restore x to the set X_i, and recover Π. Hence, this correspondence is also a bijection. $\square$

Note that the Stirling's numbers of the second kind can be tabulated in much the same way as the binomial coefficients in the well-known Pascal's triangle. Recall that $\left\{{n \atop k}\right\} = \left\{{n-1 \atop k-1}\right\} + k \cdot \left\{{n-1 \atop k}\right\}$.

$$
\begin{array}{ccccccccccccc}
&&&&&& 1 &&&&&& \\
&&&&& 1 && 1 &&&&& \\
&&&& 1 && 3 && 1 &&&& \\
&&& 1 && 7 && 6 && 1 &&& \\
&& 1 && 15 && 25 && 10 && 1 && \\
& 1 && 31 && 90 && 65 && 15 && 1 & \\
1 && 63 && 301 && 350 && 140 && 21 && 1
\end{array}
$$

For example, the number of the equivalences and partitions on a 5-element set is $1 + 15 + 25 + 10 + 1 = 52$.

In the next subsection we will consider topological spaces in which closure and interior operators have a major role. An *interior operator* $I: \wp(U) \to \wp(U)$ satisfies the conditions

(IO1) $I(X) \subseteq X$;
(IO2) $X \subseteq Y$ implies $I(X) \subseteq I(Y)$;
(IO3) $I(I(X)) = I(X)$.

An *interior system* is a family of sets closed under arbitrary unions. Since interior operator and systems are the dual notions of closure operators and systems, we get many of their properties without any work. For example, the correspondence between interior operators and interior systems is bijective. In particular, if I is an interior operator U, the family $\mathcal{N} = \{I(X) \mid X \subseteq U\}$ is an interior system, and then $(\mathcal{N}, \subseteq)$ is a complete lattice in which

$$\bigvee \mathcal{H} = \bigcup \mathcal{H} \quad \text{and} \quad \bigwedge \mathcal{H} = I\left(\bigcap \mathcal{H}\right)$$

for all $\mathcal{H} \subseteq \mathcal{N}$.

5.2   Topological Spaces

A *topological space* $(U, \mathcal{T})$ consists of a set U and a family $\mathcal{T} \subseteq \wp(U)$ such that

(TS1) $\emptyset \in \mathcal{T}$ and $U \in \mathcal{T}$,
(TS2) $X \cap Y \in \mathcal{T}$ for any sets $X, Y \in \mathcal{T}$, and
(TS3) $\bigcup \mathcal{H} \in \mathcal{T}$ for any subfamily $\mathcal{H} \subseteq \mathcal{T}$.

The family $\mathcal{T}$ is called a *topology* on U and the members of $\mathcal{T}$ are *open sets*. The complement of an open set is called a *closed set*. The family of closed sets is denoted by

$$\mathcal{L}_{\mathcal{T}} = \{X^c \mid X \in \mathcal{T}\}.$$

The union of any two closed set is closed and any intersection of closed sets is closed. Furthermore, the sets $\emptyset$ and U are closed. Clearly, all open sets form an interior system and all closed systems form a closure system in the sense of Section 5.1. It is also obvious that $\mathcal{T}$ and $\mathcal{L}_{\mathcal{T}}$ are rings of sets, and therefore they form distributive lattices.

Let $(U, \mathcal{T})$ be a topological space. The *interior* $I_{\mathcal{T}}(X)$ of a set $X \subseteq U$ in $\mathcal{T}$ is defined to be the greatest open set included in X. Similarly, the *closure* $C_{\mathcal{T}}(X)$ of a set $X \subseteq U$ in $\mathcal{T}$ is defined to be the smallest closed set containing X.

Proposition 53. *Let $(U, \mathcal{T})$ be a topological space.*

(a) $\mathcal{T}$ *is a pseudocomplemented lattice such that* $X^* = I_{\mathcal{T}}(X^c)$ *for all* $X \in \mathcal{T}$.
(b) *The ordered sets* $(\mathcal{T}, \subseteq)$ *and* $(\mathcal{L}_{\mathcal{T}}, \subseteq)$ *are dually order-isomorphic.*
(c) *For all* $X \subseteq U$, $I_{\mathcal{T}}(X)^c = C_{\mathcal{T}}(X^c)$.

Proof. (a) We have already noted that $\mathcal{T}$ is a distributive lattice in which joins are given by set unions and meets by set intersections. Further, $\emptyset$ is the least element of $\mathcal{T}$. Let $X \in \mathcal{T}$. Then $X \cap I_{\mathcal{T}}(X^c) \subseteq X \cap X^c = \emptyset$. If $X \cap Y = \emptyset$ for some $Y \in \mathcal{T}$, then $Y \subseteq X^c$ and hence $Y \subseteq I_{\mathcal{T}}(X^c)$. Thus, $X^* = I_{\mathcal{T}}(X^c)$.

(b) We show that $\varphi \colon \mathcal{T} \to \mathcal{L}_{\mathcal{T}}, X \mapsto X^c$ is a dual order-isomorphism. If $Y \in \mathcal{L}_{\mathcal{T}}$, then $Y^c \in \mathcal{T}$ and $\varphi(Y^c) = Y$, that is, Y is onto. If $X, Y \in \mathcal{T}$, then $X \subseteq Y$ is equivalent to $\varphi(Y) = Y^c \subseteq X^c = \varphi(X)$.

(c) If $X \subseteq U$, then

$$
\begin{aligned}
I_{\mathcal{T}}(X)^c &= \left(\bigcup \{Y \mid Y \in \mathcal{T} \text{ and } Y \subseteq X\} \right)^c \\
&= \bigcap \{Y^c \mid Y \in \mathcal{T} \text{ and } Y \subseteq X\} \\
&= \bigcap \{Y \mid Y \in \mathcal{L}_{\mathcal{T}} \text{ and } Y^c \subseteq X\} \\
&= \bigcap \{Y \mid Y \in \mathcal{L}_{\mathcal{T}} \text{ and } X^c \subseteq Y\} \\
&= C_{\mathcal{T}}(X^c). \qquad \square
\end{aligned}
$$

Example 54. Let us consider the topology $\mathcal{T}$ of Fig. 13. As we have noted, $\mathcal{T}$ is a pseudocomplemented distributive lattice such that $X^* = I_{\mathcal{T}}(X^c)$ for all $X \in \mathcal{T}$. For example, $\{a\}^* = I_{\mathcal{T}}(\{a\}^c) = I_{\mathcal{T}}(\{b, c\}) = \{c\}$.

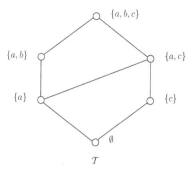

Fig. 13.

The *Kuratowski closure axioms* allow us to define a topology on U by means of an operator on U. An operator $C\colon \wp(U) \to \wp(U)$ is a *Kuratowski closure operator* if for any $X, Y \subseteq U$,

(K1) $X \subseteq C(X)$,
(K2) $C(C(X)) = C(X)$,
(K3) $C(X \cup Y) = C(X) \cup C(Y)$, and
(K4) $C(\emptyset) = \emptyset$.

It is obvious that if $C\colon \wp(U) \to \wp(U)$ is a Kuratowski closure operator, it is a closure operator in the sense of Section 5.1, because (K3) implies that if $X \subseteq Y$, then $C(Y) = C(X \cup Y) = C(X) \cup C(Y)$, that is, $C(X) \subseteq C(Y)$.

Proposition 55. *If $(U, \mathcal{T})$ is a topological space, then the operator $C_{\mathcal{T}}$ is a Kuratowski closure operator.*

Proof. Since $C_{\mathcal{T}}$ is a closure operator, it satisfies conditions (K1) and (K2). $C_{\mathcal{T}}$ is also order-preserving, which gives $C_{\mathcal{T}}(X) \cup C_{\mathcal{T}}(Y) \subseteq C_{\mathcal{T}}(X \cup Y)$. On the other hand, $X \cup Y \subseteq C_{\mathcal{T}}(X) \cup C_{\mathcal{T}}(Y) \in \mathcal{L}_{\mathcal{T}}$ implies $C_{\mathcal{T}}(X \cup Y) \subseteq C_{\mathcal{T}}(C_{\mathcal{T}}(X) \cup C_{\mathcal{T}}(Y)) = C_{\mathcal{T}}(X) \cup C_{\mathcal{T}}(Y)$. Hence, (K3) holds.

Because $U \in \mathcal{T}$ by definition, $\emptyset \in \mathcal{L}_{\mathcal{T}}$ and $C_{\mathcal{T}}(\emptyset) = \emptyset$. Thus, also (K4) is satisfied. □

By the previous lemma, each topology induces a Kuratowski closure operator. On the other hand, let $C\colon \wp(U) \to \wp(U)$ be a Kuratowski closure operator. Let us denote

$$\mathcal{T}_C = \{C(X)^c \mid X \subseteq U\}.$$

Then we can write the following proposition.

Proposition 56. *If $C\colon \wp(U) \to \wp(U)$ be a Kuratowski closure operator, then $\mathcal{T}_C$ is a topology on U.*

Proof. Since C is a closure operator, the family $\{C(X) \mid X \subseteq Y\}$ is closed under arbitrary intersections. This implies that the family $\mathcal{T}_C$ is closed under arbitrary unions. For all $X, Y \subseteq U$, $C(X)^c \cap C(Y)^c = (C(X) \cup C(Y))^c = C(X \cup Y)^c \in \mathcal{T}_C$, that is, $\mathcal{T}_C$ is closed under finite intersections.

Since $C(\emptyset) = \emptyset$ and $C(U) = U$, we have that $\emptyset = U^c = C(U)^c \in \mathcal{T}_C$ and $U = \emptyset^c = C(\emptyset)^c \in \mathcal{T}_C$. □

It should now be obvious that the correspondence between topological spaces and Kuratowski closure operators is bijective.

Let $(U, \mathcal{T})$ be a topological space. A family of sets $\mathcal{B} \subseteq \mathcal{T}$ is called a *base* for $\mathcal{T}$ if each member of $\mathcal{T}$ is the union of some members of $\mathcal{B}$. Because $\mathcal{T}$ is a complete lattice such that $\bigvee \mathcal{H} = \bigcup \mathcal{H}$ for all $\mathcal{H} \subseteq \mathcal{T}$, a base is simply a join-dense subset of $\mathcal{T}$.

If $X \subseteq Y$ and $Y \in \mathcal{T}$, then Y is called a *neighbourhood* of X. Further, any neighbourhood of the singleton set $\{x\}$ is called a *neighbourhood* of the point $x \in U$.

5.3 Alexandrov Spaces

A topology $\mathcal{T}$ on U is called an *Alexandrov topology* if the intersection of every family of open sets is also open. If $\mathcal{T}$ is an Alexandrov topology on U, the pair $(U, \mathcal{T})$ is called an *Alexandrov space*. Clearly, Alexandrov topologies are complete rings of sets, as usual topologies are just rings of sets closed under arbitrary unions.

Lemma 57. *If $(U, \mathcal{T})$ is an Alexandrov space, then*

$$C_\mathcal{T}\left(\bigcup \mathcal{H}\right) = \bigcup C_\mathcal{T}(\mathcal{H})$$

for all $\mathcal{H} \subseteq \mathcal{T}$.

Proof. Obviously, $\bigcup \{C_\mathcal{T}(X) \mid X \in \mathcal{H}\} \subseteq C_\mathcal{T}(\bigcup \mathcal{H})$ because $C_\mathcal{T}$ is order-preserving. On the other hand, $\bigcup \mathcal{H} \subseteq \bigcup \{C_\mathcal{T}(X) \mid X \in \mathcal{H}\} \in \mathcal{L}_\mathcal{T}$ since also $\mathcal{L}_\mathcal{T}$ is closed under arbitrary unions. Thus, $C_\mathcal{T}(\bigcup \mathcal{H}) \subseteq C_\mathcal{T}(\bigcup \{C_\mathcal{T}(X) \mid X \in \mathcal{H}\}) = \bigcup \{C_\mathcal{T}(X) \mid X \in \mathcal{H}\}$. □

By the previous lemma, each Alexandrov topology $\mathcal{T}$ on U defines a complete join-morphism $C_\mathcal{T} \colon \wp(U) \to \wp(U)$. We say that a closure operator is an *Alexandrov closure operator* if it satisfies

$$C\left(\bigcup \mathcal{H}\right) = \bigcup C(\mathcal{H})$$

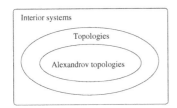

Fig. 14.

for all $\mathcal{H} \subseteq \mathcal{T}$, that is, C is a complete join-morphism. Trivially, each Alexandrov closure operator is a Kuratowski closure operator. Further, we know by Lemma 57 that the closure operator of an Alexandrov space is an Alexandrov closure operator, see Fig. 14.

Also the following lemma similar to Proposition 56 holds.

Lemma 58. *If C is an Alexandrov closure operator on U, then the family*

$$\mathcal{T}_C = \{C(X)^c \mid X \subseteq U\}$$

is an Alexandrov topology on U.

Proof. It suffices to show that $\mathcal{T}_C$ is closed under arbitrary intersections. For all $\mathcal{H} \subseteq \wp(U)$,

$$\bigcap C(\mathcal{H})^c = \left(\bigcup C(\mathcal{H})\right)^c = \left(C\left(\bigcup \mathcal{H}\right)\right)^c \in \mathcal{T}_C. \qquad \square$$

Because in an Alexandrov topology $\mathcal{T}$, the intersection of every family of open sets is open, each set $X \subseteq U$ has a smallest neighbourhood, denoted by $N_{\mathcal{T}}(X)$. Clearly,

$$N_{\mathcal{T}}(X) = \bigcap \{Y \in \mathcal{T} \mid X \subseteq Y\}.$$

It is clear that $N_{\mathcal{T}} \colon \wp(U) \to \wp(U)$ is also an Alexandrov closure operator and a complete join-morphism, because $\mathcal{T}$ is closed under arbitrary intersections and unions. Further, let us denote by $N_{\mathcal{T}}(x)$ the smallest neighbourhood of the point $x \in U$. Notice that we have already considered smallest neighbourhoods of points in Section 3.2. It is clear by Proposition 27 that for all $X \in \mathcal{T}$,

$$X = \bigcup \{N_{\mathcal{T}}(x) \mid x \in X\},$$

and that $\{N_{\mathcal{T}}(x) \mid x \in X\}$ is the smallest base of $\mathcal{T}$.

The next lemma characterizes Alexandrov spaces by means of neighbourhoods.

Lemma 59. *If $(U, \mathcal{T})$ is topological space, then the following assertions are equivalent.*

(a) *$\mathcal{T}$ is an Alexandrov topology.*
(b) *Every point $x \in U$ has a smallest neighbourhood.*

Proof. We have already shown that (a) implies (b). Suppose that (b) holds and let $\mathcal{H} \subseteq \mathcal{T}$. Then for all $x \in \bigcap \mathcal{H}$, $x \in N_{\mathcal{T}}(x) \subseteq \bigcap \mathcal{H}$, which implies $\bigcap \mathcal{H} \subseteq \bigcup \{N_{\mathcal{T}}(x) \mid x \in \bigcap \mathcal{H}\} \subseteq \bigcap \mathcal{H}$, and thus $\bigcap \mathcal{H} \in \mathcal{T}$. So, also (b) implies (a). $\qquad \square$

The following condition holds between the closures and the smallest neighbourhoods of singleton sets.

Lemma 60. *Let (U, T) be an Alexandrov space. Then for all $x, y \in U$,*

$$x \in C_T(\{y\}) \iff y \in N_T(\{x\}).$$

Proof. If $x \in C_T(\{y\})$, then

$$x \in \bigcap \{Y \mid Y \in \mathcal{L}_T \text{ and } y \in Y\} = \bigcap \{X^c \mid X \in T \text{ and } y \in X^c\}.$$

This is equivalent to the condition that for all $X \in T$, $y \in X^c$ implies $x \in X^c$, or equivalently, that for all $X \in T$, $x \in X$ implies $y \in X$. This gives $y \in N_T(\{x\})$. □

Because the open sets and the closed sets in an Alexandrov space T satisfy exactly the same axioms, they may be interchanged. So, instead of calling the elements in T open, we may call them closed, and analogously, we can call the elements of $\mathcal{L}_T$ open. Therefore, we get a new Alexandrov space

$$T^D = \{X^c \mid X \in T\}$$

which is called the *dual* of T. It is now trivial that

$$T^D = \mathcal{L}_T \quad \text{and} \quad T = \mathcal{L}_{T^D}.$$

This implies that for all $X \subseteq U$,

$$C_T(X) = N_{T^D}(X) \quad \text{and} \quad N_T(X) = C_{T^D}(X),$$

that is, the closure operator of an Alexandrov topology is the neighbourhood operator of its dual topology. Trivially, T is dually isomorphic to T^D justifying the name 'dual topology'.

Bibliographical Notes

Closure systems and operators are studied in the books [3,7,8,11,21]. Also an early paper by McKinsey and Tarski [39] deserves to be mentioned. Number of partitions and equivalences are considered, for instance, in [5]. A detailed study on topological spaces can be found in [35]. Alexandrov spaces were originally introduced in [1], where also most of the results of Section 5.3 can be found.

6   Fixpoints and Closure Operators on Ordered Sets

The topics considered in this section are:

6.1 Fixpoint Theorems
6.2 Closure Operators on Ordered Sets

6.1   Fixpoint Theorems

Given an ordered set P and a self-map f on P, an element $x \in P$ is called a *fixpoint* of f if $f(x) = x$. We denote by $\mathrm{Fix}(f)$ the set of all fixpoints of f. Recall that if $C \colon \wp(U) \to \wp(U)$ is a closure operator, then the set of its fixpoints $\mathrm{Fix}(C)$ is equal to the set $\mathcal{L}_C$ of its closed elements.

Theorem 61 (Knaster–Tarski Fixpoint Theorem). *If f is an order-preserving self-map on a complete lattice L, then*

$$\bigvee \{x \in L \mid x \leq f(x)\}$$

is the greatest fixpoint of f. Dually, f has a least fixpoint $\bigwedge \{x \in L \mid x \geq f(x)\}$.

Proof. Let $H = \{x \in L \mid x \leq f(x)\}$ and $\alpha = \bigvee H$. For all $x \in H$, we have $x \leq \alpha$ and so $x \leq f(x) \leq f(\alpha)$. This means that $f(\alpha)$ is an upper bound of H, from which we get $\alpha \leq f(\alpha)$. Because f is order-preserving, $f(\alpha) \leq f(f(\alpha))$. This means that $f(\alpha) \in H$ and hence $f(\alpha) \leq \alpha$. We have now shown that $f(\alpha) \in \mathrm{Fix}(f)$. If β is any fixpoint of f, then $\beta \in H$ implies $\beta \leq \alpha$.   □

By applying the previous theorem we can get the following.

Proposition 62. *If f is an order-preserving self-map on a complete lattice L, then $\mathrm{Fix}(f)$ is a complete lattice with respect to the order of L.*

Proof. Let $X \subseteq \mathrm{Fix}(f)$ and let Y be the set of the upper bounds of X in P. Then for all $x \in X$ and $y \in Y$, $x = f(x) \leq f(y)$ since f is order-preserving. This implies that $f(y) \in Y$ for all $y \in Y$. Let f_Y be the restriction of f to Y.

Clearly, Y is a complete lattice with respect to the order of L, because it is a complete sublattice of L. Thus, f_Y has a least fixpoint α by the Knaster–Tarski Fixpoint Theorem. Since $\alpha \in Y$, α is an upper bound of X, and if $\beta \in \mathrm{Fix}(f)$ is an upper bound of X, then $\beta \in Y$ and $\alpha \leq \beta$. Thus, $\bigvee X = \alpha$ in $\mathrm{Fix}(f)$. Since $\bigvee X$ exists in $\mathrm{Fix}(f)$ for all $X \subseteq \mathrm{Fix}(f)$, $\mathrm{Fix}(f)$ is a complete lattice by the dual of Theorem 14.   □

By the previous proposition, $\mathrm{Fix}(f)$ is always a complete lattice. Next we consider some special cases.

Proposition 63. *If f is an extensive and order-preserving self-map on a complete lattice L, then $\mathrm{Fix}(f)$ is a complete meet-sublattice of L.*

Proof. Let $S \subseteq \mathrm{Fix}(f)$. Because f is extensive, $\bigwedge S \leq f(\bigwedge S)$. For all $x \in S$, we have $\bigwedge S \leq x$ and $f(\bigwedge S) \leq f(x) = x$. Thus, also $f(\bigwedge S) \leq \bigwedge S$ holds and $\bigwedge S \in \mathrm{Fix}(f)$.   □

Let f be an extensive and order-preserving self-map on a complete lattice L. Since $\mathrm{Fix}(f)$ is closed under arbitrary meets in L, there exists a smallest fixpoint of f above any $x \in L$. Let us denote this element by $\overline{f}(x)$. Clearly,

$$\overline{f}(x) = \bigwedge \{\alpha \in \mathrm{Fix}(f) \mid x \leq \alpha\}.$$

We will study the properties of the map $\overline{f} \colon L \to L$ in Section 6.2.

Proposition 64. *If* $f: L \to L$ *is a complete join-morphism on a complete lattice* L, *then* $\mathrm{Fix}(f)$ *is a complete join-sublattice of* L.

Proof. Since f is a complete join-morphism, $f(\bigvee S) = \bigvee \{f(x) \mid x \in S\} = \bigvee \{x \mid x \in S\} = \bigvee S$. Hence, $\bigvee S \in \mathrm{Fix}(f)$.   □

Propositions 63 and 64 have the following corollary.

Corollary 65. *If* $f: L \to L$ *is an extensive complete join-morphism on a complete lattice* L, *then* $\mathrm{Fix}(f)$ *is a complete sublattice of* L.

Example 66. Let C be an Alexandrov closure operator on a set U. Since C is an extensive complete join-morphism on $\wp(U)$, the set of closed elements $\mathcal{L}_C$ is a complete sublattice of $\wp(U)$ – as we already know by Section 5.3.

Next we present a more concrete description of the smallest fixpoint of a complete join-morphism. If f is a self-map on an ordered set P, then we define for any integer $i \geq 0$, the i-fold composition $f^i(x)$ by $f^0(x) = x$ and $f^{i+1}(x) = f(f^i(x))$ for all $x \in P$.

Theorem 67 (Kleene's Fixpoint Theorem). *If* $f: L \to L$ *is a complete join-morphism on a complete lattice* L, *then*

$$\bigvee \{f^i(\bot) \mid i \geq 0\}$$

is the least fixpoint of f.

Proof. Let us denote $\alpha = \bigvee \{f^i(\bot) \mid i \geq 0\}$. Because f is a complete join-morphism,

$$\begin{aligned} f(\alpha) &= f\left(\bigvee \{f^i(\bot) \mid i \geq 0\} \right) \\ &= \bigvee \{f^{i+1}(\bot) \mid i \geq 0\} \\ &= \bigvee \{f^i(\bot) \mid i \geq 1\} \\ &= \bigvee \{f^i(\bot) \mid i \geq 0\} \\ &= \alpha. \end{aligned}$$

Thus, α is a fixpoint of f. If β is a fixpoint of f, then $f^i(\bot) \leq f^i(\beta) = \beta$ for all $i \geq 0$. Thus, $\alpha = \bigvee \{f^i(\bot) \mid i \geq 0\} \leq \beta$.   □

6.2   Closure Operators on Ordered Sets

In this section we consider closure operators on ordered sets and particularly on complete lattices. This generalizes the study carried out in Section 5.1.

For an ordered set P, a function $c: P \to P$ is called a *closure operator* on P, if for all $a, b \in P$,

(co1) $a \leq c(a)$ (extensive)
(co2) $c(c(a)) = c(a)$ (idempotent)
(co3) $a \leq b$ implies $c(a) \leq c(b)$ (order-preserving)

An element $a \in P$ is called *closed* if $c(a) = a$. *Interior operators* on ordered sets are defined dually.

In the next lemma we present some basic properties of closure operators.

Lemma 68. *If $c \colon P \to P$ is a closure operator on an ordered set P, then the following assertions hold.*

(a) *The set of c-closed elements is $c(P) = \{c(a) \mid a \in P\}$.*
(b) *For any $x \in P$, $c(x) = \bigwedge_P \{c(a) \mid x \le a\}$.*
(c) *If $S \subseteq c(P)$ and $\bigvee S$ exists in P, $\bigvee S$ exists in $c(P)$ and equals $c(\bigvee_P S)$.*
(d) *If $S \subseteq c(P)$ and $\bigwedge S$ exists in P, $\bigwedge S$ exists in $c(P)$ and equals $\bigwedge_P S$.*

Proof. (a) Assume that x is closed. Then $c(x) = x$ and so $x \in c(P)$. On the other hand, $c(c(x)) = c(x)$ for all $c(x) \in c(P)$.

(b) If $x \le a$, then $c(x) \le c(a)$, which shows that $c(x)$ is a lower bound of $\{c(a) \mid x \le a\}$. Since $c(x)$ itself is in $\{c(a) \mid x \le a\}$, this implies that $c(x) = \bigwedge_P \{c(a) \mid x \le a\}$.

(c) Suppose $S \subseteq c(P)$ and $\bigvee S$ exists in P. For all $x \in S$, $x \le \bigvee_P S \le c(\bigvee_P S) \in c(P)$. If $c(y)$ is an upper bound of S in $c(P)$, then $\bigvee_P S \le c(y)$ and $c(\bigvee_P S) \le c(c(y)) = c(y)$.

(d) Suppose $S \subseteq c(P)$ is such that $\bigwedge S$ exists in P. Then for all $x \in S$, $c(\bigwedge_P S) \le c(x) = x$ and $c(\bigwedge_P S) \le \bigwedge_P S$. Clearly, $c(\bigwedge_P S) \ge \bigwedge_P S$. Hence, $\bigwedge_P S \in c(P)$ which implies that the infimum of S in $c(P)$ is $\bigwedge_P S$. □

The previous lemma has the immediate consequence that if L is a lattice, then $c(L)$ is a lattice in which

$$a \vee b = c(a \vee_L b) \quad \text{and} \quad a \wedge b = a \wedge_L b$$

for all $a, b \in c(L)$. Similarly, if L is a complete lattice, then $c(L)$ is a complete lattice such that

$$\bigvee S = c\left(\bigvee_L S\right) \quad \text{and} \quad \bigwedge S = \bigwedge_L S$$

for all $S \subseteq c(L)$.

The map c is not always join-preserving. However, we can write the following.

Lemma 69. *If c is a closure operator on a complete lattice L, then for all $S \subseteq L$,*

$$c\left(\bigvee S\right) = c\left(\bigvee c(S)\right)$$

and especially for all $x, y \in L$,

$$c(x \vee y) = c(c(x) \vee c(y)).$$

Proof. It is clear that $c(\bigvee S) \le c(\bigvee c(S)) \le c(c(\bigvee S)) = c(\bigvee S)$. The proof for the rest is analogous. □

We have shown that in a complete lattice L, each closure operator c determines a complete meet-sublattice $c(L)$ of L. Also the opposite holds, as we see in the next lemma.

Lemma 70. *Let S be a complete meet-sublattice of a complete lattice L. Then the map*

$$x \mapsto \bigwedge \{z \in S \mid x \leq z\}$$

is a closure operator on L such that the set of its closed elements is S.

Proof. Let us denote $c(x) = \bigwedge \{z \in S \mid x \leq z\}$. It is clear that $x \leq c(x)$. If $x \leq y$, then $\{z \in S \mid x \leq z\} \supseteq \{z \in S \mid y \leq z\}$, which implies $c(x) = \bigwedge \{z \in S \mid x \leq z\} \leq \bigwedge \{z \in S \mid y \leq z\} = c(y)$. Since S is a complete meet-sublattice of L, $c(x) \in S$. This implies $c(x) = \bigwedge \{z \in S \mid c(x) \leq z\} = c(c(x))$.

If $y \in S$, then $c(y) = \bigwedge \{z \in S \mid y \leq z\} = y$, that is, y is c-closed. On the other hand, if y is c-closed, then $y = c(y) = \bigwedge \{z \in S \mid y \leq z\}$. Because S is a complete meet-sublattice of L, we have $y \in S$. □

If P is an ordered set, then we denote by $\mathrm{Clo}(P)$ the set of all closure operators on P. Because $\mathrm{Clo}(P) \subseteq P^P$, $\mathrm{Clo}(P)$ has an order inherited from P^P. Obviously, $x \mapsto x$ is the least element in $\mathrm{Clo}(P)$, and if P has a top element $\top$, then $x \mapsto \top$ is the greatest element in $\mathrm{Clo}(P)$.

Proposition 71. *If L is a complete lattice, then $\mathrm{Clo}(L)$ is a complete lattice with respect to the pointwise order.*

Proof. Suppose $\Phi \subseteq \mathrm{Clo}(L)$. We will show that $c = \bigwedge_{L^L} \Phi$ belongs to $\mathrm{Clo}(L)$.

For all $x \in L$ and $f \in \Phi$, $x \leq f(x)$. This implies $x \leq \bigwedge_{L^L} \{f(x) \mid f \in \Phi\} = c(x)$, that is, c is extensive. If $x \leq y$, then $f(x) \leq f(y)$ for all $f \in \Phi$, which implies $\bigwedge \{f(x) \mid f \in \Phi\} \leq f(y)$ for all $f \in \Phi$ and hence $c(x) = \bigwedge \{f(x) \mid f \in \Phi\} \leq \bigwedge \{f(y) \mid f \in \Phi\} = c(y)$.

It is clear that $c(x) \leq c(c(x))$ for all $x \in L$. Let $g \in \Phi$ and $x \in L$. Then $c(x) = \bigwedge \{f(x) \mid f \in \Phi\} \leq g(x)$. Because $c(c(x)) = \bigwedge \{f(c(x)) \mid f \in \Phi\} \leq g(c(x))$, we get $c(c(x)) \leq g(c(x)) \leq g(g(x)) = g(x)$. Hence, $c(c(x)) \leq \bigwedge \{f(x) \mid f \in \Phi\} = c(x)$. Thus, $c(c(x)) = c(x)$ and c is a closure operator. □

We showed in Proposition 63 that if f is an extensive and order-preserving self-map on a complete lattice L, then $\mathrm{Fix}(f)$ is closed under arbitrary meets. As before, we denote by

$$\overline{f}(x) = \bigwedge \{\alpha \in \mathrm{Fix}(f) \mid x \leq \alpha\}$$

the smallest fixpoint of f above x.

Proposition 72. *If f is an extensive and order-preserving self-map on a complete lattice L, then $\overline{f} \colon L \to L$ is the smallest closure operator above f.*

Proof. By Proposition 63 and Lemma 70, the map $\overline{f}$ is a closure operator. It is clear that $\overline{f}$ is above f with respect to the pointwise order, because $x \leq \overline{f}(x)$, and this implies $f(x) \leq f(\overline{f}(x)) = \overline{f}(x)$ for all $x \in L$. If c is a closure operator above f, then for all $x \in L$, $c(c(x)) = c(x)$ and $x \leq c(x) \in \mathrm{Fix}(f)$. This gives $\overline{f}(x) \leq c(x)$. □

In Proposition 71 we showed that if L is complete lattice, then $\mathrm{Clo}(L)$ is a complete lattice in which the meets are defined pointwise. Next we will describe joins in $\mathrm{Clo}(L)$. We need the following lemma.

Lemma 73. *If L is a complete lattice and $H \subseteq \mathrm{Clo}(L)$, then $\bigvee_{LL} \Phi$ is extensive and order-preserving for all $\Phi \subseteq \mathrm{Clo}(L)$.*

Proof. Suppose $\Phi \subseteq \mathrm{Clo}(L)$. Let us denote $c = \bigvee_{LL} \Phi$. For all $f \in \Phi$ and $x \in L$, $x \leq f(x) \leq \bigvee \{f(x) \mid f \in \Phi\} = c(x)$. If $x \leq y$, then for all $f \in \Phi$, $f(x) \leq f(y)$, which implies $c(x) = \left(\bigvee_{LL} \Phi\right)(x) \leq \left(\bigvee_{LL} \Phi\right)(y) = c(y)$. $\qquad\square$

Because $\bigvee_{LL} \Phi$ is extensive and order-preserving, by Proposition 72 the map $\overline{\bigvee_{LL} \Phi}$ is the smallest closure operator above $\bigvee_{LL} \Phi$. This implies that in the complete lattice $\mathrm{Clo}(L)$,

$$\bigvee \Phi = \overline{\bigvee_{LL} \Phi}.$$

Example 74. Let us consider the set $\mathbb{N}_\infty = \mathbb{N} \cup \{\infty\}$, in which the order relation $\leq$ is defined by

$$n \leq m \quad \Longleftrightarrow \quad n \leq m \text{ holds in } \mathbb{N} \text{ or } m = \infty.$$

It is clear that $\mathbb{N}_\infty$ is a complete lattice in which $\bigvee S = \max S$ for finite nonempty subsets S and $\bigvee S = \infty$ in case S is infinite. Furthermore, $\bigvee \emptyset = 0$.

The closure operators c_1 and c_2 are defined on $\mathbb{N}_\infty$ by

$$c_1(n) = \begin{cases} n+1 & \text{if } n \text{ is odd} \\ n & \text{if } n \text{ is even} \\ \infty & \text{if } n = \infty \end{cases}$$

and

$$c_2(n) = \begin{cases} n+1 & \text{if } n \text{ is even} \\ n & \text{if } n \text{ is odd} \\ \infty & \text{if } n = \infty. \end{cases}$$

The pointwise join f of c_1 and c_2 is the map

$$f(n) = \begin{cases} \infty & \text{if } n = \infty \\ n+1 & \text{otherwise,} \end{cases}$$

and f is obviously not a closure operator. In fact, the infinity ∞ is the only fixpoint of f, and hence the map $\overline{f} \colon x \mapsto \infty$ is the join of c_1 and c_2 in $\mathrm{Clo}(\mathbb{N}_\infty)$.

In Sections 5.2 and 5.3 we considered Kuratowski and Alexandrov closure operators of topological spaces. Here we study their counterparts on complete lattices. A closure operator c on a complete lattice L is called a *Kuratowski closure operator* if $c(\bot) = \bot$ and $c(a \vee b) = c(a) \vee c(b)$ for all $a, b \in L$. Further, if the closure operator c is also a complete join-morphism, it is an *Alexandrov closure operator*. Therefore, every Alexandrov closure operator is a Kuratowski closure operator. The corresponding *interior operators* are defined as dual concepts canonically.

We showed in Proposition 72 that for every extensive and order-preserving map f, there exists a smallest closure operator $\overline{f}$ above f. Next we show a similar result for extensive and bottom-preserving join-morphisms.

Proposition 75. *If* $f : L \to L$ *is an extensive and bottom-preserving join-morphism on a complete lattice* L, *then* $\overline{f}$ *is a Kuratowski closure operator on* L.

Proof. We know by Proposition 72 that $\overline{f}$ is a closure operator. It is also obvious that if $f(\bot) = \bot$, then $\overline{f}(\bot) = \bot$. We have to show that $\overline{f}(a) \vee \overline{f}(b) = \overline{f}(a \vee b)$. As in the proof of Proposition 64 we can show that $\overline{f}(a) \vee \overline{f}(b)$ is a fixpoint of f and clearly $\overline{f}(a) \vee \overline{f}(b) \leq \overline{f}(a \vee b)$. If α is a fixpoint of f above $a \vee b$, then $\overline{f}(a \vee b) \leq \overline{f}(\alpha) = \alpha$. Especially this implies that $\overline{f}(a \vee b) \leq \overline{f}(a) \vee \overline{f}(b)$. □

Next we present another description of $\overline{f}$ in case $f : P \to P$ is an extensive complete join-morphism on a complete lattice L by applying Kleene's Fixpoint Theorem.

Lemma 76. *If* $f : L \to L$ *is an extensive complete join-morphism on a complete lattice* L, *then*

$$\overline{f}(x) = \bigvee \{ f^i(x) \mid i \geq 0 \}$$

for all $x \in L$.

Proof. Let $x \in L$. Because f is extensive, $f([x)) \subseteq [x)$. Clearly, $[x)$ is also a complete sublattice of L with x as its bottom element. If f_x is the restriction of f into $[x)$, then f_x is a complete join-morphism on $[x)$, and the result follows from Kleene's Fixpoint Theorem. □

Finally, we show that to any extensive join-morphism we may attach a smallest Alexandrov closure operator which is above it.

Proposition 77. *If* $f : L \to L$ *is an extensive complete join-morphism on a complete lattice* L, *then* $\overline{f}$ *is an Alexandrov closure operator on* L.

Proof. We know that the map $\overline{f}$ is a Kuratowski closure operator. Because f is a complete join-morphism,

$$\overline{f}\left(\bigvee S\right) = \bigvee_{i \geq 0} f^i\left(\bigvee S\right) = \bigvee_{i \geq 0} \left(\bigvee f^i(S)\right)$$

for any $S \subseteq L$. Clearly, $f^i(x) \leq \overline{f}(x) \leq \bigvee \overline{f}(S)$ for all $i \geq 0$ and $x \in S$. Hence, $\bigvee f^i(S) \leq \bigvee \overline{f}(S)$ for all $i \geq 0$ and so $\overline{f}(\bigvee S) \leq \bigvee \overline{f}(S)$. Because $\overline{f}$ is order-preserving, we have $\bigvee \overline{f}(S) \leq \overline{f}(\bigvee S)$. □

Example 78. Let us return to the self-map $f : \mathbb{N}_\infty \to \mathbb{N}_\infty$ of Example 74 which is defined by

$$f(n) = \begin{cases} \infty & \text{if } n = \infty \\ n+1 & \text{otherwise.} \end{cases}$$

It it easy to observe that f is not a complete join-morphism, because f is not 0-preserving. Therefore, we have to do a slight modification to the definition of the map f. Let f be re-defined as follows

$$f(n) = \begin{cases} 0 & \text{if } n = 0 \\ \infty & \text{if } n = \infty \\ n+1 & \text{otherwise.} \end{cases}$$

Clearly, the new f is an extensive complete join-morphism.

For all $i \geq 0$,

$$f^i(n) = \begin{cases} 0 & \text{if } n = 0 \\ \infty & \text{if } n = \infty \\ n+i & \text{otherwise.} \end{cases}$$

Then,

- $\bigvee \{f^i(0) \mid i \geq 0\} = \bigvee \{0\} = 0$,
- $\bigvee \{f^i(\infty) \mid i \geq 0\} = \bigvee \{\infty\} = \infty$, and
- $\bigvee \{f^i(n) \mid i \geq 0\} = \bigvee \{n, n+1, \ldots\} = \infty$ for all $n \in \mathbb{N} - \{0\}$.

Thus, the map

$$\overline{f}(n) = \begin{cases} 0 & \text{if } n = 0 \\ \infty & \text{otherwise} \end{cases}$$

is the smallest closure operator above f by Lemma 76. By Proposition 77, this map is also an Alexandrov closure operator.

The previous considerations can now be summarized as follows.

- If f is extensive and order-preserving, then $\overline{f}$ is a closure operator.
- If f is extensive and bottom-preserving join-morphism, then $\overline{f}$ is a Kuratowski closure operator.
- If f is extensive complete join-morphism, then $\overline{f}$ is an Alexandrov closure operator.

Bibliographical Notes

Knaster–Tarski and Kleene's Fixpoint Theorems can be found, for instance, in [11]. In [54] it was originally proved that the set of fixpoints of an order-preserving map on a complete lattice forms a complete lattice. That fixpoints of an extensive and order-preserving map form complete meet-semilattices, and fixpoints of a complete join-morphism form complete join-sublattices originate in [19]. The basic results concerning closure operators on ordered sets appear in [7,8,11]. It should be noted that already in [56] it was proved that the pointwise meet of closure operators is a closure operator. Most of the results in the last part of Section 6.2 appear also in [25].

7   Galois Connections and Their Fixpoints

This section has the following two subsections

7.1 Galois Connections and Conjugate Functions
7.2 Fixpoints of Galois Connections

7.1   Galois Connections and Conjugate Functions

Galois connections can be found in various settings in mathematics and theoretical computer science. Galois connections are pairs of maps which enable us to move back and forth between two different structures. After an element is mapped to the other structure and back, a certain stability is reached in such a way that further mappings give the same results. Furthermore, the image sets of the maps forming the Galois connection are isomorphic.

For two ordered sets P and Q, a pair (f, g) of maps $f \colon P \to Q$ and $g \colon Q \to P$ is called a *Galois connection* between P and Q if for all $p \in P$ and $q \in Q$,

$$f(p) \leq q \iff p \leq g(q).$$

The map g is called the *adjoint* and f is called the *co-adjoint*. Moreover, if (f, g) is a Galois connection, then we say that f has an adjoint g, and g has a co-adjoint f.

It is clear by the definition that if (f, g) is a Galois connection between bounded ordered sets P and Q, then f is bottom-preserving and g is top-preserving, because $\bot_P \leq g(\bot_Q)$ implies $f(\bot_P) \leq \bot_Q$, and $f(\top_P) \leq \top_Q$ yields $\top_P \leq g(\top_Q)$. The following lemma gives a characterization of Galois connections.

Lemma 79. *Let $f \colon P \to Q$ and $g \colon Q \to P$ be maps between ordered sets P and Q. The pair (f, g) is a Galois connection if and only if*

(a) $p \leq g(f(p))$ *for all $p \in P$ and $f(g(q)) \leq q$ for all $q \in Q$;*
(b) *the maps f and g are order-preserving.*

Proof. Suppose (f, g) is a Galois connection between P and Q. If $p \in P$, then $f(p) \leq f(p)$ implies $p \leq g(f(p))$. Similarly, $g(q) \leq g(q)$ implies $f(g(q)) \leq q$. Thus, (a) holds. If $p_1 \leq p_2$ in P, then $p_1 \leq p_2 \leq g(f(p_2))$ by (a). Clearly, this is equivalent to $f(p_1) \leq f(p_2)$, which means that f is order-preserving. The other part of (b) can be proved analogously.

On the other hand, assume that (a) and (b) hold. Suppose that $f(p) \leq q$, where $p \in P$ and $q \in Q$. This implies $p \leq g(f(p)) \leq g(q)$. Similarly, if $p \leq g(q)$, then $f(p) \leq f(g(q)) \leq q$. Hence, (f, g) is a Galois connection.   □

Remark 80. In the literature can be found two ways to define Galois connections – the one adopted here, in which the maps are order-preserving, and the other, in which they are order-reversing. Originally, Galois connections were introduced with maps that reverse the order, but in this work we use the other form, since it is more natural for rough approximation operators. The two definitions are theoretically equivalent since if (f, g) is a Galois connection between P and Q of the other sense, then (f, g) is a Galois connection between P and Q^{op} of the other sense.

Example 81. (a) If L is a pseudocomplemented lattice, then the pair $(^*, ^*)$ is a Galois connection between L and its dual L^{op} by Lemma 42.

(b) Section 8 is devoted to Pawlak's information systems. They are pairs (U, A), where U is a set of objects, called the universe, and A is a set of attributes. For each attribute $a \in A$, a set V_a consisting of values of the attribute a is attached. Every attribute $a \in A$ can be viewed as a map $U \to V_a$ and the image $a(x)$ is the value of the attribute a for the object x. The fundamental idea in Pawlak's information systems is that each subset $B \subseteq A$ of attributes determines so-called indiscernibility relation $\text{ind}(B)$ which is defined so that two objects x and y of the universe U are B-indiscernible if their values for all attributes in the set B are equal. We will show that the pair of maps (ind, att), where ind is the map attaching to each subset of A its indiscernibility relation on U and att is the function giving for any binary relation R on U the greatest subset of A of which indiscernibility relation includes R, forms a Galois connection.

(c) In Section 9 we will study rough set approximations defined by means of an equivalence $\approx$ on a set U. For any subset X of U, let

$$X^{\blacktriangledown} = \{x \in U \mid [x]_{\approx} \subseteq X\} \quad \text{and} \quad X^{\blacktriangle} = \{x \in U \mid X \cap [x]_{\approx} \neq \emptyset\}.$$

The sets $X^{\blacktriangledown}$ and $X^{\blacktriangle}$ are called the *lower* and the *upper approximations* of X. We will show that the pair $(^{\blacktriangle}, ^{\blacktriangledown})$ is a Galois connection on $\wp(U)$.

The next proposition presents some basic properties of Galois connections.

Proposition 82. *Let (f, g) be a Galois connection between two ordered sets P and Q.*

(a) *The composition $f \circ g \circ f$ equals f and the composition $g \circ f \circ g$ equals g.*
(b) *The composition $g \circ f$ is a closure operator on P and the set of $g \circ f$-closed elements is $g(Q)$, that is, $(g \circ f)(P) = g(Q)$*
(c) *The composition $f \circ g$ is an interior operator on Q and the set of $f \circ g$-closed elements is $f(P)$, that is, $(f \circ g)(Q) = f(P)$.*
(d) *The image sets $f(P)$ and $g(Q)$ are order-isomorphic.*
(e) *The map f is a complete join-morphism and g is a complete meet-morphism.*
(f) *The maps f and g uniquely determine each other by the equations*

$$f(p) = \bigwedge \{q \in Q \mid p \leq g(q)\} \quad and \quad g(q) = \bigvee \{p \in P \mid f(p) \leq q\}.$$

Proof. (a) This follows easily from Lemma 79. If $p \in P$, then $p \leq g(f(p))$ implies $f(p) \leq f(g(f(p)))$ On the other hand, $f(g(f(p))) \leq f(p)$. The second part can be proved similarly.

(b) The composition $g \circ f : P \to P$ is extensive and order-preserving by Lemma 79. Further, $(g \circ f) \circ (g \circ f) = (g \circ f \circ g) \circ f = g \circ f$ by (a). If $p \in g(Q)$, then $p = g(q)$ for some $q \in Q$, implying $(g \circ f)(p) = (g \circ f \circ g)(q) = g(q) = p$. Conversely, if p is $g \circ f$-closed, then $p = g(f(p))$, that is, $p \in g(Q)$. Claim (c) can be proved as (b).

(d) If $g(q) \in g(Q)$, then $f(g(q)) \in f(P)$ and $g(f(g(q))) = g(q)$, that is, the map $f(p) \mapsto g(f(p))$ is onto $g(Q)$. If $f(p_1) \leq f(p_2)$, then trivially $g(f(p_1)) \leq$

$g(f(p_2))$. On the other hand, $g(f(p_1)) \leq g(f(p_2))$ implies $f(p_1) = f(g(f(p_1))) \leq f(g(f(p_2))) = f(p_2)$. Thus, $f(p) \mapsto g(f(p))$ is an order-isomorphism between $f(P)$ and $g(Q)$.

(e) Suppose that $S \subseteq P$ and $\bigvee S$ exists in P. Then for all $p \in S$, $f(p) \leq f(\bigvee S)$ and therefore $\bigvee \{f(p) \mid p \in S\} \leq f(\bigvee S)$. Further, if $x \in Q$ is an upper bound for $\{f(p) \mid p \in S\}$, then $p \leq g(f(p)) \leq g(x)$ for all $p \in S$, which gives $\bigvee S \leq g(x)$ and $f(\bigvee S) \leq f(g(x)) \leq x$. Thus, $f(\bigvee S) = \bigvee f(S)$.

(f) It is clear that $f(p)$ is a lower bound of $\{q \in Q \mid p \leq g(q)\} = \{q \in Q \mid f(p) \leq q\}$, and since $f(p)$ is itself in $\{q \in Q \mid p \leq g(q)\}$, the claim is obvious. The proof for the other part is analogous. $\qquad \square$

By the previous proposition, we can easily write also the following result.

Proposition 83. *Let (f, g) be a Galois connection between two complete lattices L and K.*

(a) *The ordered set $f(L)$ is a complete lattice such that for all $S \subseteq f(L)$,*

$$\bigvee S = \bigvee_K S \quad and \quad \bigwedge S = f(g(\bigwedge_K S)) = f(\bigwedge_L g(S)).$$

(b) *The ordered set $g(K)$ is a complete lattice such that for all $S \subseteq g(K)$,*

$$\bigvee S = g(f(\bigvee_L S)) = g(\bigvee_K f(S)) \quad and \quad \bigwedge S = \bigwedge_L S.$$

Proof. By Proposition 82, $g \circ f$ is a closure operator on L such that its set of closed elements is $g(K)$. Then by Lemma 68,

$$\bigvee S = g(f(\bigvee_L S)) = g(\bigvee_K f(S)) \quad and \quad \bigwedge S = \bigwedge_L S$$

for all $S \subseteq g(K)$. This proves (b), and (a) can be proved in a similar way. $\qquad \square$

The next result states when a map on a complete lattice induces a Galois connection.

Proposition 84. *Let L and K be complete lattices.*

(a) *A map $f : L \to K$ has an adjoint if and only if f is a complete join-morphism.*
(b) *A map $g : K \to L$ has a co-adjoint if and only if g is a complete meet-morphism.*

Proof. We prove (a); the proof for (b) is analogous. If f has an adjoint f^a, that is, (f, f^a) is a Galois connection, then by Proposition 82, f is a complete join-morphism.

On the other hand, if f is a complete join-morphism, then we define for all $q \in K$,

$$f^a(q) = \bigvee \{z \in L \mid f(z) \leq q\}.$$

Let $p \in L$ and $q \in K$. If $f(p) \leq q$, then trivially $p \leq \bigvee \{z \in L \mid f(z) \leq q\} = f^a(q)$. Conversely, if $p \leq f^a(q) = \bigvee \{z \in L \mid f(z) \leq q\}$, we obtain $f(p) \leq \bigvee \{f(z) \mid f(z) \leq q\} \leq q$. Thus, (f, f^a) is a Galois connection. $\qquad \square$

The previous proposition shows that for each complete join-morphism $f\colon L \to K$, the pair (f, f^a), where f^a is defined by

$$f^a(q) = \bigvee \{p \in L \mid f(p) \leq q\},$$

is a Galois connection. Similarly, for each complete meet-morphism $g\colon K \to L$, the pair (g^a, g) is a Galois connection, where for any $p \in L$,

$$g^a(p) = \bigwedge \{q \in K \mid p \leq g(q)\}.$$

Thus, in a way, complete join- or meet-morphisms and Galois connections can be regarded as the two sides of the same coin.

Example 85. The map $f\colon L \to K$ in Example 21 is a complete join-morphism between complete lattices. Therefore, it has an adjoint $g\colon K \to L$ which is depicted in Fig. 15.

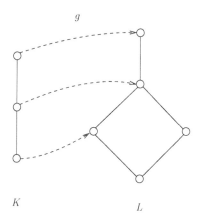

Fig. 15.

In the following we study conjugate functions on a Boolean lattice. We show, for example, that there is a correspondence between Galois connections and conjugate function pairs. Let f and g be two self-maps on a complete Boolean lattice B. We say that g is a *conjugate* of f, if for all $x, y \in B$, we have

$$x \wedge f(y) = 0 \iff y \wedge g(x) = 0.$$

It is clear that if g is a conjugate of f, then f is a conjugate of g. Therefore, in the following we shall say 'f and g are conjugate' instead of 'g is a conjugate of f'. Furthermore, each map has at most one conjugate. In particular, if a map f is the conjugate of itself, then we call f *self-conjugate*.

The next proposition characterizes self-maps on complete Boolean lattices having a conjugate. Note that this result holds only for complete Boolean lattices, not for complete lattices, as Proposition 84.

Proposition 86. *Let f be a self-map on a complete Boolean lattice. Then f has a conjugate if and only if f is a complete join-morphism.*

Proof. Let f be a self-map on a complete Boolean lattice. Suppose that f has a conjugate g. We show first that f is order-preserving which implies $f\left(\bigvee S\right) \geq \bigvee f(S)$ for all $S \subseteq B$. If $x \leq y$, then $f(x \vee y) = f(y)$ and thus $f(x \vee y) \wedge f(y)' = 0$. We obtain $g\left(f(y)'\right) \wedge (x \vee y) = 0$ and so $g\left(f(y)'\right) \wedge x = 0$. This gives $f(x) \wedge f(y)' = 0$, which is equivalent to $f(x) \leq f(y)$. Therefore, f is order-preserving.

If $S \subseteq B$, then for all $x \in S$, $f(x) \leq \bigvee f(S)$. This gives $f(x) \wedge \left(\bigvee f(S)\right)' = 0$ and $g\left(\left(\bigvee f(S)\right)'\right) \wedge x = 0$ for all $x \in S$. Hence, $g\left(\left(\bigvee f(S)\right)'\right) \wedge \bigvee S = 0$ and $f\left(\bigvee S\right) \wedge \left(\bigvee f(S)\right)' = 0$. This means that $f\left(\bigvee S\right) \leq \bigvee f(S)$.

Conversely, assume that f is a complete join-morphism. Then

$$g(y) = \left(\bigvee \{x \mid f(x) \leq y'\}\right)' = \bigwedge \{x' \mid f(x) \wedge y = 0\}$$

defines a function g on B. Now for all $x, y \in B$, $f(x) \wedge y = 0$ implies $g(y) \leq x'$, that is, $g(y) \wedge x = 0$. On the other hand,

$$f\left(g(y)'\right) = f\left(\bigvee \{x \mid f(x) \leq y'\}\right) = \bigvee \{f(x) \mid f(x) \leq y'\} \leq y'.$$

If x and y are such that $g(y) \wedge x = 0$, then $x \leq g(y)'$. Thus, $f(x) \leq f\left(g(y)'\right) \leq y'$, that is, $f(x) \wedge y = 0$. □

If f is a complete join-morphism, its conjugate also is necessarily a complete join-morphism. Next we show a natural conjugate pair for Alexandrov topologies.

Theorem 87. *If $\mathcal{T}$ is an Alexandrov topology on U, then the closure operator $C_{\mathcal{T}} \colon \wp(U) \to \wp(U)$ and the smallest neighbourhood operator $N_{\mathcal{T}} \colon \wp(U) \to \wp(U)$ are conjugate.*

Proof. We known that $C_{\mathcal{T}}(X) = \bigcup_{x \in X} C_{\mathcal{T}}(\{x\})$ and $N_{\mathcal{T}}(X) = \bigcup_{x \in X} N_{\mathcal{T}}(\{x\})$. Further, by Lemma 60,

$$x \in C_{\mathcal{T}}(\{y\}) \iff y \in N_{\mathcal{T}}(\{x\}).$$

This implies

$$
\begin{aligned}
X \cap N_{\mathcal{T}}(Y) \neq \emptyset &\iff (\exists x \in X)\,(\exists y \in Y)\, x \in N_{\mathcal{T}}(\{y\}) \\
&\iff (\exists y \in Y)\,(\exists x \in X)\, y \in C_{\mathcal{T}}(\{x\}) \\
&\iff Y \cap C_{\mathcal{T}}(X) \neq \emptyset.
\end{aligned}
$$

□

Let f and g be self-maps on a complete Boolean lattice B. We say that g is the *dual* of f, if for any $x \in B$,

$$f(x') = g(x)'.$$

For any function f, we denote by f^{∂} the dual of f. It is obvious that if $g = f^{\partial}$, then $f = g^{\partial}$. Therefore, we usually say 'f and g are dual' instead of 'g is the dual of f'. It is also obvious that each function f on B has exactly one dual. For example, if $\mathcal{T}$ is a topology on U, then by Proposition 53, the closure operator $C_{\mathcal{T}}: \wp(U) \to \wp(U)$ and the interior operator $I_{\mathcal{T}}: \wp(U) \to \wp(U)$ are dual.

The following lemma connects complete join-morphisms and complete meet-morphisms with each other through the notion of duality.

Lemma 88. *Let B be a complete Boolean lattice. A function f on B is a complete join-morphisms if and only if f^{∂} is a complete meet-morphism.*

Proof. Suppose that $f(\bigvee S) = \bigvee f(S)$ for all $S \subseteq B$. Then

$$
\begin{aligned}
f^{\partial}\left(\bigwedge S\right) &= \left(f\left(\left(\bigwedge S\right)'\right)\right)' \\
&= \left(f\left(\bigvee\{x' \mid x \in S\}\right)\right)' \\
&= \left(\bigvee\{f(x') \mid x \in S\}\right)' \\
&= \bigwedge\{f(x')' \mid x \in S\} \\
&= \bigwedge\{f^{\partial}(x) \mid x \in S\}.
\end{aligned}
$$

Thus, f^{∂} is a complete meet-morphism. The converse holds by duality. □

In Proposition 84 we showed that for complete lattices, each complete join-morphism induces a Galois connection, and a similar result holds also for complete meet-morphisms. By Proposition 86 we also know for complete Boolean lattices that each complete join-morphism has a conjugate. We end this subsection by presenting the result connecting conjugate maps and Galois connections.

Proposition 89. *Let B be a complete Boolean lattice.*

(a) *For any complete join-morphism f on B, its adjoint is the dual of the conjugate of f.*
(b) *For any complete meet-morphism g on B, its co-adjoint is the conjugate of the dual of g.*

Proof. We prove (a). Let $f: B \to B$ be a complete join-morphism. Then it has the adjoint $f^a: B \to B$ defined by $f^a(x) = \bigvee\{y \mid f(y) \le x\}$. On the other hand, the conjugate g of f is defined as $g(x) = \bigwedge\{y' \mid f(y) \le x'\}$. The dual of g is

$$
\begin{aligned}
g^{\partial}(x) &= g(x')' \\
&= \left(\bigwedge\{y' \mid f(y) \le x''\}\right)' \\
&= \bigvee\{y'' \mid f(y) \le x\} \\
&= \bigvee\{y \mid f(y) \le x\} \\
&= f^a(x).
\end{aligned}
$$

The proof for (b) is analogous. □

By the previous proposition, if B is a complete Boolean lattice and $f\colon B \to B$ is a complete join-morphism, then f has a unique adjoint and we may define the conjugate of f as the dual of the adjoint.

Note that if $\mathcal{T}$ is an Alexandrov topology on U, then by Theorem 87 the closure operator $C_{\mathcal{T}}\colon \wp(U) \to \wp(U)$ and the smallest neighbourhood operator $N_{\mathcal{T}}\colon \wp(U) \to \wp(U)$ are conjugate. Further, we know that $C_{\mathcal{T}}$ and $I_{\mathcal{T}}$ are dual. This implies the following corollary.

Corollary 90. *If $\mathcal{T}$ is an Alexandrov topology, then the pair $(N_{\mathcal{T}}, I_{\mathcal{T}})$ is a Galois connection.*

7.2 Fixpoints of Galois Connections

In this section we study fixpoints of Galois connections. Recall that if (f, g) is a Galois connection on an ordered set P, then f is a complete join-morphism, g is a complete meet-morphism, and thus, both f and g are order-preserving. Further, if P is bounded, then $f(\bot) = \bot$ and $g(\top) = \top$.

Lemma 91. *Let (f, g) be a Galois connection on an ordered set P. The following are equivalent:*

(a) $x \leq f(x)$ *for all* $x \in P$;
(b) $g(x) \leq x$ *for all* $x \in P$.

Proof. If (a) holds, then $g(x) \leq f(g(x)) \leq x$ for every $x \in P$. Conversely, if (b) holds, then for any $x \in P$, $x \leq g(f(x)) \leq f(x)$. □

Proposition 92. *Let (f, g) be a Galois connection on an ordered set P. If f is extensive, then f and g have exactly the same fixpoints.*

Proof. If x is a fixpoint of an extensive map f, then $f(x) \leq x$ implies $x \leq g(x) \leq x$. Conversely, if y is a fixpoint of g, then $y \leq g(y)$ and $f(y) \leq y \leq f(y)$. □

Example 93. That f is extensive is necessary for Proposition 92. Let us consider the interval $[0, 1]$ with its usual order. If

$$f(x) = \min\{1/2, x\},$$

then f is clearly a complete join-morphism which is not extensive. The map

$$g(x) = \bigvee\{y \in [0, 1] \mid \min\{1/2, y\} \leq x\}$$

is the adjoint of f, and we have $f(1/2) = 1/2$ and $g(1/2) = 1$. Hence, $1/2$ is a fixpoint of f, but not of g.

Corollary 94. *Let (f, g) be a Galois connection on an ordered set P. If f is extensive, then the following are equivalent:*

(a) x *is a fixpoint of* f;
(b) x *is a fixpoint of* g;
(c) $f(x) = g(x)$.

Proof. By Lemma 92, (a) and (b) are equivalent and they imply (c). If $f(x) = g(x)$ for some $x \in P$, then $x \leq f(x) = g(x) \leq x$, which means that (c) implies both (a) and (b).   □

Lemma 95. *If (f,g) is a Galois connection on an ordered set P, then the following are equivalent:*

(a) $f(f(x)) \leq f(x)$ *for all $x \in P$;*
(b) $g(x) \leq g(g(x))$ *for all $x \in P$.*

Proof. Assume that (a) holds. Now $f(g(x)) \leq x$ implies $f(f(g(x))) \leq x$, from which we get $f(g(x)) \leq g(x)$ and $g(x) \leq g(g(x))$. The other direction can be proved analogously.   □

Lemmas 91 and 95 have the following obvious corollary.

Corollary 96. *If (f,g) is a Galois connection on an ordered set P, then the following are equivalent:*

(a) *f is a closure operator;*
(b) *g is an interior operator.*

Notice that for a Galois connection (f,g), f is a complete join-morphism and g is a complete meet-morphism. Therefore, f is in fact an Alexandrov closure operator and g is an Alexandrov interior operator in Corollary 96. Note also that Proposition 92 and Corollary 96 imply that if f is a closure operator, then for all $x \in P$,

$$g(f(x)) = f(x) \text{ and } f(g(x)) = g(x).$$

We know that for an Alexandrov space $(U, \mathcal{T})$, the pair $(N_{\mathcal{T}}, I_{\mathcal{T}})$ is a Galois connection. Further, because $N_{\mathcal{T}}$ is a closure operator, we have

$$I_{\mathcal{T}}(N_{\mathcal{T}}(X)) = N_{\mathcal{T}}(X) \text{ and } N_{\mathcal{T}}(I_{\mathcal{T}}(X)) = I_{\mathcal{T}}(X)$$

for all $X \subseteq U$.

As before, we denote for a Galois connection (f,g) by $\mathrm{Fix}(f)$ the set of all fixpoints of f. Note that if f is extensive, then $\mathrm{Fix}(f)$ is also the set of fixpoints of g. In Section 6.1 we showed that if L is a complete lattice, then $\mathrm{Fix}(f)$ is a complete lattice and, in fact, a complete join-sublattice of L, since f is a complete join-morphism. Further, we know that if f is extensive, $\mathrm{Fix}(f)$ is a complete sublattice of L.

In general, as the following example shows, the set $\mathrm{Fix}(f)$ may not be closed under complementation even if L is a complete Boolean lattice and f is an extensive complete join-morphism.

Example 97. Let L be the 4-element complete Boolean lattice depicted in Fig. 11, and let the maps f and g be defined as follows:

$$f(0) = 0, \, f(a) = 1, \, f(b) = b, \, f(1) = 1,$$
$$g(0) = 0, \, g(a) = 0, \, g(b) = b, \, g(1) = 1.$$

Now f is extensive and (f,g) is a Galois connection. In this case, since $\mathrm{Fix}(f) = \{0, b, 1\}$, we have $b \in \mathrm{Fix}(f)$, but $b' = a \notin \mathrm{Fix}(f)$.

Proposition 98. *Let (f, g) be a Galois connection on a complete Boolean lattice B. If f is extensive and self-conjugate, then $\mathrm{Fix}(f)$ is a complete Boolean sublattice of B.*

Proof. By Corollary 65, $\mathrm{Fix}(f)$ is a complete sublattice of B. Since f is self-conjugate, its adjoint g is equal to the dual f^{∂} of f by Proposition 89. If $x \in \mathrm{Fix}(f)$, then x' is a fixpoint of f^{∂} because $f^{\partial}(x') = f(x)' = x'$. Now, $x' \geq f(f^{\partial}(x')) = f(x')$. Since f is extensive, also $x' \leq f(x')$ holds. So, $x' \in \mathrm{Fix}(f)$. $\qquad\square$

If (f, g) is a Galois connection on a complete lattice L such that f is extensive, then by Lemma 76,

$$\overline{f}(x) = \bigvee \{f^i(x) \mid i \geq 0\}$$

is the least fixpoint of f above x. Further, $\overline{f}$ is an Alexandrov closure operator by Proposition 77. To complete this section, let us assign

$$\Gamma = \{x' \in B \mid x \in \mathrm{Fix}(f)\},$$

where B is a complete Boolean lattice. Now Γ can be seen as the set of 'closed sets of an Alexandrov topology'.

Proposition 99. *Let (f, g) be a Galois connection on a complete Boolean lattice B. If f is extensive, then $\Gamma = \{x \in B \mid x = g^{\partial}(x)\}$, that is, Γ consists of the fixpoints of the conjugate of f.*

Proof. $\Gamma = \{x' \mid x \in \mathrm{Fix}(f)\} = \{x' \mid x = g(x)\} = \{x \mid x' = g(x')\} = \{x \mid x' = g^{\partial}(x)'\} = \{x \mid x = g^{\partial}(x)\}$. $\qquad\square$

As we have noted, for an Alexandrov topology $\mathcal{T}$, the pair $(N_{\mathcal{T}}, I_{\mathcal{T}})$ is a Galois connection, and $C_{\mathcal{T}}$ is the conjugate of $N_{\mathcal{T}}$. Now, $N_{\mathcal{T}}$ and $I_{\mathcal{T}}$ have the same fixpoints, and the fixpoints of $C_{\mathcal{T}}$, that is, the closed sets of $\mathcal{T}$, are the complements of the fixpoints of $N_{\mathcal{T}}$ and $I_{\mathcal{T}}$. Naturally, the fixpoints of $N_{\mathcal{T}}$ and $I_{\mathcal{T}}$ are the open sets of $\mathcal{T}$.

Example 100. Let us consider rough approximations introduced in Example 81(c). As we have already noted, the pair $(^{\blacktriangle}, ^{\blacktriangledown})$ is a Galois connection on $\wp(U)$. Furthermore, because $^{\blacktriangle}$ and $^{\blacktriangledown}$ will be shown to be dual, the map $^{\blacktriangle}$ is self-conjugate by Proposition 89. We will also show that $^{\blacktriangle}$ is an Alexandrov closure operator and $^{\blacktriangledown}$ is an Alexandrov interior operator. Furthermore,

$$X^{\blacktriangle\blacktriangledown} = X^{\blacktriangle} \quad \text{and} \quad X^{\blacktriangledown\blacktriangle} = X^{\blacktriangledown}$$

for all $X \subseteq U$.

Let us denote

$$\mathrm{Def} = \{X \subseteq U \mid X^{\blacktriangle} = X\},$$

that is, Def consists of the fixpoints of $^{\blacktriangle}$. The sets in Def are called definable and they are important because $X^{\blacktriangle}$ can be interpreted as a set of elements possibly belonging to X. Therefore, definable sets are 'closed with respect to possibility'. Since $\blacktriangle$ is an extensive and self-dual map, Def is a complete field of sets by Proposition 98. Further, by Corollary 94,

$$X = X^{\blacktriangle} \iff X = X^{\blacktriangledown} \iff X^{\blacktriangledown} = X^{\blacktriangle}$$

for all $X \subseteq U$. In Section 9 we give a more detailed study on rough approximations.

Bibliographical Notes

Many results concerning Galois connections can be found in [7,11], and [15,41] are early papers studying Galois connections on lattices. The paper [14] provides the rudiments of the theory of Galois connections together with many examples and applications. The definition of the conjugate of a self-map on complete Boolean lattices appeared in [34] and characterization of maps which have a conjugate can be found there. Self-conjugate functions on Boolean algebras were considered in [53]. Subsection 7.2 is based on a section of the article [31].

8   Information Systems

In this section we consider the following topics:

8.1 Armstrong Systems on Ordered Sets
8.2 Indiscernibility in Information Systems
8.3 Independent Attribute Sets and Reducts
8.4 Other Types of Information Relations

8.1   Armstrong Systems on Ordered Sets

In relational databases the notion of functional dependencies is essential. As we will see in the sequel, dependency relations play an important role in Pawlak's information systems, also. A *functional dependency*, denoted by $X \to Y$, between two attribute sets X and Y of, for example, a database table, specifies that in every row the values corresponding to the attributes in Y are uniquely determined by the values of the attributes in X. For example, the social security number uniquely determines a name, denoted by SSN $\to$ NAME. Armstrong axioms are a set of rules that enable us to infer all functional dependencies that hold on a relational database. Here we study Armstrong systems and dependencies in a more general setting of ordered sets.

Let P be an ordered set and let F be a set of ordered pairs $a \to b$, where $a, b \in P$. We say that F is an *Armstrong system* on P if the following (modified) *Armstrong axioms* hold for all $x, y, z \in P$:

(AS1) $x \geq y$ implies $x \to y \in F$;
(AS2) $x \to y \in F$ and $y \to z \in F$ imply $x \to z \in F$;
(AS3) the set $\{y \mid x \to y \in F\}$ has a greatest element x^{+}.

Usually, we write $x \to y \in F$ simply as $x \to y$ and say that y is *dependent* on x. Note that x^+ is just the greatest element dependent on x and

$$x^+ = \bigvee \{y \mid x \to y\}.$$

Let P be an ordered set. We denote by $\mathrm{Arm}(P)$ the set of all Armstrong systems on P. The set $\mathrm{Arm}(P)$ can be ordered with the usual set inclusion relation. The ordered set $(\mathrm{Arm}(P), \subseteq)$ has $\{a \to a \mid a \in P\}$ as its least element and if P has a greatest element $\top$, then $\mathrm{Arm}(P)$ has a greatest element $\{a \to b \mid a, b \in P\}$.

Proposition 101. *Let P be an ordered set.*

(a) *If F is an Armstrong system, then the map $x \mapsto x^+$ is a closure operator on P.*
(b) *If c is a closure operator on P, then the set $\{a \to b \mid c(a) \geq c(b)\}$ is an Armstrong system on P.*

Proof. Let F be an Armstrong system. We show that the map $x \mapsto x^+$ satisfies conditions (co1)–(co3). Since $x \leq x$, we obtain $x \to x$ and $x \leq x^+$ by (AS1) and (AS3). If $y \leq x$, then $x \to y$ by (AS1). The fact $y \to y^+$ implies $x \to y^+$ by (AS2) and so $y^+ \leq x^+$. Since $x \to x^+$ and $x^+ \to x^{++}$, we get $x \to x^{++}$ and $x^{++} \leq x^+$. Because $x^+ \to x^+$, we have $x^+ \leq x^{++}$. Hence, $x^+ = x^{++}$.

Let c be a closure operator. We show that the set $\{a \to b \mid c(a) \geq c(b)\}$ satisfies the Armstrong axioms (AS1)–(AS3). If $x \geq y$, then $c(x) \geq c(y)$ by (co3), and thus $x \to y$. If $x \to y$ and $y \to z$, then $c(x) \geq c(y) \geq c(z)$ and $x \to z$. Because $c(x) = c(c(x))$ by (co2), we get $x \to c(x)$. If $x \to y$, then $y \leq c(y) \leq c(x)$ by (co1). Hence, $c(x)$ is the greatest element dependent on x. $\square$

We can write the following useful lemma.

Lemma 102. *If F is an Armstrong system on an ordered set P, then the following conditions are equivalent for all $x, y \in P$:*

(a) $x \to y$;
(b) $x^+ \geq y^+$.

Proof. If $x \to y$, then $y \leq x^+$ by (AS3). Because $^+ : P \to P$ is a closure operator, we have $y^+ \leq x^{++} = x^+$. Thus, (a) implies (b).

If $x^+ \geq y^+$, then $y \leq y^+ \leq x^+$, which implies $x^+ \to y$ by (AS1). Since $x \to x^+$, we get $x \to y$ because $\to$ is transitive by (AS2). $\square$

For complete lattices, we can also write the following lemma.

Lemma 103. *If F is an Armstrong system on a complete lattice L, then $\to$ is completely join-compatible, that is, if $x_i \to y_i$ for all $i \in I$, then $\bigvee_{i \in I} x_i \to \bigvee_{i \in I} y_i$.*

Proof. Suppose that $x_i \to y_i$ for all $i \in I$. Then $y_i \leq y_i^+ \leq x_i^+ \leq (\bigvee_{i \in I} x_i)^+$ and $\bigvee_{i \in I} y_i \leq (\bigvee_{i \in I} x_i)^+$. This implies $(\bigvee_{i \in I} y_i)^+ \leq (\bigvee_{i \in I} x_i)^{++} = (\bigvee_{i \in I} x_i)^+$, that is, $\bigvee_{i \in I} x_i \to \bigvee_{i \in I} y_i$. □

We know by Proposition 71 that if L is a complete lattice, then $\mathrm{Clo}(L)$ is a complete lattice with respect to the pointwise order. Next we point out that for any ordered set P, the ordered sets of Armstrong systems and closure operators on P are isomorphic, which implies that $\mathrm{Arm}(L)$ is a complete lattice if L is a complete lattice.

Proposition 104. *If P is an ordered set, then $\mathrm{Clo}(P) \cong \mathrm{Arm}(P)$.*

Proof. For a closure operator c, we denote by F_c the induced Armstrong system. We show that $c_1 \leq c_2$ if and only if $F_{c_1} \subseteq F_{c_2}$ for all $c_1, c_2 \in \mathrm{Clo}(P)$, and that the map $\mathrm{Clo}(P) \to \mathrm{Arm}(P), c \mapsto F_c$, is onto.

Suppose that $c_1 \leq c_2$. If $x \to y \in F_{c_1}$, then $c_1(y) \leq c_1(x)$. This implies $y \leq c_1(y) \leq c_1(x) \leq c_2(x)$ and $c_2(y) \leq c_2(c_2(x)) = c_2(x)$. Thus, $x \to y \in F_{c_2}$ and so $F_{c_1} \subseteq F_{c_2}$. Conversely, assume that $F_{c_1} \subseteq F_{c_2}$. Because $x \to c_1(x) \in F_{c_1} \subseteq F_{c_2}$ for all $x \in P$, we obtain $c_1(x) \leq c_2(x)$. Hence, $c_1 \leq c_2$ in P^P.

Let $F \in \mathrm{Arm}(P)$. It is clear that $F = F_{(c_F)}$ since

$$x \to y \in F \iff c_F(y) \leq c_F(x) \iff x \to y \in F_{(c_F)}$$

for all $x, y \in P$. Thus, the map $c \mapsto F_c$ is onto $\mathrm{Arm}(P)$. □

We can also write the following observation.

Lemma 105. *If L is a complete lattice and $\mathcal{F} \subseteq \mathrm{Arm}(L)$, then $\bigcap \mathcal{F}$ is an Armstrong system on L.*

Proof. Let $\mathcal{F} \subseteq \mathrm{Arm}(L)$. It is clear that $\bigcap \mathcal{F}$ satisfies condition (AS1).

If $x \to y$ and $y \to z$ are in $\bigcap \mathcal{F}$, then $x \to y$ and $y \to z$ belong to F for all $F \in \mathcal{F}$. This implies that $x \to z \in F$ for all $F \in \mathcal{F}$. Hence, $x \to z \in \bigcap \mathcal{F}$ and (AS2) holds.

Let us write $c = \bigwedge_{L^L} \{c_F \mid F \in \mathcal{F}\}$, where c_F denotes the closure operator corresponding the Armstrong system F. Clearly, c is a closure operator. Let $x \in L$. Since $c(x) \leq c_F(x)$ and $x \to c_F(x) \in F$ for all $F \in \mathcal{F}$, we get $x \to c(x) \in F$ for all $F \in \mathcal{F}$, and hence $x \to c(x) \in \bigcap \mathcal{F}$. Moreover, if $x \to y \in \bigcap \mathcal{F}$, then $y \leq c_F(x)$ for all $F \in \mathcal{F}$. Hence, $y \leq \bigwedge_L \{c_F(x) \mid F \in \mathcal{F}\} = c(x)$. This means that $c(x)$ is the greatest element in the set $\{y \mid x \to y \in \bigcap \mathcal{F}\}$. □

Proposition 104 and Lemma 105 imply that if L is complete lattice, then $\mathrm{Arm}(L)$ is a complete lattice in which

$$\bigwedge \mathcal{F} = \bigcap \mathcal{F} \qquad \text{and} \qquad \bigvee \mathcal{F} = F_c,$$

where F_c is the Armstrong system determined by $c = \bigvee \{c_F \mid F \in \mathcal{F}\}$ formed in $\mathrm{Clo}(L)$.

Next we show how each Galois connection (f, g) between two complete lattices L and K induces an Armstrong system on L such that $x^+ = (g \circ f)(x)$ for any $x \in L$. This is done by defining a set F_f of ordered pairs of elements of L by

$$F_f = \{x \to y \mid f(x) \geq f(y)\}.$$

Theorem 106. *If (f, g) is a Galois connection between two complete lattices L and K, then F_f is an Armstrong system on L such that for all $x \in L$,*

$$x^+ = (g \circ f)(x).$$

Further, $L^+ = g(K) \cong f(L)$.

Proof. If $x \geq y$, then $f(x) \geq f(y)$, that is, $x \to y \in F_f$ and (AS1) holds. If $x \to y \in F_f$ and $y \to z \in F_f$, then $f(x) \geq f(y) \geq f(z)$, which gives $x \to z \in F_f$ and also (AS2) is satisfied.

For (AS3), let us denote

$$x^+ = \bigvee \{y \mid x \to y \in F_f\}.$$

We show first that $x \to x^+ \in F_f$. Indeed,

$$
\begin{aligned}
f(x^+) &= f\left(\bigvee\{y \mid x \to y \in F_f\}\right) \\
&= f\left(\bigvee\{y \mid f(x) \geq f(y)\}\right) \\
&= \bigvee\{f(y) \mid f(x) \geq f(y)\} \\
&\leq f(x),
\end{aligned}
$$

that is, $x \to x^+ \in F_f$. If $x \to y \in F_f$, then trivially $y \leq \bigvee\{y \mid x \to y \in F_f\} = x^+$.

Let $x \in L$. Then

$$
\begin{aligned}
g(f(x)) &= \bigvee\{y \mid y \leq g(f(x))\} \\
&= \bigvee\{y \mid f(y) \leq f(x))\} \\
&= \bigvee\{y \mid x \to y \in F_f\} \\
&= x^+.
\end{aligned}
$$

Finally, it is known by Proposition 82 that $f(L)$ and $g(K)$ are order-isomorphic and that the set of $g \circ f$-closed elements is $g(K)$. $\qquad\square$

We end this section by considering dense sets of Armstrong systems. Let P be an ordered set and $S \subseteq P$. We define a set F_S of ordered pairs of elements of P by

$$F_S = \{x \to y \mid (\forall z \in S)\, x \leq z \Longrightarrow y \leq z\}.$$

It turns out that in complete lattices every subset determines an Armstrong system, as we see in the following proposition.

Proposition 107. *If L is a complete lattice and $S \subseteq L$, then F_S is an Armstrong system on L such that*

$$x^+ = \bigwedge \{z \in S \mid x \leq z\}.$$

Proof. We will show that F_S satisfies conditions (AS1)–(AS3).

(AS1) Assume that $a \geq b$. Let $z \in S$. If $a \leq z$, then obviously $b \leq z$. Thus, $a \to b \in F_S$.

(AS2) Suppose that $a \to b \in F_S$ and $b \to c \in F_S$. Let $z \in S$. If $a \leq z$, then $b \leq z$. But this implies that also $c \leq z$ holds. Hence, $a \to c \in F_S$.

(AS3) We show that $x^+ = \bigwedge \{z \in S \mid x \leq z\}$. If $x \leq z$ for some $z \in S$, then $z \in \{z \in S \mid x \leq z\}$ and $\bigwedge \{z \in S \mid x \leq z\} \leq z$. This means $x \to \bigwedge \{z \in S \mid x \leq z\} \in F_S$. If $x \to y \in F_S$, then $\{z \in S \mid x \leq z\} \subseteq \{z \in S \mid y \leq z\}$ giving $\bigwedge \{z \in S \mid x \leq z\} \geq \bigwedge \{z \in S \mid y \leq z\} \geq y$. $\square$

Let F be an Armstrong system on a complete lattice L. We say that that a subset S of L is *dense* for F if $F_S = F$. It is clear that if S is dense for F, the following conditions are equivalent for all $x, y \in L$:

(ds1) $x \to y$;
(ds2) $x^+ \geq y^+$;
(ds3) $(\forall z \in S)\, x \leq z \Longrightarrow y \leq z$.

The following proposition connects dense sets of Armstrong systems and meet-dense subsets of the corresponding ordered set of $^+$-closed elements.

Proposition 108. *Let F be an Armstrong system on a complete lattice L. Then a subset $S \subseteq L$ is dense for F if and only if S is a meet-dense subset of L^+.*

Proof. Suppose that $S \subseteq L$ is dense for F. Hence,

$$x^+ = \bigwedge \{z \in S \mid x \leq z\}$$

for all $x \in L$. Then, S is a meet-dense subset of L^+.

Conversely, let S be a meet-dense subset of L^+. Assume that $x \to y \in F$. If $x \leq z$ for some $z \in S$, then $y \leq y^+ \leq x^+ \leq z^+ = z$, that is, $x \to y \in F_S$. On the other hand, if $x \to y \in F_S$, then $\{z \in S \mid x \leq z\} \subseteq \{z \in S \mid y \leq z\}$. Clearly, for all $a \in L$ and $z \in S$, $a \leq z$ if and only if $a^+ \leq z$. Thus, we get $\{z \in S \mid x^+ \leq z\} \subseteq \{z \in S \mid y^+ \leq z\}$. Because S is a meet-dense subset of L^+, $x^+ = \bigwedge \{z \in S \mid x^+ \leq z\} \geq \bigwedge \{z \in S \mid y^+ \leq z\} = y^+$ by the dual of Lemma 23, which gives $x \to y \in F$ by Lemma 102. $\square$

Notice that if (f, g) is a Galois connection between complete lattices L and K, then $g(K)$ is the greatest F_f-dense set.

8.2 Indiscernibility in Information Systems

In this section we study information systems introduced by Pawlak. An *information system* is a pair (U, A), where U is a set of *objects*, called the *universe*, and A is a set of *attributes*. Each attribute $a \in A$ is a map $a \colon U \to V_a$, where each V_a consists of *values* the attribute a can have.

Example 109. An information system (U, A) in which the sets U and A are finite can be represented by a table. The rows of the table are labeled by the objects and the columns by the attributes of the system. In the intersection of the row labeled by an object x and the column labeled by an attribute a we find the value $a(x)$.

Let us define an information system (U, A) such that the object set U consists of four persons, say 1, 2, 3, and 4, and the attribute set A has the attributes GENDER, MOTHER TONGUE, DEGREE, and POSITION. The corresponding value sets are

- $V_{\text{GENDER}} = \{\text{Male, Female}\}$,
- $V_{\text{MOTHER TONGUE}} = \{\text{Finnish, Swedish}\}$,
- $V_{\text{DEGREE}} = \{\text{MSc, PhD}\}$,
- $V_{\text{POSITION}} = \{\text{Assistant, Lecturer, Professor}\}$,

and the values of attributes are defined as in Table 1. We will return to this information system several times.

Table 1. A simple information system

U	GENDER	MOTHER TONGUE	DEGREE	POSITION
1	Male	Swedish	PhD	Professor
2	Male	Finnish	MSc	Assistant
3	Male	Finnish	PhD	Assistant
4	Female	Finnish	PhD	Lecturer

Let (U, A) be an information system. For any $B \subseteq A$, we can define on U the *indiscernibility relation* $\text{ind}(B)$ of B by setting:

$$(x, y) \in \text{ind}(B) \iff (\forall a \in B)\, a(x) = a(y).$$

If $(x, y) \in \text{ind(B)}$, then objects x and y are said to be *B-indiscernible*. The relation $\text{ind}(B)$ is clearly an equivalence, and the partition corresponding to $\text{ind}(B)$ can be viewed as a classification of objects, in which the equivalence classes of $\text{ind}(B)$ consist of objects which have exactly the same B-values. It can be seen easily from the definition that

$$\text{ind}(\emptyset) = U \times U$$

and

$$(\forall B, C \subseteq A)\, B \subseteq C \implies \text{ind}(B) \supseteq \text{ind}(C).$$

In the sequel, we denote $\text{ind}(\{a\})$ simply by $\text{ind}(a)$ for any $a \in B$.

Example 110. Let us consider the information system (U, A) of Example 109. Let us denote the attributes GENDER, MOTHER TONGUE, DEGREE, and POSITION simply by the letters a, b, c, and d, respectively. All indiscernibility relations determined by subsets of A are presented in Fig. 16. Notice that sets of attributes are denoted simply by sequences of their elements.

It is easy to observe that

$$\text{ind}(d) = \text{ind}(ab) = \text{ind}(ad) = \text{ind}(bd) = \text{ind}(abd)$$

and

$$\text{ind}(cd) = \text{ind}(abc) = \text{ind}(acd) = \text{ind}(bcd) = \text{ind}(A).$$

We begin our study of properties of indiscernibility relations with the following lemma.

Lemma 111. *Let (U, A) be an information system. If $\{B_i\}_{i \in I}$ is a family of subsets of A, then*

$$\bigcap_{i \in I} \text{ind}(B_i) = \text{ind}\Big(\bigcup_{i \in I} B_i\Big).$$

Proof. Let $x, y \in U$. Then

$$(x, y) \in \bigcap_{i \in I} \text{ind}(B_i) \iff (\forall i \in I)\, (x, y) \in \text{ind}(B_i)$$

$$\iff (\forall i \in I)(\forall a \in B_i)\, a(x) = a(y)$$

$$\iff (\forall a \in \bigcup_{i \in I} B_i)\, a(x) = a(y)$$

$$\iff (x, y) \in \text{ind}\Big(\bigcup_{i \in I} B_i\Big). \qquad \square$$

Let us denote by $\text{Rel}(U)^{\text{op}}$ the dual $(\text{Rel}(U), \supseteq)$ of the complete lattice $(\text{Rel}(U), \subseteq)$. By Lemma 111, the map

$$\text{ind} : \wp(A) \to \text{Rel}(U)^{\text{op}}$$

is a complete join-morphism, because the join in $\text{Rel}(U)^{\text{op}}$ is the intersection of relations. Then, by Proposition 84, the complete join-morphism $\text{ind} : \wp(A) \to \text{Rel}(U)^{\text{op}}$ has an adjoint denoted by

$$\text{att} : \text{Rel}(U)^{\text{op}} \to \wp(A).$$

By Proposition 82, the adjoint of ind is defined for all $R \in \text{Rel}(U)$ by

$$\text{att}(R) = \bigcup\{B \in \wp(A) \mid \text{ind}(B) \supseteq R\}.$$

It is easy to see that the adjoint can be written also in a simpler form:

$$\text{att}(R) = \{a \in A \mid R \subseteq \text{ind}(a)\}.$$

It is now clear that the pair (ind, att) is a Galois connection between complete lattices $\wp(A)$ and $\text{Rel}(U)^{\text{op}}$. For each $R \in \text{Rel}(U)$, the attribute set $\text{att}(R)$ can be considered as the greatest set of attributes of which indiscernibility relation contains R.

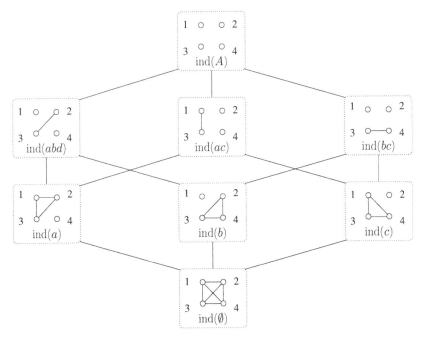

Fig. 16.

By the properties of Galois connections, the map $\mathrm{att}\colon \mathrm{Rel}(U)^{\mathrm{op}} \to \wp(A)$ is a complete meet-morphism, which means that for all $\{R_i\}_{i \in I}$,

$$\mathrm{att}\Big(\bigcup_{i \in I} R_i\Big) = \bigcap_{i \in I} \mathrm{att}(B_i),$$

Furthermore, $R_1 \subseteq R_2$ implies $\mathrm{att}(R_2) \subseteq \mathrm{att}(R_1)$.

Because $(\mathrm{ind}, \mathrm{att})$ is a Galois connection between $\wp(A)$ and $\mathrm{Rel}(U)^{\mathrm{op}}$, we may, as in Section 8.1, define a dependency relation $\to$ on $\wp(A)$ by setting

$$B \to C \iff \mathrm{ind}(B) \subseteq \mathrm{ind}(C).$$

This means that if $B \to C$ and two objects have the same B-values, they then have also the same C-values. This can be interpreted so that the values of C-attributes of objects are determined by their values of B-attributes.

By Theorem 106, we may determine for each $B \subseteq A$ the greatest set B^+ dependent on B, which can be written in several ways:

$$\begin{aligned}
B^+ &= \mathrm{att}(\mathrm{ind}(B)) \\
&= \{a \in A \mid \mathrm{ind}(B) \subseteq \mathrm{ind}(a)\} \\
&= \{a \in A \mid B \to a\},
\end{aligned}$$

where $B \to a$ means $B \to \{a\}$. Theorem 106 also gives that

$$\wp(A)^+ = \mathrm{att}(\mathrm{Rel}(U)) \cong \mathrm{ind}(\wp(A))^{\mathrm{op}}.$$

Since $^+ \colon \wp(A) \to \wp(A)$ is a closure operator on A, $\wp(A)^+$ is a closure system and a complete lattice such that

$$\bigwedge \mathcal{H} = \bigcap \mathcal{H} \qquad \text{and} \qquad \bigvee \mathcal{H} = \left(\bigcup \mathcal{H}\right)^+.$$

Furthermore, $\wp(A)^+$ is order-isomorphic to $\mathrm{ind}(\wp(A))^{\mathrm{op}}$ and the isomorphism is simply $B^+ \mapsto \mathrm{ind}(B)$. By Proposition 83, $\mathrm{ind}(\wp(A))^{\mathrm{op}}$ is a complete lattice such that

$$\bigvee_{i \in I} \mathrm{ind}(B_i) = \bigcap_{i \in I} \mathrm{ind}(B_i) = \mathrm{ind}\left(\bigcup_{i \in I} B_i\right)$$

and

$$\bigwedge_{i \in I} \mathrm{ind}(B_i) = \mathrm{ind}\left(\mathrm{att}\left(\bigcup_{i \in I} \mathrm{ind}(B_i)\right)\right) = \mathrm{ind}\left(\bigcap_{i \in I} \mathrm{att}\left(\mathrm{ind}(B_i)\right)\right) = \mathrm{ind}\left(\bigcap_{i \in I} B_i^+\right).$$

Finally, by Proposition 82, the mapping

$$R \mapsto \mathrm{ind}(\mathrm{att}(R))$$

is an interior operator on $\mathrm{Rel}(U)^{\mathrm{op}}$. Thus, it is a closure operator on $\mathrm{Rel}(U)$. It maps every relation R to the smallest indiscernibility relation containing R expressible by means of attributes in A. Note that $R \mapsto \mathrm{ind}(\mathrm{att}(R))$ is usually not the operator $^E \colon \mathrm{Rel}(U) \to \mathrm{Rel}(U)$ studied in Section 5.1.

Example 112. Let us consider the information system of Example 109. The complete lattice $\wp(A)^+ = \mathrm{att}(\mathrm{Rel}(U))$ is presented in Fig. 17. Note that $\wp(A)^+$ is isomorphic to the ordered set $\mathrm{ind}(\wp(A))^{\mathrm{op}}$ depicted in Fig. 16.

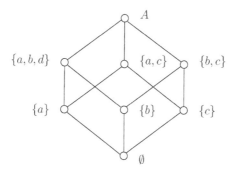

Fig. 17.

We end this subsection by considering dense families for information systems. Let $\mathcal{H}$ be family of subsets of A. We define a set of pairs $F_{\mathcal{H}}$ of ordered pairs of elements of $\wp(A)$ by by setting

$$F_{\mathcal{H}} = \{X \to Y \mid (\forall Z \in \mathcal{H})\, X \subseteq Z \Longrightarrow Y \subseteq Z\}.$$

We say that a family $\mathcal{H} \subseteq \wp(A)$ is *dense* in (U, A) if $F_{\mathcal{H}}$ is the set of dependencies that hold in (U, A). It is clear by Proposition 108 that $\mathcal{H}$ is dense in (U, A) if and only if it is $\bigcap$-dense subfamily of $\wp(A)^+$. Furthermore, the following conditions are equivalent for any dense family $\mathcal{H}$:

(DS1) $X \to Y$;
(DS2) $X^+ \supseteq Y^+$;
(DS3) $(\forall Z \in \mathcal{H}) \, X \subseteq Z \Longrightarrow Y \subseteq Z$.

Example 113. Let us consider the closure system $\wp(U)^+$ of Fig. 17. The family $\{\{a, c\}, \{b, c\}, \{a, b, d\}\}$ is the smallest $\bigcap$-dense subset of $\wp(A)^+$. This means that it is also the smallest dense family in (U, A).

Now, for example, $\{a, b\} \to d$, because $\{a, b\}$ is included only in $\{a, b, d\}$ and also $\{d\} \subseteq \{a, b, d\}$.

It is clear that there exists always at least one dense family in (U, A), namely $\wp(A)^+$. However, construction of $\wp(A)^+$ directly from the information system can be rather tedious. Next we present a simpler way to find dense families. For an information system (U, A), the *indiscernibility matrix* $\mathbf{m} = (m_{xy})$ of (U, A) is defined so that for all $x, y \in U$,

$$m_{xy} = \{a \in A \mid a(x) = a(y)\}.$$

Thus, the entry m_{xy} consists of those attributes $a \in A$ for which x and y are indiscernible. The next lemma is trivial.

Lemma 114. *If (U, A) is an information system and $\mathbf{m} = (m_{xy})$ is its indiscernibility matrix, then for all $B \subseteq A$ and $x, y \in U$,*

$$(x, y) \in \mathrm{ind}(B) \text{ if and only if } B \subseteq m_{xy}.$$

Proposition 115. *If (U, A) is an information system and $\mathbf{m} = (m_{xy})$ is its indiscernibility matrix, then the family $\{m_{xy} \mid x, y \in U\}$ is dense in (U, A).*

Proof. (a) Let us denote $\mathcal{H} = \{m_{xy} \mid x, y \in U\}$. If $B \to C$, then $\mathrm{ind}(B) \subseteq \mathrm{ind}(C)$. This implies by Lemma 114 that if $B \subseteq m_{xy}$, then $(x, y) \in \mathrm{ind}(B) \subseteq \mathrm{ind}(C)$, which is equivalent to $C \subseteq m_{xy}$. Thus, $(B, C) \in F_{\mathcal{H}}$.

Conversely, if $(B, C) \in F_{\mathcal{H}}$, then $(x, y) \in \mathrm{ind}(B)$ yields $B \subseteq m_{xy}$. This is equivalent to $C \subseteq m_{xy}$ and $(x, y) \in \mathrm{ind}(C)$. Thus, $\mathrm{ind}(B) \subseteq \mathrm{ind}(C)$ and $B \to C$. $\square$

The next example shows how we may easily obtain dense families in information systems by applying the previous proposition.

Example 116. Let us consider the information system of Example 109. Its indiscernibility matrix is the 4×4-matrix

$$\mathbf{m} = \begin{pmatrix} A & a & ac & c \\ a & A & abd & b \\ ac & abd & A & bc \\ c & b & bc & A \end{pmatrix}$$

By Proposition 115, the family

$$\mathcal{H} = \{\{a\}, \{b\}, \{c\}, \{a, c\}, \{b, c\}, \{a, b, d\}, A\}$$

consisting of the entries of $\mathbf{m}$ is dense in (U, A). This is also clear since $\mathcal{H} \subseteq \wp(A)^+$ includes the family $\{\{a, c\}, \{b, c\}, \{a, b, d\}\}$ which is known to be the smallest dense set.

8.3   Independent Attribute Sets and Reducts

In this section we consider independent attribute sets and attribute reduction. Let (U, A) be an information system. Then, a subset B of A is *dependent* if there exists $a \in B$ such that $(B - \{a\}) \to a$, that is, the values of the attribute a for objects in U are determined by the values of the other attributes in B. If B is not dependent, it is *independent*. Obviously, every subset of an independent set is independent and supersets of dependent sets are dependent. Note that for a dependent set B, there always exists a proper subset C of B such that $C \to B$.

Example 117. Let us consider the information system of Example 109. It can be easily checked that the independent subsets of A are $\emptyset$, $\{a\}$, $\{b\}$, $\{c\}$, $\{d\}$, $\{a, b\}$, $\{a, c\}$, $\{b, c\}$, $\{c, d\}$, and $\{a, b, c\}$.

A subset C of a set $B \subseteq A$ is called a *reduct* of B if $C \to B$ and C is independent. Note that if C is a reduct of B, then $\mathrm{ind}(B) = \mathrm{ind}(C)$. Reducing an attribute set is of practical importance because one can get the same classification accuracy with a smaller set of attributes. On the other words, the attributes not belonging to a reduct are superfluous with respect to classification of objects of the universe. The next proposition gives some important properties of reducts.

Proposition 118. *In an information system (U, A), the following assertions hold for all $B, C \subseteq A$.*

(i) *The set C is a reduct of B if and only if C is a minimal subset of B such that $C \to B$.*

(ii) *If C is a reduct of B, then C is a maximal independent subset of B.*

Proof. (a) Let C be a reduct of B. Suppose that there exists a proper subset D of C on which B is dependent. Then $D \to B$ and $B \to C$, which imply $D \to C$, that is, C is dependent, a contradiction!

Conversely, let C be a minimal subset of B such that $C \to B$. Assume that C is dependent. Then there exists a proper subset D of C such that $D \to C$. But since $\to$ is transitive, $D \to B$ holds, a contradiction!

(b) Let C be a reduct of B. Suppose that there exists an independent set D such that $C \subset D \subseteq B$. Then $C \to B$ and $B \to D$ imply that there exists a proper subset C of D such that $C \to D$. Hence, D is dependent, a contradiction!   □

Example 119. All independent subsets of A in the information system (U, A) are listed in Example 117. Clearly, $\{c, d\}$ and $\{a, b, c\}$ are the maximal independent subsets of A, which means that these two sets are the potential reducts of A by Proposition 118.

Clearly, $\{c,d\}^+ = \{a,b,c\}^+ = A^+ = A$ implying $\{c,d\} \to A$ and $\{a,b,c\} \to A$. Hence, $\{c,d\}$ and $\{a,b,c\}$ really are the reducts of A.

Next we present a characterization of reducts by means of dense families.

Proposition 120. *Let $\mathcal{H}$ be a dense family of an information system (U,A) and $B \subseteq A$. Then C is a reduct of B if and only if C is a minimal set with respect to the property of containing an element from each nonempty difference $B - Z$, where $Z \in \mathcal{H}$.*

Proof. Suppose that C is a reduct of B. Then $C \subseteq B$, $C \to B$, and $B \subseteq Z$ for all $Z \in \mathcal{H}$ such that $C \subseteq Z$. This means that for all $Z \in \mathcal{H}$, $B \not\subseteq Z$ implies $C \not\subseteq Z$, which is clearly equivalent to the condition of C containing an element from each nonempty difference $B - Z$, where $x \in \mathcal{H}$. Since C is a reduct, C must clearly be a minimal set satisfying this condition.

Conversely, let C be a minimal subset of A with respect to the property of containing an element from each nonempty difference $B - Z$, $Z \in \mathcal{H}$. First we show that C is a subset of B. If $C \not\subseteq B$, then $B \cap C \subset C$ and $(B \cap C) \cap (B - Z) = C \cap (B - Z) \neq \emptyset$ whenever $B - Z \neq \emptyset$, a contradiction! Thus, $C \subseteq B$. It is clear that $C \subseteq Z$ implies $B \subseteq Z$ for all $Z \in \mathcal{H}$. Thus, $C \to B$. It is also obvious that C must now be a minimal set on which B is dependent, which means that C is a reduct of B. $\qquad\square$

Example 121. Let us find the reducts of the set A in the information system (U,A) of Example 109 by applying Proposition 120. As we have noted, the family $\mathcal{H} = \{\{a,c\}, \{b,c\}, \{a,b,d\}\}$ is dense in (U,A). The differences $A - Z$, where $Z \in \mathcal{H}$, are

$$A - \{a,c\} = \{b,d\}, \quad A - \{b,c\} = \{a,d\}, \quad \text{and} \quad A - \{a,b,d\} = \{c\}.$$

Clearly, $\{c,d\}$ and $\{a,b,c\}$ are the minimal sets containing an element from each (nonempty) difference and therefore these sets are the reducts of A.

In this section we have considered independent subsets and reducts of attribute sets, and characterized the reducts by means of dense families. However, we have not considered the problem whether a subset has any reducts at all.

In fact, in information systems in which the universe and the attribute set are infinite, it may happen that there exist sets that do not possess any reducts, as the following example shows.

Example 122. Let (U,A) be an information system such that $U = \mathbb{N}$ and $A = \{a_i \mid i \in \mathbb{N}\}$, where $\mathbb{N} = \{0,1,2,\ldots\}$. For each $i \in \mathbb{N}$, the attribute a_i is defined by

$$a_i(j) = \begin{cases} j & \text{if } j \leq i \\ i+1 & \text{otherwise.} \end{cases}$$

The equivalence classes of $\text{ind}(a_i)$ are

$$\{0\}, \{1\}, \{2\}, \ldots, \{i\}, \{i+1, i+2, i+3, \ldots\}.$$

It is easy to see that the only independent subsets of A are the singletons $\{a_i\}$. So, these set are the only potential reducts of A. Obviously, $\{a_i\} \nrightarrow A$ for all $i \in \mathbb{N}$. Thus, the set A has no reducts. In fact, it is easy to see that every infinite subset of A has no reducts.

However, for finite universes we can write the following lemma.

Lemma 123. *Let (U, A) be an information system such that U is finite. Then for any subset B of A, there exists a finite subset $F \subseteq B$ such that $F \to B$.*

Proof. Let us consider the family

$$\mathcal{F} = \{\mathrm{ind}(F) \mid F \text{ is a finite subset of } B\}.$$

Because U is finite, the family $\mathcal{F} \subseteq U \times U$ is finite and it is nonempty since $\mathrm{ind}(\emptyset) \in \mathcal{F}$. This implies by Lemma 10 that $\mathcal{F}$ must with respect to inclusion have a minimal element $\mathrm{ind}(F)$ for some finite $F \subseteq B$. For all $a \in B$, $\mathrm{ind}(F \cup \{a\}) \in \mathcal{F}$ and trivially $\mathrm{ind}(F \cup \{a\}) \subseteq \mathrm{ind}(F)$. Because $\mathrm{ind}(F)$ is minimal, we have $\mathrm{ind}(F) = \mathrm{ind}(F \cup \{a\}) = \mathrm{ind}(F) \cap \mathrm{ind}(a)$ for all $a \in B$. Thus, $\mathrm{ind}(F) \subseteq \mathrm{ind}(a)$ and $F \to a$ for all $a \in B$. Because $\to$ is completely $\bigcup$-compatible by Lemma 103, we have $F \to B$. $\qquad\square$

The next proposition guarantees that in information systems in which the universe or the attribute set is finite, each subset of attributes has reducts.

Proposition 124. *If (U, A) is an information system such that U or A is finite, then every subset of A has at least one reduct.*

Proof. (a) By Lemma 123, if U is finite, then for every subset B of A there exists a finite subset F of B such that $F \to B$. Trivially, if A finite, then every subset B of A is finite, and we may choose $F = B$.

Let us assume that $F = \{a_1, a_2, \dots, a_n\}$. We define inductively the sets $F_0, F_1, \dots, F_n$ as follows:

$$F_0 = F \qquad \text{and} \qquad F_i = \begin{cases} F_{i-1} - \{a_i\} & \text{if } (F_{i-1} - \{a_i\}) \to a_i, \\ F_{i-1} & \text{otherwise.} \end{cases}$$

Obviously, $F_n \subseteq F_{n-1} \subseteq \cdots \subseteq F_1 \subseteq F_0 \subseteq B$ and $\mathrm{ind}(B) = \mathrm{ind}(F_0) = \mathrm{ind}(F_1) = \cdots = \mathrm{ind}(F_n)$. Let us assume that F_n is dependent, that is, $(F_n - \{a_i\}) \to a_i$ for some i such that $a_i \in F_n$. Because $F_n \subseteq F_{i-1}$, we have $(F_{i-1} - \{a_i\}) \to (F_n - \{a_i\}) \to a_i$. This implies $a_i \notin F_i$ and thus $a_i \notin F_n$, a contradiction! $\qquad\square$

Note that even if all value sets of the attributes in the system are finite, this does not guarantee that every subset has a reduct – consider Example 122, for instance. There, $V_{a_i} = \{0, 1, 2, \dots, i, i+1\}$ for each $i \in \mathbb{N}$.

Example 125. Let us consider the information system of Table 2 describing weather conditions and suitable actions. The table can be interpreted so that the attributes CLOUD AMOUNT, HARD WIND, TEMPERATURE, and RAIN can be viewed as *condition attributes* and the attributes READ A BOOK and PLAY TENNIS can be considered as *decision attributes*. It is clear that in the table the decision attributes are dependent on the condition attributes. Therefore, we may define the problem of finding all minimal subsets of the condition attributes on which the decision attributes are dependent.

Table 2. An information system describing weather conditions and suitable actions

	CLOUD AMOUNT	HARD WIND	TEMPERATURE	RAIN	READ A BOOK	PLAY TENNIS
1	cloudy	no	warm	yes	yes	no
2	clear	no	warm	no	no	yes
3	half cloudy	no	hot	no	yes	yes
4	cloudy	yes	cold	snow	yes	no
5	half cloudy	no	warm	no	no	yes

Next we present a proposition which characterizes in terms of dense families the minimal subsets C of B which satisfy $C \to D$ for a dependency $B \to D$. Note that Proposition 120 characterizing the reducts of a set B can be obtained as a special case of this proposition. We may obtain reducts of B simply by finding for the dependency $B \to B$ all minimal subsets C of B such that $C \to B$ holds.

Proposition 126. *Let $\mathcal{H}$ be a dense family of an information system (U, A). If $B \to D$, then C is a minimal subset of B which satisfies $C \to D$ if and only if C is a minimal set with respect to the property of containing an element from each difference $B - Z$, where $Z \in \mathcal{H}$ and satisfies $D - Z \neq \emptyset$.*

Proof. Suppose that $B \to D$ and let C be a minimal subset of B such that $C \to D$. This means that for all $Z \in \mathcal{H}$, $C \subseteq Z$ implies $D \subseteq Z$. Since $C \subseteq B$, the assumption $C \to D$ gives that $C \cap (B - Z) = (C \cap B) - Z = C - Z \neq \emptyset$ for all $Z \in \mathcal{H}$ such that $D - Z \neq \emptyset$. Assume that there exists $X \subset C$ which satisfies $X \cap (B - Z) \neq \emptyset$ for all $Z \in \mathcal{H}$ such that $D - Z \neq \emptyset$. However, $X \subset C \subseteq B$ implies $X - Z = X \cap (B - Z) \neq \emptyset$ for all $Z \in \mathcal{H}$ which satisfy $D - Z \neq \emptyset$. Thus $X \to D$, a contradiction!

Conversely, assume that $B \to D$ and let C be a minimal set containing an element from each difference $B - Z$ such that $Z \in \mathcal{H}$ satisfies $D - Z \neq \emptyset$. If $C \not\subseteq B$, then $C \cap B \subset C$ and $(C \cap B) \cap (B - Z) = C \cap (B - Z) \neq \emptyset$ for all $Z \in \mathcal{H}$ such that $D - Z \neq \emptyset$, a contradiction! Hence, $C \subseteq B$. This implies $C - Z = C \cap (B - Z) \neq \emptyset$ for all $Z \in \mathcal{H}$ which satisfy $D - Z \neq \emptyset$. This means $C \to D$. Suppose that there exists $X \subset C$ such that $X \to D$. Then $X \subset C \subseteq B$ implies $X \cap (B - Z) = X - Z \neq \emptyset$ whenever $D - Z \neq \emptyset$, a contradiction!   □

Example 127. Let us return to the information system of Example 125. Let us denote the attributes CLOUD AMOUNT, HARD WIND, TEMPERATURE, RAIN, READ A BOOK, and PLAY TENNIS simply by the letters a, b, c, d, e, and f, respectively. As we have noted, $\{a, b, c, d\} \to \{e, f\}$ and our task is to find all minimal subsets C of $\{a, b, c, d\}$ such that $C \to \{e, f\}$. This can be done by applying Proposition 126.

The indiscernibility matrix of the information system is the 5×5-matrix

$$\mathbf{m} = \begin{pmatrix} A & bc & be & aef & bc \\ bc & A & bdf & \emptyset & bcdef \\ be & bdf & A & e & abdf \\ aef & \emptyset & e & A & \emptyset \\ bc & bcdef & abdf & \emptyset & A \end{pmatrix}$$

By Proposition 115, the family

$$\mathcal{H} = \{\emptyset, \{e\}, \{b, c\}, \{b, e\}, \{a, e, f\}, \{b, d, f\}, \{a, b, d, f\}, \{b, c, d, e, f\}, A\}$$

consisting of the entries of $\mathbf{m}$ is dense in (U, A).

The differences $\{e, f\} - Z$, $Z \in \mathcal{H}$ are nonempty for $Z = \emptyset$, $\{e\}$, $\{b, c\}$, $\{b, e\}$, $\{b, d, f\}$, $\{a, b, d, f\}$. The corresponding differences $\{a, b, c, d\} - Z$ are the following:

(i) $\{a, b, c, d\} - \emptyset = \{a, b, c, d\}$;
(ii) $\{a, b, c, d\} - \{e\} = \{a, b, c, d\}$;
(iii) $\{a, b, c, d\} - \{b, c\} = \{a, d\}$;
(iv) $\{a, b, c, d\} - \{b, e\} = \{a, c, d\}$;
(v) $\{a, b, c, d\} - \{b, d, f\} = \{a, c\}$;
(vi) $\{a, b, c, d\} - \{a, b, d, f\} = \{c\}$.

Next we must find all such minimal sets that contain an element from all differences (i)–(vi). Because $\{c\}$ and $\{a, d\}$ are the minimal differences, it suffices to consider them only. Clearly, $\{a, c\}$ and $\{c, d\}$ are the minimal sets which contain an element from all of these differences. So, $\{$CLOUD AMOUNT, TEMPERATURE$\}$ and $\{$TEMPERATURE, RAIN$\}$ are the minimal subsets C of the four condition attributes which satisfy $C \to \{$READ A BOOK, PLAY TENNIS$\}$.

Similarly, we can see that in this example $\{$CLOUD AMOUNT$\}$ and $\{$RAIN$\}$ are the minimal subsets of condition attributes on which $\{$PLAY TENNIS$\}$ is dependent.

8.4 Other Types of Information Relations

In this section we shortly deal with many-valued information systems. Formally, a *many-valued information system* is a pair (U, A), where U is a set of objects and A is a set of attributes such that each attribute is a map $a : U \to \wp(V_a)$. This means that attributes attach sets of values to objects. For example, if a is the attribute 'knowledge of languages' and a person denoted by x knows English and Finnish, then $a(x) = \{$English, Finnish$\}$. Notice that it is possible that some person, say y, does not speak any of the languages belonging V_a. Then, $a(y) = \emptyset$.

In many-valued information systems it is possible to define different types of relations reflecting either indistinguishability or distinguishability of objects. In general, these kinds of relations are referred to as *information relations*.

Now we may present a list of *indistinguishability* relations derived from an information system (U, A). For every $B \subseteq A$, we define the relations:

- *indiscernibility*: $(x, y) \in \text{ind}(B) \iff (\forall a \in B)\, a(x) = a(y)$
- *similarity*: $(x, y) \in \text{sim}(B) \iff (\forall a \in B)\, a(x) \cap a(y) \neq \emptyset$
- *inclusion*: $(x, y) \in \text{inc}(B) \iff (\forall a \in B)\, a(x) \subseteq a(y)$

If a is again the attribute 'knowledge of languages', then two objects x and y are similar with respect to a if they have a common language. Further, x is $\text{inc}(a)$-related to y, if y can speak any language x speaks; and possibly some other languages as well.

The following collection of *distinguishability* relations may be defined in an information system for any $B \subseteq A$:

- *diversity*: $(x, y) \in \text{div}(B) \iff (\forall a \in B)\, a(x) \neq a(y)$
- *orthogonality*: $(x, y) \in \text{ort}(B) \iff (\forall a \in B)\, a(x) \cap a(y) = \emptyset$
- *negative similarity*: $(x, y) \in \text{nsim}(B) \iff (\forall a \in B)\, a(x) \cap a(y)^c \neq \emptyset$

For example, two objects are diverse by their knowledge of languages, if they cannot speak exactly the same languages, and they are orthogonal, if they do not share a common language. Further, they are negatively similar, if x can speak a language that y cannot.

Information relations are similar in the sense that two objects are in a certain relation with respect to an attribute set B if their values of the B-attributes are in a specified relation. Next we introduce the general notions of preimage relations and information frames allowing us to study the general properties of information relations of many-valued information systems.

Assume that U and X are nonempty sets, A is a subset of X^U of all functions from U to X, and R is a binary relation on X. Then the quadruple (U, X, A, R) is called an *information frame*. For all $B \subseteq A$, the *R-preimage relation* of B on U is defined by setting

$$(x, y) \in \text{pre}_R(B) \iff (\forall f \in B)\, f(x)\, R\, f(y).$$

Proposition 128. *If (U, X, A, R) is an information frame, then the map*

$$\text{pre}_R \colon \wp(A) \to \text{Rel}(U)^{\text{op}}$$

is a complete join-morphism.

Proof. For any $x, y \in U$ and $\{B_i\}_{i \in I} \subseteq \wp(A)$,

$$
\begin{aligned}
(x, y) \in \text{pre}_R\Big(\bigcup_{i \in I} B_i\Big) &\iff \Big(\forall f \in \bigcup_{i \in I} B_i\Big)\, f(x)\, R\, f(y) \\
&\iff (\forall i \in I)(\forall f \in B_i)\, f(x)\, R\, f(y) \\
&\iff (\forall i \in I)\, (x, y) \in \text{pre}_R(B_i) \\
&\iff (x, y) \in \bigcap_{i \in I} \text{pre}_R(B_i). \qquad \square
\end{aligned}
$$

Because $\mathrm{pre}_R\colon \wp(A) \to \mathrm{Rel}(U)^{\mathrm{op}}$ is a complete join-morphism, we may define a dependency relation $\xrightarrow{R}$ on $\wp(A)$ by

$$B \xrightarrow{R} C \iff \mathrm{pre}_R(B) \subseteq \mathrm{pre}_R(C)$$

and now $B^+ = \{a \in A \mid B \xrightarrow{R} a\}$. It is clear that the set of all R-preimage relations is a complete lattice in which

$$\bigvee_{i \in I} \mathrm{pre}_R(B_i) = \mathrm{pre}_R\Big(\bigcup_{i \in I} B_i\Big)$$

and

$$\bigwedge_{i \in I} \mathrm{pre}_R(B_i) = \mathrm{pre}_R\Big(\bigcap_{i \in I} B_i^+\Big).$$

We may also introduce matrix representations of preimage relations. Let (U, X, A, R) be an information frame. The *matrix* $\mathbf{m}_R = (m_{xy})$ of *preimage relations* with respect to R is defined so that

$$m_{xy} = \{f \in A \mid f(x)\, R\, f(y)\}$$

for all $x, y \in U$. It is clear that $(x, y) \in \mathrm{pre}_R(B)$ if and only if $B \subseteq m_{xy}$.

Because preimage relations determine dependency relations, we may define dense families for them as in Section 8.2. The following proposition can be proved in a similar way than Proposition 115.

Proposition 129. *Let (U, X, A, R) be an information frame. If $\mathbf{m}_R = (m_{xy})$ is the matrix of preimage relations with respect to R, the family $\{m_{xy} \mid x, y \in U\}$ is dense for the dependency relation $\xrightarrow{R}$.*

Since independent sets and reducts are defined by means of dependency relations, we could present similar results for preimage relation as for indiscernibility relations in Section 8.3. Finally, we show that information relations are preimage relations.

Example 130. Let (U, A) be a many-valued information system. Let us set $V = \bigcup_{a \in A} V_a$. Then every attribute can be considered as a function $a\colon U \to \wp(V)$. This means that $\wp(V)$ has the role of X in the previous considerations.

Let us define the following four relations on $\wp(V)$:

$$(X, Y) \in R_= \iff X = Y;$$
$$(X, Y) \in R_\cap \iff X \cap Y \neq \emptyset;$$
$$(X, Y) \in R_\subseteq \iff X \subseteq Y.$$

Now each $(U, \wp(V), A, R)$, where R may be any of the above-defined relations, is an information frame. It is easy to see that for all $B \subseteq A$,

$$\mathrm{ind}(B) = \mathrm{pre}_{(R_=)}(B) \quad \text{and} \quad \mathrm{div}(B) = \mathrm{pre}_{(R_=)^c}(B);$$
$$\mathrm{sim}(B) = \mathrm{pre}_{(R_\cap)}(B) \quad \text{and} \quad \mathrm{ort}(B) = \mathrm{pre}_{(R_\cap)^c}(B);$$
$$\mathrm{inc}(B) = \mathrm{pre}_{(R_\subseteq)}(B) \quad \text{and} \quad \mathrm{nsim}(B) = \mathrm{pre}_{(R_\subseteq)^c}(B).$$

Bibliographical Notes

Functional dependencies between sets of attributes in relational databases originate in [9]. Dependency relations can be also found, for instance, in formal concept analysis [17]. Armstrong axioms were introduced in [2], and in [12] functional dependencies in a more general setting of complete lattices are studied. There Armstrong systems on a complete lattice L satisfy (AS1) and (AS2), and they are complete join-sublattices of $L \times L$. Therefore, the definition here is more general than the one in [12] since it is applicable to any ordered set. However, by Lemma 103, these two definitions agree in case of complete lattices. Also in [40] an algebraic treatment of dependency is given. Several results of Section 8.1 appear in [27].

This model for information systems was introduced by Pawlak in [43], where indiscernibility relations, independent sets, and reducts are defined. Furthermore, his book [45] on theoretical aspects of rough set theory contains more detailed studies on these subjects. Joins and meets in the complete lattice of all indiscernibility relations were originally described in [23], where also many of the results of Section 8.2 appear. Preimage relations and their matrices were presented in [24] to generalize the notions of indiscernibility matrices and discernibility matrices introduced [49]. Several observations of Section 8.3 are presented in [24] for preimage relations. In addition, many results of Sections 8.2 and 8.3 appear also in [30]. Many-valued information systems are defined for the first time in [42] and different types of information relations studied in Section 8.4 can be found in [13].

9   Rough Set Approximations

This section has three subsections:

9.1 Indiscernibility and Approximations
9.2 Generalizations of Approximations
9.3 Definable Sets

9.1   Indiscernibility and Approximations

Rough set theory is a mathematical framework for dealing with uncertainty and to some extent overlapping fuzzy set theory. In fuzzy set theory vagueness is expressed by a membership function. The rough set theory approach is based on indiscernibility relations and approximations. A major advantage of rough set theory is that it needs no preliminary or additional information about data, such as membership functions in fuzzy set theory.

The basic idea of rough set theory is that knowledge about objects is represented by indiscernibility relations. Indiscernibility relations are usually assumed to be equivalences interpreted so that two objects are equivalent if we cannot distinguish them by their properties. We may observe objects only by the accuracy given by an indiscernibility relation. This means that our ability to distinguish objects is blurred – we cannot distinguish individual objects, only their equivalence classes. As we have seen, in an information system an indiscernibility relation arises naturally when one considers a given set of attributes: two objects are equivalent when their values of all attributes in the set are the same.

Let us consider the situation in Fig. 18. Let $\approx$ be an equivalence, called *indiscernibility relation*, on a universe U. The relation $\approx$ enables us to divide the objects of U into three disjoint sets with respect to any given subset $X \subseteq U$:

(a) the objects that surely are in X;
(b) the objects that are surely not in X;
(c) the objects that possibly are in X.

The objects in class (a) form the lower approximation of X, and the objects of type (a) and (c) form together its upper approximation. The boundary of X consists of the objects in class (c).

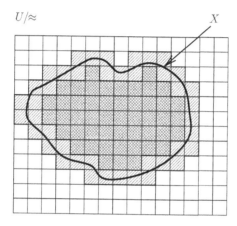

Fig. 18.

Next we define formally the upper and the lower approximations of an indiscernibility relation $\approx$ on U. As before, we denote for any $x \in U$, the equivalence class of x by $[x]_\approx = \{y \in U \mid x \approx y\}$. Thus, $[x]_\approx$ consists of the elements which cannot be discerned from x. For any subset X of U, let

$$X^\blacktriangledown = \{x \in U \mid [x]_\approx \subseteq X\}$$

and

$$X^\blacktriangle = \{x \in U \mid X \cap [x]_\approx \neq \emptyset\}.$$

The sets $X^\blacktriangledown$ and $X^\blacktriangle$ are called the *lower* and the *upper approximation* of X, respectively. The set $B(X) = X^\blacktriangle - X^\blacktriangledown$ is the *boundary* of X.

The above definitions mean that $x \in X^\blacktriangle$ if there is an element in X to which x is $\approx$-related. Similarly, $x \in X^\blacktriangledown$ if all the elements to which x is $\approx$-related are in X. Furthermore, $x \in B(X)$ if both in X and outside X there are elements which cannot be discerned from x. If $B(X) = \emptyset$ for some $X \subseteq U$, this means that for any object $x \in U$, we can with certainty decide whether $x \in X$ just by knowing x 'modulo $\approx$'.

Example 131. Let us consider the information system of Table 3 describing some symptoms of persons vising a doctor's reception on one afternoon. The attributes FEVER, HEADACHE, and FATIGUE can be regarded as the condition attributes and let us denote the set consisting of these three attributes by C. The attribute FLU can be considered as a decision attribute. Note that the condition attribute is not dependent on the decision attributes, that is, $C \not\rightarrow$ FLU.

Table 3. An information system describing flu symptoms

U	FEVER	HEADACHE	FATIGUE	FLU
1	no	yes	no	no
2	yes	no	yes	yes
3	yes	yes	yes	yes
4	no	no	yes	no
5	yes	no	yes	no
6	yes	yes	no	yes
7	yes	yes	yes	yes

Let us denote by X the set of persons that have a flu, that is,

$$X = \{x \mid \text{FLU}(x) = \text{'yes'}\} = \{2, 3, 6, 7\}.$$

Further, let us denote the C-indiscernibility relation $\text{ind}(C)$ simply by $\approx$. The equivalence classes of $\approx$ are $\{1\}$, $\{2, 5\}$, $\{3, 7\}$, $\{4\}$, and $\{6\}$. The lower approximation of X is $X^{\blacktriangledown} = \{3, 6, 7\}$, X's upper approximation is $X^{\blacktriangle} = \{2, 3, 5, 6, 7\}$, and the boundary of X is $\{2, 5\}$. Thus, based on this 'training example', persons indiscernible with 3, 6, or 7 have certainly a flu by their symptoms, persons indiscernible with 2 and 5 possibly have a flu, and persons indiscernible with 1 or 4 surely not have a flu.

As we already mentioned, indiscernibility relations are commonly assumed to be equivalences. The literature, however, contains studies in which rough approximations are defined also by other types of relations. In the following we argue that there exist indiscernibility relations which are not reflexive, symmetric, or transitive.

Reflexivity. It may seem reasonable to assume that every object is indiscernible from itself. But on some occasions this is not true, since it is possible that our information is so imprecise. For example, we may discern persons by comparing photographs taken of them. But it may happen that we are unable to recognize that a same person appears in two different photographs.

Symmetry. Usually it is supposed that indiscernibility relations are symmetric, which means that if we cannot discern x from y, then we cannot discern y from x either. But indiscernibility relations may be directional. For example, if a person x speaks English and Finnish, and a person y speaks English, Finnish, and German, then x cannot distinguish y from himself by the property 'knowledge of languages' since y can communicate with x in any language that x speaks. On the other hand, y can distinguish x from himself by asking a simple question in German, for example.

Transitivity. Transitivity is the least obvious of the three properties usually associated with indiscernibility relations. For example, we may have a sequence $x_1, \ldots, x_n$ of objects such that $x_1 \approx x_2$, $x_2 \approx x_3, \ldots, x_{n-1} \approx x_n$, but $x_1 \not\approx x_n$. This can be interpreted so that always two consecutive objects x_i and x_{i+1} are so similar that there is no way to distinguish them, but if we take objects at the utmost ends, the objects are already distinguishable by their properties.

9.2   Generalizations of Approximations

This section is devoted to the properties of approximations determined by arbitrary binary relations. We start by defining the approximations. Let R be a binary relation on U, and let us denote for all $x \in U$,

$$R(x) = \{y \in U \mid x\,R\,y\}.$$

The *upper approximation* of $X \subseteq U$ is

$$X^{\blacktriangle} = \{x \in U \mid R(x) \cap X \neq \emptyset\}$$

and the *lower approximation* of X is

$$X^{\blacktriangledown} = \{x \in U \mid R(x) \subseteq X\}.$$

The set $B(X) = X^{\blacktriangle} - X^{\blacktriangledown}$ is the *boundary* of X.

The next proposition lists basic properties of rough approximations. Note that in the sequel we denote

$$\wp(U)^{\blacktriangledown} = \left\{ X^{\blacktriangledown} \mid X \subseteq U \right\} \quad \text{and} \quad \wp(U)^{\blacktriangle} = \left\{ X^{\blacktriangle} \mid X \subseteq U \right\}.$$

Proposition 132. *If R is a binary relation on U, then following assertions hold.*

(a) *The maps $^{\blacktriangledown} \colon \wp(U) \to \wp(U)$ and $^{\blacktriangle} \colon \wp(U) \to \wp(U)$ are mutually dual.*
(b) *The boundary of any set is equal to the boundary of its complement.*
(c) *The map $^{\blacktriangledown} \colon \wp(U) \to \wp(U)$ is a complete meet-morphism.*
(d) *The map $^{\blacktriangle} \colon \wp(U) \to \wp(U)$ is a complete join-morphism.*
(e) *The maps $^{\blacktriangledown} \colon \wp(U) \to \wp(U)$ and $^{\blacktriangle} \colon \wp(U) \to \wp(U)$ are order-preserving.*
(f) *The family $\wp(U)^{\blacktriangledown}$ is a closure system and $\wp(U)^{\blacktriangle}$ is an interior system.*
(g) *The complete lattices $\wp(U)^{\blacktriangle}$ and $\wp(U)^{\blacktriangledown}$ are dually isomorphic.*

Proof. (a) $x \in X^{\blacktriangledown c} \iff x \notin X^{\blacktriangledown} \iff R(x) \not\subseteq X \iff R(x) \cap X^c \neq \emptyset \iff x \in X^{c\blacktriangle}$. Further, $X^{\blacktriangle c} = X^{cc\blacktriangle c} = X^{c\blacktriangledown cc} = X^{c\blacktriangledown}$.

(b) $B(X) = X^{\blacktriangle} - X^{\blacktriangledown} = X^{\blacktriangle} \cap X^{\blacktriangledown c} = X^{c\blacktriangledown c} \cap X^{c\blacktriangle} = X^{c\blacktriangle} - X^{c\blacktriangledown} = B(X^c)$.

(c) $x \in (\bigcap \mathcal{H})^{\blacktriangledown} \iff R(x) \subseteq \bigcap \mathcal{H} \iff (\forall X \in \mathcal{H})\, R(x) \subseteq X \iff (\forall X \in \mathcal{H})\, x \in X^{\blacktriangledown} \iff x \in \bigcap \mathcal{H}^{\blacktriangledown}$.

(d) $x \in (\bigcup \mathcal{H})^{\blacktriangle} \iff R(x) \cap \bigcup \mathcal{H} \neq \emptyset \iff (\exists X \in \mathcal{H})\ R(x) \cap X \neq \emptyset \iff (\exists X \in \mathcal{H})\ x \in X^{\blacktriangle} \iff x \in \bigcup \mathcal{H}^{\blacktriangle}$.

(e) If $X \subseteq Y$, then $X^{\blacktriangledown} \cap Y^{\blacktriangledown} = (X \cap Y)^{\blacktriangledown} = X^{\blacktriangledown}$ and $X^{\blacktriangle} \cup Y^{\blacktriangle} = (X \cup Y)^{\blacktriangle} = Y^{\blacktriangle}$. Thus, $X^{\blacktriangledown} \subseteq Y^{\blacktriangledown}$ and $X^{\blacktriangle} \subseteq Y^{\blacktriangle}$.

(f) Claim is obvious by (c) and (d).

(g) We show that the map $\phi \colon X^{\blacktriangle} \mapsto X^{c\blacktriangledown}$ is an order-isomorphism between $(\wp(U)^{\blacktriangle}, \subseteq)$ and $(\wp(U)^{\blacktriangledown}, \supseteq)$. Clearly, $X^{\blacktriangle} \subseteq Y^{\blacktriangle}$ if and only if $\phi(X^{\blacktriangle}) = X^{c\blacktriangledown} = X^{\blacktriangle c} \supseteq Y^{\blacktriangle c} = Y^{c\blacktriangledown} = \phi(Y^{\blacktriangle})$. Thus, ϕ is an order-embedding. Further, if $X^{\blacktriangledown} \in \wp(U)^{\blacktriangledown}$, then $\phi(X^{c\blacktriangle}) = X^{cc\blacktriangledown} = X^{\blacktriangledown}$, that is, ϕ is onto. □

In the previous proposition, (a) is interpreted so that if an element does not belong with certainty to a set, it belongs possibly to the complement of that set, and if an element does not belong possibly to a set, then it belongs with certainty to the complement. Assertion (b) means that if we cannot decide whether an element belongs to a set, we cannot decide whether the element is in the set's complement either. Claim (c) says that elements belong possibly to the union of some sets if they belong possibly to at least one of the sets in question. An element belongs with certainty to the intersection of sets if it is with certainty in all sets; this is stated in (d). Furthermore, because $^{\blacktriangle}$ is a complete join-morphism, it is bottom-preserving, that is, $\emptyset^{\blacktriangle} = \emptyset$. Similarly, $^{\blacktriangledown}$ is top-preserving, meaning $U^{\blacktriangledown} = U$. Concerning (f), notice that $X \mapsto X^{\blacktriangledown}$ is not necessarily a closure operator; the closure operator corresponding $\wp(U)^{\blacktriangledown}$ is defined by

$$X \mapsto \bigcap \{Y^{\blacktriangledown} \mid X \subseteq Y^{\blacktriangledown}\}.$$

Example 133. Let $U = \{a, b, c, d\}$ and assume that R is a binary relation on U of Fig. 19. The dually order-isomorphic complete lattices $\wp(U)^{\blacktriangledown}$ and $\wp(U)^{\blacktriangle}$ are also presented there. For simplicity, sets are denoted by sequences of their elements. As we have noted, $\wp(U)^{\blacktriangledown}$ is a closure system and $\wp(U)^{\blacktriangle}$ is an interior system. The lattices $\wp(U)^{\blacktriangledown}$ and $\wp(U)^{\blacktriangle}$ are not distributive, because they contain $\mathbf{M_3}$ as a sublattice. It is also easy to observe that these lattices are not complemented.

Since $^{\blacktriangle} \colon \wp(U) \to \wp(U)$ is a complete join-morphism, it induces a Galois connection by Proposition 84. Similarly, the complete meet-morphism $^{\blacktriangledown} \colon \wp(U) \to \wp(U)$

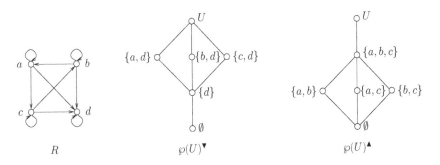

Fig. 19.

must determine a Galois connection. Next we will find their adjoint and co-adjoint, respectively.

As before, we denote by R^{-1} the inverse relation of R and

$$R^{-1}(x) = \{y \mid y\,R\,x\}.$$

Next we define the upper and the lower approximations of R^{-1} canonically. Let R be a binary relation on U and let $X \subseteq U$. Then,

$$X^{\vartriangle} = \{x \in U \mid R^{-1}(x) \cap X \neq \emptyset\}$$

and

$$X^{\triangledown} = \{x \in U \mid R^{-1}(x) \subseteq X\}.$$

Notice that if R is symmetric, then $R = R^{-1}$ which means that $X^{\blacktriangle} = X^{\vartriangle}$ and $X^{\blacktriangledown} = X^{\triangledown}$ for all $X \subseteq U$. It is also trivial that the operators $^{\vartriangle}$ and $^{\triangledown}$ have the properties of Proposition 132.

Now we can write the following important observation.

Proposition 134. *For any binary relation on U, the pairs $(^{\blacktriangle}, ^{\triangledown})$ and $(^{\vartriangle}, ^{\blacktriangledown})$ are Galois connections on $\wp(U)$.*

Proof. We show that $(^{\blacktriangle}, ^{\triangledown})$ is a Galois connection. The other part can be proved analogously. The maps $X \mapsto X^{\blacktriangle}$ and $X \mapsto X^{\triangledown}$ are order-preserving by Proposition 132. If $x \in X^{\triangledown\blacktriangle}$, then there exists $y \in X^{\triangledown}$ such that $(x, y) \in R$. Because $y \in X^{\triangledown}$ and $(y, x) \in R^{-1}$, we have $x \in X$. Hence, $X^{\triangledown\blacktriangle} \subseteq X$. This also gives $X^{\vartriangle\blacktriangledown c} = X^{c\triangledown\blacktriangle} \subseteq X^c$, that is, $X \subseteq X^{\vartriangle\blacktriangledown}$. Hence, by Lemma 79, $(^{\blacktriangle}, ^{\triangledown})$ is a Galois connection. $\qquad\square$

Since $^{\vartriangle}$ is the dual of $^{\triangledown}$, we can write the following corollary by Proposition 89.

Corollary 135. *For any binary relation, $^{\blacktriangle}$ and $^{\vartriangle}$ are conjugate.*

Because the pairs $(^{\vartriangle}, ^{\blacktriangledown})$ and $(^{\blacktriangle}, ^{\triangledown})$ form Galois connections on $\wp(U)$, the maps $X \mapsto X^{\vartriangle\blacktriangledown}$ and $X \mapsto X^{\blacktriangle\triangledown}$ are closure operators, and $X \mapsto X^{\triangledown\blacktriangle}$ and $X \mapsto X^{\blacktriangledown\vartriangle}$ are interior operators. Furthermore,

$$X^{\blacktriangle} = X^{\blacktriangle\triangledown\blacktriangle}, \quad X^{\triangledown} = X^{\triangledown\blacktriangle\triangledown}, \quad X^{\vartriangle} = X^{\vartriangle\blacktriangledown\vartriangle}, \quad \text{and} \quad X^{\blacktriangledown} = X^{\blacktriangledown\vartriangle\blacktriangledown}$$

for all $X \subseteq U$. It also follows from the general properties of Galois connections that the map $X^{\vartriangle} \mapsto X^{\vartriangle\blacktriangledown}$ is an order-isomorphism between $(\wp(U)^{\vartriangle}, \subseteq)$ and $(\wp(U)^{\blacktriangledown}, \subseteq)$, and $X^{\blacktriangle} \mapsto X^{\blacktriangle\triangledown}$ is an order-isomorphism between $(\wp(U)^{\blacktriangle}, \subseteq)$ and $(\wp(U)^{\triangledown}, \subseteq)$. Thus,

$$(\wp(U)^{\vartriangle}, \subseteq) \cong (\wp(U)^{\blacktriangledown}, \subseteq) \cong (\wp(U)^{\blacktriangle}, \supseteq) \cong (\wp(U)^{\triangledown}, \supseteq).$$

Note that if R is symmetric, then $X^{\blacktriangle} = X^{\vartriangle}$ and $X^{\blacktriangledown} = X^{\triangledown}$ for all $X \subseteq U$, which yields

$$(\wp(U)^{\blacktriangledown}, \subseteq) = (\wp(U)^{\triangledown}, \subseteq) \cong (\wp(U)^{\blacktriangle}, \subseteq) = (\wp(U)^{\vartriangle}, \subseteq).$$

Hence, for a symmetric R, the ordered sets $(\wp(U)^{\blacktriangledown}, \subseteq) = (\wp(U)^{\triangledown}, \subseteq)$ and $(\wp(U)^{\blacktriangle}, \subseteq) = (\wp(U)^{\triangle}, \subseteq)$ are self-dual. Furthermore, because the conjugate of $\blacktriangle$ is $\triangle$, and $\triangle$ is equal to $\blacktriangle$ for symmetric relations, we obtain that $\blacktriangle$ is self-conjugate in case R is symmetric.

It is clear that for any $x \in U$, $\{x\}^{\blacktriangle} = R^{-1}(x)$ and $\{x\}^{\triangle} = R(x)$. Hence, for all $X \subseteq U$ and for any arbitrary binary relation R on U,

$$X^{\triangle} = \bigcup_{x \in X} R(x) \quad \text{and} \quad X^{\blacktriangle} = \bigcup_{x \in X} R^{-1}(x).$$

Below we show how properties of binary relations are expressed by rough approximations, and conversely.

Proposition 136. *If R is a binary relation on U, then the following assertions are equivalent:*

(a) R *is connected;*
(b) $X^{\blacktriangledown} \subseteq X^{\blacktriangle}$ *for all $X \subseteq U$.*

Proof. (a) $\Longrightarrow$ (b): Let $x \in X^{\blacktriangledown}$. Then $R(x) \subseteq X$, which gives $R(x) \cap X = R(x) \neq \emptyset$, that is, $x \in X^{\blacktriangle}$.
(b) $\Longrightarrow$ (a): Assume that R is not connected, that is, $R(x) = \emptyset$ for some $x \in U$. This means that $x \in X^{\blacktriangledown}$ and $x \notin X^{\blacktriangle}$ for this particular x and for any set $X \subseteq U$, a contradiction! $\qquad\square$

Each set is bounded by its approximations determined by a reflexive relation, as seen in the next proposition.

Proposition 137. *If R is a binary relation on U, then the following assertions are equivalent:*

(a) R *is reflexive;*
(b) $X \subseteq X^{\blacktriangle}$ *for all $X \subseteq U$;*
(c) $X^{\blacktriangledown} \subseteq X$ *for all $X \subseteq U$.*

Proof. (a) $\Longrightarrow$ (b): If $x \in X$, then $x \in R(x) \cap X \neq \emptyset$, that is, $x \in X^{\blacktriangle}$.
(b) $\Longrightarrow$ (c): Obviously, $X^c \subseteq X^{c\blacktriangle} = X^{\blacktriangledown c}$, which is equivalent to $X^{\blacktriangledown} \subseteq X$.
(c) $\Longrightarrow$ (a): If R is not reflexive, then there exists $x \in U$ such that $(x, x) \notin R$. Let us consider the set $X = U - \{x\}$. For all $y \in U$, $(x, y) \in R$ implies $y \in X$. Thus, $x \in X^{\blacktriangledown}$ and $x \notin X$, a contradiction! $\qquad\square$

Proposition 138. *If R is a binary relation on U, then the following assertions are equivalent:*

(a) R *is symmetric;*
(b) $(\blacktriangle, \blacktriangledown)$ *is a Galois connection on $(\wp(U), \subseteq)$.*

Proof. (a) $\Longrightarrow$ (b): If R is symmetric, then $X^{\blacktriangle} = X^{\triangle}$ and $X^{\blacktriangledown} = X^{\triangledown}$ for all $X \subseteq U$. Thus, the implication is clear by Proposition 134.
(b) $\Longrightarrow$ (a): Assume that R is not symmetric. Then, for some $x, y \in U$, $(x, y) \in R$, but $(y, x) \notin R$. Let us consider the set $X = \{x\}$. For all $z \in U$, $(y, z) \in R$ implies $z \notin X$. This gives $y \notin X^{\blacktriangle}$. Hence, $x \in X$ and $x \notin X^{\blacktriangle\blacktriangledown}$, a contradiction! $\qquad\square$

Proposition 139. *If R is a binary relation on U, then the following assertions are equivalent:*

(a) R *is transitive;*
(b) $X^{\blacktriangle\blacktriangle} \subseteq X^{\blacktriangle}$ *for all $X \subseteq U$;*
(c) $X^{\blacktriangledown} \subseteq X^{\blacktriangledown\blacktriangledown}$ *for all $X \subseteq U$.*

Proof. (a) $\Longrightarrow$ (b): Let $x \in X^{\blacktriangle\blacktriangle}$. This means that there exists $y \in X^{\blacktriangle}$ such that $(x, y) \in R$. Because $y \in X^{\blacktriangle}$, there is $z \in X$ such that $(y, z) \in R$. Now, $(x, z) \in R$ by the transitivity of R. Hence, $x \in X^{\blacktriangle}$.

(b) $\Longrightarrow$ (c): Obviously, $X^{\blacktriangledown\blacktriangledown c} = X^{c\blacktriangle\blacktriangle} \subseteq X^{c\blacktriangle} = X^{\blacktriangledown c}$, which means $X^{\blacktriangledown} \subseteq X^{\blacktriangledown\blacktriangledown}$.

(c) $\Longrightarrow$ (a): Assume that R is not transitive. Then there exist $x, y, z \in U$ such that $(x, y) \in R$ and $(y, z) \in R$, but $(x, z) \notin R$. Let us consider the set $X = U - \{z\}$. Then for all $w \in U$, (x, w) implies $w \in X$. This implies $x \in X^{\blacktriangledown}$. Obviously, $y \notin X^{\blacktriangledown}$ and hence $x \notin X^{\blacktriangledown\blacktriangledown}$, a contradiction! $\qquad\square$

Note also that R is reflexive if and only if R^{-1} is reflexive, and similar conditions hold also for symmetry and transitivity. Therefore, we could state similar correspondences between R and the operators $X \mapsto X^{\triangle}$ and $X \mapsto X^{\triangledown}$. However, the connectedness of R does not imply the connectedness of R^{-1}. Therefore,

$$(\forall X \subseteq U)\, X^{\triangledown} \subseteq X^{\triangle} \iff R^{-1} \text{ is connected.}$$

Propositions 137 and 139 have the following corollary:

Corollary 140. *If R is a binary relation on U, then the following assertions are then equivalent:*

(a) R *is a preorder;*
(b) *The map $X \mapsto X^{\blacktriangle}$ is a closure operator;*
(c) *The map $X \mapsto X^{\blacktriangledown}$ is an interior operator.*

In fact, since $^{\blacktriangle}: \wp(U) \to \wp(U)$ is a complete join-morphism and $^{\blacktriangledown}: \wp(U) \to \wp(U)$ is a complete meet-morphism, we may write that the following are equivalent:

(a) R is a preorder;
(b) The map $X \mapsto X^{\blacktriangle}$ is an Alexandrov closure operator;
(c) The map $X \mapsto X^{\blacktriangledown}$ is an Alexandrov interior operator.

We end this subsection by considering how rough approximation operators relate to fuzzy sets.

Example 141. Let $\leq$ be a preorder on a set L. Then the pair $(L, \leq)$ is called a *preordered set.* An *L-fuzzy set* φ on U is a mapping $\varphi : U \to L$. We may order the family of all L-fuzzy sets on U by the pointwise order:

$$\varphi \leq \psi \iff (\forall x \in U)\, \varphi(x) \leq \psi(x).$$

If L is equal to $\{0, 1\}$, then each L-fuzzy set is simply the characteristic function of some conventional subset of U and the pointwise ordered set of all $\{0, 1\}$-sets on U can be identified with $(\wp(U), \subseteq)$. Each L-fuzzy set $\varphi: U \to L$ determines naturally also a preorder $\lesssim$ on U is defined by setting for all $x, y \in U$,

$$x \lesssim y \iff \varphi(x) \leq \varphi(y).$$

Note that $\lesssim$ can be regarded as the $\leq$-preimage relation of φ.

Typically, L may consists of adjectives such as 'good', 'excellent', 'poor', and 'adequate', for example. Assume that $\varphi: U \to L$ is an L-fuzzy set describing what is the ability of persons in U to speak Finnish. For example, there may exist persons x and y in U such that $\varphi(x) =$ 'adequate' and $\varphi(y) =$ 'excellent'. Now it is clear that $x \lesssim y$.

For any L-fuzzy set $\varphi: U \to L$, we may define the operators $^{\blacktriangle}$, $^{\blacktriangledown}$, $^{\vartriangle}$, and $^{\triangledown}$ determined by the relation $\lesssim$ by

$$X^{\blacktriangledown} = \{x \in U \mid x \lesssim y \text{ implies } y \in X\};$$
$$X^{\blacktriangle} = \{x \in U \mid \text{ there is } y \in X \text{ such that } x \lesssim y\};$$
$$X^{\triangledown} = \{x \in U \mid x \gtrsim y \text{ implies } y \in X\};$$
$$X^{\vartriangle} = \{x \in U \mid \text{ there is } y \in X \text{ such that } x \gtrsim y\}.$$

For example, if φ is the L-fuzzy set describing what is the ability of persons to speak Finnish, then $x \in X^{\blacktriangledown}$ if all persons in U that can speak Finnish at least as well as x are in X, and $x \in X^{\vartriangle}$ if there is $y \in X$ such that x can speak Finnish as well as y.

9.3   Definable Sets

In this section we consider sets $X \subseteq U$ such that $X^{\blacktriangle} = X$, that is, the fixpoints of the map $X \mapsto X^{\blacktriangle}$. These are important because the set $X^{\blacktriangle}$ is interpreted as a set of elements possibly belonging to X when objects are observed by the accuracy given by an indiscernibility relation. A fixpoint $X = X^{\blacktriangle}$ is called *definable*, because the set X and the set of elements possibly in X are equal. Let us denote

$$\mathrm{Def} = \{X \subseteq U \mid X^{\blacktriangle} = X\}.$$

Recall that

$$X^{\blacktriangle} = \{x \in U \mid x \, R \, y \text{ for some } y \in X\}.$$

Thus, each definable set $X = X^{\blacktriangle}$ is such that

- Each element of X is R-related to some element of X.
- Elements outside X are not R-related to elements in X.

In fact, if X is definable, then

$$X = \bigcup_{x \in X} \{x\}^{\blacktriangle} = \bigcup_{x \in X} R^{-1}(x).$$

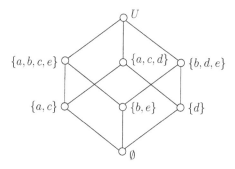

Fig. 20.

Example 142. Let $\approx$ be an equivalence on $\{a, b, c, d, e\}$ such that its equivalence classes are $\{a, c\}$, $\{b, e\}$, and $\{d\}$. The family Def is depicted in Fig. 20.

Because $\approx$ is an equivalence, the map $^{\blacktriangle}$ is self-conjugate and it can be considered as the smallest-neighborhood operator of the Alexandrov topology

$$\mathrm{Def} = \{X^{\blacktriangle} \mid X \subseteq U\}.$$

Further, by Proposition 98, Def is a complete Boolean sublattice of $\wp(U)$. This means that Def is a complete field of sets and thus, by Proposition 41, it is a complete atomic Boolean lattice.

Let $X \subseteq U$. Since $\approx$ is symmetric,

$$X^{\blacktriangle} = \bigcup_{x \in X} [x]_{\approx}.$$

Furthermore, the fact that Def is a complete field of sets gives

$$X^{\blacktriangle} = \bigcap \{Y \in \mathrm{Def} \mid X \subseteq Y\}.$$

By Proposition 41 we have that the atoms of Def are the $\approx$-equivalence classes. Notice that if $\approx$ is the indiscernibility relation determined by some subset B of the attribute set A in an information system (U, A), the definable sets can be described by using the values of B-attributes.

The previous example shows that Def is a complete atomic Boolean lattice whenever indiscernibility relations are equivalences. Next we give a systematic study on the properties of definable sets determined by arbitrary relations. We begin with the following proposition which follows directly from Proposition 62.

Proposition 143. *For any binary relation, the set* Def *is a complete lattice with respect to the set-inclusion relation.*

The previous proposition does not guarantee that Def is a sublattice of $\wp(U)$. However, we can also present the following stronger result by Proposition 64, since $^{\blacktriangle}$ is always a complete $\cup$-morphism.

Proposition 144. *For any binary relation, the set* Def *is a closed under arbitrary joins.*

For an arbitrary relation, Def is not closed under intersections. For example, it may happen that an element x belonging to the intersection $X \cap Y$ of two definable sets X and Y is related to only one element in $X - (X \cap Y)$ and to just one element in $Y - (X \cap Y)$. This means that x is not R-related to any element is $X \cap Y$, see Fig. 21.

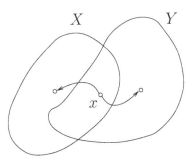

Fig. 21.

Next point out that for reflexive relations, the definable sets are also closed under arbitrary intersections.

Proposition 145. *For a reflexive relation,* Def *is a complete sublattice of* $\wp(U)$.

Proof. If R is reflexive, then by Proposition 137, $X \subseteq X^{\blacktriangle}$ for all $X \subseteq U$, that is, $X \mapsto X^{\blacktriangle}$ is extensive. Since $X \mapsto X^{\blacktriangle}$ is also a complete join-morphism, the claim is clear by Corollary 65.   □

Thus, definable sets determined by reflexive relations form an Alexandrov topology, that is, a complete ring of sets. For a reflexive relation, we may also write by Corollary 94 the following:

$$(\forall X \subseteq U) \ X = X^{\blacktriangle} \iff X^{\triangledown} = X \iff X^{\triangledown} = X^{\blacktriangle}.$$

This means that Def consists of elements satisfying one of these conditions.

It is known that for tolerances the pair $(^{\blacktriangle}, ^{\triangledown})$ is a Galois connection. Further, $^{\blacktriangle}$ is self-conjugate and by Proposition 98 we can get the following fact.

Proposition 146. *For a tolerance,* Def *is a complete field of sets.*

Proposition 145 states that Def is an Alexandrov topology when the relation R is reflexive. However, by the correspondence of Corollary 140, Def does not equal $\{X^{\blacktriangle} \mid X \subseteq U\}$ unless the relation is reflexive and transitive. Therefore, it is natural to ask what are the interior, closure, and smallest-neighbourhood operators in the Alexandrov topology Def determined by a reflexive relation.

In the following we consider approximations and definable sets determined by reflexive indiscernibility relations. We denote

$$X^{\blacktriangle(i)} = X^{\overbrace{\blacktriangle\blacktriangle\cdots\blacktriangle}^{i \text{ times}}}$$

for all $X \subseteq U$ and $i \geq 0$. Because $\blacktriangle$ is an extensive complete join-morphism, the map $X^{\overline{\blacktriangle}} : \wp(U) \to \wp(U)$ defined by

$$X^{\overline{\blacktriangle}} = \bigcup\left\{X^{\blacktriangle(i)} \mid i \geq 0\right\}$$

is such that $X^{\overline{\blacktriangle}}$ is the least fixpoint of $\blacktriangle$ above $X \subseteq U$ by Lemma 76. Therefore, by Proposition 77, we can get the following.

Proposition 147. *For any reflexive relation on U, the map $\overline{\blacktriangle}$ is the smallest-neighbourhood operator of the Alexandrov topology* Def. *Furthermore,* $\left\{\{x\}^{\overline{\blacktriangle}}\right\}_{x \in X}$ *is the smallest base of* Def.

Notice that by Proposition 146, Def is a complete atomic Boolean lattice for tolerances. Clearly, $\left\{\{x\}^{\overline{\blacktriangle}}\right\}_{x \in X}$ is the set of its atoms.

The set $\{X^c \mid X \in \text{Def}\}$ of closed elements of the topology Def consists of the fixpoints of the conjugate $\triangle$ of $\blacktriangle$ by Lemma 99. Let us denote

$$X^{\overline{\triangle}} = \bigcup\left\{X^{\triangle(i)} \mid i \geq 0\right\},$$

where

$$X^{\triangle(i)} = X^{\overbrace{\triangle\triangle\cdots\triangle}^{i \text{ times}}}.$$

The next proposition is now obvious.

Proposition 148. *For a reflexive relation on U, the map $\overline{\triangle}$ is the Alexandrov closure operator of* Def.

Let us define the map $\overline{\nabla} : \wp(U) \to \wp(U)$ by

$$X^{\overline{\nabla}} = \bigcap\{X^{\nabla(i)} \mid i \geq 0\},$$

where

$$X^{\nabla(i)} = X^{\overbrace{\nabla\nabla\cdots\nabla}^{i \text{ times}}}.$$

It is clear that $X^{\nabla(i)}$ is the dual of $X^{\triangle(i)}$ for all $i \geq 0$, which implies easily that $X^{\overline{\triangle}}$ and $X^{\overline{\nabla}}$ are dual. Therefore, we can write the following proposition, since the dual of the closure operator of a topology is the topology's interior operator.

Proposition 149. *For a reflexive relation on U, the map $\overline{\nabla}$ is the Alexandrov interior operator of* Def.

Let us consider approximation determined by preorders. Then it is clear that $X^{\overline{\blacktriangle}} = X^{\blacktriangle}$, $X^{\overline{\vartriangle}} = X^{\vartriangle}$, and $X^{\overline{\triangledown}} = X^{\triangledown}$ for all $X \subseteq U$. This implies the following proposition.

Proposition 150. *For a preorder,* $\blacktriangle$ *is the smallest-neighbourhood operator,* $\vartriangle$ *is the closure operator, and* $\triangledown$ *is the interior operator of* Def.

The previous proposition gives that for preorders on U,

$$\text{Def} = \wp(U)^{\blacktriangle} = \wp(U)^{\triangledown}.$$

Then for all $X \subseteq U$,

$$X^{\blacktriangle\triangledown} = X^{\blacktriangle} \quad \text{and} \quad X^{\triangledown\blacktriangle} = X^{\triangledown}$$

and analogously,

$$X^{\vartriangle\blacktriangledown} = X^{\blacktriangle} \quad \text{and} \quad X^{\blacktriangledown\vartriangle} = X^{\blacktriangledown}.$$

It is also clear that $\wp(U)^{\blacktriangle} = \wp(U)^{\triangledown}$ and $\wp(U)^{\blacktriangle} = \wp(U)^{\triangledown}$ are dual topologies in the sense of Section 5.3.

Notice that for an equivalence, Def is a complete atomic Boolean lattice such that $\left\{\{x\}^{\blacktriangle}\right\}_{x \in X}$ is the set of its atoms, as we already mentioned in Example 142. Furthermore,

$$\wp(U)^{\blacktriangle} = \wp(U)^{\triangledown} = \wp(U)^{\vartriangle} = \wp(U)^{\blacktriangledown}.$$

Finally, we present an example considering definable sets of L-fuzzy sets introduced in Example 141.

Example 151. If $\varphi : U \to L$ is an L-fuzzy set, then $\lesssim$ is a preorder, and thus

$$\text{Def} = \wp(U)^{\blacktriangle} = \wp(U)^{\triangledown}.$$

Recall that

$$X^{\blacktriangle} = \{x \in U \mid \text{there is } y \in X \text{ such that } x \lesssim y\}$$

and

$$X^{\triangledown} = \{x \in U \mid x \gtrsim y \text{ implies } y \in X\}.$$

Let $X \in$ Def, that is, $X = X^{\triangledown}$. If $x \in X$ and $x \gtrsim y$, then necessarily $y \in X$. Thus, Def consists of down-sets of $\lesssim$. So, also for L-fuzzy sets, definability has a nice interpretation.

Bibliographical Notes

Rough sets defined by equivalences were introduced in [44], where also the essential properties of rough set approximations and definable sets were given.

In the literature can be found several papers which consider rough approximations determined by relations that are not necessarily equivalences. For instance, rough approximations defined by tolerances are studied in [26,47,50]. The paper

[51] considers approximations induced by reflexive binary relations and in [60] approximations based on arbitrary binary relations are investigated. Furthermore, in [13] operators determined by frames of information relations reflecting distinguishability or indistinguishability of objects of a many-valued information system are examined. Additionally, motivated by the fact that $R(x)$ can be regarded as a neighbourhood of x, relational interpretations of neighbourhood operators were considered in [58]. Interestingly, in [10] so-called knowledge representation algebras are studied, where the usual powerset algebra of the universe is equipped with an upper approximation operator.

In [28] and [31], approximations are studied in a more general setting of complete atomic Boolean lattices. The fact that $(^{\blacktriangle},^{\triangledown})$ and $(^{\triangle},^{\blacktriangledown})$ are Galois connections appears already in [14], and in [34] it is noted that $^{\blacktriangle}$ and $^{\triangle}$ are conjugate. In [55] correspondence results for modal logic are given. Definable sets determined by preorders are studied in [32], and in [36] it is shown that definable sets determined by reflexive relations form an Alexandrov topology. It should be noted that already Birkhoff noticed in [6] that if R is a preorder, then $^{\blacktriangle}$ is a closure operator in a unique Alexandrov topology. It was also shown in [52] that the ordered sets of all Alexandrov topologies and preorders on a given set are dually isomorphic. Note also that in [57] generalizations of Pawlak's approximation operators are reviewed from lattice-theoretical and topological point of view.

Fuzzy sets were defined originally by Zadeh in [61] as mappings from a nonempty set U into the unit interval $[0, 1]$. Goguen generalized fuzzy sets to L-fuzzy sets in [20], where L is a 'transitive partially ordered set'. The relational view of fuzzy sets considered here was introduced in [37,38]. The approximations and Alexandrov spaces determined by L-fuzzy sets are also studied in [33].

Finally, some authors have studied connections between rough sets and formal concepts. In particular, modal-like operators are considered with respect to two universes; see [59], where more references can be found.

10   Lattices of Rough Sets

In this final section we study the following issues:

10.1 Orders for Rough Sets
10.2 Rough Sets Determined by Equivalences
10.3 Rough Sets Determined by Arbitrary Binary Relations

10.1   Orders for Rough Sets

Let R be any binary relation on a universe U. Let us define a binary relation $\equiv$ on the powerset $\wp(U)$ of U by setting

$$X \equiv Y \text{ if and only if } X^{\blacktriangledown} = Y^{\blacktriangledown} \text{ and } X^{\blacktriangle} = Y^{\blacktriangle}.$$

Obviously, $\equiv$ is an equivalence on U called *rough equality relation*. The equivalence classes of $\equiv$ are called *rough sets*. The set of all rough sets if denoted by $\mathcal{R}$.

The idea is that if subsets of U are observed within the limitation given by the knowledge represented by R, then the sets in the same rough set look the same; $X \equiv Y$ means that exactly the same elements belong *certainly* to X and to Y, and that precisely the same elements are *possibly* in X and in Y. Therefore, $\equiv$ can also be viewed as an indiscernibility relation, but now between subsets of the universe.

In the sequel, we denote the $\equiv$-class of $X \subseteq U$ simply by $[X]$. Next we study the structure of rough sets $\mathcal{R} = \{ [X] \mid X \subseteq U \}$ more carefully. Let us begin with defining the *rough inclusion* relation $\subseteq$ on $\mathcal{R}$ by setting

$$[X] \subseteq [Y] \text{ if and only if } X^{\blacktriangledown} \subseteq Y^{\blacktriangledown} \text{ and } X^{\blacktriangle} \subseteq Y^{\blacktriangle}.$$

The relation is well-defined, because the $\equiv$-classes consist of elements which have the same lower and upper approximations.

Lemma 152. *The relation $\subseteq$ is an order on $\mathcal{R}$.*

Proof. We have to show that $\subseteq$ is reflexive, antisymmetric, and transitive. It is trivial that $\subseteq$ is reflexive. Suppose $[X] \subseteq [Y]$ and $[Y] \subseteq [X]$ for some $X, Y \subseteq U$. Then, $X^{\blacktriangledown} \subseteq Y^{\blacktriangledown} \subseteq X^{\blacktriangledown}$ and $X^{\blacktriangle} \subseteq Y^{\blacktriangle} \subseteq X^{\blacktriangle}$, which means $X \equiv Y$ and $[X] = [Y]$. If $[X] \subseteq [Y]$ and $[Y] \subseteq [Z]$ for some $X, Y, Z \subseteq U$, then $X^{\blacktriangledown} \subseteq Y^{\blacktriangledown} \subseteq Z^{\blacktriangledown}$ and $X^{\blacktriangle} \subseteq Y^{\blacktriangle} \subseteq Z^{\blacktriangle}$, that is, $[X] \subseteq [Z]$. □

Next we try to find out whether $\mathcal{R}$ is a lattice with respect to the order $\subseteq$. At first glance it seems tempting to define the operators $\vee$ and $\wedge$ in $\mathcal{R}$ pointwise by

$$[X] \vee [Y] = [X \cup Y] \text{ and } [X] \wedge [Y] = [X \cap Y].$$

Unfortunately, this definition is not well-defined since it depends in general on the choice of representatives of $\equiv$-classes. For instance, let us consider the equivalence on $\{a, b, c, d\}$ which has the equivalence classes $\{a, d\}$ and $\{b, c\}$. Then obviously

$$\{a, b\} \equiv \{a, c\} \equiv \{b, d\} \equiv \{c, d\},$$

because they all have the lower approximation $\emptyset$ and the upper approximation $\{a, b, c, d\}$. This would imply $[\{a, b\}] \wedge [\{c, d\}] = [\{a, b\} \cap \{c, d\}] = [\emptyset]$, which is of course senseless. However, in case of equivalences, we are able to find a uniform well-behaving family of representatives of $\equiv$-classes, as we will see in the sequel.

We may also define another order for $\mathcal{R}$. Let R be a binary relation on U. Then the pair $(X^{\blacktriangledown}, X^{\blacktriangle})$ is called the *approximation* of X. Let us denote by $\mathcal{A}$ the set of all approximations, that is,

$$\mathcal{A} = \{ (X^{\blacktriangledown}, X^{\blacktriangle}) \mid X \subseteq U \}.$$

Because $\mathcal{A} \subseteq \wp(U) \times \wp(U)$, the set $\mathcal{A}$ may be ordered by the same order as $\wp(U) \times \wp(U)$. A natural question is, whether $\mathcal{A}$ is a sublattice of $\wp(U) \times \wp(U)$? We will study this question in the next section. By the next, lemma the presented two orders are essentially the same.

Lemma 153. *For any binary relation, $(\mathcal{R}, \subseteq) \cong (\mathcal{A}, \leq)$.*

Proof. We show that the map $[X] \mapsto (X^{\blacktriangledown}, X^{\blacktriangle})$ is an order-isomorphism. For all $X, Y \subseteq U$, it is trivial that $[X] \subseteq [Y]$ if and only if $(X^{\blacktriangledown}, X^{\blacktriangle}) \leq (Y^{\blacktriangledown}, Y^{\blacktriangle})$. Further, if $(X^{\blacktriangledown}, X^{\blacktriangle}) \in \mathcal{A}$, then clearly it is the image of $[X]$, that is, the map is onto. □

This section is ended by a lemma showing that $\mathcal{R} \cong \mathcal{R}^{\mathrm{op}}$ and $\mathcal{A} \cong \mathcal{A}^{\mathrm{op}}$.

Lemma 154. *For any binary relation, $\mathcal{R}$ and $\mathcal{A}$ are self-dual.*

Proof. We show that $\mathcal{A}$ is self-dual. Let us define a map $\phi: \mathcal{A} \to \mathcal{A}$ by

$$(X^{\blacktriangledown}, X^{\blacktriangle}) \mapsto (X^{c\blacktriangledown}, X^{c\blacktriangle}).$$

The map ϕ is onto $\mathcal{A}$, because $\phi(X^{c\blacktriangledown}, X^{c\blacktriangle}) = (X^{\blacktriangledown}, X^{\blacktriangle})$ for all $(X^{\blacktriangledown}, X^{\blacktriangle}) \in \mathcal{A}$. Clearly,

$$
\begin{aligned}
(X^{\blacktriangledown}, X^{\blacktriangle}) \leq (Y^{\blacktriangledown}, Y^{\blacktriangle}) &\iff X^{\blacktriangledown} \subseteq Y^{\blacktriangledown} \text{ and } X^{\blacktriangle} \subseteq Y^{\blacktriangle} \\
&\iff Y^{\blacktriangledown c} \subseteq X^{\blacktriangledown c} \text{ and } Y^{\blacktriangle c} \subseteq X^{\blacktriangle c} \\
&\iff Y^{c\blacktriangle} \subseteq X^{c\blacktriangle} \text{ and } Y^{c\blacktriangledown} \subseteq X^{c\blacktriangledown} \\
&\iff (Y^{c\blacktriangledown}, Y^{c\blacktriangle}) \leq (X^{c\blacktriangledown}, X^{c\blacktriangle}) \\
&\iff \phi(Y^{\blacktriangledown}, Y^{\blacktriangle}) \leq \phi(X^{\blacktriangledown}, X^{\blacktriangle}).
\end{aligned}
$$

Since $\mathcal{R}$ is isomorphic to $\mathcal{A}$, it is also self-dual. □

In the following subsections, we study the order-theoretical properties of $\mathcal{R}$ and $\mathcal{A}$ with respect to the properties of binary relations inducing these approximations and rough sets.

10.2 Rough Sets Determined by Equivalences

We study here approximations and rough sets defined by equivalences. We show that for equivalences, we can for any $\equiv$-class pick a representative of that class such that the family of representatives forms a complete sublattice of $\wp(U)$. This will imply also that $\mathcal{R}$ and $\mathcal{A}$ can be embedded into a complete lattice of sets. We also show that they are Stone lattices. First we recall the Axiom of Choice by Zermelo:

Axiom of Choice. Let P be any set. Then there exists a function f which selects from each nonempty subset $S \subseteq P$ a member $f(S)$ of S.

Let E be an equivalence on U. We denote by U/E the set of all equivalence classes of E. By the Axiom of Choice, there exists a function

$$f: U/E \to U$$

such that $f(C) \in C$ for every E-class C. Any such function f is called a *choice function for* E. The range $\{f(C) \mid C \in U/E\}$ of f is denoted by $\mathrm{Rg}(f)$.

Let us denote for any $X \subseteq U$,

$$\overline{X} = X^{\blacktriangledown} \cup (X^{\blacktriangle} \cap \mathrm{Rg}(f)).$$

The definition means that $\overline{X}$ contains all equivalence classes included in X, and from the classes intersecting X only one element is chosen.

Let us also denote

$$\wp^f(U) = \{\overline{X} \mid X \subseteq U\}.$$

It is clear that for any definable set X, $\overline{X} = X$ because $X^{\blacktriangledown} = X^{\blacktriangle}$.

Example 155. Let E be an equivalence on $U = \{a, b, c, d\}$ with the equivalence classes $\{a, d\}$ and $\{b, c\}$. Further, let $f : U/E \to U$ be a choice function such that $f(\{a, d\}) = a$ and $f(\{b, c\}) = b$. The ordered set $\wp^f(U)$ is depicted in Fig. 22. Clearly, $\wp^f(U)$ is order-isomorphic to $\mathbf{3} \times \mathbf{3}$.

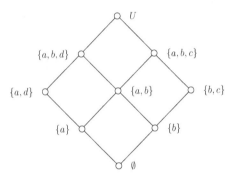

Fig. 22.

Lemma 156. *Let f be a choice function for an equivalence on U.*

(a) *For any $X \subseteq U$, $X \equiv \overline{X}$.*
(b) *For each $X \subseteq U$, the set $\overline{X}$ is the unique member of $\wp^f(U)$ in $[X]$.*
(c) *$\mathcal{R} \cong \wp^f(U)$.*
(d) *$\overline{\overline{X}} = \overline{X}$.*
(e) *$X \subseteq Y$ implies $\overline{X} \subseteq \overline{Y}$.*

Proof. (a) By definition, $X^{\blacktriangledown} \subseteq \overline{X}$. On the other hand, $\overline{X} - X^{\blacktriangledown}$ cannot contain any complete equivalence classes, because $\mathrm{Rg}(f)$ contains just one element from each class, and there cannot be any singleton classes included in $\overline{X} - X^{\blacktriangledown}$. Therefore, $X^{\blacktriangledown} = \overline{X}^{\blacktriangledown}$. If $x \in X^{\blacktriangle}$, then the equivalence class of x intersects with X, but this implies that the equivalence class of x intersects also with $\overline{X}$. Thus, $X^{\blacktriangle} \subseteq \overline{X}^{\blacktriangle}$. Because $\overline{X} \subseteq X^{\blacktriangle}$, $\overline{X}^{\blacktriangle} \subseteq X^{\blacktriangle\blacktriangle} = X^{\blacktriangle}$.

(b) By (a), $\overline{X}$ is always in the same $\equiv$-class as X. If $X \equiv Y$, then $\overline{Y} = Y^{\blacktriangledown} \cup (Y^{\blacktriangle} \cap \mathrm{Rg}(f)) = X^{\blacktriangledown} \cup (X^{\blacktriangle} \cap \mathrm{Rg}(f)) = \overline{X}$.

(c) We show that the map $[X] \mapsto \overline{X}$ is an order-isomorphism. It is now obvious that the map is onto $\wp^f(U)$. If $[X] \sqsubseteq [Y]$, then $X^{\blacktriangledown} \subseteq Y^{\blacktriangledown}$ and $X^{\blacktriangle} \subseteq Y^{\blacktriangle}$. This gives,

$$\overline{X} = X^{\blacktriangledown} \cup (X^{\blacktriangle} \cap \mathrm{Rg}(f)) \subseteq Y^{\blacktriangledown} \cup (Y^{\blacktriangle} \cap \mathrm{Rg}(f)) = \overline{Y}.$$

Conversely, if $\overline{X} \subseteq \overline{Y}$, then

$$X^{\blacktriangledown} = \overline{X}^{\blacktriangledown} \subseteq \overline{Y}^{\blacktriangledown} = Y^{\blacktriangledown} \quad \text{and} \quad X^{\blacktriangle} = \overline{X}^{\blacktriangle} \subseteq \overline{Y}^{\blacktriangle} = Y^{\blacktriangle}.$$

Claim (d) is obvious by (a) and (b). If $X \subseteq Y$, then $X^{\blacktriangledown} \subseteq Y^{\blacktriangledown}$ and $X^{\blacktriangle} \subseteq Y^{\blacktriangle}$. This gives $\overline{X} \subseteq \overline{Y}$ which proves (e). □

Note that $X \to \overline{X}$ is not a closure operator, because $X \subseteq \overline{X}$ does not usually hold. The following lemma will also be useful.

Lemma 157. *Let E be an equivalence on U.*

(a) *If X is definable, then for all $Y \subseteq U$,*

$$(X \cup Y)^{\blacktriangledown} = X^{\blacktriangledown} \cup Y^{\blacktriangledown} \quad \text{and} \quad (X \cup Y)^{\blacktriangle} = X^{\blacktriangle} \cap Y^{\blacktriangle}.$$

(b) *If $E(x) = \{x\}$, then for all $X \subseteq U$, $x \in X^{\blacktriangledown}$ if and only if $x \in X^{\blacktriangle}$.*

Proof. It is known that $X^{\blacktriangledown} \cup Y^{\blacktriangledown} \subseteq (X \cup Y)^{\blacktriangledown}$. Let $x \in (X \cup Y)^{\blacktriangledown}$. If $E(x) \cap X \neq \emptyset$, then $E(x) \subseteq X$ and $x \in X^{\blacktriangledown}$, because X is definable. If $E(x) \cap X = \emptyset$, then $E(x) \subseteq Y$ and $x \in Y^{\blacktriangledown}$. Hence, in both cases $x \in X^{\blacktriangledown} \cup Y^{\blacktriangledown}$.

It is also clear that $(X \cap Y)^{\blacktriangle} \subseteq X^{\blacktriangle} \cap Y^{\blacktriangle}$. Let $x \in X^{\blacktriangle} \cap Y^{\blacktriangle}$. Then $E(x) \cap X \neq \emptyset$ and $E(x) \cap Y \neq \emptyset$. Since X is definable, $E(x) \subseteq X$, and $E(x) \cap (X \cap Y) = (E(x) \cap X) \cap Y = E(x) \cap Y \neq \emptyset$. Thus, $x \in (X \cap Y)^{\blacktriangle}$.

Claim (b) is obvious by the definition of approximations. □

The following lemma presents important properties of representatives.

Lemma 158. *Let f be a choice function for an equivalence on U. Then for all families $\{X_i \mid i \in I\} \subseteq \wp(U)$,*

$$\left(\bigcup_{i \in I} \overline{X_i}\right)^{\blacktriangledown} = \bigcup_{i \in I} X_i^{\blacktriangledown} \quad \text{and} \quad \left(\bigcap_{i \in I} \overline{X_i}\right)^{\blacktriangle} = \bigcap_{i \in I} X_i^{\blacktriangle}.$$

Proof. Let us omit the subscripts $i \in I$ from the unions. Then

$$\left(\bigcup \overline{X_i}\right)^{\blacktriangledown} = \left(\bigcup (X_i^{\blacktriangledown} \cup (X_i^{\blacktriangle} \cap \mathrm{Rg}(f)))\right)^{\blacktriangledown}$$

$$= \left(\bigcup X_i^{\blacktriangledown} \cup \bigcup (X_i^{\blacktriangle} \cap \mathrm{Rg}(f))\right)^{\blacktriangledown}$$

$$= \left(\bigcup X_i^{\blacktriangledown}\right)^{\blacktriangledown} \cup \left(\bigcup (X_i^{\blacktriangle} \cap \mathrm{Rg}(f))\right)^{\blacktriangledown}$$

$$= \bigcup X_i^{\blacktriangledown} \cup \left(\bigcup X_i^{\blacktriangle} \cap \mathrm{Rg}(f)\right)^{\blacktriangledown}.$$

Recall that $\bigcup X_i^{\blacktriangledown}$ is definable. If $x \in \left(\bigcup X_i^{\blacktriangle} \cap \mathrm{Rg}(f)\right)^{\blacktriangledown}$, then necessarily $E(x) \subseteq \mathrm{Rg}(f)$ by Lemma 157, which implies $E(x) = \{x\}$. It is also clear that $x \in X_i^{\blacktriangle}$

for some $i \in I$, which implies $x \in X_i^{\blacktriangledown}$ with any such i. So, $x \in \bigcup X_i^{\blacktriangledown}$. Hence, $\left(\bigcup X_i^{\blacktriangle} \cap \mathrm{Rg}(f)\right)^{\blacktriangledown} \subseteq \bigcup X_i^{\blacktriangledown}$ which implies the desired equality.

For the other part,

$$
\begin{aligned}
\left(\bigcap \overline{X_i}\right)^{\blacktriangle} &= \left(\bigcap (X_i^{\blacktriangledown} \cup (X_i^{\blacktriangle} \cap \mathrm{Rg}(f)))\right)^{\blacktriangle} \\
&= \left(\bigcap ((X_i^{\blacktriangledown} \cup X_i^{\blacktriangle}) \cap (X_i^{\blacktriangledown} \cup \mathrm{Rg}(f)))\right)^{\blacktriangle} \\
&= \left(\bigcap (X_i^{\blacktriangle} \cap (X_i^{\blacktriangledown} \cup \mathrm{Rg}(f)))\right)^{\blacktriangle} \\
&= \left(\bigcap X_i^{\blacktriangle} \cap \bigcap (X_i^{\blacktriangledown} \cup \mathrm{Rg}(f))\right)^{\blacktriangle} \\
&= \left(\bigcap X_i^{\blacktriangle}\right)^{\blacktriangle} \cap \left(\bigcap (X_i^{\blacktriangledown} \cup \mathrm{Rg}(f))\right)^{\blacktriangle} \\
&= \bigcap X_i^{\blacktriangle} \cap \left(\bigcap X_i^{\blacktriangledown} \cup \mathrm{Rg}(f)\right)^{\blacktriangle} \\
&= \bigcap X_i^{\blacktriangle} \cap U \\
&= \bigcap X_i^{\blacktriangle}. \qquad \qquad \square
\end{aligned}
$$

The previous lemma has several consequences.

Proposition 159. *Let f be a choice function for an equivalence on U. Then $\wp^f(U)$ is a complete ring of sets.*

Proof. Let $\{\overline{X_i} \mid i \in I\}$ be a subfamily of $\wp^f(U)$. Then by Proposition 132 and Lemma 158,

$$
\begin{aligned}
\bigcup \overline{X_i} &= \bigcup (X_i^{\blacktriangledown} \cup (X_i^{\blacktriangle} \cap \mathrm{Rg}(f))) \\
&= \bigcup X_i^{\blacktriangledown} \cup \bigcup (X_i^{\blacktriangle} \cap \mathrm{Rg}(f)) \\
&= \left(\bigcup \overline{X_i}\right)^{\blacktriangledown} \cup \left(\bigcup X_i^{\blacktriangle} \cap \mathrm{Rg}(f)\right) \\
&= \left(\bigcup \overline{X_i}\right)^{\blacktriangledown} \cup \left(\bigcup \overline{X_i}^{\blacktriangle} \cap \mathrm{Rg}(f)\right) \\
&= \left(\bigcup \overline{X_i}\right)^{\blacktriangledown} \cup \left(\left(\bigcup \overline{X_i}\right)^{\blacktriangle} \cap \mathrm{Rg}(f)\right) \\
&= \overline{\left(\bigcup \overline{X_i}\right)}.
\end{aligned}
$$

Hence, $\bigcup \overline{X_i} \in \wp^f(U)$. The other part can be proved in a similar way:

$$
\begin{aligned}
\bigcap \overline{X_i} &= \bigcap (X_i^{\blacktriangledown} \cup (X_i^{\blacktriangle} \cap \mathrm{Rg}(f))) \\
&= \bigcap ((X_i^{\blacktriangledown} \cup X_i^{\blacktriangle}) \cap (X_i^{\blacktriangledown} \cup \mathrm{Rg}(f))) \\
&= \bigcap X_i^{\blacktriangle} \cap \bigcap (X_i^{\blacktriangledown} \cup \mathrm{Rg}(f)) \\
&= \bigcap X_i^{\blacktriangle} \cap \left(\bigcap X_i^{\blacktriangledown} \cup \mathrm{Rg}(f)\right) \\
&= \left(\bigcap \overline{X_i}\right)^{\blacktriangle} \cap \left(\bigcap \overline{X_i}^{\blacktriangledown} \cup \mathrm{Rg}(f)\right)
\end{aligned}
$$

$$= \left(\bigcap \overline{X_i}\right)^{\blacktriangle} \cap \left(\left(\bigcap \overline{X_i}\right)^{\blacktriangledown} \cup \mathrm{Rg}(f)\right)$$

$$= \left(\left(\bigcap \overline{X_i}\right)^{\blacktriangle} \cap \left(\bigcap \overline{X_i}\right)^{\blacktriangledown}\right) \cup \left(\left(\bigcap \overline{X_i}\right)^{\blacktriangle} \cup \mathrm{Rg}(f)\right)$$

$$= \left(\bigcap \overline{X_i}\right)^{\blacktriangledown} \cup \left(\left(\bigcap \overline{X_i}\right)^{\blacktriangle} \cup \mathrm{Rg}(f)\right)$$

$$= \overline{\left(\bigcap \overline{X_i}\right)}.$$

□

Because $\wp^f(U)$ is a complete sublattice of $\wp(U)$, and $\wp^f(U) \cong \mathcal{R} \cong \mathcal{A}$ by Lemmas 153 and 156, then also $\mathcal{R}$ and $\mathcal{A}$ are distributive lattices that can be embedded into $\wp(U)$. Further, since $\mathcal{R}$ and $\mathcal{A}$ are isomorphic to a complete ring of sets, their elements can be represented as a join of some – or none – completely join-irreducible elements.

Further, $\wp^f(U)$ is an Alexandrov topology, which means that there is a smallest neighbourhood operator $N_f\colon \wp(U) \to \wp(U)$ defined by

$$N_f(X) = \bigcap \{Y \in \wp^f(U) \mid X \subseteq Y\}.$$

Obviously, the set $\{N_f(x) \mid x \in U\}$ is the smallest base for $\wp^f(U)$ and these are the completely join-irreducible elements of $\wp^f(U)$ by Proposition 27. For example, in case of Example 155,

$$N_f(a) = \{a\}, \ N_f(b) = \{b\}, \ N_f(c) = \{b,c\} \ \text{and} \ N(d) = \{a,d\}.$$

Note that the 'representative operator' $X \mapsto \overline{X}$ is not the neighbourhood operator of $\wp^f(U)$.

Example 160. Let us again consider the equivalence on the set $\{a,b,c,d\}$ having the equivalence classes $\{a,d\}$ and $\{b,c\}$. The ordered set $\mathcal{A}$ is presented in Fig. 23. Completely join-irreducible elements are marked with filled circles.

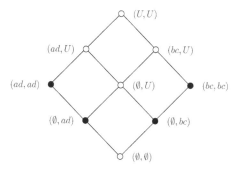

Fig. 23.

It is clear that only the elements $(\emptyset,\emptyset)$, (ad,ad), (bc,bc), and (U,U) have complements, and notice that these sets can be identified with the family of definable sets Def.

Now we are able to answer the question stated in Section 10.1.

Theorem 161. *For any equivalence on U, $\mathcal{A}$ is a complete sublattice of $\wp(U) \times \wp(U)$.*

Proof. Let $\{(X_i^\blacktriangledown, X_i^\blacktriangle) \mid i \in I\}$ be a subset of $\mathcal{A}$. Then by Lemmas 156 and 158,

$$\left(\bigcup_{i \in I} X_i^\blacktriangledown, \bigcup_{i \in I} X_i^\blacktriangle\right) = \left(\bigcup_{i \in I} X_i^\blacktriangledown, \bigcup_{i \in I} \overline{X_i}^\blacktriangle\right) = \left(\left(\bigcup_{i \in I} \overline{X_i}\right)^\blacktriangledown, \left(\bigcup_{i \in I} \overline{X_i}\right)^\blacktriangle\right).$$

and

$$\left(\bigcap_{i \in I} X_i^\blacktriangledown, \bigcap_{i \in I} X_i^\blacktriangle\right) = \left(\bigcap_{i \in I} \overline{X_i}^\blacktriangledown, \bigcap_{i \in I} X_i^\blacktriangle\right) = \left(\left(\bigcap_{i \in I} \overline{X_i}\right)^\blacktriangledown, \left(\bigcap_{i \in I} \overline{X_i}\right)^\blacktriangle\right).$$

Thus,

$$\bigvee_{i \in I} (X_i^\blacktriangledown, X_i^\blacktriangle) = \left(\bigcup_{i \in I} X_i^\blacktriangledown, \bigcup_{i \in I} X_i^\blacktriangle\right) \quad \text{and} \quad \bigwedge_{i \in I} (X_i^\blacktriangledown, X_i^\blacktriangle) = \left(\bigcap_{i \in I} X_i^\blacktriangledown, \bigcap_{i \in I} X_i^\blacktriangle\right). \qquad \square$$

Recall from Lemma 46 that for any complete Boolean lattice B, $B^{[2]} = \{(a, b) \in B \times B \mid a \le b\}$ is a complete Stone lattice such that the pseudocomplement is $(a, b)^* = (b', b')$. It is also known that $B^{[2]}$ is a complete sublattice of $B \times B$. Since Def is a complete Boolean lattice, $\text{Def}^{[2]}$ is a complete Stone lattice in which $(X, Y)^* = (Y^c, Y^c)$.

Now for all $X \subseteq U$, $X^\blacktriangledown, X^\blacktriangle \in \text{Def}$ and $X^\blacktriangledown \subseteq X^\blacktriangle$, which implies $\mathcal{A} \subseteq \text{Def}^{[2]}$. It is also clear that $\mathcal{A}$ is a complete sublattice of $\text{Def}^{[2]}$. However, $\mathcal{A}$ is not always equal to $\text{Def}^{[2]}$, because for elements $x \in U$ such that $E(x) = \{x\}$, $\mathcal{A}$ does not contain $(\emptyset, \{x\})$.

Proposition 162. *For any equivalence on U, $\mathcal{A}$ is a Stone lattice such that $(X^\blacktriangledown, X^\blacktriangle)^* = (X^{\blacktriangle c}, X^{\blacktriangle c})$ for all $X \subseteq U$.*

Proof. It is clear that for any X, $(X^{\blacktriangle c}, X^{\blacktriangle c})$ is in $\mathcal{A}$, because it is the approximation pair of the definable set $X^{\blacktriangle c}$. Now, $X^\blacktriangledown \cap X^{\blacktriangle c} \subseteq X^\blacktriangle \cap X^{\blacktriangle c} = \emptyset$, and this gives $(X^\blacktriangledown, X^\blacktriangle) \wedge (X^{\blacktriangle c}, X^{\blacktriangle c}) = (X^\blacktriangledown \cap X^{\blacktriangle c}, X^\blacktriangle \cap X^{\blacktriangle c}) = (\emptyset, \emptyset)$. Further, if $(X^\blacktriangledown, X^\blacktriangle) \wedge (Y^\blacktriangledown, Y^\blacktriangle) = (X^\blacktriangledown \cap Y^\blacktriangledown, X^\blacktriangle \cap Y^\blacktriangle) = (\emptyset, \emptyset)$, then $Y^\blacktriangledown \subseteq Y^\blacktriangle \subseteq X^{\blacktriangle c}$, that is, $(Y^\blacktriangledown, Y^\blacktriangle) \le (X^{\blacktriangle c}, X^{\blacktriangle c})$, which completes the proof. $\qquad \square$

The skeleton of the lattice $\mathcal{A}$ is

$$S(\mathcal{A}) = \{(X^\blacktriangledown, X^\blacktriangle) \mid X \in \text{Def}\} \cong \text{Def}$$

and the dense set is

$$S(\mathcal{A}) = \{(X^\blacktriangledown, X^\blacktriangle) \mid X^\blacktriangle = U\}.$$

We shall now describe the structure of the lattices of rough sets. It turns out that the lattices $\mathcal{R}$ and $\mathcal{A}$ are determined up to isomorphism by the number of singleton equivalence classes and the number of non-singleton equivalence classes.

Proposition 163. *For any equivalence E, the set of approximations $\mathcal{A}$ is isomorphic to $\mathbf{2}^I \times \mathbf{3}^J$, where I is the set of singleton E-classes and J is the set of the non-singleton E-classes.*

Proof. Let us define a mapping

$$\varphi \colon \mathcal{A} \to \mathbf{2}^I \times \mathbf{3}^J, (X^{\blacktriangledown}, X^{\blacktriangle}) \mapsto (f, g),$$

such that $f \colon I \to \{0, 1\}$ and $g \colon J \to \{0, u, 1\}$ are defined for any $[a] \in I$ and $[b] \in J$ by

$$f([a]) = \begin{cases} 1 & \text{if } a \in X \\ 0 & \text{if } a \notin X \end{cases} \qquad \text{and} \qquad g([b]) = \begin{cases} 1 & \text{if } b \in X^{\blacktriangledown} \\ u & \text{if } b \in X^{\blacktriangle} - X^{\blacktriangledown} \\ 0 & \text{if } b \notin X^{\blacktriangle}. \end{cases}$$

First we show that φ is an order-embedding. Let us denote $\varphi(X^{\blacktriangledown}, X^{\blacktriangle}) = (f_1, g_1)$ and $\varphi(Y^{\blacktriangledown}, Y^{\blacktriangle}) = (f_2, g_2)$. Assume that $(X^{\blacktriangledown}, X^{\blacktriangle}) \leq (Y^{\blacktriangledown}, Y^{\blacktriangle})$. We will show that this implies $(f_1, g_1) \leq (f_2, g_2)$, that is, $f_1([a]) \leq f_2([a])$ and $g_1([b]) \leq g_2([b])$ for all $[a] \in I$ and $[b] \in J$. If $f_1([a]) = 1$ for some $[a] \in I$, then $a \in X$, and since $[a] = \{a\}$, we get $a \in X^{\blacktriangledown} \subseteq Y^{\blacktriangledown} \subseteq Y$. Hence, $f_2([a]) = 1$ and $f_1 \leq f_2$. Further, if $g_1([b]) = 1$ for some $[b] \in J$, then $b \in X^{\blacktriangledown} \subseteq Y^{\blacktriangledown}$ and $g_2([b]) = 1$. If $g_1([b]) = u$, then necessarily $b \in X^{\blacktriangle} \subseteq Y^{\blacktriangle}$, which gives $g_2([b]) \geq u$. Thus, also $g_1 \leq g_2$.

Conversely, assume that $(f_1, g_1) \leq (f_2, g_2)$. We will show that $(X^{\blacktriangledown}, X^{\blacktriangle}) \leq (Y^{\blacktriangledown}, Y^{\blacktriangle})$. Let $x \in X^{\blacktriangledown}$. If $[x] \in I$, then $[x] = \{x\}$ and $1 = f_1([x]) \leq f_2([x])$. Thus, we get $x \in Y$ and $x \in Y^{\blacktriangledown}$. If $[x] \in J$, then $1 = g_1([x]) \leq g_2([x])$ gives $x \in Y^{\blacktriangledown}$. We have shown that $X^{\blacktriangledown} \subseteq Y^{\blacktriangledown}$. Assume that $x \in X^{\blacktriangle}$. If $[x] \in I$, then $x \in X$, and $1 \leq f_1([x]) \leq f_2([x])$ gives $x \in Y \subseteq Y^{\blacktriangle}$. If $[x] \in J$, then $u \leq g_1([x]) \leq g_2([x])$ implies $x \in Y^{\blacktriangle}$. Thus, also $X^{\blacktriangle} \subseteq Y^{\blacktriangle}$ and $(X^{\blacktriangledown}, X^{\blacktriangle}) \leq (Y^{\blacktriangledown}, Y^{\blacktriangle})$.

We still have to show that φ is a surjection. For that we need the Axiom of Choice. Let $(f, g) \in \mathbf{2}^I \times \mathbf{3}^J$ and let $F \colon U/E \to U$ be any choice function. Let

$$X = \bigcup \{[a] \in I \mid f([a]) = 1\} \cup \bigcup \{[b] \in J \mid g([b]) = 1\}$$
$$\cup \bigcup \{[c] \cap \mathrm{Rg}(F) \mid [c] \in J \text{ and } g([c]) = u\},$$

and denote $\varphi(X^{\blacktriangledown}, X^{\blacktriangle}) = (f', g')$. It should now be clear that $(f, g) = (f', g')$. For example, if $g([x]) = u$, then X contains only one element from $[x]$ picked by the choice function F. This implies $x \in X^{\blacktriangle}$, but since $[x]$ has at least two elements, we have $[x] \not\subseteq X$ and $x \notin X^{\blacktriangledown}$. This gives $g'([x]) = u$. Conversely, if $g'([x]) = u$, then $x \in X^{\blacktriangle} - X^{\blacktriangledown}$ and $[x] \not\subseteq X$, which means $g([x]) < 1$. If $g([x]) = 0$, then $[x]$ cannot intersect with X, which gives $x \notin X^{\blacktriangle}$ and $g'([x]) = 0$, a contradiction. Thus, also $g([x])$ must be u. $\qquad \square$

10.3   Rough Sets Determined by Arbitrary Binary Relations

Here we study ordered sets of rough sets defined by arbitrary binary relations. Recall that for a binary relation R on U, we denote $R(x) = \{y \in U \mid x \, R \, y\}$, and the approximation operators are defined by

$$X^{\blacktriangledown} = \{x \in U \mid R(x) \subseteq X\} \quad \text{and} \quad X^{\blacktriangle} = \{x \in U \mid R(x) \cap X \neq \emptyset\}$$

for any $X \subseteq U$. The relation $\equiv$ and the ordered sets $\mathcal{A}$ and $\mathcal{R}$ are defined as before.

As shown in the previous section, the ordered set of rough sets defined by equivalences is a complete Stone lattice. We begin our study on ordered sets of rough sets by an example showing that for tolerances and transitive relations, $\mathcal{A}$ is not always a lattice.

Example 164. (a) Let us consider the tolerance depicted in Fig. 24(i). Surprisingly, if we omit the transitivity, the structure of rough sets changes quite dramatically. The Hasse diagram of $\mathcal{A}$ is given in Fig. 25. For instance, the elements (a, abc) and $(\emptyset, abcd)$ do not have a least upper bound. Similarly, the elements $(ab, abcd)$ and (a, U) do not have a greatest lower bound. This means that $\mathcal{A}$ and $\mathcal{R}$ are not lattices.

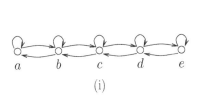

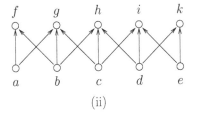

Fig. 24.

(b) The set of approximations $\mathcal{A}$ determined the transitive relation depicted in Fig. 24(ii) is the 22-element set

$$\{(fghik, \emptyset), (fghik, ab), (fghik, abc), (fghik, bcd), (fghik, cde),$$
$$(fghik, de), (afghik, abc), (fghik, abcd), (fghik, abcde), (fghik, abde),$$
$$(fghik, bcde), (efghik, cde), (abfghik, abcd), (afghik, abcde),$$
$$(cfghik, abcde), (defghik, bcde), (abcfghik, abcde), (abfghik, abcde),$$
$$(aefghik, abcde), (defghik, abcde), (cdefghik, abcde),$$
$$(abcdefghik, abcde)\}.$$

Note that since the relation is not connected, $X^{\blacktriangledown} \not\subseteq X^{\blacktriangle}$ for all $X \subseteq U$. It is easy to see that the ordered set of all approximations is isomorphic to the ordered set depicted in Fig. 25. Now, for example, $(abfghik, abcd) \wedge (afghik, abcde)$ does not exist; the set of lower bounds of this pair is $\{(afghik, abc), (fghik, abcd), (fghik, abc), (fghik, ab), (fghik, bcd), (fghik, \emptyset)\}$, which does not have a greatest element. Similarly, $(afghik, abc) \vee (fghik, abcd)$ does not exist because this pair of elements has two minimal upper bounds. Therefore $\mathcal{A}$ and $\mathcal{R}$ are not lattices.

Our next proposition shows that the rough sets defined by a symmetric and transitive binary relation form a complete Stone lattice.

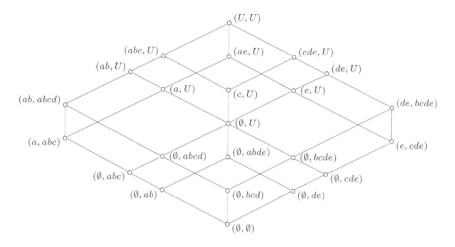

Fig. 25.

Proposition 165. *For a symmetric and transitive binary relation R, the ordered set of rough sets $\mathcal{A}$ is a complete Stone lattice isomorphic to $\mathbf{2}^I \times \mathbf{3}^J$, where I is the set of singleton R-classes and J is the set of R-classes that have at least two elements.*

Proof. Let R be a symmetric and transitive binary relation on a set U. Let us denote $U^* = \{x \in U \mid R(x) \neq \emptyset\}$. It is obvious that R is connected, symmetric, and transitive relation on U^*. As we noted in Section 2.2, R is therefore an equivalence on U^*. The resulting ordered set of rough sets on U^* is a complete Stone lattice by Proposition 163. Further, it is isomorphic to $\mathbf{2}^I \times \mathbf{3}^J$, where I is the set of singleton R-classes and J is the set of R-classes that have at least two elements.

Let us denote by $\mathcal{A}$ the set of rough sets on U, and by $\mathcal{A}^*$ the set of rough sets on U^*. We show that $\mathcal{A}^* \cong \mathcal{A}$. Let $\Sigma = U - U^*$ and let us define a map $\varphi \colon \mathcal{A}^* \to \mathcal{A}$ by setting

$$(X^{\blacktriangledown}, X^{\blacktriangle}) \mapsto (X^{\blacktriangledown} \cup \Sigma, X^{\blacktriangle}).$$

Assume that $x \in \Sigma$. Because $R(x) = \emptyset$, $R(x) \subseteq X$ and $R(x) \cap X = \emptyset$ hold for all $X \subseteq U$. By applying this it is easy to see that the map φ is an order-isomorphism, and hence $\mathcal{A}$ is a complete Stone lattice. $\square$

Note that if R is symmetric and transitive, but not reflexive, the elements that are not related to any element behave quite absurdly: they belong to every lower approximation, but not in any upper approximation, as shown in the previous proof.

Example 166. Let us consider the preorder of Fig. 26(i). The corresponding ordered set $\mathcal{A}$ is given in Fig. 26(ii).

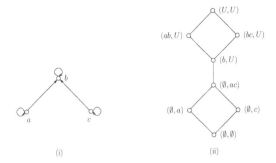

Fig. 26.

It is clear that we cannot find any family of representatives of ≡-classes isomorphic to $\mathcal{A}$, because the cardinality of each ≡-class is one. However, in this case $\mathcal{A}$ is a sublattice of $\wp(U) \times \wp(U)$.

In this section we have considered rough sets determined by indiscernibility relations which are not necessarily reflexive, symmetric, or transitive. Our studies can be summarized as follows:

- For any symmetric and transitive relation, $\mathcal{A}$ and $\mathcal{R}$ are Stone lattices (see Propositions 162 and 165).
- For tolerances, $\mathcal{A}$ and $\mathcal{R}$ are not always lattices (see Example 164(a)).
- For transitive relations, $\mathcal{A}$ and $\mathcal{R}$ are not always lattices (see Example 164(b)).
- Is is not known whether $\mathcal{A}$ and $\mathcal{R}$ determined by preorders are lattices (cf. Example 166).

These observations are depicted in Fig. 27.

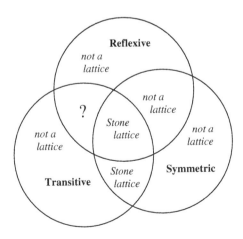

Fig. 27.

Bibliographical Notes

The idea of defining rough sets as equivalence classes of the equivalence $\equiv$ appears already in [44]. In [22] it was proposed that rough sets can be ordered coordinatewise by approximation pairs. The fact that the ordered set of approximations forms a Stone lattice was proved in [46], where also was mentioned without proof that the family of representatives is a sublattice of the powerset of the universe. In fact, that representatives form a complete ring of sets was proved in [25]. Finally, in [18] it was shown that this lattices has the structure of the form $\mathbf{2}^I \times \mathbf{3}^J$, where I is the set of singleton equivalence classes and J is the set of the non-singleton equivalence classes. The work [4] provides a good survey on lattice structures of rough sets determined by equivalences.

The result that the ordered set of rough sets determined by a tolerance does not always form a lattice appeared in [25]. It is shown in [29] that rough sets determined by transitive relations do not necessarily form lattices, and the fact that for symmetric and transitive relations, $\mathcal{A}$ and $\mathcal{R}$ are Stone Lattices can be also found there.

Acknowledgements

The author is very grateful to Ewa Orłowska, Magnus Steinby, Jari Kortelainen, and Michiro Kondo for the careful reading of the manuscript and for their valuable comments and suggestions. Special thanks are also due to Andrzej Skowron and James Peters for offering the opportunity to do this work.

References

1. P. Alexandroff. Diskrete räume. *Matematičeskij Sbornik*, 2:501–518, 1937.
2. W. W. Armstrong. Dependency structures of data base relationships. In *Information Processing 74, Proceedings of IFIP Congress 74, Stockholm, Sweden, August 5–10, 1974*, pages 580–583. North-Holland, 1974.
3. R. Balbes and P. Dwinger. *Distributive Lattices*. University of Missouri Press, Columbia, Missouri, 1974.
4. M. Banerjee and M. K. Chakraborty. Algebras from rough sets. In S. K. Pal, L. Polkowski, and A. Skowron, editors, *Rough-Neural Computing: Techniques for Computing with Words*, pages 157–184. Springer-Verlag, Berlin, 2004.
5. N. L. Biggs. *Discrete Mathematics*. Oxford University Press, New York, 1985.
6. G. Birkhoff. Rings of sets. *Duke Mathematical Journal*, 3:443–454, 1937.
7. G. Birkhoff. *Lattice Theory*, volume XXV of *Colloquim publications*. American Mathematical Society (AMS), Providence, Rhode Island, third edition, 1995.
8. S. Burris and H. P. Sankappanavar. *A Course in Universal Algebra*, volume 78 of *Graduate Texts in Mathematics*. Springer-Verlag, New York, Heidelberg, Berlin, 1981.
9. E. Codd. A relational model of data for large shared data banks. *Communications of the ACM*, 13:377–387, 1970.
10. S. D. Comer. An algebraic approach to the approximation of information. *Fundamenta Informaticae*, 14:492–502, 1991.
11. B. A. Davey and H. A. Priestley. *Introduction to Lattices and Order*. Cambridge University Press, second edition, 2002.

12. A. Day. The lattice theory of functional dependencies and normal decompositions. *International Journal of Algebra and Computation*, 2:409–431, 1992.
13. S. P. Demri and E. S. Orlowska. *Incomplete Information: Structure, Inference, Complexity.* Monographs in Theoretical Computer Science. An EATCS Series. Springer-Verlag, Berlin, Heidelberg, 2002.
14. M. Erné, J. Koslowski, A. Melton, and G. E. Strecker. A primer on Galois connections. In *Proceedings of the 1991 Summer Conference on General Topology and Applications in Honor of Mary Ellen Rudin and Her Work*, volume 704 of *Annals of the New York Academy of Sciences*, pages 103–125. New York Academy of Sciences, 1993.
15. C. J. Everett. Closure operators and Galois theory in lattices. *Transactions of the American Mathematical Society*, 55:514–525, 1944.
16. T. Feil and J. Krone. *Essential Discrete Mathematics for Computer Science.* Prentice Hall, Upper Saddle River, NJ, 2003.
17. B. Ganter and R. Wille. *Formal Concept Analysis. Mathematical Foundations.* Springer-Verlag, Berlin, 1999. Translated from the 1996 German original by Cornelia Franzke.
18. M. Gehrke and E. Walker. On the structure of rough sets. *Bulletin of Polish Academy of Sciences. Mathematics*, 40:235–245, 1992.
19. G. Gierz, K. H. Hofmann, K. Keimel, J. D. Lawson, M. Mislove, and D. S. Scott. *A Compendium of Continuous Lattices.* Springer-Verlag, Berlin, Heidelberg, New York, 1980.
20. J. A. Goguen. *L*-fuzzy sets. *Journal of Mathematical Analysis and Applications*, 18:145–174, 1967.
21. G. Grätzer. *General Lattice Theory. New appendices with B. A. Davey, R. Freese, B. Ganter, M. Greferath, P. Jipsen, H. A. Priestley, H. Rose, E. T. Schmidt, S. E. Schmidt, F. Wehrung, R. Wille.* Birkhäuser, Basel, second edition, 1998.
22. T. Iwinski. Algebraic approach to rough sets. *Bulletin of Polish Academy of Sciences. Mathematics*, 35:673–683, 1987.
23. J. Järvinen. *Representations of Information Systems and Dependence Spaces, and Some Basic Algorithms.* Ph.Lic. thesis, Department of Mathematics, University of Turku, Turku, Finland, 1997.
24. J. Järvinen. Preimage relations and their matrices. In L. Polkowski and A. Skowron, editors, *Rough Sets and Current Trends in Computing, First International Conference, RSCTC'98, Warsaw, Poland, June 22–26, 1998, Proceedings*, volume 1424 of *Lecture Notes in Computer Science*, pages 139–146. Springer-Verlag, Berlin, Heidelberg, 1998.
25. J. Järvinen. *Knowledge Representation and Rough Sets.* Ph.D. dissertation, Department of Mathematics, University of Turku, Turku, Finland, 1999. TUCS Dissertations 14.
26. J. Järvinen. Approximations and rough sets based on tolerances. In W. Ziarko and Y. Y. Yao, editors, *Rough Sets and Current Trends in Computing, Second International Conference, RSCTC 2000 Banff, Canada, October 16–19, 2000, Revised Papers*, volume 2005 of *Lecture Notes in Computer Science*, pages 182–189. Springer-Verlag, Berlin, Heidelberg, 2001.
27. J. Järvinen. Armstrong systems on ordered sets. In C. S. Calude, M. J. Dinneen, and S. Sburlan, editors, *Combinatorics, Computability and Logic*, Springer Series in Discrete Mathematics and Theoretical Computer Science, pages 137–149. Springer-Verlag, London, 2001.
28. J. Järvinen. On the structure of rough approximations. *Fundamenta Informaticae*, 53:135–153, 2002.

29. J. Järvinen. The ordered set of rough sets. In S. Tsumoto, R. Slowinski, H. J. Komorowski, and J. W. Grzymala-Busse, editors, *Rough Sets and Current Trends in Computing, 4th International Conference, RSCTC 2004, Uppsala, Sweden, June 1–5, 2004, Proceedings*, volume 3066 of *Lecture Notes in Computer Science*, pages 49–58. Springer-Verlag, Berlin, Heidelberg, 2004.

30. J. Järvinen. Pawlak's information systems in terms of Galois connections and functional dependencies. *Fundamenta Informaticae*, 75, 2007.

31. J. Järvinen, M. Kondo, and J. Kortelainen. Modal-like operators in Boolean algebras, Galois connections and fixed points. *Fundamenta Informaticae*, 76, 2007.

32. J. Järvinen and J. Kortelainen. A note on definability in rough set theory. In B. De Baets, R. De Caluwe, G. De Tré, J. Fodor, J. Kacprzyk, and S. Zadrożny, editors, *Current Issues in Data and Knowledge Engineering*, Problemy Współczesnej Nauki, Teoria i Zastosowania, Informatyka, pages 272–277. EXIT, Warsaw, Poland, 2004.

33. J. Järvinen and J. Kortelainen. A unifying study between modal-like operators, topologies, and fuzzy sets. Technical Report 642, Turku Centre for Computer Science (TUCS), Turku, Finland, 2004.

34. B. Jónsson and A. Tarski. Boolean algebras with operators. Part I. *American Journal of Mathematics*, 73:891–939, 1951.

35. J. L. Kelley. *General Topology*, volume 27 of *Graduate Texts in Mathematics*. Springer-Verlag, 1975. Reprint of the 1955 edition, published by Van Nostrand.

36. M. Kondo. On the structure of generalized rough sets. *Information Sciences*, 176:589–600, 2006.

37. J. Kortelainen. On relationship between modified sets, topological spaces and rough sets. *Fuzzy Sets and Systems*, 61:91–95, 1994.

38. J. Kortelainen. *A Topological Approach to Fuzzy Sets*. Ph.D. dissertation, Lappeenranta University of Technology, Lappeenranta, Finland, 1999. Acta Universitatis Lappeenrantaensis 90.

39. J. C. C. McKinsey and A. Tarski. The algebra of topology. *Annals of Mathematics*, 45:141–191, 1944.

40. M. Novotný and Z. Pawlak. Algebraic theory of independence in information systems. *Fundamenta Informaticae*, 14:454–476, 1991.

41. O. Ore. Galois connexions. *Transactions of American Mathematical Society*, 55:493–513, 1944.

42. E. Orłowska and Z. Pawlak. Representation of nondeterministic information. *Theoretical Computer Science*, 29:27–39, 1984.

43. Z. Pawlak. Information systems theoretical foundations. *Information Systems*, 6:205–218, 1981.

44. Z. Pawlak. Rough sets. *International Journal of Computer and Information Sciences*, 11:341–356, 1982.

45. Z. Pawlak. *Rough Sets. Theoretical Aspects of Reasoning About Data*. Kluwer Academic Publishers Group, Dordrecht, 1991.

46. J. Pomykała and J. A. Pomykała. The Stone algebra of rough sets. *Bulletin of Polish Academy of Sciences. Mathematics*, 36:495–512, 1988.

47. J. A. Pomykała. About tolerance and similarity relations in information systems. In J. J. Alpigini, J. F. Peters, A. Skowron, and N. Zhong, editors, *Rough Sets and Current Trends in Computing, Third International Conference, RSCTC 2002, Malvern, PA, USA, October 14–16, 2002, Proceedings*, volume 2475 of *Lecture Notes in Computer Science*, pages 175–182. Springer-Verlag, Berlin, Heidelberg, 2002.

48. G. N. Raney. Completely distributive complete lattices. *Proceedings of American Mathematical Society*, 3:677–680, 1952.

49. A. Skowron and C. Rauszer. The discernibility matrices and functions in information systems. In R. Slowinski, editor, *Intelligent Decision Support*, pages 331–362. Kluwer Academic Publishers, Dordrecht, The Netherlands, 1992.

50. A. Skowron and J. Stepaniuk. Tolerance approximation spaces. *Fundamenta Informaticae*, 27:245–253, 1996.

51. R. Slowinski and D. Vanderpooten. A generalized definition of rough approximations based on similarity. *IEEE Transactions on Knowledge and Data Engineering*, 12:331–336, 2000.

52. A. K. Steiner. The lattice of topologies: Structure and complementation. *Transactions of the American Mathematical Society*, 122:379–398, 1966.

53. T. A. Sudkamp. Self-conjugate functions on Boolean algebras. *Notre Dame Journal of Formal Logic*, 19:504–512, 1978.

54. A. Tarski. A lattice-theoretical fixpoint theorem and its applications. *Pacific Journal of Mathematics*, 5:285–309, 1955.

55. J. van Benthem. Correspondence theory. In D. Gabbay and F. Guenthner, editors, *Handbook of Philosophical Logic. Volume II: Extensions of Classical Logic*, pages 167–247. Reidel, Dordrecht, 1984.

56. M. Ward. The closure operators of a lattice. *Annals of Mathematics. Second Series*, 43:191–196, 1942.

57. Y. Y. Yao. On generalizing Pawlak approximation operators. In L. Polkowski and A. Skowron, editors, *Rough Sets and Current Trends in Computing, First International Conference, RSCTC'98, Warsaw, Poland, June 22–26, 1998, Proceedings*, volume 1424 of *Lecture Notes in Computer Science*, pages 298–307. Springer-Verlag, Berlin, Heidelberg, 1998.

58. Y. Y. Yao. Relational interpretations of neighborhood operators and rough set approximation operators. *Information Sciences*, 111:239–259, 1998.

59. Y. Y. Yao. A comparative study of formal concept analysis and rough set theory in data analysis. In S. Tsumoto, R. Slowinski, H. J. Komorowski, and J. W. Grzymala-Busse, editors, *Rough Sets and Current Trends in Computing, 4th International Conference, RSCTC 2004, Uppsala, Sweden, June 1–5, 2004, Proceedings*, volume 3066 of *Lecture Notes in Computer Science*, pages 59–68. Springer-Verlag, Berlin, Heidelberg, 2004.

60. Y. Y. Yao and T. Y.Lin. Generalization of rough sets using modal logics. *Intelligent Automation and Soft Computing*, 2:103–120, 1996.

61. L. A. Zadeh. Fuzzy sets. *Information and Control*, 8:338–353, 1965.

Index of Notation

Author Index

Lecture Notes in Computer Science

For information about Vols. 1–4310

please contact your bookseller or Springer

Vol. 4360: W. Dubitzky, A. Schuster, P.M.A. Sloot, M. Schroeder, M. Romberg (Eds.), Distributed, High-Performance and Grid Computing in Computational Biology. X, 192 pages. 2007. (Sublibrary LNBI).

Vol. 4358: R. Vidal, A. Heyden, Y. Ma (Eds.), Dynamical Vision. IX, 329 pages. 2007.

Vol. 4357: L. Buttyán, V. Gligor, D. Westhoff (Eds.), Security and Privacy in Ad-Hoc and Sensor Networks. X, 193 pages. 2006.

Vol. 4355: J. Julliand, O. Kouchnarenko (Eds.), B 2007: Formal Specification and Development in B. XIII, 293 pages. 2006.

Vol. 4354: M. Hanus (Ed.), Practical Aspects of Declarative Languages. X, 335 pages. 2006.

Vol. 4353: T. Schwentick, D. Suciu (Eds.), Database Theory – ICDT 2007. XI, 419 pages. 2006.

Vol. 4352: T.-J. Cham, J. Cai, C. Dorai, D. Rajan, T.-S. Chua, L.-T. Chia (Eds.), Advances in Multimedia Modeling, Part II. XVIII, 743 pages. 2006.

Vol. 4351: T.-J. Cham, J. Cai, C. Dorai, D. Rajan, T.-S. Chua, L.-T. Chia (Eds.), Advances in Multimedia Modeling, Part I. XIX, 797 pages. 2006.

Vol. 4349: B. Cook, A. Podelski (Eds.), Verification, Model Checking, and Abstract Interpretation. XI, 395 pages. 2007.

Vol. 4348: S.T. Taft, R.A. Duff, R.L. Brukardt, E. Ploedereder, P. Leroy (Eds.), Ada 2005 Reference Manual. XXII, 765 pages. 2006.

Vol. 4347: J. Lopez (Ed.), Critical Information Infrastructures Security. X, 286 pages. 2006.

Vol. 4346: L. Brim, B. Haverkort, M. Leucker, J. van de Pol (Eds.), Formal Methods: Applications and Technology. X, 363 pages. 2007.

Vol. 4345: N. Maglaveras, I. Chouvarda, V. Koutkias, R. Brause (Eds.), Biological and Medical Data Analysis. XIII, 496 pages. 2006. (Sublibrary LNBI).

Vol. 4344: V. Gruhn, F. Oquendo (Eds.), Software Architecture. X, 245 pages. 2006.

Vol. 4342: H. de Swart, E. Orłowska, G. Schmidt, M. Roubens (Eds.), Theory and Applications of Relational Structures as Knowledge Instruments II. X, 373 pages. 2006. (Sublibrary LNAI).

Vol. 4341: P.Q. Nguyen (Ed.), Progress in Cryptology - VIETCRYPT 2006. XI, 385 pages. 2006.

Vol. 4340: R. Prodan, T. Fahringer, Grid Computing. XXIII, 317 pages. 2007.

Vol. 4339: E. Ayguadé, G. Baumgartner, J. Ramanujam, P. Sadayappan (Eds.), Languages and Compilers for Parallel Computing. XI, 476 pages. 2006.

Vol. 4338: P. Kalra, S. Peleg (Eds.), Computer Vision, Graphics and Image Processing. XV, 965 pages. 2006.

Vol. 4337: S. Arun-Kumar, N. Garg (Eds.), FSTTCS 2006: Foundations of Software Technology and Theoretical Computer Science. XIII, 430 pages. 2006.

Vol. 4335: S.A. Brueckner, S. Hassas, M. Jelasity, D. Yamins (Eds.), Engineering Self-Organising Systems. XII, 212 pages. 2007. (Sublibrary LNAI).

Vol. 4334: B. Beckert, R. Hähnle, P.H. Schmitt (Eds.), Verification of Object-Oriented Software. XXIX, 658 pages. 2007. (Sublibrary LNAI).

Vol. 4333: U. Reimer, D. Karagiannis (Eds.), Practical Aspects of Knowledge Management. XII, 338 pages. 2006. (Sublibrary LNAI).

Vol. 4332: A. Bagchi, V. Atluri (Eds.), Information Systems Security. XV, 382 pages. 2006.

Vol. 4331: G. Min, B. Di Martino, L.T. Yang, M. Guo, G. Ruenger (Eds.), Frontiers of High Performance Computing and Networking – ISPA 2006 Workshops. XXXVII, 1141 pages. 2006.

Vol. 4330: M. Guo, L.T. Yang, B. Di Martino, H.P. Zima, J. Dongarra, F. Tang (Eds.), Parallel and Distributed Processing and Applications. XVIII, 953 pages. 2006.

Vol. 4329: R. Barua, T. Lange (Eds.), Progress in Cryptology - INDOCRYPT 2006. X, 454 pages. 2006.

Vol. 4328: D. Penkler, M. Reitenspiess, F. Tam (Eds.), Service Availability. X, 289 pages. 2006.

Vol. 4327: M. Baldoni, U. Endriss (Eds.), Declarative Agent Languages and Technologies IV. VIII, 257 pages. 2006. (Sublibrary LNAI).

Vol. 4326: S. Göbel, R. Malkewitz, I. Iurgel (Eds.), Technologies for Interactive Digital Storytelling and Entertainment. X, 384 pages. 2006.

Vol. 4325: J. Cao, I. Stojmenovic, X. Jia, S.K. Das (Eds.), Mobile Ad-hoc and Sensor Networks. XIX, 887 pages. 2006.

Vol. 4323: G. Doherty, A. Blandford (Eds.), Interactive Systems. XI, 269 pages. 2007.

Vol. 4322: F. Kordon, J. Sztipanovits (Eds.), Reliable Systems on Unreliable Networked Platforms. XIV, 317 pages. 2007.

Vol. 4320: R. Gotzhein, R. Reed (Eds.), System Analysis and Modeling: Language Profiles. X, 229 pages. 2006.

Vol. 4319: L.-W. Chang, W.-N. Lie (Eds.), Advances in Image and Video Technology. XXVI, 1347 pages. 2006.

Vol. 4318: H. Lipmaa, M. Yung, D. Lin (Eds.), Information Security and Cryptology. XI, 305 pages. 2006.

Vol. 4317: S.K. Madria, K.T. Claypool, R. Kannan, P. Uppuluri, M.M. Gore (Eds.), Distributed Computing and Internet Technology. XIX, 466 pages. 2006.

Vol. 4316: M.M. Dalkilic, S. Kim, J. Yang (Eds.), Data Mining and Bioinformatics. VIII, 197 pages. 2006. (Sublibrary LNBI).

Vol. 4314: C. Freksa, M. Kohlhase, K. Schill (Eds.), KI 2006: Advances in Artificial Intelligence. XII, 458 pages. 2007. (Sublibrary LNAI).

Vol. 4313: T. Margaria, B. Steffen (Eds.), Leveraging Applications of Formal Methods. IX, 197 pages. 2006.

Vol. 4312: S. Sugimoto, J. Hunter, A. Rauber, A. Morishima (Eds.), Digital Libraries: Achievements, Challenges and Opportunities. XVIII, 571 pages. 2006.

Vol. 4311: K. Cho, P. Jacquet (Eds.), Technologies for Advanced Heterogeneous Networks II. XI, 253 pages. 2006.